Natural Systems

The organisation of life

Natural Systems

The organisation of life

Markus P. Eichhorn
The University of Nottingham

WILEY Blackwell

This edition first published 2016 © 2016 by John Wiley & Sons Ltd

Registered office: John Wiley & Sons, Ltd, The Atrium, Southern Gate, Chichester, West Sussex, PO19 8SQ, UK

Editorial offices: 9600 Garsington Road, Oxford, OX4 2DQ, UK
The Atrium, Southern Gate, Chichester, West Sussex, PO19 8SQ, UK
111 River Street, Hoboken, NJ 07030-5774, USA

For details of our global editorial offices, for customer services and for information about how to apply for permission to reuse the copyright material in this book please see our website at www.wiley.com/wiley-blackwell.

Library of Congress Cataloging-in-Publication Data

Names: Eichhorn, Markus P., author.
Title: Natural systems : the organization of life / Markus P. Eichhorn.
Description: Chichester, UK ; Hoboken, NJ : John Wiley & Sons, 2016. |
Includes index.
Identifiers: LCCN 2015040946| ISBN 9781118905883 (cloth) | ISBN 9781118905920
(pbk.)
Subjects: LCSH: Biotic communities. | Life cycles (Biology)
Classification: LCC QH541 .E3542 2016 | DDC 571.8–dc23 LC record available at
http://lccn.loc.gov/2015040946

A catalogue record for this book is available from the British Library.

Wiley also publishes its books in a variety of electronic formats. Some content that appears in print may not be available in electronic books.

Cover image: © Igor Shpilenok/naturepl.com

Typeset in 9/13pt MeridienLTStd by SPi Global, Chennai, India

1 2016

Contents

Preface

Ecology is the study of how the living world works. As a scientific field it has advanced in great strides over recent years, driven by a recognition of its central role in tackling some of the most pressing problems in the modern world. At the same time, however, the conventional ecology syllabus has remained relatively static, with a focus on theories dating from half a century ago. Even higher-level undergraduates can struggle to comprehend ideas under debate in the current literature, and the leap between undergraduate and postgraduate levels has become ever wider. This book is an attempt to bridge the gap.

The overall aim is to introduce the processes determining the structure and organisation of natural systems. The core questions can be expressed in two ways. In purely academic terms, they are to

- Understand patterns of species richness
- Interpret the composition of species in any given area
- Explain how processes at local (bottom-up) and regional (top-down) scales interact

It is perhaps better to reframe these in terms that capture more practical aspects related to current global concerns and are therefore more enticing to a general audience:

- What is biodiversity, and how can we measure it?
- If we wanted to create a natural system, how would we go about it?
- Can we predict what might happen to the natural world in the future?

The text builds sequentially from the concept and importance of species, through patterns of diversity, the interactions of natural systems with their abiotic environment and how species are organised within communities. This leads to consideration of global patterns of biogeography, concluding with the topic of islands, which are the closest analogues in nature to sealed systems. Standard ecology courses take a bottom-up approach, focussing on core phenomena such as population dynamics and simple interactions among a few species. Meanwhile biogeographers tend to stress the importance of speciation, extinction and dispersal in generating broadscale patterns. This book attempts to unite the two perspectives. Specialist terms highlighted in bold on first usage are defined in the glossary at the end.

To students

In the coming years, ecology—and ecologists—have a crucial role to play. The great challenges facing humanity include feeding a growing global population, dealing with the consequences of climate change and controlling the effects of a mass extinction event triggered by our own activities. The recent formation of the Intergovernmental Platform on Biodiversity and Ecosystem Services (IPBES[1]), intended as a parallel to the similar panel on climate change, reflects how seriously the world is beginning to take such concerns. This is long overdue—in the time since the Convention on Biological Diversity was signed by 193 countries in 1992, matters have only become worse (Butchart et al., 2010). These are all problems that will require an understanding of ecology and particularly how processes interact from local to regional and global scales. Learning how natural systems operate is the first step towards making a difference.

Much of the material in this book is recent, and some of the content remains controversial, so you are encouraged to follow up with wider reading and see how the debates are proceeding in the literature. Part of becoming a scientist lies in forming your own opinions and not taking everything you're told for granted, even in textbooks. Scientific understanding advances constantly and posterity will no doubt judge some of what is contained here to be incorrect; finding out which parts are wrong will be up to you. To learn more about any of the debates presented here and to keep pace with the field, I recommend scanning issues of the journals Ecology Letters and Trends in Ecology and Evolution. The most helpful articles will be short reviews. Many other journals include or focus entirely upon ecological studies.

As ecology grows as a science, it becomes increasingly important to understand the quantitative aspects. The best ecologists combine the enthusiasm of a natural historian with the mind of a statistician, and I entreat you to not skip over mathematical sections that might initially appear 'difficult'. Every effort has been made to make these as accessible as possible. A central set of skills to develop are techniques for the assessment and measurement of diversity, which will prove essential if you have aspirations to work in conservation, environmental consultancy or natural resource management. An appendix describes how to calculate and interpret a range of diversity measures using a real dataset from a butterfly conservation project in Colorado. These metrics are used routinely in the academic literature and applied fields. Remember that without strong numerical evidence, it is almost impossible to make a convincing scientific argument.

Finally, an occasional complaint from students on my courses is that there's too much material. If you feel this way then you're missing the point! Try to focus on underlying theories and concepts, rather than attempting to memorise specific information. You're unlikely to ever need to know the exact species of plant that make up a particular succession, but you should be able to explain how and why succession occurs. There are many examples contained in the book, but these are provided to illustrate ideas, not because the details themselves are essential. Take these concepts and see whether you can match them to natural systems wherever you find yourself—on holiday in an exotic country, walking in the park or even in your own garden. A true ecological rule should apply anywhere.

To instructors

Since 2008 I have taught an undergraduate module which has formed the spine of this book. My aim is to provide a bridge between conventional ecological teaching, covering the behaviour of individuals and populations, and global patterns of life. As such it begins where most ecology textbooks end, and takes a broad perspective on the organisation of natural systems. These themes make the book relevant to students of ecology, environmental science, geography and conservation. My hope is that it will be easy to use as a course text since each chapter is derived from a single 1-hour lecture (albeit expanded). Instructors should therefore be able to readily convert the text into a teaching resource, and students will be able to use it to enhance their

[1] http://www.ipbes.net.

overall understanding and support their learning. Each chapter commences by framing the big questions and concludes by outlining outstanding problems and avenues for additional enquiry along with some suggestions for additional reading. These could be used as starting points for class discussions or to stimulate interest in active areas of controversy.

Lately there have been several attempts to provide links from core ecology through to biogeography, usually via the nascent field of macroecology. Recent books targeting an exclusively academic audience include Scheiner and Willig (2011) and Loreau (2010). A growing drive to improve the connectivity between these research fields has however yet to be represented at an accessible level for a student audience. My hope is that this book will help to guide advanced undergraduates and postgraduates towards this exciting and vital issue.

In a single text it is impossible to cover all aspects of a topic, and therefore it is worth outlining what this book does not contain. There is a relatively limited emphasis on conservation, though many of the ideas and principles lie at the heart of conservation biology. The same can be said of restoration ecology, the field devoted to rebuilding natural systems where they have been altered by human activities. It has also not been possible to include much on abiotic processes operating at the ecosystem scale; a number of excellent texts already exist in this area though (e.g. Chapin III et al., 2012). A final known omission is the relatively limited coverage of the impacts of diseases and parasites (and parasitoids) on communities. This is a topic on which there has been a growing focus in the ecological literature, but as yet little synthesis, which means that it will hopefully be included in a future edition.

The book is intended to be a summary of the present state of the field which focusses on the most promising ideas for continued investigation. It is not written as a history of the development of ideas within ecology. My own experience as both a student and teacher is that diverting attention towards old arguments or superceded theories only serves to distract or confuse. There are many great names from the history of ecology, founding figures even, who are not mentioned in these pages. This is not meant as a slight, nor is it an accidental oversight, but a deliberate choice to avoid bloating the text with outdated arguments.

Finally, please note that a number of the concepts and theories that are referenced remain subject to dispute. I do not shy away from giving a personal opinion on the more contentious points, based on the balance of current evidence and in full awareness that some will disagree. The reference list has to stop somewhere, and I have not included anything published since the end of 2014. It is inevitable that by the time this book reaches your hands, something will have been contested or overturned. If you should find yourself disagreeing with anything, then I hope this can be turned into a productive means of introducing students to difficult questions. No textbook should ever be treated as absolute truth; my goal is to provide a reasonable starting point. Should you notice any errors or omissions, then please let me know.

Markus P. Eichhorn
Nottingham, UK
12 October 2015

References

Butchart, S. H. M., M. Walpole, B. Collen, A. van Strien, J. P. W. Scharlemann, R. E. A. Almond, J. E. M. Baillie, B. Bomhard, C. Brown, J. Bruno, K. E. Carpenter, G. M. Carr, J. Chanson, A. M. Chenery, J. Csirke, N. C. Davidson, F. Dentener, M. Foster, A. Galli, J. N. Galloway, P. Genovesi, R. D. Gregory, M. Hockings, V. Kapos, J.-F. Lamarque, F. Leverington, J. Loh, M. A. McGeoch, L. McRae, A. Minasyan, M. H. Morcillo, T. E. E. Oldfield, D. Pauly, S. Quader, C. Revenga, J. R. Sauer, B. Skolnik, D. Spear, D. Stanwell-Smith, S. N. Stuart, A. Symes, M. Tierney, T. D. Tyrrell, J.-C. Vié, and R. Watson, 2010. Global biodiversity: indicators of recent declines. *Science* 328:1164–1168.

Chapin, F. S. III, P. A. Matson, and P. M. Vitousek, 2012. *Principles of Terrestrial Ecosystem Ecology*. Springer-Verlag, second edition.

Loreau, M., 2010. Linking biodiversity and ecosystems: towards a unifying ecological theory. *Philosophical Transactions of the Royal Society Series B* 365:49–60.

Scheiner, S. M. and M. R. Willig, 2011. *The Theory of Ecology*. The University of Chicago Press.

Acknowledgements

While originally planning on writing a textbook, I consulted another author for advice. It was 'don't'. I would like to thank him for the warning (anonymously of course) but also my supportive colleagues for their encouragement. Apologies are offered to the many collaborators, students, colleagues, family and friends whom I have let down while concentrating on this project. *Mega biblion, mega kakon*[1]. My wife, Sarah, deserves special gratitude for her forbearance, particularly as reading ever more papers always seemed more appealing than domestic chores.

I am especially indebted to Anne Chao for inspiring the content of Chapter 5, Jonathan Chase and Mathew Leibold for Chapter 6, Robert Whittaker and José María Fernández-Palacios for Chapters 19 and 20, and Sharon Collinge, Jeff Oliver and Katherine Prudic for the data featured in the appendix. Richard Field provided a complete and thorough review of the entire text. Further comments on sections of the manuscript were received from Robert Bagchi, Anne Chao, Francis Gilbert, Chris Lavers, Vojtech Novotny and Eleanor Slade. Ward Cooper and Kelvin Matthews at Wiley were encouraging, helpful and patient throughout.

[1] Big book, big evil—Callimachus

Abbreviations

AET actual evapotranspiration: a measure of energy describing the total amount of water that evaporates from a fixed area of plant foliage given the available supply (distinct from PET)

asl above sea level

Bya billion years ago

EMIB equilibrium model of island biogeography, the theory developed by MacArthur and Wilson (1967); see Chapter 19

LDG latitudinal diversity gradient: the observation that more species occur at lower latitudes, with richness peaking in the tropics

LGM last glacial maximum (around 20,000 years ago)

Mya million years ago

NGO non-governmental organisation

NMDS Non-Metric Multidimensional Scaling: an ordination method which calculates dissimilarity indices between all pairs of samples in a dataset then represents these differences in two-dimensional space

NPP net primary productivity: a measure of the total energy fixed by producers (e.g. plants) minus any used in respiration, usually measured in mass of carbon per unit area per year (e.g. g $C/m^2/y$)

OTT out-of-the-tropics: a hypothesis to explain the latitudinal diversity gradient. See Chapter 16.8.

PAR photosynthetically active radiation: the portion of the spectrum of sunlight that can be used in photosynthesis; spans wavelengths from 400–700 nm

PCA principal components analysis: an ordination method which reduces a large dataset of potentially correlated variables (e.g. abundances of individual species across samples) to a small number of composite axes which capture the greatest amount of variance possible.

PCR polymerase chain reaction: a biochemical technique to take small amounts of template DNA and create vast numbers of copies of a particular region of interest, which can then be sequenced

PET potential evapotranspiration: a measure of energy describing the potential amount of water that would evaporate from a fixed area of plant foliage given unlimited supply of water (contrast AET)

SAD species abundance distribution

SAR species-area relationship

SE standard error: a measure of the uncertainty of an estimate; specifically it is the standard deviation of the sampling distribution of a statistic

ya years ago

ZNGI Zero Net-Growth Isocline

CHAPTER 1
Introduction: defining nature

1.1 How little we know

Understanding the organisation of nature has never been so important. The UN declared 2010 to be International Year of Biodiversity in advance of a meeting in Nagoya to discuss the Convention on Biological Diversity. Back in April 2002 the member countries had pledged

> ... to achieve by 2010 a significant reduction of the current rate of biodiversity loss at the global, regional and national level as a contribution to poverty alleviation and to the benefit of all life on Earth.[1]

It is safe to say that this target was not met; if anything the rate of extinctions increased over this time period (Butchart et al., 2010), and the continued trajectory is not promising. But was it ever an achievable goal? Problems with the statement include the very definition of biodiversity itself—what should we be counting, how do we go about it, and when will we know that the trend is reversing? How can we begin to collect the necessary information when fewer than 14% of all species have been formally identified (Mora et al., 2011)? A major theme of this book involves trying to answer these questions.

The concatenation of linked issues facing humanity, which include overpopulation, global climate change and an ongoing mass extinction (May, 2010), has prompted some to suggest that the only future for humanity is to leave the planet and take to the stars. It has long been a trope of science fiction that natural systems could be exported beyond Earth's atmosphere. Such a bold aspiration poses immense technical challenges, but these are at least equalled by the ecological problems. Is such an achievement—or salvation—within our capabilities?

The ultimate test of our understanding of natural systems is whether we are able to construct them ourselves. This was first attempted on a realistic scale by the Biosphere 2 project (Biosphere 1 being the Earth). A sealed glasshouse was constructed in Arizona between 1987 and 1991 covering 1.27 ha; it remains the largest ever constructed. It originally contained a series of different habitats along with agricultural land. The total cost of the project was $200 million (around $0.5 billion at today's value). Two attempts were made to completely seal groups of scientists inside. One of the major problems turned out to be atmospheric control; carbon dioxide levels fluctuated wildly both daily and seasonally, and oxygen levels fell by 30% over the first 16 months, leading to an injection of oxygen on medical grounds. All pollinator species and most vertebrates went extinct, while pests such as cockroaches became superabundant. Much was learnt from these studies, but in terms of the grand ambition—conducting a pilot study for future space stations—it must be considered an abject failure. No one has tried again since the last mission

[1] For the complete text see http://www.cbd.int/decision/cop/?id=7200.

Natural Systems: The organisation of life, First Edition. Markus P. Eichhorn.

was abandoned in 1994. For all our knowledge and understanding, we still cannot build a closed, functioning ecosystem.

1.2 Pressing questions

There are several profound gaps in our understanding of the natural world. As in any branch of science, asking the questions can seem deceptively simple, but arriving at the answers is more challenging. This book attempts to address the following:

- What governs the number of species present in any one location?
- What determines the identity of these species?
- How do local and broad-scale ecological processes interact with one another?

In order to reach an appropriate level of understanding to tackle these questions, we must draw from a number of fields including diversity theory, community ecology, ecosystem functioning and biogeography.

1.3 The hierarchy of nature

First it is important to identify the major scales of organisation in nature (Figure 1.1). Knowing the differences between these is essential. Each term has a very specific meaning and conflating concepts can lead to confusion. A crucial point is that processes which operate at one scale (e.g. the local community) might be irrelevant at another (e.g. the ecoregion). For example, competition is a central structuring force in explaining species interactions within a grassland but tells us little about species distributions on the scale of an entire continent.

Ecology begins with **individuals**, typically recognised as independent reproductive organisms. This definition is less simple to enforce than it sounds and can occasionally be arbitrary in its application. Colonial organisms such as sponges and bryozoans are composed of multiple individuals which depend

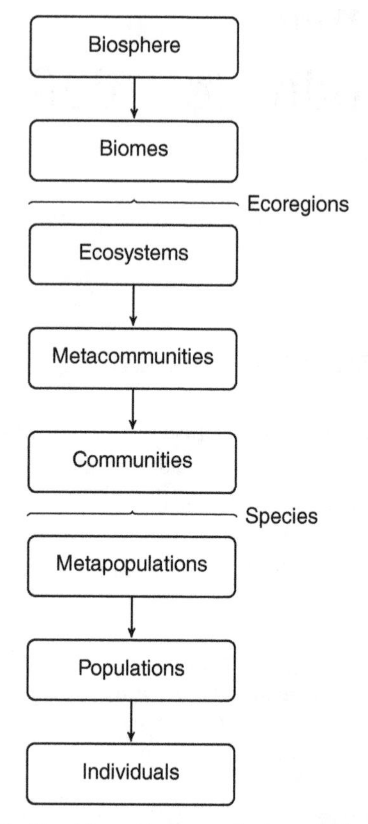

Figure 1.1 The hierarchical organisation of life on Earth. Components are linked by arrows where each level is a spatially-nested element of that above. Biomes are divided into ecoregions which are spread across the globe; likewise communities are made up of species which are not exclusive to any single community.

on their membership of a single structure to reproduce; hence it is common to count the whole colony as an individual. In some social species, such as ants, most colony members are unable to reproduce but instead support the reproduction of a single queen. Here it would make most biological sense to count each colony as an individual, yet for practical reasons (ants are easier to find and count than their nests), it is more common to count the sterile workers and treat them as individuals, which is also a more reasonable means of estimating their wider ecological impacts. Even at the individual level we have to recognise that complications can arise.

One way of identifying an individual might be to say that it is the product of a single fertilisation event, known as a genet. This is fine for sexually reproducing organisms, but not for those which are asexual, where offspring are identical copies of their parent. In some cases both can live side by side. Strawberry plants can be grown from seed, each the result of the pollination of a single ovule in a flower. As they grow, however, they send out runners which develop their own roots and can detach to become separate plants. These are known as ramets—despite being genetically identical to their parent, they still compete for all the same resources. A patch of wild strawberries could contain any proportion of genets and ramets.

Individuals can be highly variable in their behaviour, physiology and genetics; these have profound implications for the dynamics of **populations**, which are collections of individuals of the same species linked by reproduction. Once again, defining a population is simpler than recognising one; it is difficult to determine where the boundaries of reproductive links are, and therefore for convenience we typically demarcate populations based on sensible habitat features rather than by assessing gene flow (e.g. the edge of a lake would mark the boundary for a single population of fish). In more recent ecological theory it has been recognised that populations are linked by dispersal of individuals into **metapopulations** which have their own higher-order dynamics (Hanski, 1999). In truth there is often a continuum between the two, though in cases where discrete units can be identified (e.g. islands), with a discontinuity in dispersal, the concept can be vital in appreciating how local and regional dynamics are connected.

In conventional ecological theory, the population size is the sum of all the actively (or potentially) reproducing individuals. This therefore excludes juveniles (eggs, larvae and young) and—perhaps more surprisingly—males, since they do not control the rate of reproduction. Here again the theoretical and the practical are not easily reconciled. Consider the moon jellyfish (*Aurelia aurita*). It reproduces sexually, forming larvae which sink to the sea floor and form polyps. Normally these will wait for suitable conditions before budding to generate around 20 floating jellyfish. When the environment is unfavourable, however, these polyps can instead choose to create new polyps or form resilient long-lasting cysts. Apparent explosions in jellyfish abundance occur as soon as conditions allow. A species like this foils all our idealised concepts of how we determine population size.

The sum of all individuals makes up the totality of a **species**. The precise definition used to decide where the boundaries lie between species has further implications for how we interpret patterns in nature. This is the subject of the next chapter and the starting point for thinking about natural systems.

When multiple species occur in a single location, and show stability through time, it is referred to as a **community**. Typically these species are linked by feeding relationships into a food web and through interactions including competition, mutualism and parasitism. Interpreting the dynamics of any single population requires an understanding of these linkages. In an analogous fashion to populations, communities can be joined together in **metacommunities**, which are networks of communities connected by the dispersal of species (Holyoak et al., 2005). This is a relatively recent concept in ecology but has great explanatory power when linking the processes occurring at small scales to those on a regional level and *vice versa*.

At greater scales of study we recognise another entity emerging, the ecosystem. This oft-misused term actually refers to a combination of interacting living and abiotic components (Chapin et al., 2012). Rather than organisms simply responding to their environment, they also change it. Examples include how **transpiration** of a forest generates clouds and influences regional climate or peat bogs absorb carbon and store it in the soil. A hot topic in current ecological research focusses on how the components of ecosystems contribute to the resultant processes, which is the focus of Chapter 9.

Finally we can view natural systems at the biogeographical scale, where new levels of organisation become apparent. Broad patterns of life can be identified as **ecoregions**, large patches of Earth with consistent biotic characteristics in terms of their constituent communities and ecosystems. These can be further grouped into **biomes** which indicate the forms of the major natural systems as determined by their vegetation, general climate and dominant organisms within them. These are familiar as deserts, rain forests and other dominant communities on land and in the oceans. They are considered in more detail in Chapter 14. Ultimately these make up the whole Earth system, known as the **biosphere**. The controversial theory that life on Earth interacts with the abiotic environment to form a self-regulating complex system at a planetary scale is referred to as **Gaia** (*sensu* Lovelock, 1979). While most ecologists would hesitate to include such a maligned idea in this hierarchy (see Chapter 11), a global perspective can be useful, and problems such as climate change necessitate a scale of thinking at the level of the biosphere, representing all life on Earth.

1.4 Biodiversity

I have generally avoided the use of the word 'biodiversity' throughout the text, despite its prevalence in the media and (increasingly) the scientific literature. This may seem ironic; this is after all a book entirely about biodiversity in its most inclusive sense. As will become clear, however, such a simple word serves to obscure a vast array of important information and variation and is therefore a barrier to a full appreciation of how natural systems are constructed and operate. The term has a disputed history but is most commonly used to refer to the variety of life at all levels, which, according to the 1992 Convention on Biological Diversity,

> ... includes diversity within species, between species and of ecosystems.[2]

Unfortunately this means that in practical terms it has no units; one cannot point to two lakes and state quantitatively which has the greatest biodiversity. Often it is used to imply species richness, a term with a precise definition as the number of species, and therefore the latter should be used in preference since it is less prone to confusion. Only a few years after the term was coined there were already at least 85 different published definitions of biodiversity (DeLong, 1996), prompting some wags to refer to it as 'biological diversity with the logical part removed'. Its origins in the legalistic language of a political treaty mean it is of little help in resolving scientific questions.

A further difficulty with the term is that it is inherently value laden—more biodiversity is assumed to be a good thing. This is often not the case; invasive species increase species richness (at least temporarily), while many important habitats (e.g. mangrove swamps) have relatively low numbers of species. On its own, therefore, biodiversity cannot be used as a criterion for making assessments for conservation purposes.

1.5 Myths to bust

In the process of building an understanding of the organisation of natural systems, we must begin from firm foundations, which means dispensing with several beliefs that are commonly held by the naïve observer of nature. The first, and most egregious, is the idea of 'the balance of nature'. One of the recurrent themes in the text will be to demonstrate that there is no such thing: everything is in flux, usually with no clear end point, and constant change is an ecological rule. Stability is often a transient illusion. This is equally important in conservation; we should guard against superficial attempts to return systems to a 'natural' state as it is seldom possible to decide with any confidence what this ought to be.

A related idea which was long ago driven from scientific theory is the principle of providence, through which it was believed that a benevolent creator would not allow any part of nature to come to harm. Yet this pattern of thinking can insidiously creep back into our reasoning when we assume

[2] See http://www.cbd.int/convention/text.

that internal checks and balances will automatically restore natural systems to some default state after being perturbed. The fallacy of this will be revealed in due course. That natural systems respond to disruption in predictable ways is demonstrable but can be explained through more prosaic, ground-level processes without the need to invoke numinous forces. Moreover there is no guarantee that systems will return to their original starting point.

A final common belief is that natural systems act as finely balanced machines in which every component is harmoniously linked and removal of any part will inevitably lead to decay or collapse. This is related in ecological thought to the idea of the 'superorganism' (Clements, 1916), whereby species are tightly and obligately connected as *gestalt* units. We will return to this theme in Chapter 10, but for now it should be stated that the concept is discredited, and it turns out that many species are replaceable or expendable. This is not to say that each is not important, but rather that extinctions do not always imply imminent disaster, and natural systems prove to be remarkably resilient.

1.6 Further information

To set the wider political context in which the ideas presented here gain their greatest importance, you might want to follow up on the work of the various non-governmental organisations (NGOs) and intergovernmental bodies tasked with addressing the challenges of our changing world. For information on the Convention on Biological Diversity (and subsequent developments), there is a wealth of information at http://www.cbd.int. More facts and figures can be obtained from the World Resources Institute (http://wri.ogc.org/wri/biodiv) or the World Conservation Monitoring Centre (http://www.unep-wcmc.org).

1.6.1 Recommended reading

Groombridge B. & Jenkins M.D. (2002). *World Atlas of Biodiversity: Earth's Living Resources in the 21st Century*. University of California Press.

Pimm S.L., Jenkins C.N., Abell R., Brooks T.M., Gittleman J.L., Joppa L.N., Raven P.H., Roberts C.M. & Sexton J.O. (2014). The biodiversity of species and their rates of extinction, distribution and protection. *Science* 344, 1246–7532.

References

Butchart, S. H. M., M. Walpole, B. Collen, A. van Strien, J. P. W. Scharlemann, R. E. A. Almond, J. E. M. Baillie, B. Bomhard, C. Brown, J. Bruno, K. E. Carpenter, G. M. Carr, J. Chanson, A. M. Chenery, J. Csirke, N. C. Davidson, F. Dentener, M. Foster, A. Galli, J. N. Galloway, P. Genovesi, R. D. Gregory, M. Hockings, V. Kapos, J.-F. Lamarque, F. Leverington, J. Loh, M. A. McGeoch, L. McRae, A. Minasyan, M. H. Morcillo, T. E. E. Oldfield, D. Pauly, S. Quader, C. Revenga, J. R. Sauer, B. Skolnik, D. Spear, D. Stanwell-Smith, S. N. Stuart, A. Symes, M. Tierney, T.D. Tyrrell, J.-C. Vié, and R. Watson, 2010. Global biodiversity: indicators of recent declines. *Science* 328:1164–1168.

Chapin, F. S. III, P. A. Matson, and P. M. Vitousek, 2012. *Principles of Terrestrial Ecosystem Ecology*. Springer-Verlag, second edition.

Clements, F. E., 1916. *Plant Succession: Analysis of the Development of Vegetation*. Carnegie Institute of Washington.

DeLong, D. C. Jr., 1996. Defining biodiversity. *Wildlife Society Bulletin* 24:738–749.

Hanski, I., 1999. *Metapopulation Ecology*. Oxford University Press.

Holyoak, M., M. A. Leibold, and R. D, Holt editors, 2005. *Metacommunities: Spatial Dynamics and Ecological Communities*. The University of Chicago Press.

Lovelock, J., 1979. *Gaia: A New Look at Life on Earth*. Oxford University Press.

May, R. M., 2010. Ecological science and tomorrow's world. *Philosophical Transactions of the Royal Society Series B: Biological Sciences* 365:41–47.

Mora, C., D. P. Tittensor, S. Adl, A. G. B. Simpson, and B. Worm, 2011. How many species are there on earth and in the ocean? *PLoS Biology* 9:e1001–127.

PART I
Species

CHAPTER 2
What is a species?

2.1 The big question

This may seem, on first impression, like a strange place to start. After all, everyone knows what a species is, don't they? Even Darwin, in his preamble to *On The Origin of Species*, claimed that he did, though in truth he dodged the question:

> ... Nor shall I discuss here the various definitions which have been given of the term species. No one definition has yet satisfied all naturalists; yet every naturalist knows vaguely what he means when he speaks of a species. (Darwin, 1859)

In other words, he would know one when he saw it. For such a broadly used concept it proves surprisingly difficult to decide exactly what comprises a species and where the boundaries of classification ought to lie. What separates species from mere varieties? Should decisions be based upon their traits or their reproduction? This debate stretches back to the earliest days of natural history, and neither approach was considered satisfactory:

> External resemblance and community of descent are both defective and liable to break down if rigidly applied. (Nicholson, 1872)

An attempt to firmly grasp the nettle was made by Mayr (1942) in his book *Systematics and the Origin of Species*. Mayr recognised five distinct versions of the species concept at large in the scientific literature. Since publication of this book, rather than a process of consolidation, ever more definitions have arisen, such that there are now at least 20 in circulation (Hey, 2001).

The majority of diversity studies, and most of the chapters in this book, revolve around counting the number of species, or asking how many species there ought to be, or how many species are necessary. In order to be sure we're counting the right thing, we need a clear idea about what really comprises a species. Counting species is becoming an ever more pressing exercise in the midst of a mass extinction, combined with a decline in the number of professional taxonomists (Godfray, 2002) and when the overwhelming majority of species still remain to be described (Mora et al., 2011).

Species also have political implications, and without clear rules the concept is liable to abuse, leading to a problem known as species inflation, whereby increasing numbers of doubtful species are named. Often these are populations which have long been known as subspecies or varieties but which become elevated to full species status based on criteria which need to be carefully scrutinised. The predicament arises when counting species becomes a competitive endeavour, for example in deciding who has the most species-rich rain forest in the world. Some countries have used taxonomy as a means of emphasising their distinctiveness: Soviet Russia was especially prone to recognising dubious endemics in an effort to 'make Russia special'. There are often career incentives for individuals to describe new species; a

Natural Systems: The organisation of life, First Edition. Markus P. Eichhorn.

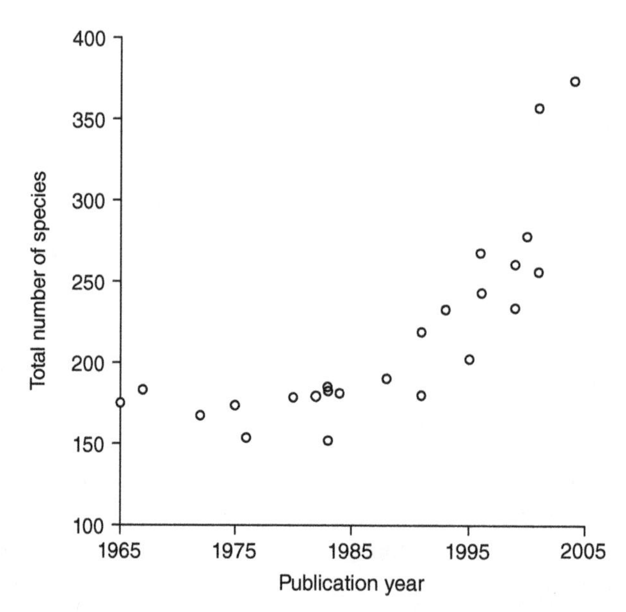

Figure 2.1 Global lists of primate species published between 1965 and 2005. (*Source*: Isaac, Mallet and Mace (2004). Reproduced with permission from Elsevier.)

taxonomist receives greater acclaim for the number of new taxa they have disovered than the number of spurious species they have corrected or quashed. Often the best possible motives are involved. In tropical conservation it is widely believed that the way to ensure an area becomes gazetted as a nature reserve is to discover a new species. These social and political pressures cannot be ignored.

The problem of species inflation is well illustrated by the global numbers of primates (Isaac et al., 2004). Primates are probably the best studied of all animal groups, certainly in terms of the resources expended relative to the number of species involved. Between 1965 and 1985 most authorities agreed that there were around 175 species. Yet thereafter up to the present day, this number appears to have doubled (Fig. 2.1). Have we really found so many new species in 50 years within a well-known group of large and charismatic vertebrates? If so, then how could we have missed them for so long? In truth, most of these 'new' species are divisions of existing taxa.

It is therefore worthwhile to review the history of the species concept and its many interpretations to see how we ended up in this state, and perhaps find a way to chart a path to a clearer idea of what we really mean by the word 'species'. For this section it should be assumed that all ideas refer to multicellular forms of life; definitions become even less certain when faced with microorganisms.

2.2 Species concepts

2.2.1 Nominalistic

A cynical observer, on the basis of all the confusion, might come to the conclusion that species are simply social constructs, artificial entities with no real biological meaning. Humans certainly have a need to classify things; every culture has named the animals and plants around them, paralleling our modern desire to give everything a taxonomic category. Even in the Garden of Eden, Adam is supposed to have spent his time naming the animals:

> Now the Lord God had formed out of the ground all the wild animals and all the birds in the sky. He brought them to the man to see what he would name them; and whatever the man called each living creature, that was its name. So the man gave names to all the livestock, the birds in the sky and all the wild animals. (Genesis 2:19–20)

If taxonomy is genuinely an innate human drive, then we would expect it to be consistent across cultures. If it matches biological realities, we might anticipate that (supposedly) impartial scientific classifications match the distinctions in common parlance.

Evidence to test these hypotheses was collected from native Tzeltal languages in the mountains of Chiapas in south Mexico (Berlin et al., 1966). Of 200 native names that were recognised, 41% covered more than a single taxonomic species, while 25% were more specific and matched what would be viewed as subspecies or varieties. Only 34% of native names corresponded with international taxonomy, and of these, more than 60% were derived from Hispanic names and therefore—indirectly—influenced by European views.

This lack of concordance is unsurprising, as cultural influences tend to cause us to be more specific in our classification of things that are more useful or interesting to us. As a simple exercise, first think how many breeds of dog you can name. These are all clearly within the same taxon—*Canis lupus familiaris*—itself a subspecies of the grey wolf. Yet the English language has many names for these types that are far below the level of a species. By way of contrast, how many types of grass can you name? Non-biologists would probably be surprised to learn that there are 113 native grass species in the British Isles and at least 220 regularly found in the wild (Cope and Gray, 2009). Most people struggle to name more than a few, yet all have evocative names, including Yorkshire fog (*Holcus lanatus*), creeping bent (*Agrostis stolonifera*) and holy grass (*Anthoxanthum nitens*). In past times, when the quality of grazing land was a matter of great importance, such names were widely known and used. Our more recent urbanised culture has allowed these names to be confined to the pages of floristic manuals while vernacular taxonomy has atrophied.

In conclusion, there is no common cultural level at which humans recognise species, and our system of classification changes over time to match present needs. Whatever we mean by the word 'species', it is more than just the representation of a common human bias.

2.2.2 Morphological

Morphology is the foundation of traditional scientific taxonomy. It is based on the visible characteristics of organisms, focussing on those presumed to be heritable, that is to say, those with a strong genetic component rather than overly influenced by the environment. For this reason plant taxonomy tends to concentrate on reproductive parts (i.e. flowers) rather than more plastic traits such as height or shape. A miniature bonsai tree is the same species as its relatives in the forest; despite the reduction in overall size, it will still produce flowers with an identical structure.

What comprises sufficient grounds to denote a new species depends on the group involved, which leads to accusations that some taxonomists have united entities that should be distinct ('lumpers'), while others may have divided things too finely ('splitters'). In this context Bickford et al. (2007) cite the examples of moss and marine molluscs. While most species of moss are widely distributed, often across several continents and encompassing a large degree of morphological variation, marine molluscs are finely differentiated on the basis of local variation in colour and pattern which modern molecular analyses have tended to show to be unreliable characters. In general, groups of organisms classified by professional taxonomists tend to be more 'lumped' than those that attract wider interest and large numbers of amateurs.

In recent years efforts have been made to minimise the subjective nature of judgements by use of a quantitative method known as Morphology-Based Alpha Taxonomy (MOBAT). This can be extremely revealing. For example, *Tetramorium* is a genus of Palæarctic ants of which two species have been recognised for over 150 years and the subject of more than 500 published papers. Recent detailed MOBAT-based examination of specimens revealed that there were in fact at least seven hidden or 'cryptic' species that had not been noted before (Schlick-Steiner et al., 2006). How could this have happened? Put simply, of over 200 myrmecologists working in Europe, only two were using MOBAT techniques to classify species. It's not that no one had looked, but rather that they hadn't looked in the right way.

In general it is believed that taxonomy as a science is under threat, with declining numbers of specialists and inadequate funding (Godfray, 2002). With so few people using these tools, it is likely that there are more species to be discovered simply by closer examination of specimens, while we may find that others would be better joined together.

The principles of taxonomy require that any named species should be traceable back to a single type specimen—the first to be given that name—which is kept safely in a recognised museum or herbarium. This specimen is in a literal sense the official definition of the species, though, in order for the name to be accepted, a detailed description must also be published. This allows it to be provided with an official name in the form of a Linnaean binomial comprising its genus (e.g. *Quercus* for the oaks) and a specific epithet (e.g. *robur* for the pedunculate oak). The full name includes a suffix denoting the name of the person who first named it (e.g. *Quercus robur* L., where the 'L.' refers to Linnæus). There are strict rules regarding the names that can be applied and how conflicts in nomenclature are resolved; these differ somewhat between plants and animals and are entirely different for fungi and microbes.

2.2.3 Biological

In contrast to morphological species concepts, the other major means of recognising species is by their reproduction. The biological species concept is the most widely known approach and was defined by Ernst Mayr as

> Groups of actually or potentially interbreeding natural populations which are reproductively isolated from other such groups. (Mayr, 1957)

This definition captures the true ecological sense in which we would like to recognise species and seems straightforward enough if you walk around a zoo. Nevertheless, even with large vertebrates, the weakness of the definition is revealed on closer inspection. It requires that each species be 'reproductively isolated', a simple phrase that disguises great complexity. The mechanisms for this isolation differ among taxa, meaning that detailed knowledge of the natural history of the organisms concerned is essential and often informed judgements must be made since it is impossible to attempt to mate every putative species with each other. Sometimes genetic evidence can be obtained that indicates which individuals are breeding with one another (e.g. Lagache et al., 2013), though this still takes considerable effort and expense to achieve.

As an example of the problems that can arise, the American red wolf *Canis rufus* is the recipient of large amounts of conservation funding, despite its unpopularity with livestock owners. An examination of its genetic heritage indicates that it is likely to be a hybrid of the grey wolf (*Canis lupus*) and the coyote (*Canis latrans*; Allendorf et al., 2001). The cause and timing of this hybridisation event are uncertain. There is an active and ongoing debate over the true taxonomic status of the red wolf, and naïve application of the biological species concept has done little to resolve this. Should we care about saving the red wolf?

Hybrids are widespread in nature. In a few cases, when the common ancestor of two species remains alive, they can be linked without the need for interbreeding. Some special cases of this are known as ring species, where populations have diverged in two directions, their descendents evolving into distinct and reproductively isolated lineages.

In central Siberia, two forms of greenish warbler coexist without interbreeding (Irwin et al., 2005). Were we to visit the region, we would have no qualms about deciding that they were different species. But once their wider distribution is examined, some doubts emerge (Fig. 2.2). The origin of the warblers was in the Himalayas, and they dispersed northwards as the glaciers retreated. The high-altitude deserts of the Tibetan plateau are inhospitable to warblers, and as a result the species

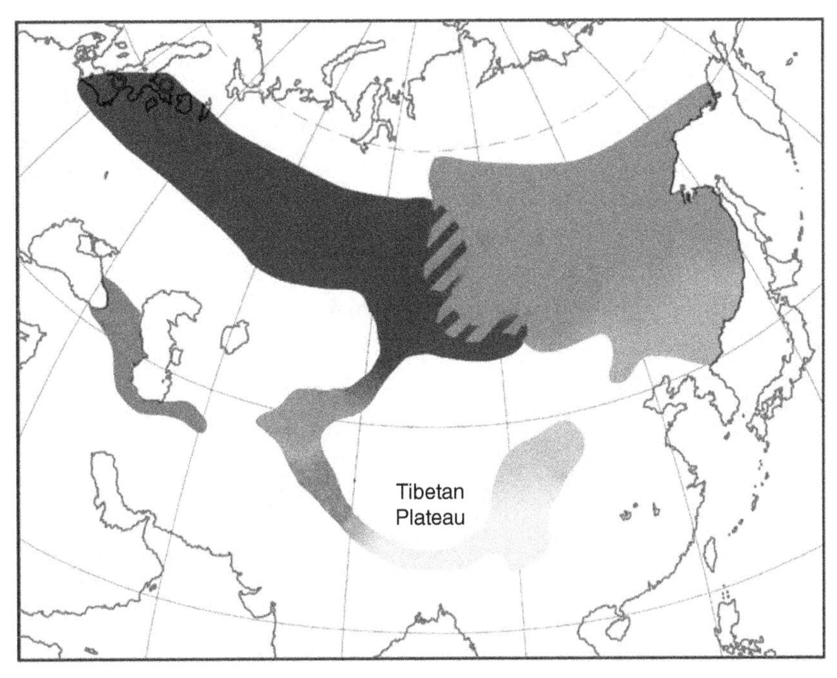

Figure 2.2 Breeding range of greenish warblers in Asia. Shades represent distinct subspecies, with gradations indicating change in morphology. The hatched area in central Siberia is the overlap zone between two distinct subspecies. Sampling sites are indicated by symbols corresponding to mitochondrial clades. (*Source*: Irwin et al. (2005). Reproduced with permission from American Association for the Advancement of Science.) (*See colour plate section for the colour representation of this figure.*)

split in two directions. One form reached west Siberia (*Phylloscopus trochiloides viridanus*), while the other branch spread to the east (*P. t. plumbeitarsus*; the gap in north China is probably the result of recent habitat loss). Once they met again in central Siberia, they had evolved along separate trajectories, with differing songs, calls and plumage, and had become reproductively isolated. This is known as speciation by distance. Yet their ancestral populations maintain a smooth transition in their morphology and genetics[1]. A gene which arose in one subspecies could plausibly reach the other by spreading through the linked populations. Where should the line between species be drawn?

We have clearly missed an important element out of our definitions of species so far: evolution. Without appreciating the origin of these two types of warbler from a common southern ancestor, we cannot understand their current status. Can this lesson be applied more generally?

[1] In fact the genetics of these subspecies is more complicated than this simplified account suggests, and there are discontinuities in the ring (Alcaide et al., 2014).

2.2.4 Phylogenetic

In one of the most controversial developments in the history of taxonomy, Cracraft (1989) proposed an entirely new species concept to bring morphological taxonomy and evolution together. In order to distinguish a species using the phylogenetic species concept, a population must share a common ancestor and exhibit at least one defining and heritable character. The crucial concept is that of **monophyly**: to be confirmed as members of the same species, individuals must be from the same lineage. Reproductive isolation can take place but is not necessary to recognise species as distinct. It is this latter element which has led to the most vociferous objections. Slight differences can always be found among populations, meaning that it is often possible to divide taxa into ever smaller sections.

The concept has been widely applied since its publication, especially to vertebrates, and aided by advances in genetic techniques that allow phylogenies to be rapidly resolved. An early demonstration took on the birds of paradise (Paradisaeidae), a well-known group of Australasian birds, whose males possess brightly coloured plumage and distinctive mating behaviour (Cracraft, 1992). Classical accounts using the biological species concept had recognised 40–42 species. Cracraft raised this to 90. In a large number of other studies, use of the phylogenetic species concept has increased numbers of species by around 50%. This is one element in apparent species inflation since the 1980s.

A notorious example was the discovery of a 'new' species of clouded leopard (*Neofelis diardi*) in Borneo and Sumatra (Meiri and Mace, 2007), much fêted in the media thanks to being a photogenic feline (Fig. 2.3). The times even declared it to be the first new big cat in almost 200 years—sadly stretching the truth, as it had been described by Cuvier in 1823 and since merged with its peninsular sister species *Neofelis nebulosa*. What was the basis for elevating this long-known variant? It boiled down to some differences in coat pattern, though this is hardly surprising in a population which had been isolated for a few thousand years. Use of similar reasoning led in the past to some truly outrageous assertions,

Figure 2.3 The Sundaland clouded leopard *Neofelis diardi*. (*Source*: WWF—Canon/Alain Compost. Reproduced with permission from Alain Compost.)

such as the identification of 82 'species' of North American brown bear (Merriam, 1918). Before embarking on a similar path, we must consider whether this new line of thinking leads us to an improved understanding of the nature of species and whether this is consistent for both ecologists and population geneticists.

Phylogenetic evidence has permeated conventional taxonomy, leading to revolutions at higher levels of classification than the species. For instance, the Angiosperm Phylogeny Group (APG III, 2009) has overturned our understanding of the relationships among groups of plants by basing their scheme upon evolutionary relationships rather than taxonomic characters. This has been an important development but comes no closer to resolving the species problem.

2.2.5 Genetic

The increasing affordability and usage of genetic tools has prompted some reconsideration of the essentials of taxonomy. Could we even dispense with visual taxonomic characters altogether and simply base our classification schemes on the genetics of the organisms? There are many advantages to such an approach: it becomes ever more rapid and affordable to provide the full genome of an individual, and no specialist knowledge of the natural history (or even appearance) of the organisms is required to produce a **phylogeny**. Efforts to find a universal DNA barcode for species are ongoing and may ultimately provide a shortcut for rapid identification without the need to laboriously work through keys or consult museum and herbarium specimens; it could even prove more accurate in the field as genetic tests make fewer errors than taxonomists (Dexter et al., 2010).

The chief problem, as with every other species concept, emerges when we decide when to draw the line between entities. A simple rule of genetic differentiation doesn't work (what would it be: 1%? 0.1%?) as it causes havoc with established taxonomy and the number of base pair substitutions is a poor correlate of morphological variation among species or even of reproductive isolation. The genetic variation within agreed species runs from 0.01% in the Eurasian lynx (*Lynx lynx*) up to a staggering 8.01% in the sea squirt *Ciona savignyi* (Leffler et al., 2012).

Once again human judgement has to be invoked. When there is evidence that there is genetic exchange among groups of individuals, they should be considered to be the same species. This works best for sympatric groups, those which occur in the same location, but performs poorly for allopatric groups, those in different locations, since in the latter case the lack of genetic exchange may be the result of geographical barriers or limited dispersal of individuals among groups.

As a case study to illustrate the power of genetic tools, Smith et al. (2008) examined the diversity of parasitoid wasps from Costa Rica. Based on morphology there were 171 provisional species, then genetic analyses suggested a further 142 should be added to this list. How could they know these were genuine species? One example is the species *Apanteles leucostigmus*, long recognised morphologically and thought to be a generalist parasitoid of over 30 species of hesperiid caterpillars. Barcoding of specimens indicated that the true number of wasp species was 36, but the most important step was to return to the natural history records. It turned out that each species revealed by the genetic analyses was in fact a specialist parasitoid of a small number of closely related caterpillar species: this was better seen as a set of separate wasp species rather than a single generalist, even though by sight alone the adults were identical (Fig. 2.4).

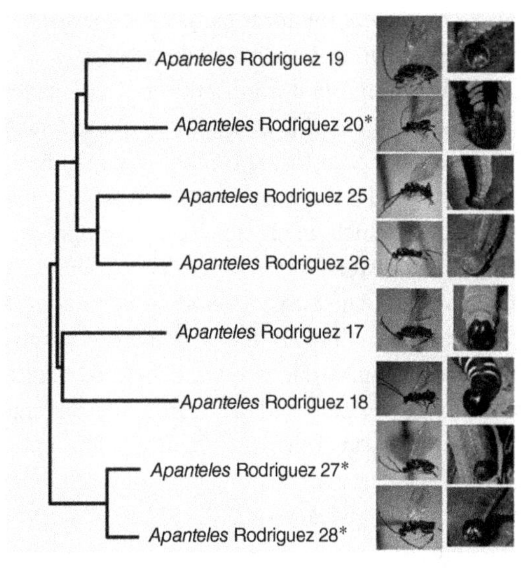

Figure 2.4 Provisional *Apanteles* wasp species within the morphological species *Apanteles leucostigmus*. Asterisks indicate wasps with more than one closely related host species. (*Source*: Smith et al. (2008). Reproduced with permission from the National Academy of Sciences, USA.)

This highlights a growing awareness of the existence of **cryptic species**, which are not detectable via normal taxonomic characters (i.e. things that humans can see) but might be apparent through other aspects of the species' natural history. These have been increasingly identified since 1986 when polymerase chain reaction (PCR) became a widespread technique, though it should also be noted that the majority of cases described so far come from temperate biomes, where the affluent universities who can afford genetics labs are based. It is likely that there are many more cryptic species in the tropics, only to be found if time and resources are invested in looking for them.

The availability of genetic data has revolutionised taxonomy in several fields, particularly fungi, where species were traditionally separated on the basis of their reproductive fruiting bodies, which were seldom visible and often difficult to tell apart. Another important use comes in the identification of **biological control** agents. Rather than attempting to introduce a generalist predator of a pest species, identification of the correct cryptic species could lead to more effective control.

Mosquito taxonomy is particularly prone to cryptic species, with many taxa recognised as species complexes. For example, the main malaria vector in Africa is *Anopheles gambiae*, a complex which contains at least seven constituent species. Each of these varies in its larval habitats, feeding preferences and behaviour and therefore—critically—in its role in the transmission of malaria (White et al., 2011). The most efficient control of mosquito populations is usually through their larvae; thus by identifying where the most dangerous species breed and targeting them, lives could be saved.

The most important contributions from genetics have come when combined with traditional taxonomy and natural history. Morphological studies are hampered by the biases inherent in human sensory capabilities, which make us highly competent in recognising distinct taxa in groups where vision is important (e.g. birds) but less so when other mechanisms of reproductive isolation are involved. Often in extreme habitats quite unrelated taxa end up looking broadly similar through convergent evolution and stasis because there are few options available for variation in traits (e.g. all tundra plants have small, tough leaves).

Genetics can direct investigations down new and unexpected avenues. Meegaskumbura et al. (2002) studied a group of Sri Lankan frogs, where 18 species were known, but many more were suspected to be present. Reproductive isolation mechanisms were more often the result of olfactory cues and acoustic signals than appearance, and variation in these traits was shown to correspond to the phylogeny as revealed through genetics. Combining these various strands of evidence in an integrative approach allowed over 100 species to be recognised, each of which would satisfy at least one of the main concepts.

The problem still remains that no consistent definition has been applied to cryptic species, which gets us no closer to an agreed definition for species themselves, even while the idea has proved immensely helpful. The incorporation of genetic evidence has led to a range of new and ever more refined species concepts (Hausdorf, 2011), though these direct us ever further away from a practical definition that can be applied to identify organisms in the field. Cryptic species are also another major contributor to recent patterns of species inflation. Cases include Lücking et al., (2014), where a lichen which 10 years previously was thought of as a single species was split into at least 126 species (and an estimated total of over 400). This forces us to reconsider our use of the word 'species'.

2.3 Solving the riddle

Where have we left go to? We began with the old question of whether species were best defined by appearance or reproduction and then found that we were omitting important information by neglecting their evolutionary background and genetics. Each of these has provided crucial information that illustrates real differences in the biology which we would like to capture. Many attempts to unite the multiple definitions of species have failed as pluralistic approaches quickly fall apart when faced with contentious cases. In response to this, taxonomists working within different clades have developed idiosyncratic approaches that work well within their own species groups, only exacerbating the lack of comparability across the natural world.

The simplest way to split this Gordian knot is to break free from our tendency to view species as fixed, immutable entities, locked to the labels attached to dessicated specimens in museums or herbaria. Nature does not draw clear lines around its constituent parts, and we should treat any attempts to do so as necessarily limited. That is not to say that the concept of a species is useless; instead we should view a species name as a hypothesis, a statement about how we believe a particular population (or metapopulation) to be behaving (Hey et al., 2003). To state that something is a species is to say that it derives from a unique lineage, following a separate evolutionary path, and that this marks it as distinct from its relatives and ancestors. Taxonomic characters are merely indicators that we can use as starting points to help identify them in the field.

Some authors have suggested that instead of viewing nature as divided into species, we should instead identify Evolutionarily Significant Units (ESUs), populations which are evolving along separate trajectories. This also captures an important feature of species: they are dynamic in time, continuously evolving, and a taxonomy based solely on specimens in drawers will become as ossified as the fossils it preserves. The analogy of the evolutionary tree is useful here; the tips of the twigs continue to grow, and it would be foolish to mark any location as the end point. Our definitions of species will need to be constantly revised and updated as new pathways of evolution emerge. Whether we call them species or ESUs, this dynamic view of the tree of life is a more helpful way to view the components of natural systems than a simple list of ingredients.

2.4 Coda: Species richness

Counting species matters because, even though common perception doesn't match up to reality, there is a broad public awareness of the term and they are an agreed currency in ecology and conservation. When people say 'biodiversity', what they often mean is species richness. Fortunately the number of species correlates well with other features of natural systems. In terms of broad taxonomy, larger numbers of species correspond to greater numbers of higher taxonomic classes (families, genera), as well as the range of traits and ecological functions they perform. For example, a higher species richness of leaf-chewing insects correlates with the range of damage types found on leaves (Carvalho et al., 2014). Clearly species matter for understanding ecological processes. Notably, however, the number of species is a poor predictor of phylogenetic structures or genetic richness. When we use species as a metric, we need to remember what they do—and do not—represent.

2.5 Conclusions

At the heart of all ecological studies is the species, but this masks an inherent uncertainty about how we define and differentiate among them. This debate stretches back long before Darwin and has spawned a plethora of competing ideologies. There are multiple definitions of what constitutes a species, differing in the relative importance of morphology, reproduction and phylogeny. Those approaching from an evolutionary perspective tend to emphasise the question of how to decide whether two populations have become reliably distinct; ecologists, on the other hand, are usually more concerned with determining what the minimum unit is for describing the composition of natural systems. Resolving the conflict requires us to move on from a restrictive view of species as immutable entities and embrace a more dynamic viewpoint. Species names are really hypotheses for identifying evolutionarily significant units in the field. This provides an appreciation of the constant change within species as well as the linkages between them.

2.5.1 Recommended reading

Bickford D., Lohmann D.J., Sodhi N.S., Ng P.K.L., Meier R., Winker K., Ingram K.K. & Das I. (2006). Cryptic species as a window on diversity and conservation. *Trends in Ecology and Evolution* 22, 148–155.

Hey J., Waples R.S., Arnold M.L., Butlin R.K. & Harrison R.G. (2003). Understanding and confronting species uncertainty in biology and conservation. *Trends in Ecology and Evolution* 18, 597–603.

Kunz W. (2012). *Do Species Exist? Principles of Taxonomic Classification*. Wiley-Blackwell.

2.5.2 Questions for the future

- Is it possible to come up with a truly unified and general definition of species that satisfies ecologists, taxonomists and evolutionary biologists?
- Species designations are seldom comparable among groups of organisms, and taxonomy rarely represents consistent evolutionary timescales. Are all species equal?
- Might there be something better we could count instead?

References

Alcaide, M., E. S. C. Scordato, T. D. Price, and D. E. Irwin, 2014. Genomic divergence in a ring species complex. *Nature* 511:83–85.

Allendorf, F. W., R. F. Leary, P. Spruell, and J. K. Wenburg, 2001. The problem with hybrids: setting conservation guidelines. *Trends in Ecology & Evolution* 16:613–622.

APG III, 2009. An update of the Angiosperm Phylogeny Group classification for the orders and families of flowering plants: APG III. *Botanical Journal of the Linnaean Society* 161:105–121.

Berlin, B., D. E. Breedlov, and P. H. Raven, 1966. Folk taxonomies and biological classification. *Science* 154:273.

Bickford, D., D. J. Lohman, N. S. Sodhi, P. K. L. Ng, R. Meier, K. Winker, K. K. Ingram, and I. Das, 2007. Cryptic species as a window on diversity and conservation. *Trends in Ecology & Evolution* 22:148–155.

Carvalho, M. R., P. Wilf, H. Barrios, D. M. Windsor, E. D. Currano, C. C. Labandeira, and C. A. Jaramillo, 2014. Insect leaf-chewing damage tracks herbivore richness in modern and ancient forests. *PLoS ONE* 9:e94950.

Cope, T. and A. Gray, 2009. *Grasses of the British Isles*. Botanical Society of the British Isles.

Cracraft, J., 1989. Speciation and its ontology: the empirical consequences of alternative species concepts for understanding patterns and processes of differentiation. In D. Otte and J. Endler, editors, *Speciation and its Consequences*, pages 28–59. Sinauer.

Cracraft, J., 1992. The species of the birds-of-paradise (Paradisaeidae): applying the phylogenetic species concept to a complex pattern of diversification. *Cladistics* 8:1–43.

Darwin, C., 1859. *The Origin of Species by Means of Natural Selection*. John Murray, first edition.

Dexter, K. G., T. D. Pennington, and C. W. Cunningham, 2010. Using DNA to assess errors in tropical tree identifications: how often are ecologists wrong and when does it matter? *Ecological Monographs* 80:267–286.

Godfray, H. C. J., 2002. Challenges for taxonomy. *Nature* 417:17–19.

Hausdorf, B., 2011. Progress toward a general species concept. *Evolution* 65:923–931.

Hey, J., 2001. The mind of the species problem. *Trends in Ecology & Evolution* 16:326–329.

Hey, J., R. S. Waples, M. L. Arnold, R. K. Butlin, and R. G. Harrison, 2003. Understanding and confronting species uncertainty in biology and conservation. *Trends in Ecology & Evolution* 18:597–603.

Irwin, D. E., S. Bensch, J. H. Irwin, and T. D. Price, 2005. Speciation by distance in a ring species. *Science* 307:414–416.

Isaac, N. J. B., J. Mallet, and G. M. Mace, 2004. Taxonomic inflation: its influence on macroecology and conservation. *Trends in Ecology & Evolution* 19:464–469.

Lagache, L., J.-B. Leger, J.-J. Daudin, R. J. Petit, and C. Vacher, 2013. Putting the Biological Species Concept to the test: using mating networks to delimit species. *PLoS ONE* 8:e68267.

Leffler, E. M., K. Bullaughey, D. R. Matute, W. K. Meyer, L. Ségurel, A. Venkat, P. Andolfatto, and M. Przeworski, 2012. Revisiting an old riddle: what determines genetic diversity levels within a species? *PLoS Biology* 10:e1001388.

Lücking, R., M. Dal-Forno, M. Sikaroodi, P. M. Gillevet, F. Bungartz, B. Moncada, A. Yánez-Ayabaca, J. L. Chaves, L. F. Coca, and J. D. Lawrey, 2014. A single macrolichen constitutes hundreds of unrecognized species. *Proceedings of the National Academy of Sciences of the United States of America* 111:11091–11096.

Mayr, E., 1942. *Systematics and the Origin of Species*. Columbia University Press.

Mayr, E., 1957. Species concepts and definitions. In E. Mayr, editor, *The Species Problem*. American Association for the Advancement of Science.

Meegaskumbura, M., F. Bossuyt, R. Pethiyagoda, K. Manamendra-Arachchi, M. Bahir, M. C. Milinkovitch, and C. J. Schneider, 2002. Sri Lanka: an amphibian hot spot. *Science* 298:379.

Meiri, S. and G. M. Mace, 2007. New taxonomy and the origin of species. *PLoS Biology* 5:e194.

Merriam, C. H., 1918. Review of the grizzly and big brown bears of North America (genus *Ursus*) with the description of a new genus *Vetularctos*. *North American Fauna* 41:1–136.

Mora, C., D. P. Tittensor, S. Adl, A. G. B. Simpson, and B. Worm, 2011. How many species are there on earth and in the ocean? *PLoS Biology* 9:e1001127.

Nicholson, H. A., 1872. *A Manual of Zoology*. Appleton & Company.

Schlick-Steiner, B. C., F. M. Steiner, K. Moder, B. Seifert, M. Sanetra, E. Dyreson, C. Stauffer, and E. Christian, 2006. A multidisciplinary approach reveals cryptic diversity in Western Palearctic *Tetramorium* ants (Hymenoptera: Formicidae). *Molecular Phylogenetics and Evolution* 40:259–273.

Smith, M. A., J. J. Rodriguez, J. B. Whitfield, A. R. Deans, D. H. Janzen, W. Hallwachs, and P. D. N. Hebert, 2008. Extreme diversity of tropical parasitoid wasps exposed by iterative integration of natural history, DNA barcoding, morphology, and collections. *Proceedings of the National Academy of Sciences of the United States of America* 105:12359–12364.

White, B. J., F. H. Collins, and N. J. Besansky, 2011. Evolution of *Anopheles gambiae* in relation to humans and malaria. *Annual Review of Ecology, Evolution, and Systematics* 42:111–132.

CHAPTER 3

The history of life

3.1 The big question

In order to place current patterns of species richness and diversity into context, we need to look back at the patterns of life throughout Earth's history and ask whether the natural systems we see today are typical. When assessing current rapid rates of human-driven extinction, it is also valuable to see what previous natural rates have been and how life has responded to extreme global events. To do so requires an understanding of how life evolved since its first emergence and depends upon evidence recovered from the fossil record as well as more modern molecular data.

3.2 Sources of evidence

3.2.1 The fossil record

Our primary source of information about past forms of life is the fossil record. While much knowledge on the pathways of evolution has come from fossils, as a record of diversity they are difficult to interpret (Holland, 2010). The main problem is that the record is incomplete: at best around 1% of species have been preserved. Moreover, the representation of taxa is biased towards species that were abundant, large and widespread, all of which increase the likelihood that their bodies would end up in the rare conditions necessary for fossilisation to take place and that we would subsequently find them. Only species with hard body parts—bones, shells or wood—tend to be fossilised (though some soft-bodied organisms have been identified) and mostly those from environments which were likely to preserve fossils. Finally, a major difficulty arises when we try to use the data for extrapolation of species richness in the past since 95% of fossils are of marine species, whereas in the modern world 85% of species are terrestrial, most of which are insects, which seldom produce fossils (notwithstanding occasional preservation in amber). This reflects a genuine difference; for most of the history of life, the oceans contained a greater number of species (Vermeij and Grosberg, 2010).

As an example of how sparse the record can be, we need only look to the dinosaurs, whose dominance falls within a relatively narrow period of Earth's history from 230 Mya until their eventual extinction 65.5 Mya. A study by Dodson (1990) estimated that there were between 900 and 1200 genera, of which at the time 285 had been recovered as fossils, from which 336 species had been described. The problem with making an effective estimate of the total species richness of these large animals, prime candidates for discovery, was that around 50% of genera were known from only a single fossil, and a complete skeleton had been found for only a fifth of those described. This issue is not exclusive to ancient creatures; the primates are a much more recent clade, diversifying only in the Palæogene, but scientists have still only managed to find an estimated 7% of species despite their charismatic status as our own ancestors (Tavaré et al., 2002).

Natural Systems: The organisation of life, First Edition. Markus P. Eichhorn.

3.2.2 Molecular evidence

Fortunately the paucity of fossils can be compensated for by examining their descendants and creating phylogenies that link extant species together. These can be combined with molecular clocks, which estimate the rate at which genetic changes accumulate in lineages and project backwards to predict the timing of the most recent common ancestor for any set of species. Typically the divergence times generated by this technique are earlier than the first recorded fossils in any given clade; in the case of primates the molecular evidence points towards their arising from an ancestor 90 Mya, while the first fossils appear about 55 Mya (Tavaré et al., 2002). This is exactly as expected since at the time of origin a group will necessarily be rare and restricted in distribution. It may take millions of years for them to spread and become numerous enough that it becomes likely that they will be preserved and then recovered as fossils. Once this expected time lag is taken into account, the timings of splits in the evolutionary tree often show concordance between fossil and molecular evidence, though the field of molecular dating has yet to develop the same consistent standards as are used for fossils (Wilf and Escapa, 2015).

3.3 A brief history of diversity

From the available evidence we can piece together an account of how life on Earth progressed (see Cowen, 2013, for a detailed account). The planet formed around 4,570 Mya, with continental crusts and liquid water forming by 4,400 Mya. It's possible that life evolved in these very first oceans, but we may never find out—a catastrophic asteroid bombardment around 3,900 Mya destroyed much evidence of the early Earth. Nonetheless, by 3,800 Mya there is evidence from the chemical composition of sedimentary rocks that the first primitive forms of life had already appeared. There followed a long, slow process of refinement and diversification, whose traces have largely been obscured since more advanced forms replaced their ancestors. For most of the history of Earth, life was entirely unicellular. The first stromatolites, fossilised microbial mats, appear 3,500 Mya. Considerable controversy surrounds which fossil traces represent the first evidence for metazoans (multicellular eukaryotes), though the question may not deserve the attention it receives, since lab experiments have demonstrated that evolving multicellularity is relatively straightforward (Ratcliff et al., 2012). It might have happened multiple times.

The real excitement began 520 Mya during the Cambrian, when metazoans suddenly and dramatically diversified into a range of bizarre and exquisite forms. The causes of this sudden outbreak of evolutionary innovation are contested. One possibility is that increased oxygen concentrations permitted the evolution of carnivores, bringing new selection pressures and more complex food webs (Sperling et al., 2013), though it is likely that a number of abiotic and biotic factors were involved in kick-starting the process (Smith and Harper, 2013). The end result was that, by 500 Mya, all the major animal phyla were present. The story of evolution since the Cambrian has largely been one of consolidation and diversification of these standard life forms, while the most peculiar and experimental phyla went extinct.

It is tricky to estimate numbers of species through geological time, though possible to gain a reasonable impression of genus-level diversity for marine invertebrates, which have the best fossil record (Fig. 3.1; Alroy et al., 2008). Estimating the diversity of life in previous eras points towards several further periods when bouts of evolutionary innovation took place. Following the Cambrian explosion there was a continued surge through to the mid-Devonian. Another spike occurred in the Permian with a final rise in the Jurassic and Cretaceous. Since then there has been relatively little change in total diversity. On land the patterns are harder to discern, though there appears to have been an approximately exponential increase in the number of families since life first emerged on land in the Silurian (Benton, 1995).

The accumulation of diversity has not been smooth. Extinction is a natural process, and global species richness trends represent a balance between

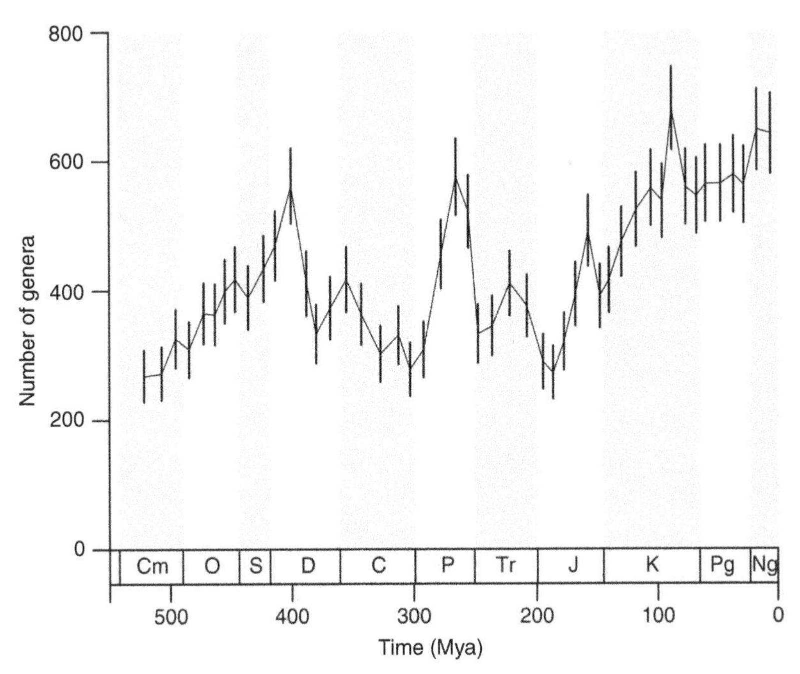

Figure 3.1 Number of genera of marine invertebrates since the Cambrian, grouped into 11 My bands with 95% confidence intervals. Periods: Cm, Cambrian; O, Ordovician; S, Silurian; D, Devonian; C, Carboniferous; P, Permian; Tr, Triassic; J, Jurassic; K, Cretaceous; Pg, Paleogene; Ng, Neogene. (*Source*: Alroy et al. (2008). Reproduced with permission from American Association for the Advancement of Science.)

speciation and extinction. Even static diversity levels mask continual turnover in the identity of those species. As a result, almost all the species that have ever existed are extinct. Those seen in the modern world are not survivors, but descendants and replacements.

Five periods in Earth's history are classically recognised as mass extinctions (Raup, 1994). These took place at the end of each of the Ordovician, Devonian, Permian, Triassic and Cretaceous. The causes of these were varied and still uncertain, though dramatic events such as asteroid impacts and volcanism no doubt played a part. In three of these events 65–85% of all marine species went extinct (generic richness was more robust; see Fig. 3.1). Though the most famous mass extinction occurred at the end of the Cretaceous, when the dinosaurs died out 66 Mya, the most dramatic was at the end of the Permian 250 Mya, coinciding with a period of massive volcanic activity in the region of modern Siberia. At this point 95% of marine species are thought to have vanished from the Earth (Sepkoski, 1984); mineral deposits after this point look eerily empty compared to the abundance of life that had been present. It seems that life itself was fortunate to survive. Each event was followed by a marked change in the dominant species groups.

Among plants there has been a turnover in the major groups that dominate the Earth. When plants evolved roots allowing them to colonise land, the **pteridophytes** (ferns and horsetails) were first to spread. These plants possess neither flowers nor seeds and reproduce solely through spores. **Gymnosperms** (seed-producing plants) evolved from them in the Carboniferous, but still in this period the pteridophytes were the main form of plant life. Despite their present diminished status, at the time they formed gigantic forests. A coal mine in Pennsylvania was discovered to have preserved an entire standing forest of lycopsid trees, a group known

Figure 3.2 An artistic reconstruction of a Carboniferous rain forest dominated by lycopsids. (*Source*: DiMichele et al. (2007). Reproduced with permission from the Geological Society of America.) (*See colour plate section for the colour representation of this figure.*)

today as the club mosses, allowing the structure of the forest to be reconstructed (Fig. 3.2; DiMichele et al., 2007). These did not look like the forests of the modern world; the trees were around 40 m tall but had open canopies with no branches and green, photosynthesising bark.

By the Cretaceous the **angiosperms**, modern flowering plants, had evolved, but the gymnosperms were at their peak. In dinosaur movies it would be more accurate to show them walking among gigantic coniferous trees than the lush foliage that film studios more often resort to (and they were certainly not browsing on plains of grass). It was only after the extinction of the dinosaurs that angiosperms reached their current level of dominance.

Similar trends can be seen among animals, with richness of various taxa rising and falling through time (Fig. 3.3). The trilobites, a mainstay of the fossil record up until their disappearance at the end of the Permian, were succeeded by the articulate Brachiopoda, which then declined as the Gastropoda and Bivalvia began to rise. Nevertheless, all these groups shared the planet for hundreds of millions of years, and there was no inevitability that the patterns of dominance seen in the modern world would arise.

How did this occur in practice? There is limited evidence for competitive replacement—in other words, it is not simply that a new set of species evolved that were superior to their forebears and rapidly drove them extinct. Instead it is more likely that shifting environmental conditions were responsible for changes within groups, over time favouring a varying set of adaptations and the species that bear

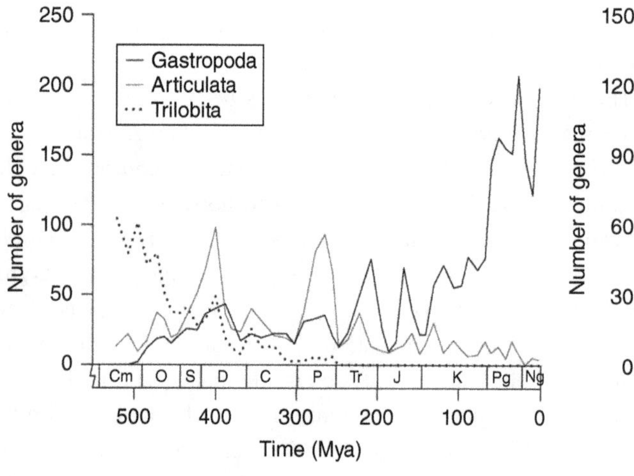

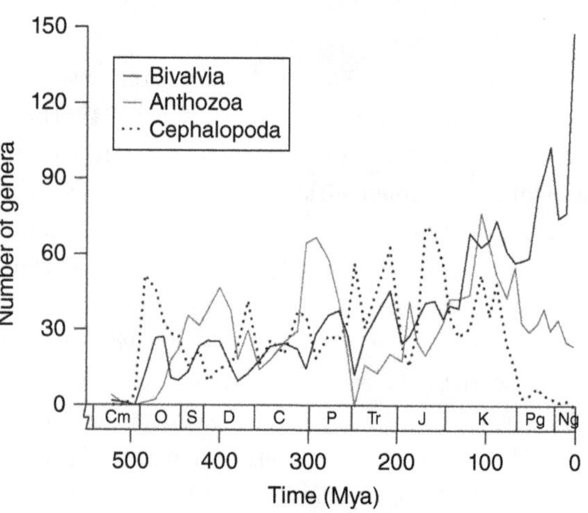

Figure 3.3 Trends in generic richness of six important groups of marine invertebrates; geological periods as in Fig. 3.1. (*Source*: Alroy et al. (2010). Reproduced with permission from the American Association for the Advancement of Science.)

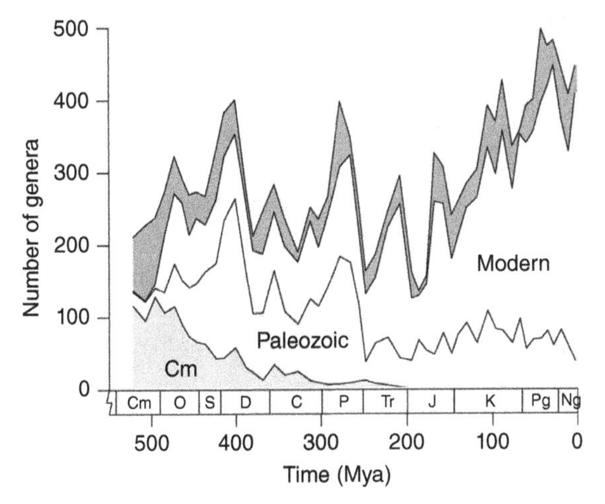

Figure 3.4 Generic diversity trends for three main marine evolutionary faunas; unlabelled area represents genera with no definite association (Alroy, 2010). Geological periods as in Fig. 3.1.

them. The faunas of the three dominant periods can be combined to show this gradual succession (Fig. 3.4). Most of the species in genera that emerged in the Cambrian went extinct at the end of the Permian, but had long been in decline, dominated by Palæozoic genera. Note that genera comprising the modern fauna have been present since the Cambrian, though only reached their present dominance in the Triassic.

It appears that rates of speciation and extinction are linked on global scales. Falls in diversity tend to be followed by reduced extinction rates, which would effectively fall to zero if diversity dropped by 90%, and by higher rates of origination (Alroy, 2008). Many ecologists now believe we are in the midst of the sixth mass extinction, with the present high rate driven by human pressure. The length of time it may take life to recover from our impacts is sobering—perhaps 10–40 My, and with a very different composition thereafter. Life will quite literally never be the same again.

The dynamic rates of origination and extinction of genera over the history of life are shown in Fig. 3.5. Note that there is no evidence that extinction rates are any more volatile than origination, and though the mass extinctions stand out, they fall along a continuum, making up only 4% of all extinctions in the past 400 My.

Whether a limit to global diversity exists remains contentious. Some argue that there must be one, although it is probable that the ceiling has changed through time (Alroy et al., 2008). Others have contested this. While evidence exists that diversification rates tend to decline within groups over time, continued speciation (especially on land) suggests

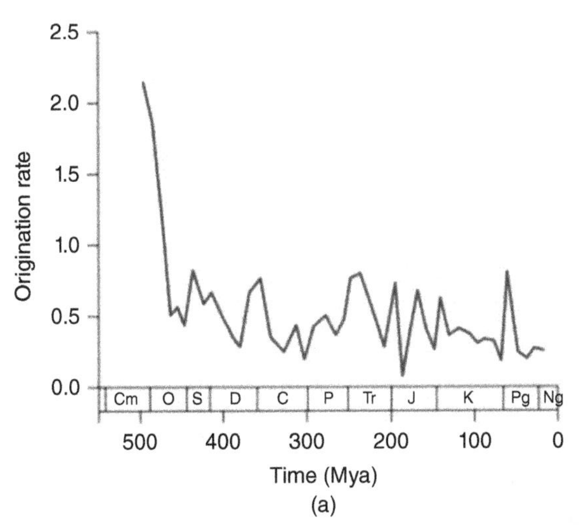

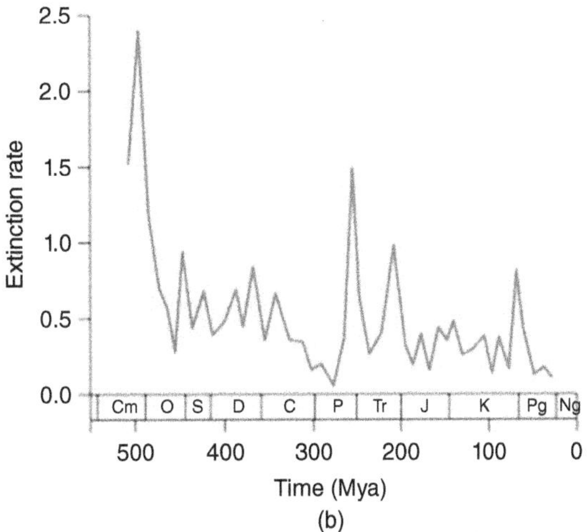

Figure 3.5 Instantaneous rates of (a) origination and (b) extinction of genera in the fossil record of marine invertebrates; geological periods as in Fig. 3.1. (*Source*: Alroy et al. (2008). Reproduced with permission from the American Association for the Advancement of Science.)

either that there is no limit or we are a long way from reaching it (Morlon et al., 2010). There seems to be no relationship between the age of a clade and the number of species it contains, indicating that individual groups are not limited, even if an overall cap on global species richness occurs (Rabosky et al., 2012). We will return to the causes of variation in global species richness through time in Chapter 17.

On average, given normal background rates of extinction, most species persist for 5–10 My, though the majority fall on the lower end of this range and many shorter-lived species have probably left no trace (May et al., 1995). Among marine animals, the longest-lived genus spans 160 My in the fossil record, less than 5% of the history of life, and even this was an exception (Raup, 1994). Note that anatomically modern humans only appeared about 200,000 years ago; we may yet become one of the shorter lived of species.

As with the major groups of life, we can ask questions about the dynamics of species' rise and fall. The pattern for both genera and species is that they tend to increase in abundance, peak briefly and then decline at the same rate (Fig. 3.6; Foote et al., 2007). The trajectory is gradual in both directions. There is no evidence of either an initial explosive increase in numbers when a new species arises, a plateau as they become the incumbent in a particular niche or a sudden crash when they are displaced. Species that go extinct tend to have been in decline for some time, and within clades their diminution is caused by both higher extinction rates and reduced speciation (Quental and Marshall, 2013).

3.4 Uneven diversity

A consistent pattern seen throughout modern nature, mirrored in the fossil record, is that there are many phyla of life, but most of these are not particularly diverse. A majority of species come from a small number of groups. For instance, most animals are arthropods, of which most are insects, and within insects the dominant group is the Coleoptera (beetles). This pattern supposedly prompted the great geneticist J.B.S. Haldane to joke that the creator must have had 'an inordinate fondness for beetles'. Similarly, most of the Mollusca (molluscs) are in the

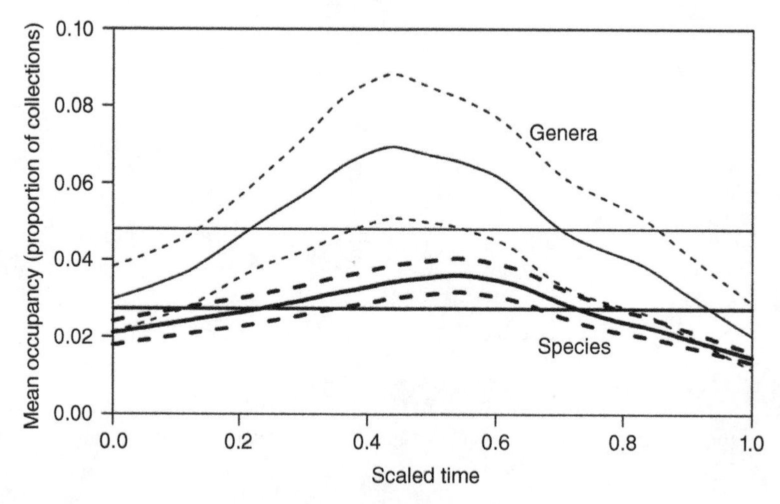

Figure 3.6 Occupancy in collections of Cenozoic (66 Mya to present) molluscan species (thick lines) and genera (thin lines) of New Zealand, with durations scaled to unit length from time of first appearance in the fossil record to their last appearance, average ± 1 SE. Horizontal lines show the overall mean occupancy of all taxa. The amplitude of the genera curve is higher as each contains multiple species. (*Source*: Foote et al. (2007). Reproduced with permission from the American Association for the Advancement of Science.)

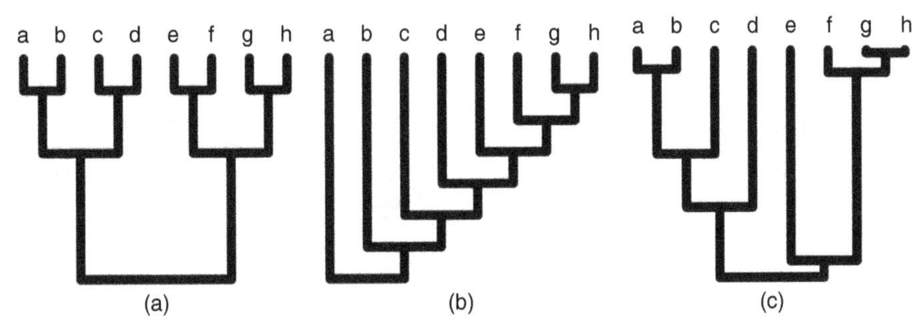

Figure 3.7 Evolutionary trees which are (a) perfectly balanced, (b) completely imbalanced (pectinate) or (c) based on random speciation and extinction. (*Source*: From Harmon (2012). Used under CC-BY-SA 2.5 http://creativecommons.org/licenses/by-sa/2.5/.)

class Gastropoda, comprising the slugs and snails. This pattern of uneven diversity occurs at all levels of life, even within groups. The 28,000 species of bony fish (Osteichthyes) almost all fall in the Actinopterygii (ray-finned fish), of which the order Perciformes (perch-like fish) makes up around 10,000 species, of which at least 1,650 are cichlids (family Cichlidae). Even within the 220 cichlid genera, 228 species are placed in a single genus *Haplochromis*. Similar fractal patterns of diversity occur throughout the reptiles (Pincheira-Donoso et al., 2013).

What is the cause of this widespread pattern? Three general explanations have been proposed:

- **It's an artefact**: When given a selection of objects and instructed to place them into groups, humans employ arbitrary criteria that end up with a few large sets and lots of small ones. This is an inadequate explanation as we know that genuine evolutionary trends have created these patterns, so there must be a biological process at work.
- **Random processes**: If you create an evolutionary tree with random speciation events, it's highly unlikely that you will produce one that is symmetrical (Fig. 3.7). With random extinction there's even less chance of a phylogeny being balanced. Even assuming such a model, real groups are still more clustered than random expectation.
- **Group traits**: Perhaps the evolution of new traits allows particular groups to diversify? An innovation such as an organ or body form might permit species to evolve into novel niche space. Nevertheless, determining why such processes start is a matter of guesswork and usually ends up as a 'just-so' story that is eminently plausible with hindsight but untestable and makes it no easier to predict where the next source of diversity will come from. It has, for instance, been suggested that the evolution of the elytra (wing cases) was the defining innovation that allowed beetles to diversify, but how could you possibly test such a proposition? One way is to examine an innovation that has evolved independently on numerous occasions. Extrafloral nectaries are a feature of over 100 families of plants which have mutualistic relationships with arthropods that defend them in return for sugar. Weber and Agrawal (2014) found that families with extrafloral nectaries had diversification rates twice those without.

In the end, we are forced to conclude that we don't really know why diversity is so unbalanced within the tree of life. It might be due to a combination of these factors, many of which are inherently stochastic and unpredictable.

3.5 Conclusions

Examining historical patterns of diversity suggests that it exploded in the Cambrian and has fluctuated ever since, though it appears to be at a relatively high point in the modern world. Within this broad trend there have been many cases of groups radiating, stabilising and then gradually declining to extinction. It is difficult to find clear evidence of a maximum number of species on Earth, nor of a 'correct' number which we should aspire to maintain. The structure of life indicates that most groups aren't very diverse, with the bulk of global species richness coming from a few taxonomic groups. This appears to have been a common pattern throughout Earth's history. Extinctions occur continually, with occasional mass extinction events leading to shifts in clade dominance and allowing new groups to emerge.

3.5.1 Recommended reading

Alroy J. (2008). Dynamics of origination and extinction in the marine fossil record. *Proceedings of the National Academy of Sciences of the United States of America* 105, 11536–11542.

Alroy J. (2010). The shifting balance of diversity among major marine animal groups. *Science* 329, 1191–1194.

Benton M.J. & Emerson B.C. (2007). How did life become so diverse? The dynamics of diversification according to the fossil record and molecular phylogenetics. *Palaeontology* 50, 23–40.

3.5.2 Questions for the future

- Is there a limit to global species richness?
- Why are some groups more diverse than others and can we predict which will diversify in the future?
- How relevant is the marine fossil record when 85% of modern species are terrestrial?

References

Alroy, J., 2008. Dynamics of origination and extinction in the marine fossil record. *Proceedings of the National Academy of Sciences of the United States of America* 105:11536–11542.

Alroy, J., 2010. The shifting balance of diversity among major marine animal groups. *Science* 329:1191–1194.

Alroy, J., M. Aberhan, D. J. Bottjer, M. Foote, F. T. Fürsich, P. J. Harries, A. J. W. Hendy, S. M. Holland, L. C. Ivany, W. Kiessling, M. A. Kosnik, C. R. Marshall, A. J. McGowan, A. I. Miller, T. D. Olszewski, M. E. Patzkowsky, S. E. Peters, L. Villier, P. J. Wagner, N. Bonuso, P. S. Borkow, B. Brenneis, M. E. Clapham, L. M. Fall, C. A. Ferguson, V. L. Hanson, A. Z. Krug, K. M. Layou, E. H. Leckey, S. Nürnberg, C. M. Powers, J. A. Sessa, C. Simpson, A. Tomašových, and C. C. Visaggi, 2008. Phanerozoic trends in the global diversity of marine invertebrates. *Science* 321:97–100.

Benton, M. J., 1995. Diversification and extinction in the history of life. *Science* 268:52–58.

Cowen, R., 2013. *History of Life*. Wiley-Blackwell, fifth edition.

DiMichele, W. A., H. J. Falcon-Lang, W. J. Nelson, S. D. Elrick, and P. R. Ames, 2007. Ecological gradients within a Pennsylvanian mire forest. *Geology* 35:415–418.

Dodson, P., 1990. Counting dinosaurs: how many kinds were there? *Proceedings of the National Academy of Sciences of the United States of America* 87: 3706–3711.

Foote, M., J. S. Crampton, A. G. Beu, B. A. Marshall, R. A. Cooper, P. A. Maxwell, and I. Matcham, 2007. Rise and fall of species occupancy in Cenozoic fossil mollusks. *Science* 318:1131–1134.

Harmon, L. J., 2012. An inordinate fondness for eukaryotic diversity. *PLoS Biology* 10:e1001382.

Holland, S. M., 2010. Additive diversity partitioning in palaeobiology: revisiting Sepkoski's question. *Palaeontology* 53:1237–1254.

May, R. M., J. H. Lawton, and N. E. Stork, 1995. Assessing extinction rates. In J. H. Lawton and R. M. May, editors, *Extinction Rates*, pages 1–24. Oxford University Press.

Morlon, H., M. D. Potts, and J. B. Plotkin, 2010. Inferring the dynamics of diversification: a coalescent approach. *PLoS Biology* 8:e1000493.

Pincheira-Donoso, D., A. M. Bauer, S. Meiri, and P. Uetz, 2013. Global taxonomic diversity of living reptiles. *PLoS ONE* 8:e59741.

Quental, T. B. and C. R. Marshall, 2013. How the Red Queen drives terrestrial mammals to extinction. *Science* 341:290–292.

Rabosky, D. L., G. J. Slater, and M. E. Alfaro, 2012. Clade age and species richness are decoupled across the eukaryotic tree of life. *PLoS Biology* 10:e1001381.

Ratcliff, W. C., R. F. Denison, M. Borrello, and M. Travisano, 2012. Experimental evolution of multicellularity. *Proceedings of the National Academy of Sciences of the United States of America* 109:1595–1600.

Raup, D. M., 1994. The role of extinction in evolution. *Proceedings of the National Academy of Sciences of the United States of America* 91:6758–6763.

Sepkoski, J. J., 1984. A kinetic model of Phanerozoic taxonomic diversity. 3. Post-Paleozoic families and mass extinctions. *Paleobiology* 10:246–267.

Smith, M. P. and D. A. T. Harper, 2013. Causes of the Cambrian Explosion. *Science* 341:1355–1356.

Sperling, E. A., C. A. Frieder, A. V. Raman, P. R. Girguis, L. A. Levin, and A. H. Knoll, 2013. Oxygen, ecology, and the Cambrian radiation of animals. *Proceedings of the National Academy of Sciences of the United States of America* 110:13446–13451.

Tavaré, S., C. R. Marshall, O. Will, C. Soligo, and R. D. Martin, 2002. Using the fossil record to estimate the last common ancestor of extant primates. *Nature* 416:726–729.

Vermeij, G. J. and R. K. Grosberg, 2010. The great divergence: when did diversity on land exceed that in the sea? *Integrative and Comparative Biology* 50:675–682.

Weber, M. G. and A. A. Agrawal, 2014. Defense mutualisms enhance plant diversification. *Proceedings of the National Academy of Sciences of the United States of America* 111:16442–16447.

Wilf, P. and I. H. Escapa, 2015. Green Web or megabiased clock? Plant fossils from Gondwanan Patagonia speak on evolutionary radiations. *New Phytologist*, 207: 283–290.

CHAPTER 4

How many species are there?

4.1 The big question

How many species exist? This question has been tackled by many people using a variety of approaches, but a resolution is still a long way off, and estimates have fluctuated widely over recent years. Many biologists believe that we are in the midst of the sixth mass extinction event in Earth history. Current rates are around 1,000 times above background levels and are expected to increase in future by another 10-fold at least (Fig. 4.1; May 2010). Placed in the context of the previous chapter, this is equivalent to the obliteration of the dinosaurs. Yet while it would seem difficult to exaggerate the importance of such a dramatic alteration in the structure of life, some nevertheless manage to do so. A cursory scan of the websites of the more strident conservation NGOs quickly reveals claims such as that 100 species per day are being lost due to destruction of tropical rain forests alone. Can this really be true, and where is the evidence? No one doubts that extinctions are happening, but how many, and whether it matters, are still open for discussion. What chance do we have of saving biological diversity if we have no reasonable measure of how much there is?

This chapter assesses estimates of the total number of species present on Earth. Three main lines of evidence are followed: discovery rates, scaling relationships, and sample-based approaches. The limitations and assumptions underlying each make a large difference to the final figure.

4.2 How can we not know?

It is often a source of amazement that such an apparently simple question—how many species are there?—has no clear answer. Confusion over what exactly comprises a species notwithstanding, the fact that until very recently no recognised catalogue existed is remarkable. By contrast, the US Library of Congress states that at the end of 2013 it contained 23,592,066 catalogued books among a total collection of 158,007,115 items. We have better mechanisms for listing the number of stars than for the species that share our planet. In recent years some amends have been made, including the Catalogue of Life[1] and the World Register of Marine Species,[2] but there is much catching up to be done.

It took a long time for the sheer scale of the problem to be recognised. The naturalist John Ray (1627–1705) was one of the first to attempt a catalogue of species, working at a time when the boundaries of the known world were still increasing. He wrote:

> ... in consequence of having discovered a greater number of English moths and butterflies I am induced to consider that the total number of British insects might be about 2,000; and those of the whole earth 20,000.

This was a radical underestimate. The naturalist Carl Linnæus (1707–1778) attempted a catalogue of

[1] http://www.catalogueoflife.org.
[2] http://www.marinespecies.org.

Natural Systems: The organisation of life, First Edition. Markus P. Eichhorn.

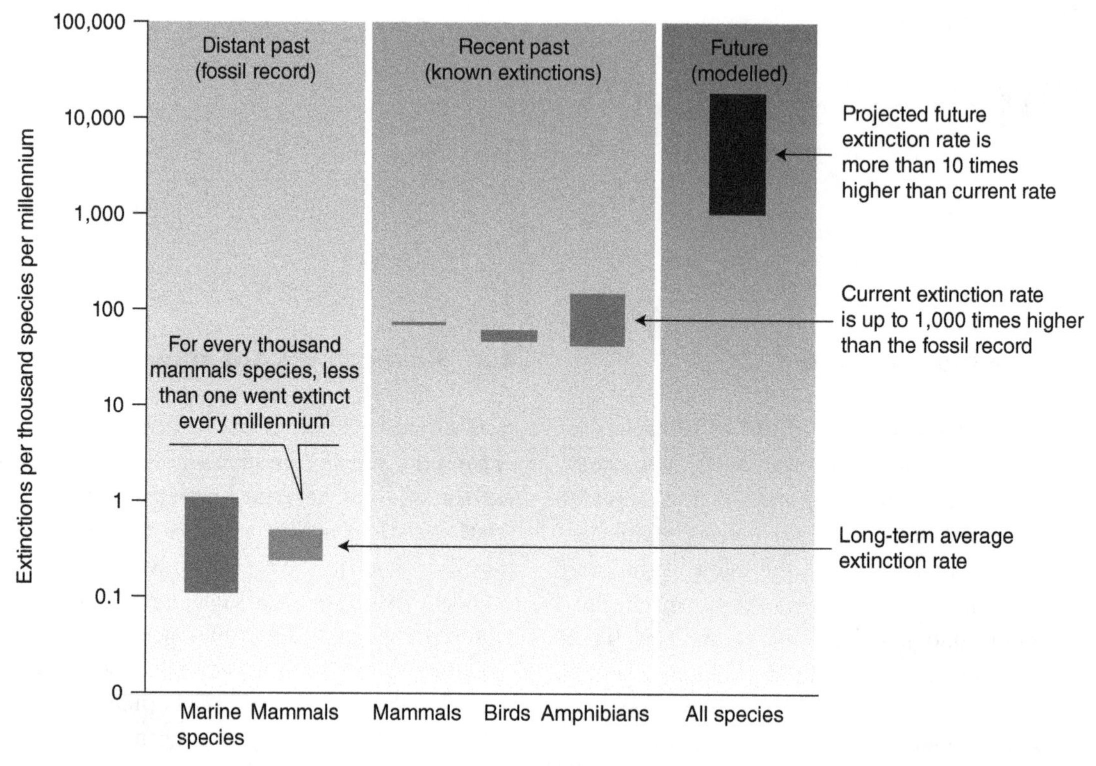

Figure 4.1 Estimated extinction rates (per thousand species per millenium) in the fossil record, the present day and the near future. (*Source*: From May (2010). https://creativecommons.org/licenses/by/4.0/.)

all known forms of life, though within his lifetime found that the influx of new species being discovered around the world outstripped his ability to keep pace. In the 1758 edition of his classic book *Systema Naturæ*, which set the standards for nomenclature in taxonomy, only around 9,000 species were included, most of which were from temperate climes. The tropics remained a mystery. Since that time there has been a seemingly endless stream of new discoveries, though the pattern differs among groups of life (Fig. 4.2).

The lack of a centralised recording scheme makes it difficult to combine estimates of known species from different sources into a single total. An attempt was made by Stork (1988), who came up with a ballpark figure of 1.8 million named species, but this is subject to a great deal of uncertainty. The Catalogue of Life, as of April 2014, recognised 1,578,063 species, yet still falls far short of the true total. There are many 'new' species that lack names and formal descriptions, languishing in museums and herbaria due to lack of time and resources on the part of those who are trained to complete the process (Bebber et al. 2010, Appeltans et al., 2012). Other problems haunt these figures, such as the common issues of homonyms (different species given the same name) and synonyms (different names given to the same species) that are bound to plague any endeavour without a unified organisation.

If we don't even know how many species have been named, what hope do we have for the rest, since the majority have yet to be documented?

4.3 Discovery rates

Some groups of organisms are better described than others, a phenomenon that is clear from Fig. 4.2b.

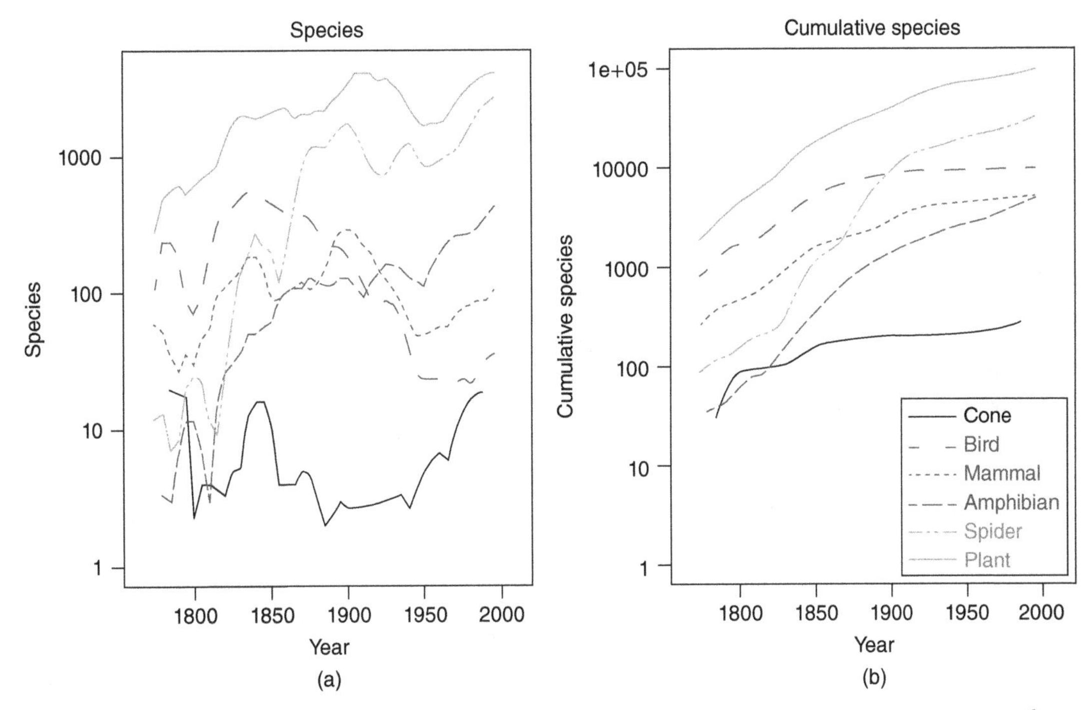

Figure 4.2 Number of species discovered (a) per 5-year interval and (b) cumulatively since 1800. (*Source*: Joppa et al. (2011). Reproduced with permission from Elsevier.) (*See colour plate section for the colour representation of this figure.*)

The numbers of known birds and mammals have been relatively stable for many years, while those of spiders or amphibians continue to climb. If we assume that these discovery curves are consistent across groups, we can project from well-known taxa to estimate species richness for those which await detailed study. Estimates using this approach tend to converge on there being about 6–7 million animal species, though there are important caveats. It is only possible to obtain accurate estimates from those groups which are already close to being complete; otherwise a large degree of error is introduced (Bebber et al., 2007). In addition, since the arrival of PCR in 1984 and novel species concepts, it is only fair to look at curves from before this time period, when a more consistent species definition was applied (see Chapter 2). This has the unfortunate side effect of meaning that the tropics are especially poorly covered since taxonomic effort in these regions has only recently accelerated. It is also suspected that most species being described today are cryptic (Scheffers et al., 2012). Such caveats severely limit the accuracy of this technique.

Attempts have been made to compensate for undersampling in the tropics. For example, Raven (1985) took well-studied groups of insects and assumed that for every temperate species there would be roughly two in the tropics. Given that two thirds of described insects were temperate, simple multiplication led to an estimate of 3–5 million species of insects worldwide.

An alternative approach used the description rates of braconid wasps, a well-resolved group of parasitoids thanks to their use as both model organisms and biological control agents (Dolphin and Quicke, 2001). They are also largely Palæarctic in distribution (see Fig. 14.3b), meaning that the number of likely species awaiting discovery in the tropics is probably small. By extrapolating discovery rates of less well-known insect groups using the pattern seen in braconids, Dolphin and Quicke arrived at an estimate of 2–3.5 million insect species.

This is probably an underestimate as a result of using a temperate group, but it nonetheless agrees approximately with Raven's study.

One assumption underlying all these attempts is the reliability of conventional taxonomy, but here some additional uncertainty creeps in. Are all the names given to species valid? We can get an impression by looking at the stability of names through time. Alroy (2002) assessed temporal patterns in the taxonomy of fossil mammals of the United States. These are intensively studied, and not particularly species rich, yet new species continue to be described. Some of these, however, turn out to be synonyms or *nomina dubia*, names that are not helpful in practice when identifying new specimens. The trouble is that there is a lag in the revision of nomenclature such that often a new species name will not be reconsidered for at least 10 years and sometimes as much as a century. This stems from the previously mentioned problem that recognising new species seems glamorous while revising old ones is fusty and dull.

Alroy's work revealed an unfortunate complication: over time, revisions meant that many species names fall out of use. Of those named up to 1,880, fewer than 60% were still considered valid. This trend does not appear to have improved over time either, meaning that somewhere in the region of

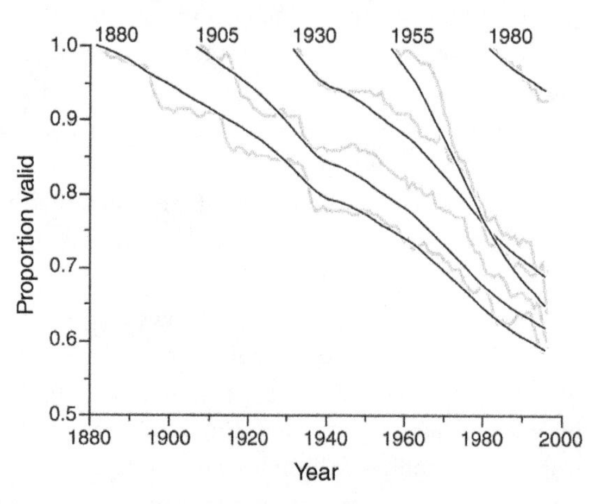

Figure 4.3 Decay in validity of species names of fossil mammals in the United States over time. (*Source*: Alroy (2002). Reproduced with permission from the National Academy of Sciences, USA.)

20–33% of all taxonomic names might turn out to be incorrect (Fig. 4.3), and perhaps even more in poorly studied groups where revisions are infrequent. Among plants upwards of 60% of names could be synonyms (Scotland and Wortley, 2003). For marine groups the average is around 40%, though with some extreme cases; a recent database assessment found 1,271 names for only 87 valid cetacean species (Appeltans et al., 2012). Any estimates based on species description rates are bound to be biased, but it is hard to know by how much.

4.4 Scaling

An entirely different technique is to use simple scaling rules in nature to extrapolate for species that have yet to be sampled. One pattern that can be exploited is in body sizes (May 1988). There are more small species than large, and this seems to follow a simple linear trend, whereby for every 10-fold decrease in body length, there are approximately 100 times as many species. Moreover, the smaller species are the ones that have been less comprehensively studied, so we can use this to estimate total numbers. Assuming that the relationship holds down to 0.2 cm, this arrives at a global estimate of 10 million species.

Sadly the rule seems to break down for organisms below 1 cm, for reasons that are poorly understood, and for microscopic organisms it is probably of little use. We also don't have a proper explanation for the existence of the relationship in the first place, and without a full mechanistic understanding, we should be careful of assuming that extrapolations are defensible.

Another more recent paper took a different tack (Mora et al., 2011). The first stage was the realisation that there was a consistent relationship between the asymptotic number of animal taxa at each level in the classification hierarchy and the level itself. In other words, the study estimated the total number of animal phyla, classes, orders, and so forth, based upon their accumulation over time, and used the relationship between these estimates to come up with a figure for the number of animal species (Fig. 4.4). This came to around 8.7 million, in tune

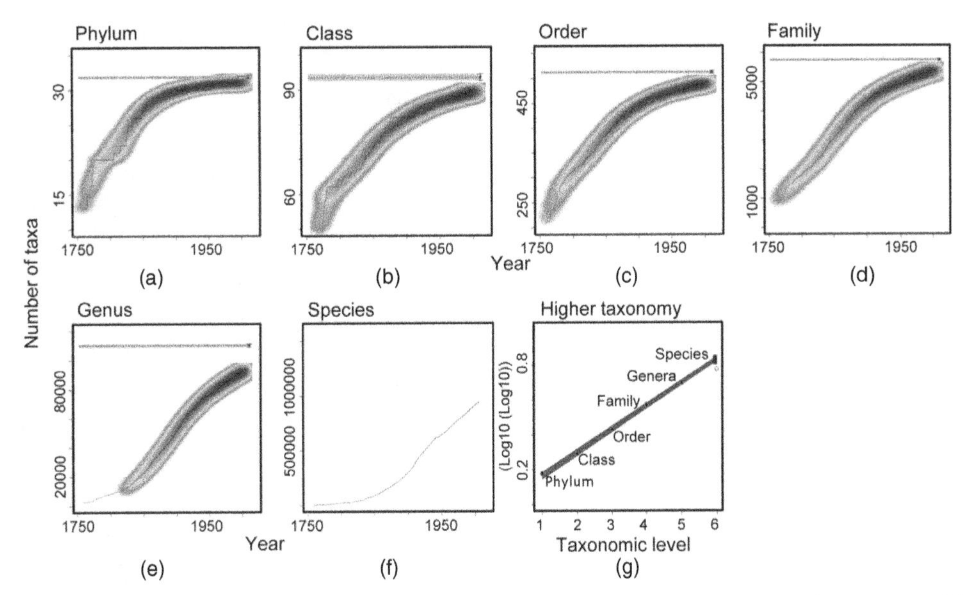

Figure 4.4 Temporal accumulation of animal taxa (solid lines) and model fits to obtain asymptotes for richness at higher taxonomic levels (a–e). Shading represents the frequency of estimates for each year across multiple models; horizontal dashed lines indicate the estimated asymptotic number of taxa, and the horizontal grey area its standard error. (f) shows the number of species described. In (g) asymptotes are extrapolated across taxonomic levels to estimate the total number of species. Black circles represent the consensus asymptotes, gray circles the catalogued number of taxa, and the box at the species level indicates the 95% confidence interval around the predicted number of species. (*Source*: From Mora et al. (2011). Used under CC-BY-SA 2.5 http://creativecommons.org/licenses/by-sa/2.5/.) (*See colour plate section for the colour representation of this figure.*)

with previous estimates, though once again we remain uncertain about the underlying mechanism. It works—but why?

4.5 Sampling-based methods

A third way to arrive at an estimate of the total number of species in the world is to scale up from a smaller and more manageable sample. This technique was famously pioneered by Terry Erwin (1982). Having recognised that the majority of species are, to a first approximation, beetles in tropical rain forests, he focussed his efforts there. This employed a technique known as canopy fogging, which involves winching an insecticide sprayer into the top of a tree then collecting and sorting the thousands of dead insects that fall to the ground. Rain forest canopies contain a greater abundance and species richness of insects than the forest floor since that is where the majority of leaves, flowers, and fruits are to be found. Canopy fogging yields vast numbers of insects, most of which are herbivores, the majority unknown to science. If you want to discover new species, then you can find scores in an afternoon.

Erwin focussed on a single species of tree in Panama, *Luehea seemannii*. His team fogged a total of 19 trees, collecting over 1,100 species. Based on this sample he scaled up to the forest, and thence to the entire world, to estimate the global total number of insect species. It is worth noting that his paper was published in a very minor journal (The Coleopterists Bulletin) but has become one of the most highly cited in ecology. This is largely because it became the fountainhead for a number of extraordinary estimates of global species richness and the more extreme claims about extinction rates. Yet the original paper had at its heart the aim of simply demonstrating how diverse and understudied were the beetles of tropical forests. We should examine this most misused of studies in more detail, including the original data and the assumptions at its heart (Table 4.1).

Table 4.1 Number of arthopod species found in various taxonomic classes, estimated host specificity and subsequent estimates of numbers of host-specific arthropod species on a single tree species.

Group	Number of species	% host specific	Number of host specialists
Herbivores	682	20	136
Predators	296	5	15
Fungivores	69	10	7
Scavengers	96	5	5
Total	1,143		163

Source: Erwin (1982). Reproduced with permission from the author.

The proportions of host-specific species were based upon educated guesswork, but these are a crucial element in what follows, especially the supposition that 20% of herbivores are host specialists, that is, will only feed on a single species of tree. The other estimates are smaller and refer to a reduced fraction of the total, so the 20% figure is the one that makes the greatest difference.

What followed was a simple back-of-a-beer mat scaling exercise. Erwin made the following apparently reasonable assumptions:

- 40% of arthropods are beetles.
- There are twice as many species in the canopy as in the forest understorey.
- In total there are 50,000 species of tropical trees.

This wound up with an estimate of 30 million species of arthropods in tropical rain forests alone, which with further extrapolation ends up at around 100 million species globally. This figure is over-quoted and was never expected to be taken entirely seriously. It did, however, focus attention on insect herbivores in tropical rain forests, as solving the mystery of their diversity was evidently key to the wider picture.

Further work has revisited this calculation and critically examined the assumptions at its core. Some served to underestimate species richness. For example, it would seem that in fact 23% of tropical rain forest arthropods are beetles and that the ratio of canopy to understorey species richness is substantially less than 2:1. Both of these amendments act to increase estimated species richness (Ødegaard, 2000).

There are however two major flaws in the reasoning that led to Erwin's estimate, and they deserve greater exposure since they have subsequently been repeated elsewhere (e.g. to give inflated estimates of the richness of parasites, fungi or deep sea fauna).

The first consideration is the relative geographical ranges of species. If tree species have broad geographical ranges and carry different insect species at various points in their range (i.e. herbivores have smaller ranges), then this will tend to increase the level of herbivore specialisation and thus species richness. On the other hand, if herbivores have broad ranges but only specialise locally (i.e. trees have smaller ranges), then overall species richness will be much lower.

Evidence in support of the latter pattern comes from a study of the plant genus *Passiflora* in South America by Thomas (1990). There are 360 species of *Passiflora* which are host to caterpillars of *Heliconius* butterflies. These are specialised feeders, and at any one site a species of butterfly will only lay eggs on a single plant species. So what happens when you compare multiple sites? Thomas surveyed 12 locations and found that on average each contained 7.2 plant species and 9.7 butterfly species. Using the logic of simple multiplication, this would imply a total of $(9.7/7.2) \times 360 = 485$ species of *Heliconius*.

How many species are there? The answer is 66, which tells us immediately that something is wrong with the rudimentary approach. In fact it turns out that the same butterfly species are present in multiple sites, but in any one place they specialise on a single plant species. The actual degree of specialisation is lower than it first appears. There may still be more going on here than meets the eye; perhaps there are cryptic species within *Heliconius* that would be revealed by genetic analyses, or else the species of *Passiflora* might be excessively divided. Without further evidence it is impossible to assess these possibilities though.

Based on an increased amount of evidence and data, Ødegaard (2000) revisited Erwin's sample-based method, accounting for variation in the known host ranges of herbivores, and came up with a figure of effective specialisation around 3–5% (much reduced from the original 20%). This resulted in an estimate of 5 million arthropod species (between 2.5 and 10 million), consistent with estimates from other methods.

In recent years there have been some extraordinary attempts to characterise the composition of tropical rain forests; we will hear more about one particular project in Chapter 5 but for now will mention only one aspect of it. In Papua New Guinea, Vojtech Novotny and a large group of researchers have collected insects from an array of tropical tree species, not simply a single species as chosen by Erwin, and also ensured that each species is actually feeding on the tree concerned (fogging tends to catch vast numbers of incidental or 'tourist' species which are probably only passing through).

Novotny et al. (2002) took 51 plant species and collected a total of 50,734 insect herbivores from them. These made up 935 species, for which they performed feeding trials to work out their actual diets. If you start by sampling a single tree species, then you have to assume that all the species on it are specialists. On moving to a second tree species, some of the insects overlap in their feeding, which brings down the observed level of dietary specialisation. The more tree species are examined, a greater the proportion of herbivore species are found on more than one of them. Eventually it transpires that there are few monophages (herbivores that can only feed on one plant species), and the majority of insect herbivores are capable of feeding on many plant species (Fig. 4.5). The average proportion of herbivorous species feeding on a particular host plant that was unique (effectively specialised) to that plant was estimated as $F_T = S_T/H_T$, that is, as the ratio of the total number of herbivorous species found on all T hosts studied (S_T), to the number of host plant records involving these hosts (H_T).

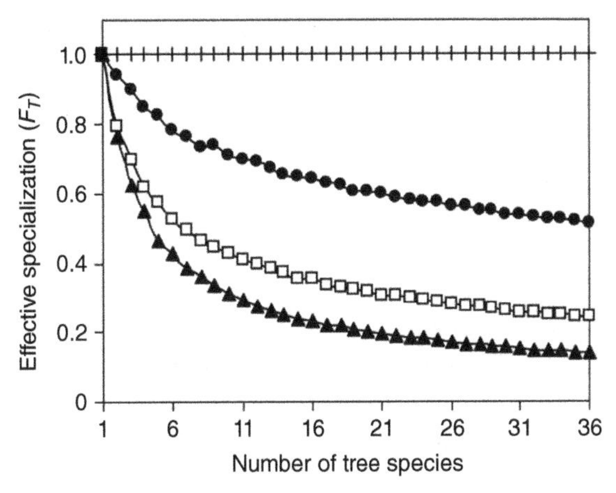

Figure 4.5 Effective specialisation F_T of insect herbivores on tropical rain forest trees in Papua New Guinea. F_T shows the average proportion of herbivore species feeding on a given plant species which are unique, and is calculated as the ratio of the total number of herbivore species found on all hosts to the number of host plant records. Samples of insect herbivore communities were taken from individual tree species, each from a different genus, then combined in random order; the mean of 100 random sequences is shown. Triangles: Orthopteroids (crickets and stick insects); squares: Coleoptera (beetles); circles: Lepidoptera (butterflies and moths); crosses: butterflies, for which all species were strict monophages. (*Source*: Novotny et al. (2002). Reproduced with permission from Nature Publishing Group.)

Interesting differences among groups emerge: Orthopteroids (crickets and stick insects) are rather unfussy, Coleoptera (beetles) fall in the middle, while Lepidoptera (moths and butterflies) are relatively specialised (Fig. 4.5). From this improved estimate of feeding specialisation it was possible to estimate a global species richness of 4–6 million arthropods, again within the range reported by others.

Still, in the back of our mind, we must retain an element of doubt—are these species really generalist feeders, or might they hide cryptic species? A study in Costa Rica by Hebert et al. (2004) gives some cause for concern. A single species of butterfly, *Astraptes fulgerator*, had been known since 1775. When genetic analyses suggested that there might be cryptic diversity, a more complex picture emerged. Even though adults looked almost identical in the eyes of entomologists, there were a range of caterpillar morphs, each of which was specialised to a particular food plant or habitat. Reconsideration suggested that there were actually 10 species present. If this is true elsewhere, then perhaps our estimates of global species richness will have to rise again.

The second, more general problem with the Erwin paper and related studies comes from the relationship between the number of species found versus the time or effort expended to collect them. This invariably forms a curve, even if it never quite reaches an asymptote in the majority of cases. The danger of a linear extrapolation from small samples is the assumption that the rate of species accumulation from a small sample will hold across all samples, which is evidently not the case. Any attempt to fit a straight line will inevitably overestimate total species richness (Fig. 4.6). This applies on all spatial scales when estimating the numbers of species present, even in a single location.

4.6 Other organisms

When assessing global species richness, most studies tend to focus purely on metazoans. Currently animals make up over 70% of named species, and plants form almost a quarter, despite there being only an estimated 350,000 plant species in total

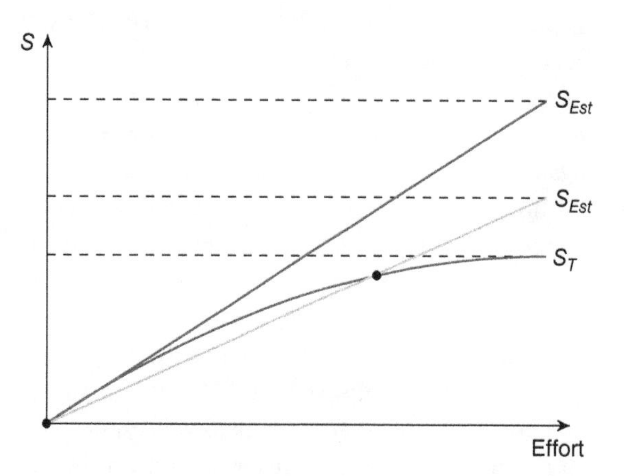

Figure 4.6 Species richness (S) against sampling effort. Asymptotic true species richness (S_T) is inevitably lower than estimates from linear extrapolations based on reduced sample sizes (S_{Est}) even when the sample size is large.

(Scheffers et al., 2012). Even among animals there is a great deal of taxonomic bias, and some groups, for example, nematodes and mites, have barely begun to be described. Bacteria comprise 5% of named species, though some have suggested that there might be as many as 10^7–10^9 species of bacteria. Is this reasonable, given that there is so much genetic exchange among them? What about viruses, which only exist alongside the organisms they infect? Species concepts designed for metazoans are not readily applicable to microorganisms, and it is safer to exclude them from global estimates. As for fungi, current estimates vary from 0.8 to 5.1 million (or higher), though many of these are based on dubious scaling relationships akin to that used by Erwin and are likely to be revised downwards (Tedersoo et al., 2014).

4.7 Wrapping up

There have been a variety of approaches taken to the problem of estimating the total species richness of the world. Many others exist; alternatives include using the proportion of known species in a sample of organisms or mechanistic models that treat taxonomists as predators and species as their prey. Over time they have coalesced into a similar range (Fig. 4.7), even though they have yet to converge (Caley et al., 2014). From early estimates in the order of tens of thousands of species to daring naturalists in the 19th century who broached the idea that there may be as many as a million, to the ludicrous extremes that exceed 100 million, we see that most methods now fall around 8–10 million species. All the techniques contain inaccuracies, assumptions and imply a certain margin of error, but it is reassuring to see that, despite their quite different methods, there is broad agreement, at least to within an order of magnitude. We may never know the true number but there is good evidence that we are close to the right answer. (A cynic, on the other hand, might suspect that herd mentality is leading estimates in the same direction.)

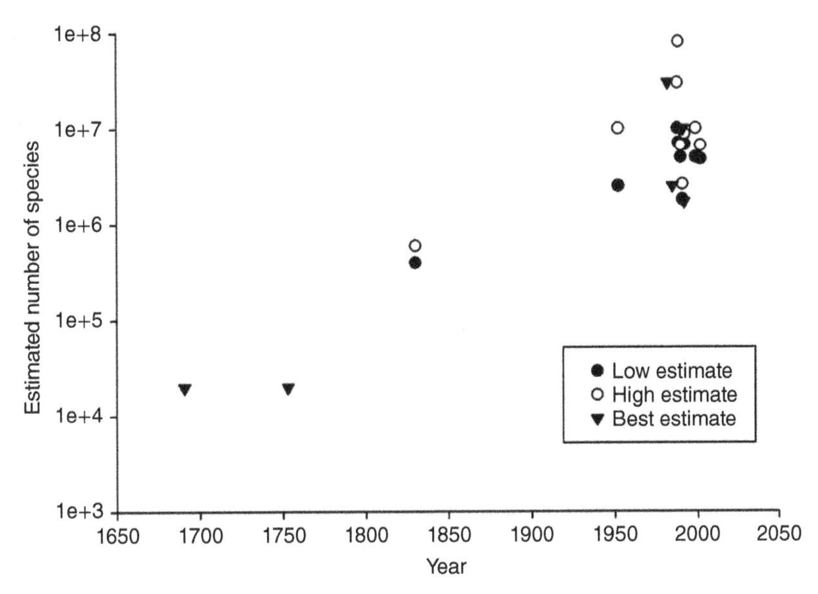

Figure 4.7 Summary of published global species richness estimates. (*Source*: Dobson et al. (2008). Reproduced with permission from the National Academy of Sciences, USA.)

Mora et al. (2011) predicted that some 86% of species (91% in the ocean) still remained to be described. Assuming current rates of progress and numbers of taxonomists, the process might take 360 years and run up a bill of US$263 billion (Carbayo and Marques, 2011). Other estimates are more or less pessimistic, but the central message is that it's never going to happen, at least not in time to find every species before it goes extinct. Increasing numbers of species continue to be discovered even in well-known areas such as Europe (Fontaine et al., 2012b), and even once collected it takes on average more than 20 years for a new species to be formally described (Bebber et al., 2010; Fontaine et al., 2012a). Improving our estimates is therefore essential.

4.8 Conclusions

Extreme estimates of global species richness (>100 million) are likely to be nonsense. More conservative estimates have been made using techniques based on discovery rates, scaling rules and sample-based approaches. Not enough data exist for any method to be truly accurate, but a consensus of around 8–10 million is forming, and the use of multiple methods increases our confidence in this. Recent reviews (Mora et al., 2011; Scheffers et al., 2012) suggest around 8 million animals and 350,000 plants, but the estimates for fungi and smaller life forms remain very uncertain. The figures will require further refinement once a greater amount of taxonomic information is available, including more sampling in the tropics and an agreed global catalogue with rigorous rules to prevent incorrect naming.

There are dangers in applying simplistic approaches to global species richness estimation. These include the problems of extrapolating from small samples, single groups or restricted locations. Assuming parameters in nature such as 'effective specialisation' also carries risks if relationships on local and regional scales are not equivalent. These are lessons with more general implications that will become important in the next chapter.

4.8.1 Recommended reading

Mora, C., D. P. Tittensor, S. Adl, A. G. B. Simpson, and B. Worm, 2011. How many species are there on earth and in the ocean? *PLoS Biology* 9:e1001127.

Scheffers, B. R., L. N. Joppa, S. L. Pimm, and W. F. Laurance, 2012. What we know and don't know about Earth's missing biodiversity. *Trends in Ecology & Evolution* 27:501–510.

4.8.2 Questions for the future

- Should we invest more resources in describing species or focus on conservation so they can be found later?
- Does the number of species we maintain on Earth matter, and if so, why?
- Is it meaningful to count species of bacteria and fungi in the same way as animals and plants?

References

Alroy, J., 2002. How many named species are valid? *Proceedings of the National Academy of Sciences of the United States of America* 99:3706–3711.

Appeltans, W., S. T. Ahyong, G. Anderson, M. V. Angel, T. Artois, N. Bailly, R. Bamber, A. Barber, I. Bartsch, A. Berta, M. Błæwicz-Paszkowycz, P. Bock, G. Boxshall, C. B. Boyko, S. N. Brandão, R. A. Bray, N. L. Bruce, S. D. Cairns, T.-Y. Chan, L. Cheng, A. G. Collins, T. Cribb, M. Curini-Galletti, F. Dahdouh-Guebas, P. J. F. Davie, M. N. Dawson, O. D. Clerck, W. Decock, S. D. Grave, N. J. de Voogd, D. P. Domning, C. C. Emig, C. Erséus, W. Eschmeyer, K. Fauchald, D. G. Fautin, S. W. Feist, C. H. Fransen, H. Furuya, O. Garcia-Alvarez, S. Gerken, D. Gibson, A. Gittenberger, S. Gofas, L. Gómez-Daglio, D. P. Gordon, M. D. Guiry, F. Hernandez, B. W. Hoeksema, R. R. Hopcroft, D. Jaume, P. Kirk, N. Koedam, S. Koenemann, J. B. Kolb, R. M. Kristensen, A. Kroh, G. Lambert, D. B. Lazarus, R. Lemaitre, M. Longshaw, J. Lowry, E. Macpherson, L. P. Madin, C. Mah, G. Mapstone, P. A. McLaughlin, J. Mees, K. Meland, C. G. Messing, C. E. Mills, T. N. Molodtsova, R. Mooi, B. Neuhaus, P. K. L. Ng, C. Nielsen, J. Norenburg, D. M. Opresko, M. Osawa, G. Paulay, W. Perrin, J. F. Pilger, G. C. B. Poore, P. Pugh, G. B. Read, J. D. Reimer, M. Rius, R. M. Rocha, J. I. Saiz-Salinas, V. Scarabino, B. Schierwater, A. Schmidt-Rhaesa, K. E. Schnabel, M. Schotte, P. Schuchert, E. Schwabe, H. Segers, C. Self-Sullivan, N. Shenkar, V. Siegel, W. Sterrer, S. Stöhr, B. Swalla, M. L. Tasker, E. V. Thuesen, T. Timm, M. A. Todaro, X. Turon, S. Tyler, P. Uetz, J. van der Land, B. Vanhoorne, L. P. van Ofwegen, R. W. M. van Soest, J. Vanaverbeke, G. Walker-Smith, T. C. Walter, A. Warren, G. C. Williams, S. P. Wilson, and M. J. Costello, 2012. The magnitude of global marine species diversity. *Current Biology* 22:2189–2202.

Bebber, D. P., M. A. Carine, J. R. I. Wood, A. H. Wortley, D. J. Harris, G. T. Prance, G. Davidse, J. Paige, T. D. Pennington, N. K. B. Robson, and R. W. Scotland, 2010. Herbaria are a major frontier for species discovery. *Proceedings of the National Academy of Sciences of the United States of America* 107:22169–22171.

Bebber, D. P., F. H. C. Marriot, K. J. Gaston, S. A. Harris, and R. W. Scotland, 2007. Predicting unknown species numbers using discovery curves. *Proceedings of the Royal Society Series B* 274:1651–1658.

Caley, M. J., R. Fisher, and K. Mengersen, 2014. Global species richness estimates have not converged. *Trends in Ecology & Evolution* 29:187–188.

Carbayo, F. and A. C. Marques, 2011. The costs of describing the entire animal kingdom. *Trends in Ecology & Evolution* 26:154–155.

Dobson, A., K. D. Lafferty, A. M. Kuris, R. F. Hechinger, and W. Jetz, 2008. Homage to Linnaeus: how many parasites? How many hosts? *Proceedings of the National Academy of Sciences of the United States of America* 105:11482–11489.

Dolphin, K. and D. L. J. Quicke, 2001. Estimating the global species richness of an incompletely described taxon: an example using parasitoid wasps (Hymenoptera: Braconidae). *Biological Journal of the Linnaean Society* 73:279–286.

Erwin, T. H., 1982. Tropical forests: their richness in Coleoptera and other arthropod species. *The Coleopterists Bulletin* 36:74–82.

Fontaine, B., A. Perrard, and P. Bouchet, 2012a. 21 years of shelf life between discovery and description of new species. *Current Biology* 22:R943–R944.

Fontaine, B., K. van Achterberg, M. A. Alonso-Zarazaga, R. Araujo, M. Asche, H. Aspöck, U. Aspöck, P. Audisio, B. Aukema, N. Bailly, M. Balsamo, R. A. Bank, C. Belfiore, W. Bogdanowicz, G. Boxshall, D. Burckhardt, P. Chylarecki, L. Deharveng, A. Dubois, H. Enghoff, R. Fochetti, C. Fontaine, O. Gargominy, M. S. G. Lopez, D. Goujet, M. S. Harvey, K.-G. Heller, P. van Helsdingen, H. Hoch, Y. De Jong, O. Karsholt, W. Los, W. Magowski, J. A. Massard, S. J. McInnes, L. F. Mendes, E. Mey, V. Michelsen, A. Minelli, J. M. N. Nafrıa, E. J. van Nieukerken, T. Pape, W. De Prins, M. Ramos, C. Ricci, C. Roselaar, E. Rota, H. Segers, T. Timm, J. van Tol, and P. Bouchet, 2012b. New species in the Old World: Europe as a frontier in biodiversity exploration, a test bed for 21st Century taxonomy. *PLoS ONE* 7:e36881.

Hebert, P. D. N., E. H. Penton, J. M. Burns, D. H. Janzen, and W. Hallwachs, 2004. Ten species in one: DNA barcoding reveals cryptic species in the neotropical skipper butterfly *Astraptes fulgerator*. *Proceedings of the National Academy of Sciences of the United States of America* 101:14812–14817.

Joppa, L. N., D. L. Roberts, and S. L. Pimm, 2011. The population ecology and social behaviour of taxonomists. *Trends in Ecology & Evolution* 26:551–553.

May, R. M., 1988. How many species are there on earth? *Science* 241:1441–1449.

May, R. M., 2010. Ecological science and tomorrow's world. *Philosophical Transactions of the Royal Society of London, Series B: Biological Sciences* 365:41–47.

Mora, C., D. P. Tittensor, S. Adl, A. G. B. Simpson, and B. Worm, 2011. How many species are there on earth and in the ocean? *PLoS Biology* 9:e1001127.

Novotny, V., Y. Basset, S. E. Miller, G. D. Weiblen, B. Bremer, L. Cizek, and P. Drozd, 2002. Low host specificity of herbivorous insects in a tropical forest. *Nature* 416:841–844.

Ødegaard, F., 2000. How many species of arthropods? Erwin's estimate revised. *Biological Journal of the Linnaean Society* 71:583–597.

Raven, P. H., 1985. Disappearing species: a global tragedy. *The Futurist* 19:8–14.

Scheffers, B. R., L. N. Joppa, S. L. Pimm, and W. F. Laurance, 2012. What we know and don't know about Earth's missing biodiversity. *Trends in Ecology & Evolution* 27:501–510.

Scotland, R. W. and A. H. Wortley, 2003. How many species of seed plants are there? *Taxon* 52:101–104.

Stork, N. E., 1988. Insect diversity: facts, fiction and speculation. *Biological Journal of the Linnaean Society* 35:321–337.

Tedersoo, L., M. Bahram, S. P olme, U. K oljalg, N. S. Yorou, R. Wijesundera, L. V. Ruiz, A. M. Vasco-Palacios, P. Q. Thu, A. Suija, M. E. Smith, C. Sharp, E. Saluveer, A. Saitta, M. Rosas, T. Riit, D. Ratkowsky, K. Pritsch, K. P oldmaa, M. Piepenbring, C. Phosri, M. Peterson, K. Parts, K. Pärtel, E. Otsing, E. Nouhra, A. L. Njouonkou, R. H. Nilsson, L. N. Morgado, J. Mayor, T. W. May, L. Majuakim, D. J. Lodge, S. S. Lee, K.-H. Larsson, P. Kohout, K. Hosaka, I. Hiiesalu, T. W. Henkel, H. Harend, L.-D. Guo, A. Greslebin, G. Grelet, J. Geml, G. Gates, W. Dunstan, C. Dunk, R. Drenkhan, J. Dearnaley, A. De Kesel, T. Dang, X. Chen, F. Buegger, F. Q. Brearley, G. Bonito, S. Anslan, S. Abell, and K. Abarenkov, 2014. Global diversity and geography of soil fungi. *Science* 346:1256688.

Thomas, C. D., 1990. Estimating the number of tropical arthropod species. *Nature* 347:237.

PART II
Diversity

CHAPTER 5

Measuring diversity

5.1 The big question

We have already met the concept of biodiversity, the critical weakness of which is that no agreed measurement fully captures this property of natural systems. Such a measure would clearly be desirable, so how could we go about describing diversity in an appropriate fashion? This requires us to think carefully about the different issues at play. The first is one of scale: how large an area should be sampled? Next, as it's impossible to record absolutely everything, robust means of estimating what hasn't been directly observed are required. Raw species numbers reveal relatively little, since some species are more abundant than others, and therefore their contribution to ecosystem processes will vary. Finally, should it be all about the species? There are other ways to measure diversity, including taxonomy or ecological interactions. The methods recommended in this chapter are demonstrated in detail through a worked example of butterflies in Colorado in Appendix A.

5.2 Scales of diversity

There are three principal levels at which diversity can be measured. The first is at the local scale, that is, at the scale of the community in front of you. This is known as α diversity. Communities vary from one to another, and the extent of these differences can be expressed as β diversity. Finally, adding together all the available communities leads to a landscape-level measure which is called γ diversity. It's possible to separate these levels (and those at both higher and lower scales) into either inventory or differentiation measures—those that record absolute values or those that record differences (Table 5.1). In practice only α, β and γ are commonly referred to, but it's important to be aware of the broader concept behind them.

5.3 Species richness

We have already seen that estimating global species richness is difficult, but determining local species richness also turns out to be surprisingly problematic, and involves many of the same issues. To illustrate the scale of the problem, Lawton et al. (1998) attempted to comprehensively sample a single hectare (100 × 100 m) of rain forest in Cameroon. They chose eight taxa for which good taxonomic information was available and collected around 2,000 species in total. Collecting was the easy part; subsequent identification took much longer, and in total the project consumed five 'scientist years'. They estimated that in order to carry out a complete survey of all species in a single hectare within one year, it would require the full-time efforts of between 10% and 20% of the 7,000 professional taxonomists in the world. Effectively it would be impossible. Add to this that African rain forests are relatively species poor compared to those in Southeast Asia and the

Natural Systems: The organisation of life, First Edition. Markus P. Eichhorn.

Table 5.1 Scales of diversity following.

Scale	Inventory	Differentiation
Sample	Point	
n Samples		Pattern
Habitat	α	
n habitats		β
Landscape	γ	
n landscapes		δ
Province	ϵ	

Source: Whittaker (1977). Reproduced with permission from Springer Science+Business Media.

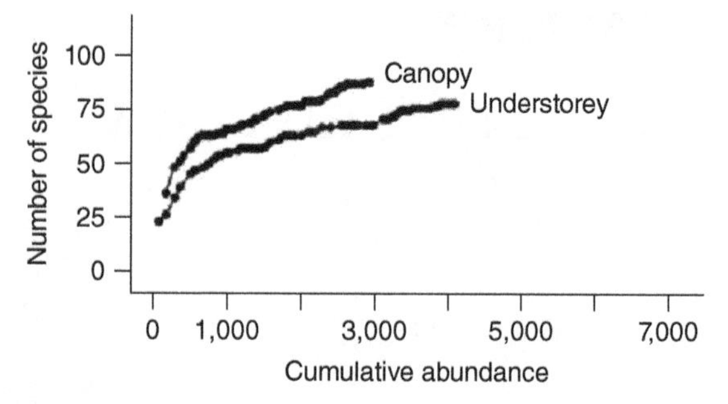

Figure 5.1 Accumulation of species by number of butterfly individuals sampled in a Costa Rican rain forest divided between canopy and understorey species. (*Source*: De Vries (2012). Reproduced with permission from John Wiley & Sons Ltd.)

Amazon (Corlett and Primack, 2011) and a complete global species list seems even further out of reach.

We therefore need to consider estimation techniques. Even outside the tropics, ecological samples are almost always incomplete; attempt to count all the species in any habitat and you will quickly discover this. The chief difficulty is that there are many uncommon species, and the longer you look, the more you find. A rule is needed to decide when to stop once we have enough to make a good estimate.

Another problem worth acknowledging is that sampling methods are all, invariably, biased. This is a crucial caveat for what follows since most statistical techniques for estimating diversity assume a random sample of the organisms present whereas what is actually collected depends on the method used. Quadrat sampling may be appropriate for finding plants, but it's a terrible way to count mammals. Similarly, a pitfall trap is suitable for ground-mobile invertebrates but not for those which fly (unless it has bait). Even the time of day or season in which sampling takes place can bias results since it may make some species more or less likely to be counted. The end result is that some species considered 'rare' might actually be very common, but they can appear to be rare as an artefact because sampling was not designed to capture them.

The starting point of any study of species richness should be to draw a **species accumulation curve** which reflects the number of species caught per unit effort. In this context effort can mean almost anything: number of individuals or samples, length of transect, time spent in the field or whatever seems most appropriate.

An example comes from a study by DeVries and colleagues (2012) who recorded the species richness of butterflies in Costa Rican rain forests in either the canopy or understorey (Fig. 5.1). The rate of increase of the species accumulation curves is declining towards the level at which all species will have been found. Are there more species in the canopy or understorey? More species have been found so far in the canopy, but it's not certain that the understorey curve has finished rising, and if more samples were collected then a different answer might be arrived at. Even after 7,000 butterflies have been counted, this simple question can't be answered without estimating.

The simplest approach might be to predict where the curve becomes flat, known as the **asymptote**. In order to do this, contiguous samples are required, which means they are connected in some way, that is, sampled over time or an increasing area. Normally we also randomise the order of samples in the curve to obtain smooth lines (or log-linear relationships).

There are a number of techniques, known as parametric estimators, that can be used to fit equations to these curves and predict the total that would be obtained with a complete ecological sample (i.e. catching every butterfly in the forest). Some are

Table 5.2 Abundance distributions for two virtual communities.

Species	A	B	C	D	E
Habitat 1	20	20	20	20	20
Habitat 2	96	1	1	1	1

outlined in Magurran (2004). These have several advantages: they are easy to interpret and work for all sampling types. Here, however, I do not recommend them, for three reasons. The first is that many models can fit a particular curve and give widely varied estimates. The fit of any particular equation is a poor reflection of its ability to tell you the true species richness, and it is impossible to compare multiple communities if different models are used. Secondly, unless the habitat has been well sampled (i.e. most of the curve is revealed), asymptotic estimates are extremely inaccurate. In most cases we haven't even come close to finding all the species. Third, and most importantly, we have to compensate for unevenness in communities.

Up until this point we have only considered species richness and ignored the abundances of those species. This misses out a crucial distinction. Imagine two habitats, each containing a total of 100 individuals from five species A–E (Table 5.2). Both have identical species richness, but in the second, one species completely dominates. For estimating species richness this poses an immediate problem. When taking random samples from these fake communities, in the first you will rapidly discover all the species, probably after taking only 10 individuals. In the second habitat you have to sample almost the entire community to discover them all (Fig. 5.2). If you tried to fit curves to these communities, you would get completely different answers, not because the species richness differed but because of unevenness. This is a problem because almost all natural systems are uneven in their species abundances, a pattern we will return to shortly. For now we need to know how to handle this from a species richness perspective.

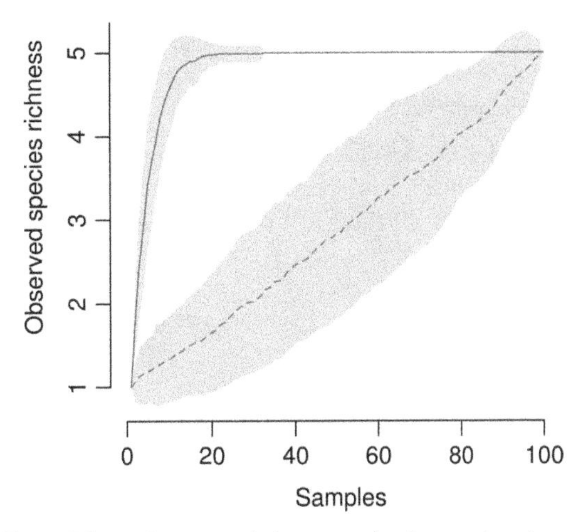

Figure 5.2 Species accumulation curves for the two virtual communities in Table 5.2 with differing degrees of evenness based on random subsampling without replacement, means ± S.E. Solid line: habitat 1; dashed line: habitat 2.

The answer is to use what are called non-parametric estimators. These are so named because they don't assume any particular form for the distribution of abundance; they can be calculated using properties of the raw data. The most important and widely used of these is the method developed by Anne Chao. This estimates the minimum species richness of an ecological community based solely on the number of species found either only once or twice:

$$S_{Chao} = S_{Obs} + \frac{f_1^2}{2f_2}$$

It requires you to know how many species were found in your sample (S_{Obs}), the number of **singletons** f_1 and doubles f_2 (you can also use the number of species found in only one or two samples if you have taken many samples). Note that the basic form of the equation is not solvable when $f_2 = 0$. For this reason a slightly more complicated bias-corrected form is also available, which can always be obtained:

$$S_{Chao-bc} = S_{Obs} + \frac{f_1(f_1 - 1)}{2(f_2 + 1)}$$

More sophisticated methods based on this general framework exist and are implemented in most diversity software packages.

As an example, we can use the Chao estimator to calculate the number of invertebrate species present on the forest floor alongside streams in forests in Oregon (data from Rykken et al., 2007). They counted a total number N of 13,348 invertebrates from 192 species (S_{Obs}). Of these, 29 were only seen once and 15 twice. This means that

$$S_{Chao} = 192 + 29^2/(2 \times 15) = 192 + 28 = 220$$

In other words, they failed to collect around 28 species that were actually present. Note that point estimates can be misleading if taken at face value, and you should always present some measure of uncertainty. It is possible to calculate a 95% confidence interval for this estimate, which should therefore be presented as 220 (203.3–261.6). The maths behind this is more complex but most diversity packages will do it for you (see Chao, 1984). Note that the lower bound estimate is still above S_{Obs}. Since accurate point estimates of species richness are so difficult to obtain, a reasonable minimum estimate is often of more practical use.

Some might be sceptical about there being at least 11 species which had not been found, especially after observing over 13,000 individuals. This credibility gap is a common problem when presenting estimates.

5.4 Believing in estimates

Imagine that you are working for a conservation non-governmental organisation (NGO) and are asked to find out how many species of bird are present in a forest that has been scheduled for timber harvesting. After doing your best to count the number of species (and their abundances!), you come up with an estimate and present it to the inquiry panel. On the one hand, the timber company may not trust you. After all, you saw only S_{Obs} species—surely you're making the others up? On the other side, a conservation NGO might put pressure on you in the opposite direction. The Chao formula gives a minimum estimate, but what is the maximum? The more species you can claim are in the forest, the stronger their case will be. If you're going to make up 50 species, then why not 100? Such conflicts are common and vexing.

Your first argument in defence, speaking to the timber company, is to point out the inadequacy of any sample. For instance, the Hong Kong Bird Race is an annual competition where teams of four birdwatchers have 24 hours to count as many species as they can on the island.[1] Strict rules apply to enforce accurate identification and honesty. In 2010 the winners spotted 136 species—an impressive total. Summing across all teams that participated gave a complete number of bird species found of 194. Even the top birdwatchers in the country managed to miss at least 58 species in the time available. Just because they didn't see them doesn't mean they weren't there.

Turning to the conservation NGO, the resolution to their plea for higher estimates requires you to learn why this deceptively simple piece of maths works. The root comes from a statistical property of a sample known as coverage:

$$C = \Sigma_{i=1}^{S} p_i \times X_i$$

This formula appears complicated, but you only need to understand what it conveys. Coverage C is calculated by taking each species i in the community of S species, multiplying its proportion in the sample p by the number of individuals of that species X and then adding these up for all species. Mathematically this obtains a number with an interesting property: $1 - C$ is the conditional (i.e. given your data) probability that if you were to take one more observation (or collect one more individual) it would be a new species. This provides an excellent and objective measure of sample completeness. C itself tells you the proportion of all individuals in the full community that are represented by the species in your sample.

This equation doesn't belong in the dry tomes of statistical theory—in WW2 it became a matter of life and death. The great mathematician Alan Turing was

[1] See www.wwf.org.hk/en/getinvolved/hkbbr/ for full details.

employed as a codebreaker by the British government and was trying to crack the Enigma code, used by enemy forces to conceal their communications. His problem was identical to that faced by ecologists. Given a sample of the code, how likely was it that he had seen all the possible elements? His moment of genius, along with his colleague Jack Good, was to demonstrate that 1-C could be estimated by the proportion of singletons in the sample, that is, the elements that had only been seen once (f_1). To put it mathematically, for a sample of N individuals, $\hat{C}_N = 1 - \frac{f_1}{N}$. A slightly improved estimator has since been proposed by Chao and Jost (2012) but remains easy to calculate:

$$\hat{C}_N = 1 - \frac{f_1}{N}\left[\frac{(N-1)f_1}{(N-1)f_1 + 2f_2}\right]$$

This is why Chao's simple non-parametric estimator (and its derivatives) works. The most valuable information is contained in the abundances of those species that have hardly ever been seen, rather than those which are common. A number of similar tools have been developed using the same principle, such as the Abundance Coverage Estimator (ACE; Chao, 2005); see Magurran and McGill (2011) for more details.

Another important outcome is that we can use this property to come up with a stopping rule for when we have sampled enough of the habitat. Turing's formula suggests that, provided you have taken a random sample, and the number of individuals n is sufficiently large, you can stop when there are no more singletons. Once you have found every species twice, you should have found all the species. Alas this doesn't happen very often! A more rigorous mathematical approach to finding out how far you are from reaching the total is given in Chao et al. (2009), along with a spreadsheet to carry out the calculations.

5.5 A SAD story

Having met the problem of unevenness of species abundance in communities, we should now turn to methods for describing this, through what are known as species abundance distributions (SADs). Typically there are a few highly abundant (dominant) species and many which are rare, though you must remember that they might be methodologically rare rather than actually rare, that is, sampling might not be designed to catch them. Figure 5.3a and b shows some examples from natural assemblages. Note that the majority of species fall on the left-hand side of these plots, indicating that they are uncommon, while relatively few exist in high numbers. This is true even on large scales; across the entirety of Amazonia, just 227 tree species out of an estimated 16,000 constitute more than half of all individuals, whereas the rarest third of species make up only 0.0003% of trees (ter Steege et al., 2013).

Two methods are commonly used to more clearly represent the abundances of species. The first is the Preston plot, named after Frank Preston, the great mathematician who noticed a particular pattern (Preston, 1948). He spotted that for large samples, if you take abundance classes on the x-axis which follow a log scale (usually base 2, otherwise known as octaves; each successive class is double the size), then the number of species that fall into each class often follows a normal distribution. This is called the **lognormal** abundance distribution and is a central assumption of much ecological theory (Fig. 5.3c and d). We will return to the reasons why this pattern emerges so often in Chapter 10.

A frequent observation in Preston plots is that the left-hand side of the curve may not be entirely visible, a phenomenon known as the **veil line**. Both the assemblages in Fig. 5.3 are large and therefore the curve is almost complete, but with small samples the less-common species might be absent.

The second common way to represent SADs is as Whittaker plots, sometimes also known as rank/abundance or dominance/diversity curves (Fig. 5.3e and f). Their essential feature is that species are ranked in terms of abundance along the x-axis, and then either their absolute or relative abundance (i.e. the proportion of the total sample made up of each species) is plotted on the y-axis,

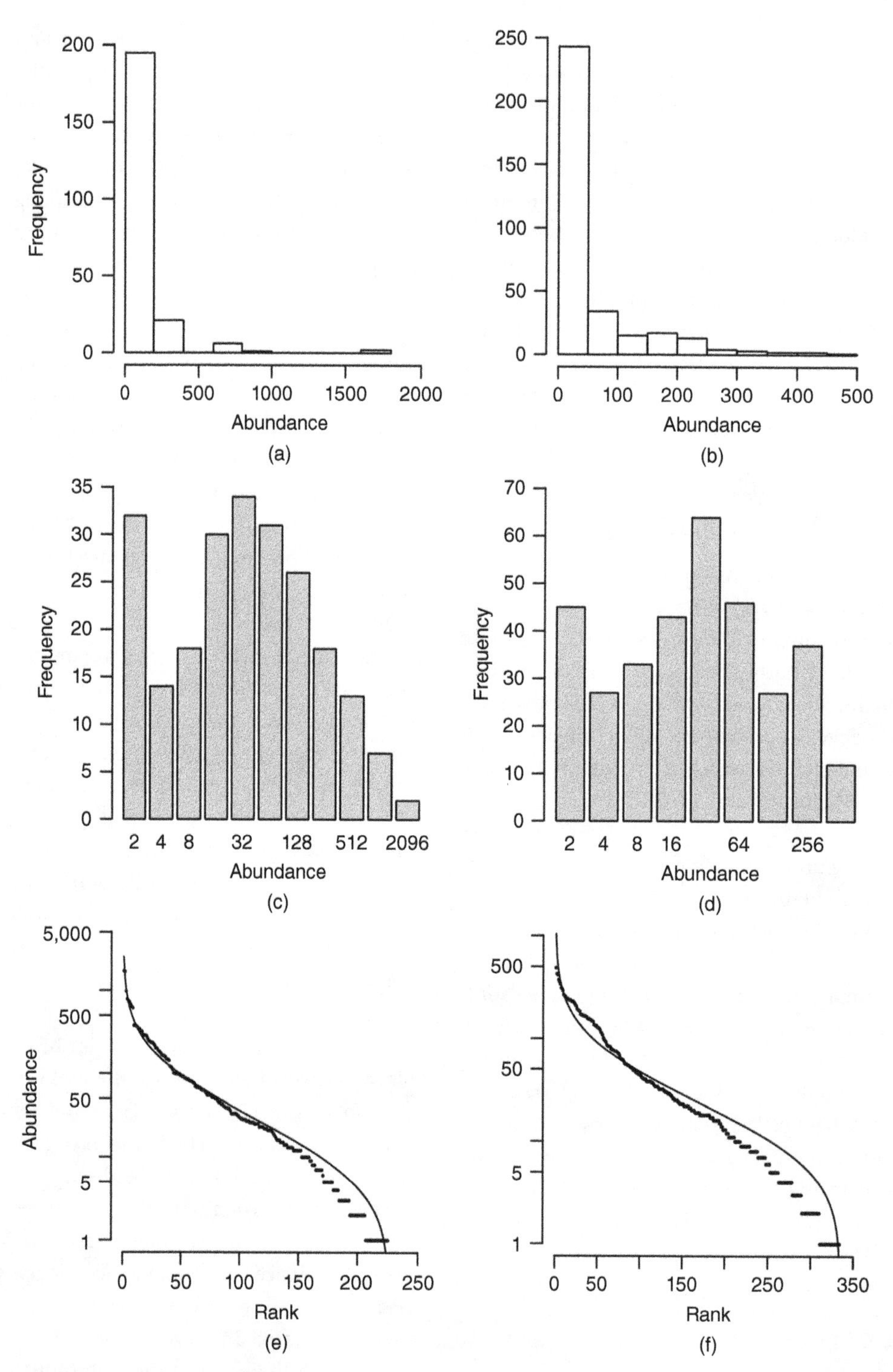

Figure 5.3 Data from tree species in a 50 ha plot on Barro Colorado Island, Panama (left column; http://www.ctfs.si.edu/), and rodents in the Chihuahuan Desert (right column; Ernest et al., 2009); (a,b) cross-species abundance distributions, (c,d) Preston plots and (e,f) Whittaker plots with fitted lognormal distributions (lines).

which is log scaled. These are a good place to start when presented with a new dataset as they illustrate species richness, dominance, evenness and structure of samples in a clear fashion. They can also be used to make comparisons, for example, examining human impacts, the effects of ecological disturbance or changes through time. In Fig. 5.3e and f we can also see the fitted model for the lognormal distribution, and although the real data don't exactly follow this pattern, it's a good first approximation, while failing to accurately represent the rare species.

5.6 Diversity of species

Having decided that species richness isn't sufficient to describe what's really going on within communities, we need to come up with a way of describing the variation in abundance among species empirically. It's not enough given the assemblages in Fig. 5.3 to simply say that the number of species S is 225 and 334, respectively.

We accomplish this by using **diversity indices.** These are methods to combine both species richness and their abundances into a single metric. There are many available—a diversity of diversity indices—and you can explore the options at your leisure (Southwood, 1978; Magurran, 2004; Maurer and McGill, 2011). Be careful though as not all are useful, and in many cases their continued application owes more to familiarity and tradition than to their effectiveness in conveying the information that people might want them to. This is an area of ecology in which a mess has been allowed to develop, though a consensus has begun to form, so you are recommended to see the latest treatments for advice (Tuomisto, 2010a; Maurer and McGill, 2011).

The best indices are those which scale in a sensible way, that is, if you were to take two assemblages of equal species richness but with no shared species and merge them together in the same proportions, then the value of the diversity index should double. This doubling property (otherwise known as the replication principle) captures how we would intuitively expect indices to behave. Indices which satisfy this criterion have been referred to as 'true' diversity to set them apart (Jost, 2007). One measure which already does this is species richness; if you have twice as many species, then S doubles. Others are less straightforward. A widely used index is Shannon's entropy:

$$H' = -\Sigma p_i \ln p_i$$

Here p_i refers to the proportion of the sample made up of species i, and the calculation is summed across all species. In mathematical terms it represents the uncertainty in the identity of a random individual taken from the sample. Since samples are invariably incomplete, the value for the whole community has to be estimated from the data (in the same way as for species richness). This involves complicated maths but many packages exist that can perform it; several are listed in Appendix A.1. The trouble with H' is that it does not scale linearly as new species are added, which makes comparisons between communities problematic. This is easily fixed by calculating its exponential, $e^{H'}$, which has important properties which we will come to in a moment.

Before that we should define another common metric, Simpson's index:

$$D = \Sigma p_i^2$$

In mathematical terms this represents the probability that two randomly sampled individuals will be of the same species. The simple formula above is for a community sampled with replacement of individuals, and a slightly more complicated version is required in the common situation when sampling without replacement takes place:

$$D = \Sigma \frac{n_i[n_i - 1]}{N[N - 1]}$$

Despite its appearance, this is also very simple to calculate, as n_i is simply the number of individuals of species i, and N is the total number of individuals across all species. This scales in a peculiar fashion, with values between 0 and 1, and the most diverse communities are closer to 0. This makes little intuitive sense so it is usually presented as $1/D$.

Table 5.3 Hill's family of indices ^qD, their derivation and interpretation.

q	Derivation	Interpretation
0	S	Total number of species
1	$e^{H'}$	Effective number of common species
2	$1/D$	Effective number of abundant species

Source: Hill (1973). Reproduced with permission from the Ecological Society of America.

Though these indices of diversity seem very different, they are actually related to one another mathematically, as was demonstrated in a key paper by Hill (1973). They can be seen as members of a family of indices of order ^qD, each representing something slightly different. Species richness is heavily skewed by the number of rare species; if you have lots of singletons, S will rise rapidly. The exponential of Shannon's entropy, $e^{H'}$, is more resistant to this and instead reflects the effective number of moderately common species. Finally, the inverse of Simpson's index $1/D$ shows the effective number of highly abundant or dominant species (Table 5.3). They are referred to as 'effective' numbers of species because they show the number of equally abundant species that would give the same value of the diversity index concerned (MacArthur, 1965).

When assessing a community it is a good idea to calculate all three of these (^0D–^2D) and compare them. There are other possible values; in fact, the family can be expanded to incorporate any value of q, but others are less intuitive:

$$^qD = \left(\sum_{i=1}^{S} p_i^q\right)^{(1/(1-q))}$$

The steepness of the relationship between values of ^qD is a good indication of the evenness of the community. This is easier to appreciate with an example (Table 5.4). Imagine four assemblages, each of which contains a hundred species (S = 100) and 500 individuals (N = 500), but a different blend of each. In the first, all species have five individuals, which we can present as $f_5 = 100$. The others vary in the pattern of relative abundance of species.

Note how species richness always remains the same, and in the case where all species are equally abundant, $^0D = {}^1D = {}^2D$. As the communities get more uneven, the difference between them becomes more pronounced. We can illustrate this by plotting the full series of q for each assemblage (Fig. 5.4). This gives a simple and intuitive way of illustrating the variation in diversity at all levels, from the rarest to the most abundant sets of species.

Calculating diversity indices can be much more efficient in terms of research effort than species richness. This is because estimating them with a reasonable degree of accuracy requires less data. For example, Lande et al. (2000) collected data from two butterfly communities, one of which had been disturbed. This tends to increase species richness, as species tolerant of more open conditions are able to invade, whereas it usually decreases evenness. This is one of those cases when species richness on its own might give a misleading impression of the conservation value of a community.

Their question was how large a sample would be required to correctly ascertain the difference in diversity between a disturbed and an undisturbed site (Table 5.5). In order to get an accurate estimate

Table 5.4 Example communities and values of Hill's numbers.

Assemblage	Species abundance distribution	^0D	^1D	^2D
1	$f_5 = 100$	100	100	100
2	$f_7 = 50, f_3 = 50$	100	92.10	86.21
3	$f_{10} = 22, f_5 = 28, f_3 = 40, f_2 = 10$	100	86.44	75.76
4	$f_{120} = 1, f_{80} = 1, f_{70} = 1, f_{50} = 1, f_{20} = 3, f_{10} = 3, f_1 = 90$	100	17.83	8.39

Source: Based on an example in Gotelli and Chao (2013).

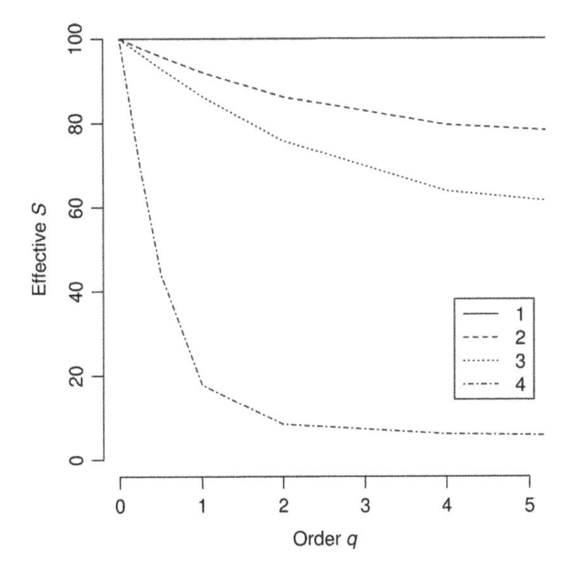

Figure 5.4 Effective number of species S at each level of the Hill series ${}^qD$ for the four virtual assemblages in Table 5.4. (*Source*: Adapted from Gotelli and Chao (2013), Fig. 6.)

Table 5.5 Butterfly assemblage metrics from disturbed and intact forests and the minimum sample required to rank them correctly.

Diversity measure	Disturbed	Intact	Minimum sample
S	142	101	1,801
$1/D$	12.0	21.7	81

Source: Lande et al. (2000). Reproduced with permission from John Wiley and Sons Ltd.

of species richness, you would need to collect at least 1,800 butterflies, but to correctly spot the higher Simpson's diversity in the intact forest, you would need to sample fewer than 100. This makes diversity indices of great importance in conservation where time and resources are always limited. Of all the measures of diversity, species richness is the hardest to measure accurately because estimates are so sensitive to the numbers of individuals or samples collected.

5.7 Other measures of diversity

We have already learnt in Chapter 2 to be cautious of uncritically using species as our measure of diversity. This has been recognised at least since Pielou (1975), who suggested that measures of diversity could encompass variation among species as well, unlike the metrics above which treat all species as being equally distinct. An assemblage containing many types of species will be more diverse than one in which all are very similar, even if both have the same richness. One reason for this is that species can vary greatly in their evolutionary background. Instead of ignoring this, we can incorporate it in our assessments. When you have a phylogeny, one approach is to calculate the taxonomic distinctness of an assemblage. This innovative method was suggested by Clarke and Warwick (1998) who proposed the metric Δ^+ which represents the average taxonomic distance between any two species in the sample (Fig. 5.5). A range of other measures of phylogenetic diversity have since been proposed (Vellend et al., 2011). These can capture information that would not otherwise be accounted for by an approach that treated all species as equivalent.

Another type of measure is functional diversity. This considers the values of species **traits** and forms a composite index capturing how varied the assemblage is. A wide range of such indices exist in the literature (Weiger, 2011). Some researchers have even tried to come up with unified measures of diversity that include abundance, phylogeny and function at the same time (e.g. Scheiner, 2012); this is ambitious as there are few systems where all the information is available. It is also possible to extend the Hill diversity series to functional and phylogenetic diversity (Chao et al., 2010).

A further and relatively new approach is intended to help with those communities with such incredible species richness that grasping the scale of them becomes difficult. Instead of counting species, Dyer et al. (2010) suggest calculating interaction diversity, the number of different interactions among species and their frequency. For example, they took sample plots in Ecuador and Louisiana and counted the number of different pairings they found between plants, lepidopteran caterpillars collected from them and parasitoids which emerged from the caterpillars in the laboratory (Table 5.6).

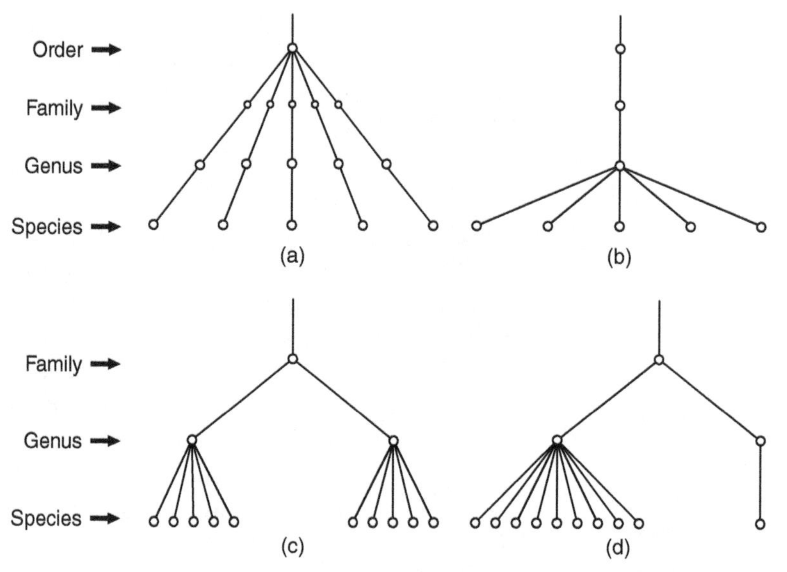

Figure 5.5 Illustrative phylogenies for calculating Δ^+ for groups of (a,b) 5 or (c,d) 10 species. Based on equal weighting of links, Δ^+ values are (a) 3.0, (b) 1.0, (c) 1.56 and (d) 1.2. (*Source*: Clarke and Warwick (1998). Reproduced with permission from John Wiley & Sons Ltd.)

Table 5.6 Diversity estimates from 10 m diameter plots in Ecuador and Louisiana, sampling all plants and caterpillars, and rearing the latter for adults and parasitoids (Dyer et al., 2010). Species from all three trophic levels are combined to produce means per plot ± SE.

	Ecuador	Louisiana
Sample size (number of plots)	429	222
S per plot	20.5 ± 0.8	20.8 ± 0.8
Interaction richness per plot	10.2 ± 0.3	5.7 ± 0.2
Interaction diversity (^2D) per plot	5.6 ± 0.3	3.3 ± 0.3

Source: Dyer (2010). Reproduced with permission from the Association for Tropical Biology and Conservation.

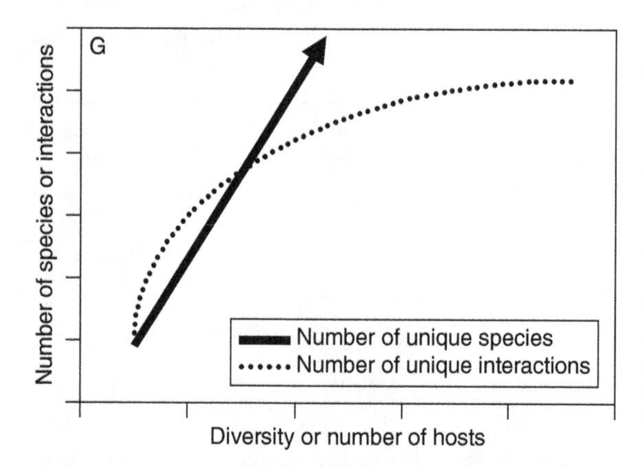

Figure 5.6 Accumulation rates of species and interactions. (*Source*: Dyer et al. (2010). Reproduced with permission from the Association for Tropical Biology and Conservation.)

In these small, although species richness didn't appear to differ between regions, the number of interactions was approximately doubled in Ecuador, and the diversity of interactions was also greater.

There are many advantages to using interactions as the basic unit of diversity. The number recorded increases more slowly than species (Fig. 5.6), which makes them a more tractable target for sampling than species themselves. Many interactions are rare, allowing similar estimation techniques and indices to be used to describe them. It also helps to deal with the large number of species collected as tourists: those that are passing through and don't really belong in the community being sampled. By focussing only on species that are actually doing something, a much clearer picture emerges of the actual structure of the community.

This approach has great potential. Consider a study from Ecuador where Tylianakis et al. (2007) created interaction webs representing the links between parasitoids (on the top row) and their insect hosts (bottom row; Fig. 5.7). The width of the bars reflects the abundance of each species, whereas the width of the grey links shows the number of times each interaction occurred. Each community has around 33 host species (bees and wasps) and 9 parasitoids. The distribution of abundance within each group responds to a gradient of human disturbance moving from intact forest to an abandoned coffee plantation, to an active coffee agroforest, to a pasture and finally to a rice paddy. Note how in the forest (the top web) there are many interactions which occur commonly. The same is true of the agroforests. In the pasture and rice paddy this has changed dramatically; one interaction dominates each of these communities. Their interaction diversity is clearly lower. We will return to this theme in the Communities section and consider what it means for their stability and functioning.

In short, there are more ways to look at biological diversity than just species; they are becoming increasingly common but further work is needed to understand what they might tell us and their relevance to real problems in the field.

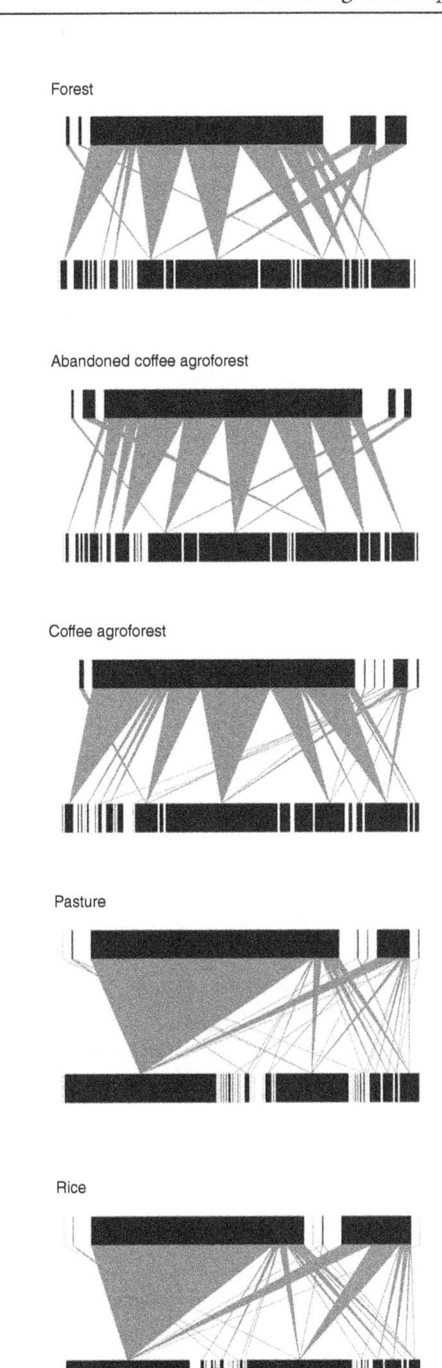

Figure 5.7 Host–parasitoid interaction webs in Ecuador. Lower bars represent host (bee and wasp) abundance and upper bars represent parasitoid abundance, drawn to different scales. Linkage width indicates frequency of each trophic interaction. (*Source*: Tylianakis et al. (2007). Reproduced with permission from Nature Publishing Group.)

5.8 β diversity

Having spent some time thinking about α diversity, we need to place it into context by examining the link with landscape-level γ diversity. Comparing between one area and another, some new species appear in communities, while others drop out. This phenomenon, known as **turnover**, represents another form of diversity, described in Table 5.1 as a measure of differentiation known as β diversity. The degree of this will depend on the scale at which both α and γ diversities are measured. When the samples used to measure α diversity are small, there will be a large amount of variation among them. As they make up a larger proportion of the available area, α diversity will tend to increase towards γ, while β will fall.

The simplest measure of β diversity, and the one which best represents the original intention, is Whittaker's (1972) index:

$$\beta_W = S/\bar{\alpha} - 1$$

Here S is the total species richness of the region, that is, the γ richness, divided by the *average* species richness for any individual sample $\bar{\alpha}$. β_W can take values from 0 (all species present in the region are present in all samples) to a maximum of $N - 1$ (where N is the total number of samples) when there is no overlap between samples whatsoever. An important criterion before applying Whittaker's index is that all samples taken of α richness are of equal size (or sampling effort).

As with diversity indices, there are a large number of other ways of approaching β diversity (Tuomisto, 2010b), though not all of these are consistent with the original meaning, so should be used only if you are certain they are more appropriate to the question at hand. One important and related class of metrics is dissimilarity indices, of which Sørenson's index $\beta_{SØ}$ is the most widely used:

$$\beta_{SØ} = \frac{b + c}{2a + b + c}$$

Here a is the number of shared species between two samples, b is the number in one sample and c is the number in the other. The output is identical to β_W for two samples. Dissimilarity indices are used to measure the degree of difference between any pair of samples and can be very useful when interpreting ecological patterns. You can find out more about them in Magurran (2004) or Jost et al. (2011). They are also frequently referred to as similarity indices; this is merely a matter of perspective and the literature is not consistent. In the case of $\beta_{SØ}$ above, it increases the more different two samples are.

5.9 Case study: The Binatang project

To finish we will revisit the interactions between tropical trees and their insect herbivores, which we found in Chapter 4 to be crucial for estimating global species richness. This leads us to the work of one of the greatest modern ecologists, Vojtech Novotny, and his pioneering studies in the rain forests of Papua New Guinea. These forests pose a special problem for study as they are vast in size, impenetrable to vehicles and filled with an incredible diversity of species, almost all of which remain unknown to taxonomists. Where to start? Novotny's innovation has been to train indigenous people— parataxonomists—to collect samples and data on his behalf.

The first problem for understanding these systems is the astonishing level of α richness. How many species occur in any one place? This was a question that Novotny and his team turned to first, selecting a single hectare of lowland forest and trying to find as many species of insect as they could within it (Novotny et al., 2004). For 59 species of woody plants they located every stem greater than 5 cm diameter at breast height and spent an entire year collecting every insect they could find on them. One important feature of their work was that they took each insect back to their 'lab'—actually a makeshift shed filled with bags of leaves—and tested to make sure that it really did feed on the plant they had collected it from. This ensured that every single individual was truly part of the community, not simply passing through.

There was no shortage of insects to be collected. By the end of the year they had found 58,483 individuals, all confirmed as feeding, making up a grand total of 940 species. Like any ecologists engaged in a study of diversity, they drew species accumulation curves, which demonstrated the true scale of the problem (Fig. 5.8). Even with all that work, none of the curves, for any plant species, looked like it had reached an asymptote.

Their one hectare plot contained a total of 152 species of woody plant. By projecting their curves and extrapolating to the plant species they hadn't examined, they estimated that there might be 1,567 to 2,559 insect herbivore species per hectare (the wide range indicates the inherent uncertainty in using species accumulation curves as a means

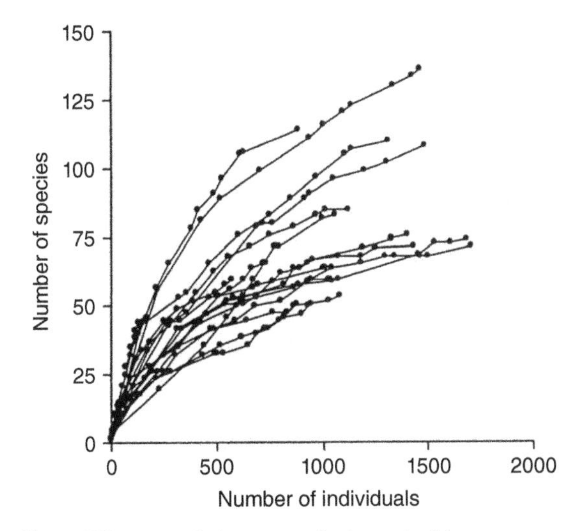

Figure 5.8 Accumulation curves for insect herbivores on representative plant species in Papua New Guinea. Each line represents a different plant species. (*Source*: Novotny et al. (2004). Reproduced with permission from John Wiley & Sons Ltd.)

of prediction). This is an intimidating number of species in an area about the same as two football pitches.

Given the challenge of describing the α richness of this forest, working out the regional number of species might seem an insurmountable task, but Novotny was undaunted. The key was to work out how much turnover there was between different patches of forest, giving an insight into the scale of β richness.

The forest was extensive and unbroken, though occasionally dotted with landing strips for light aircraft, originally built to allow missionaries or medical aircraft to reach the remotest of settlements. The team selected eight sites spread across an area of continuous forest stretching over 75,000 km^2, roughly the size of Ireland (Fig. 5.9). Picking four widespread plant genera as their focus, the parataxonomists set about collecting yet more insects. This endeavour occupied about 34 person-years of effort.

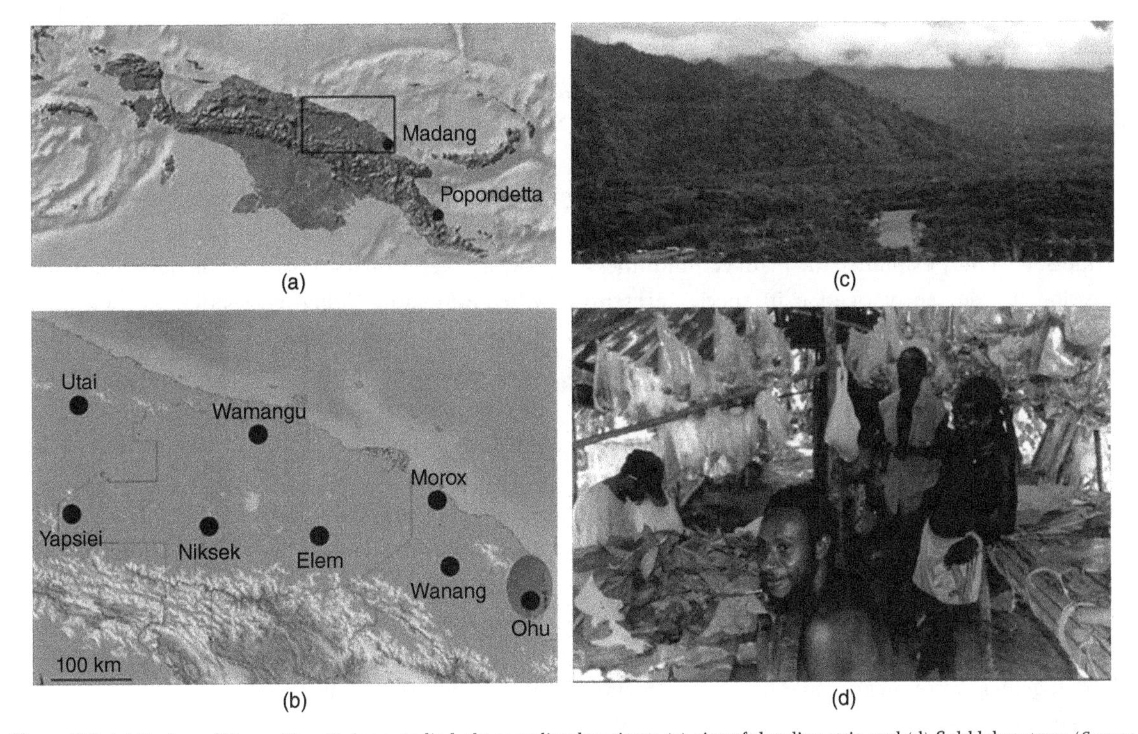

Figure 5.9 (a) Region of Papua New Guinea studied; (b) sampling locations; (c) aircraft landing strip and (d) field laboratory. (*Source*: Novotny et al. (2007). Reproduced with permission from Nature Publishing Group.) (*See colour plate section for the colour representation of this figure.*)

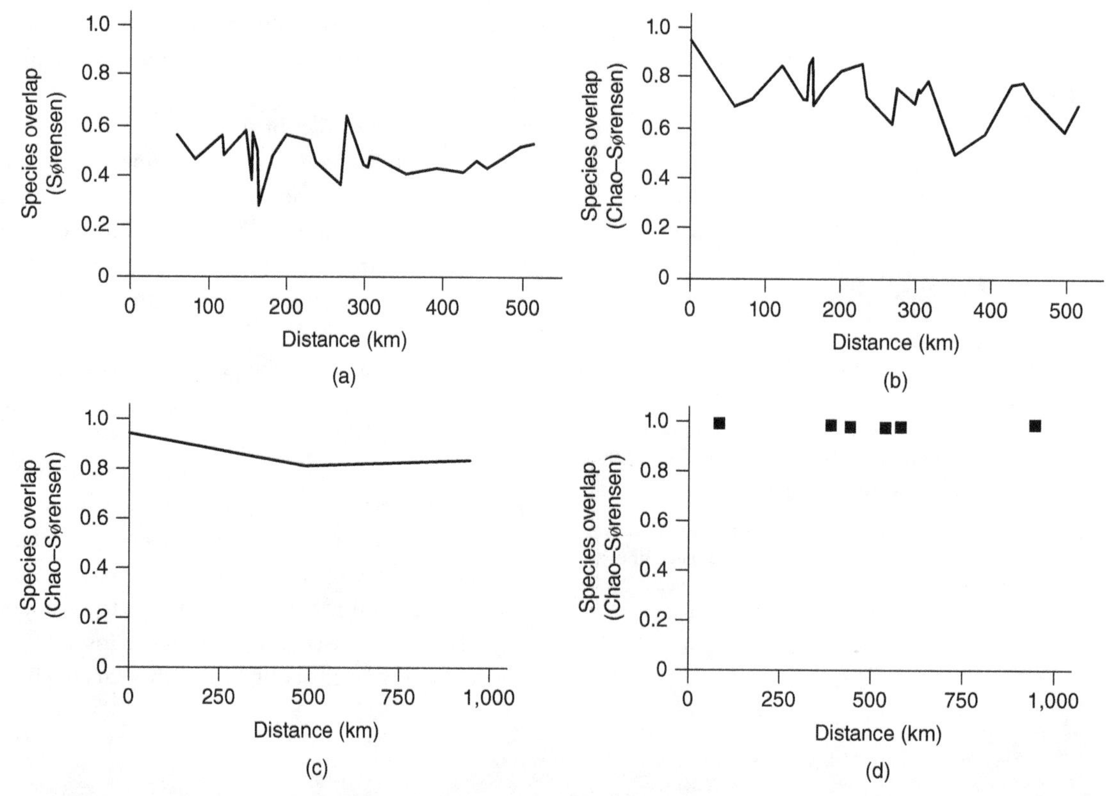

Figure 5.10 Turnover in species composition with distance for (a) plants, (b) caterpillars, (c) *Ambrosia* beetles and (d) fruit flies. (*Source*: Novotny et al. (2007). Reproduced with permission from Nature Publishing Group.)

Novotny and his team were now in a position to determine the link between local and regional richness. First they examined the plant communities, assessing the degree of overlap in species present between sites up to 500 km apart (Fig. 5.10a). On average only around 50% of plant species were shared between any two sites.

Having collected—and reared to adulthood—in the order of 75,000 caterpillars, they could also see how the herbivores responded on the same scales. This found something remarkable. There was a high degree of overlap, and though it declined with distance, it remained at 75% even for communities separated by 500 km, while those that were close together were nearly identical (Fig. 5.10b; the measure shown, Chao-Sørensen, is a similarity index). Other groups showed even less turnover; for *Ambrosia* beetles similarity remained above 80% (Fig. 5.10c), while for fruit flies almost exactly the same species were found everywhere (Fig. 5.10d).

This takes us back to one of the questions we met when estimating global species richness of insects in Chapter 4.5 do trees have large ranges, with specialist insect herbivores in different locations, or are the insects widespread, but specialising on a particular plant species in each location? The implication here is clear: turnover rates are much lower for insects than plants, and typically even the tiniest of species have ranges over 500 km in breadth. Moreover, even though they might only feed on a single plant species in any given patch, the majority turn out to have broad diets or at least to feed on a range of related plants within the same genus (Fig. 5.11).

We can now state with confidence that those who predicted vast numbers of insect species in tropical rain forests, based on projecting from small samples,

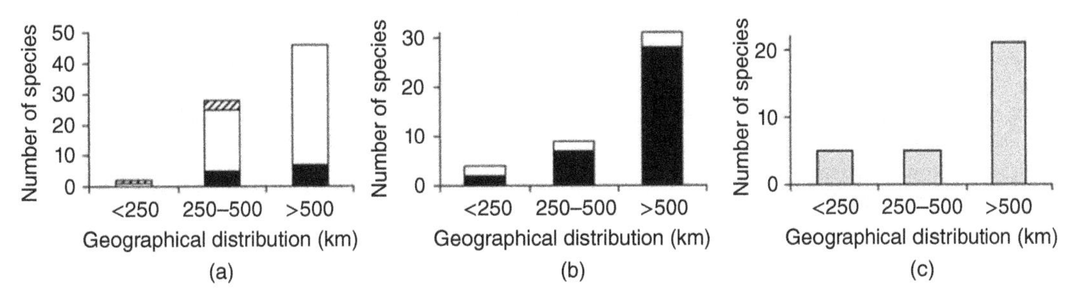

Figure 5.11 Proportion of insect species in three range size categories classified as generalists (black) or specialised on a single plant genus (white) or species (hatched) for (a) caterpillars, (b) *Ambrosia* beetles and (c) fruit flies (diet breadth untested). (*Source*: Novotny et al. (2007). Reproduced with permission from Nature Publishing Group.)

were wrong. There is an intimidating degree of α richness, but the same species tend to be found over large areas. This finding has since been replicated elsewhere; just 1 ha of rain forest in Panama contains more than 60% of regional arthropod species richness (Basset et al., 2012). We must turn our attention to the larger regional scale to work out why this is the case, but for now we have achieved our aim of demonstrating how simple collections of insects and calculation of diversity metrics can answer big questions.

5.10 Conclusions

There are three commonly recognised levels of diversity, measured as α at local scales, β as differentiation between patches and γ at the regional scale. Species richness can seldom be determined absolutely and instead requires some form of estimation. It also fails to capture variation in abundance, a key feature of the structure of communities, and so diversity indices have been developed which include this information, as well as being more efficient to obtain in terms of effort in the field. A family of indices can be used to convey multiple aspects of assemblage structure. Still other measures bring in salient features such as taxonomic (Δ^+), functional or interaction diversity. Finally, measuring turnover is essential to answer questions about how natural systems change between locations. Rather than seeking a single number that means 'biodiversity', the complexity of natural systems necessitates a range of approaches and metrics that reveal different kinds of information, and you should pick those which suit the question at hand. For a worked example illustrating how to use these techniques and interpret the findings, turn to Appendix A.

5.10.1 Recommended reading

Magurran, A. E., 2004. *Measuring Biological Diversity.* Blackwell Publishing.

Magurran, A. E. and B. J. McGill, 2011. *Biological Diversity: Frontiers in Measurement and Assessment.* Oxford University Press.

For more on the Binatang research station in Papua New Guinea and the work of Vojtech Novotny see http://www.entu.cas.cz/png/parataxoweb.htm

5.10.2 Questions for the future

- Are species the best units for describing diversity?
- What matters more for ecological processes—rare or common species?
- Why is β diversity so much higher for tropical plants than insects? What about for other groups of organisms?

References

Basset, Y., L. Cizek, P. Cuénoud, R. K. Didham, F. Guilhaumon, O. Missa, V. Novotny, F. Ødegaard, T. Roslin, J. Schmidl, A. K. Tishechkin, N. N. Winchester, D. W. Roubik, H.-P. Aberlenc, J. Bail, H. Barrios, J. R. Bridle, G. Casta no Meneses, B. Corbara, G. Curletti, W. Duarte da Rocha, D. De Bakker, J. H. C. Delabie, A. Dejean,

L. L. Fagan, A. Floren, R. L. Kitching, E. Medianero, S. E. Miller, E. Gama de Oliveira, J. Orivel, M. Pollet, M. Rapp, S. P. Ribeiro, Y. Roisin, J. B. Schmidt, L. Sørensen, and M. Leponce, 2012. Arthropod diversity in a tropical forest. *Science* 338:1481–1484.

Chao, A., 1984. Non-parametric estimation of the number of classes in a population. *Scandinavian Journal of Statistics* 11:265–270.

Chao, A., 2005. Species estimation and applications. In S. Kotz, N. Balakrishnan, C. B. Read, and B. Vidakovic, editors, *Encyclopedia of Statistical Sciences*, pages 7907–7916. John Wiley & Sons, Inc., second edition.

Chao, A., C.-H. Chiu, and L. Jost, 2010. Phylogenetic diversity measures based on Hill numbers. *Philosophical Transactions of the Royal Society of London, Series B: Biological Sciences* 365:3599–3609.

Chao, A., R. K. Colwell, C. W. Lin, and N. J. Gotelli, 2009. Sufficient sampling for asymptotic minimum species richness estimators. *Ecology* 90:1125–1133.

Chao, A. and L. Jost, 2012. Coverage-based rarefaction and extrapolation: standardizing samples by completeness rather than size. *Ecology* 93:2533–2547.

Clarke, K. R. and R. M. Warwick, 1998. A taxonomic distinctness index and its statistical properties. *Journal of Applied Ecology* 35:523–531.

Corlett, R. T. and R. B. Primack, 2011. *Tropical Rain Forests: An Ecological and Biogeographical Comparison*. Wiley-Blackwell, second edition.

DeVries, P. J., L. G. Alexander, I. A. Chacon, and J. A. Fordyce, 2012. Similarity and difference among rainforest fruit-feeding butterfly communities in Central and South America. *Journal of Animal Ecology* 81:472–482.

Dyer, L. A., T. R. Walla, H. F. Greeney, J. O. Stireman III, and R. F. Hanzen, 2010. Diversity of interactions: a metric for studies of biodiversity. *Biotropica* 42:281–289.

Ernest, S. K. M., T. J. Valone, and J. H. Brown, 2009. Long-term monitoring and experimental manipulation of a Chihuahuan Desert ecosystem near Portal; Arizona; USA. *Ecological Archives* 90:1708.

Gotelli, N. J. and A. Chao, 2013. Measuring and estimating species richness, species diversity, and biotic similarity from sampling data. In S. A. Levin, editor, *Encyclopedia of Biodiversity*, volume 5, pages 195–211. Academic Press, second edition.

Hill, M. O., 1973. Diversity and evenness: a unifying notation and its consequences. *Ecology* 54:427–432.

Jost, L., 2007. Partitioning diversity into independent alpha and beta components. *Ecology* 88:2427–2439.

Jost, L., A. Chao, and R. L. Chazdon, 2011. Compositional similarity and β (beta) diversity. In A. E. Magurran and B. J. McGill, editors, *Biological Diversity: Frontiers in Measurement and Assessment*, pages 66–84. Oxford University Press.

Lande, R., P. J. DeVries, and T. R. Walla, 2000. When species accumulation curves intersect: implications for ranking diversity using small samples. *Oikos* 89:601–605.

Lawton, J. H., D. E. Bignell, B. Bolton, G. F. Bloemers, P. Eggleton, P. M. Hammond, M. Hodda, R. D. Holt, T. B. Larsen, N. A. Mawdsley, N. E. Stork, D. S. Srivastava, and A. D. Watt, 1998. Biodiversity inventories, indicator taxa and effects of habitat modification in tropical forest. *Nature* 391:72–76.

MacArthur, R. H., 1965. Patterns of species diversity. *Biological Reviews* 40:510–533.

Magurran, A. E., 2004. *Measuring Biological Diversity*. Blackwell Publishing.

Magurran, A. E. and B. J. McGill, editors, 2011. *Biological Diversity: Frontiers in Measurement and Assessment*. Oxford University Press.

Maurer, B. A. and B. J. McGill, 2011. Measurement of species diversity. In A. E. Magurran and B. J. McGill, editors, *Biological Diversity: Frontiers in Measurement and Assessment*, pages 55–65. Oxford University Press, Oxford.

Novotny, V., Y. Basset, S. E. Miller, R. L. Kitching, M. Laidlaw, P. Drozd, and L. Cicek, 2004. Local species richness of leaf-chewing insects feeding on woody plants from one hectare of a lowland rainforest. *Conservation Biology* 18:227–237.

Novotny, V., S. E. Miller, J. Hulcr, R. A. I. Drew, Y. Basset, M. Janda, G. P. Setliff, K. Darrow, A. J. A. Stewart, J. Auga, B. Isua, K. Molem, M. Manumbor, E. Tamtiai, M. Mogia, and G. D. Weiblen, 2007. Low beta diversity of herbivorous insects in tropical forests. *Nature* 448: 692–695.

Pielou, E. C., 1975. *Ecological Diversity*. John Wiley & Sons, Inc.

Preston, F. W., 1948. The commonness, and rarity, of species. *Ecology* 29:254–283.

Rykken, J. J., A. R. Moldenke, and D. H. Olson, 2007. Headwater riparian forest-floor invertebrate communities associated with alternative forest management practices. *Ecological Applications* 17:1168–1183.

Scheiner, S. M., 2012. A metric of biodiversity that integrates abundance, phylogeny, and function. *Oikos* 121:1191–1202.

Southwood, T. R. E., 1978. *Ecological Methods*. Chapman and Hall.

ter Steege, H., N. C. A. Pitman, D. Sabatier, C. Baraloto, R. P. Salom ao, J. E. Guevara, O. L. Phillips, C. V. Castilho, W. E. Magnusson, J.-F. Molino, A. Monteagudo, P. Nú nez Vargas, J. C. Montero, T. R. Feldpausch, E. N. H. Coronado, T. J. Killeen, B. Mostacedo, R. Vasquez, R. L. Assis, J. Terborgh, F. Wittmann, A. Andrade, W. F. Laurance, S. G. W. Laurance, B. S. Marimon, B.-H. Marimon, I. C. Guimar aes Vieira, I. L. A. Amaral, R. Brienen, H. Castellanos, D. Cárdenas López, J. F.

Duivenvoorden, H. F. Mogollón, F. D. D. A. Matos, N. Dávila, R. García-Villacorta, P. R. Stevenson Diaz, F. Costa, T. Emilio, C. Levis, J. Schietti, P. Souza, A. Alonso, F. Dallmeier, A. J. D. Montoya, M. T. Fernandez Piedade, A. Araujo-Murakami, L. Arroyo, R. Gribel, P. V. A. Fine, C. A. Peres, M. Toledo, G. A. Aymard C., T. R. Baker, C. Cerón, J. Engel, T. W. Henkel, P. Maas, P. Petronelli, J. Stropp, C. E. Zartman, D. Daly, D. Neill, M. Silveira, M. R. Paredes, J. Chave, D. D. A. Lima Filho, P. M. Jørgensen, A. Fuentes, J. Schöngart, F. Cornejo Valverde, A. Di Fiore, E. M. Jimenez, M. C. Pe nuela Mora, J. F. Phillips, G. Rivas, T. R. van Andel, P. von Hildebrand, B. Hoffman, E. L. Zent, Y. Malhi, A. Prieto, A. Rudas, A. R. Ruschell, N. Silva, V. Vos, S. Zent, A. A. Oliveira, A. C. Schutz, T. Gonzales, M. Trindade Nascimento, H. Ramirez-Angulo, R. Sierra, M. Tirado, M. N. Uma na Medina, G. van der Heijden, C. I. A. Vela, E. Vilanova Torre, C. Vriesendorp, O. Wang, K. R. Young, C. Baider, H. Balslev, C. Ferreira, I. Mesones, A. Torres-Lezama, L. E. Urrego Giraldo, R. Zagt, M. N. Alexiades, L. Hernandez, I. Huamantupa-Chuquimaco, W. Milliken, W. Palacios Cuenca, D. Pauletto, E. Valderrama Sandoval, L. Valenzuela Gamarra, K. G. Dexter, K. Feeley, G. Lopez-Gonzalez, and M. R. Silman, 2013. Hyperdominance in the Amazonian tree flora. *Science* 342:1243092.

Tuomisto, H., 2010a. A consistent terminology for quantifying species diversity? Yes, it does exist. *Oecologia* 164:853–860.

Tuomisto, H., 2010b. A diversity of beta diversities: straightening up a concept gone awry. Part 1. Defining beta diversity as a function of alpha and gamma diversity. *Ecography* 33:2–22.

Tylianakis, J. M., T. Tscharntke, and O. T. Lewis, 2007. Habitat modification alters the structure of tropical host-parasitoid food webs. *Nature* 445:202–205.

Vellend, M., W. K. Cornwell, K. Magnusen-Ford, and A. O. Mooers, 2011. Measuring phylogenetic biodiversity. In A. E. Magurran and B. J. McGill, editors, *Biological Diversity: Frontiers in Measurement and Assessment*, Chapter 14, pages 194–207. Oxford University Press.

Weiger, E., 2011. A primer of trait and functional diversity. In A. E. Magurran and B. J. McGill, editors, *Biological Diversity: Frontiers in Measurement and Assessment*, Chapter 13, pages 175–193. Oxford University Press.

Whittaker, R. H., 1972. Evolution and measurement of species diversity. *Taxon* 1972:213–251.

Whittaker, R. H., 1977. Evolution of species diversity in land communities. *Evolutionary Biology* 10:1–67.

CHAPTER 6

Niches

6.1 The big question

The niche of an organism describes its place within a natural system, and an oft-stated maxim in ecology is that no two species can occupy the same niche. That said, there is probably no term in ecology that is so broadly used, and so frequently misunderstood, as that of the niche. As with species, it's an idea that seems intuitively obvious but which proves remarkably slippery once we try to create a firm definition. A variety of inconsistent definitions of the niche have led to confusion and no clear method for measuring their sizes or boundaries. Our understanding is excessively influenced by a bias for studying large animals, where it is easy to imagine species coexisting by using distinct food types, nesting sites or other obvious factors. In the vast majority of cases this is untrue and resources are shared among many competing species.

Fortunately major progress has been made in recent years, particularly through a book by Jonathan Chase and Mathew Leibold (2003), building on previous models by MacArthur (1972) and Tilman (1982). They propose an empirical way of looking at the niche which makes predictions that can be tested in the real world. Whether or not two species coexist in a given location can be predicted from a set of simple rules. This chapter summarises their arguments, but if you want to know more, then it's one of the most important books in ecology in recent years and worth reading in full. The fundamental problem we wish to resolve is: how can so many species coexist when they require basically the same resources? Why doesn't one always outcompete the others? To see why their insight is so important, it is necessary to first see how we got into a mess over niches.

6.2 Historical background

From the earliest years of ecological science, the niche concept was widespread, but two different schools of thought rapidly developed. The first clear definition was given by Grinnell (1917), later expanded by Hutchinson (1957). Essentially this saw the niche as describing the requirements of a species for its survival and persistence. A rival set of authors, led initially by Elton (1927) and built upon by MacArthur and Levins (1967), instead preferred to focus on the impacts of a species upon its environment and resources. These parallel definitions existed in the literature for some time. Hutchinson's concept remains the classic textbook definition. He made the distinction between the fundamental niche, which was the region of environmental space a species could potentially occupy, and its realised niche, which was what was left over following competition with other species (Fig. 6.1). The fundamental niche was expected to be hardly ever seen since species seldom occur in isolation, so the distribution of species should instead match their realised niche. An underlying assumption was that

Natural Systems: The organisation of life, First Edition. Markus P. Eichhorn.

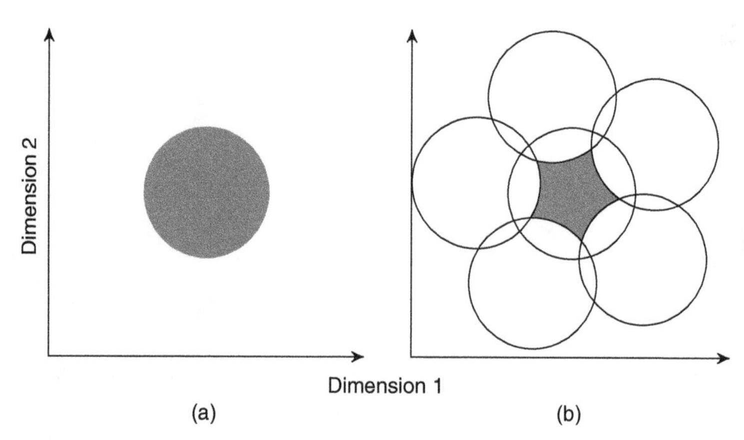

Figure 6.1 (a) The fundamental niche and (b) realised niche (grey shading) following competition with other species as envisaged by Hutchinson (1957).

competition was the main driving force in determining the difference between the fundamental and realised niche. Though represented here along two axes, which you might think of as environmental clines (e.g. temperature and rainfall) or resource gradients (e.g. nitrogen and water), Hutchinson realised that in fact many different axes could be influential and acting simultaneously. This means that these shapes, which are not necessarily circular, could also be expressed in three, four or many dimensions, creating a virtual niche in what he called an 'n-dimensional hypervolume'. The niche is therefore not a particular location or habitat but a theoretical concept.

This points towards the main problem with Hutchinson's formulation—how do you go about measuring something in n dimensions? When do you stop adding new variables? It was a nice idea but in practice proved fiendishly difficult to apply, and it is only with the advent of more powerful computers and new statistical methods that its analysis has even become tractable (Blonder et al., 2014). A deeper question, however, is that if the fundamental niche is never seen in nature, can you ever know what it is?

An attempt to take Hutchinson's vague ideas and embed them in a more mathematical framework was made by MacArthur and Levins (1967). They tried to link it with other well-known theories that were circulating in ecology at the time, including Gause's principle that no two species could occupy the same niche (Gause, 1936), in combination with the models for competition among species produced by Lotka (1924) and Volterra (1926). It's worth mentioning that this paragraph has just name-checked many of the greatest figures in the early history of ecological thought, which means the approach comes with a fine pedigree.

MacArthur and Levins (1967) came up with a model in virtual space in which, it was presumed, species would distribute themselves such that there was a maximum degree of overlap before niches became too small for each species to persist. This was referred to as 'niche packing' (Fig. 6.2). The degree of overlap could be predicted mathematically as a function of d, the distance between the mean niches of any two species in environmental space, and the radius of those niches σ. The ratio between d and σ determined the limiting similarity of two species, or how much two species could overlap without one excluding the other (May and MacArthur, 1972).

While a beautiful and elegant model, it has since found little support, as few studies have demonstrated regular spacing of species along environmental or resource axes (assuming all niches to be equal in size, which is a leap of faith). Another difficulty is that it isn't properly falsifiable; failure to find niche packing can simply provoke the defence

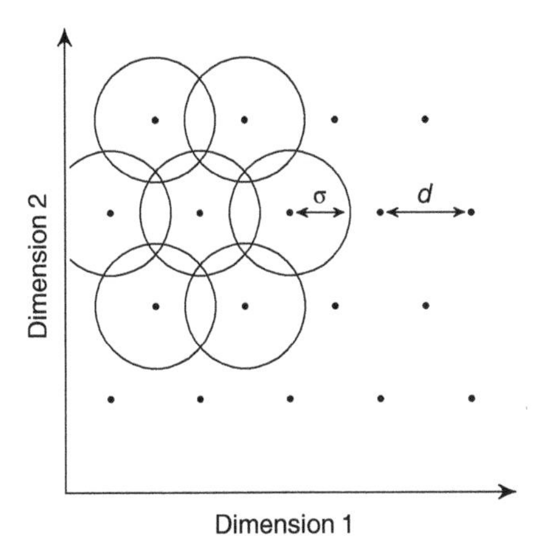

Figure 6.2 Niche packing model based on distance between niche centres d and niche radius σ. (*Source*: Adapted from MacArthur and Levins (1967).)

that there must be another, as yet unmeasured, axis along which they are arranged. Other problems include the assumption that species must trade off along axes, and that once again it's all about competition, with other vital ecological processes such as predation or stress excluded.

As a result of these competing views of niches, none of which made exclusive and falsifiable predictions that permitted proper scrutiny, the idea of the niche fell into disregard. This led to some critical comments from leading lights in ecology:

> No concept in ecology has been more variously defined or more universally confused than 'niche'.
>
> *Real and Levin (1991)*

> I believe that community ecology will have to rethink completely the classical niche-assembly paradigm from first principles.
>
> *Hubbell (2001)*

Nevertheless, it's impossible to dismiss the idea of niches entirely, mainly because (like species) much of ecology fails to make sense without them (Turnbull, 2014). The early attempts to come up with a satisfactory definition may have ultimately proven inadequate, but they were important in framing the debate that followed.

6.3 Back to basics

The way out of this mess is to step back and think about the original problem more clearly. Starting with plants, it is obvious that many species coexist in the same place, yet essentially all of them need only three main resources—light, water and soil nutrients. Is it really possible to differentiate species along axes of these resources when they basically require the same thing?

A well-known principle of agricultural production, Liebig's law, states that plant growth is restricted by the most limiting resource (von Liebig, 1840). There's no point in giving your crops more light or fertiliser if they don't have sufficient water. A similar principle applies to animals; in most cases, animal growth rates are limited by the availability of nitrogen for making proteins, and this is often required in a specific form. Other single forces can be seen to restrict population growth rates, including predation or environmental stress. In most cases, when you consider the evidence, only one or two factors are responsible for restricting the growth and reproduction of individuals and (by extension) populations. The identity of limiting resources might change through time depending on the environment or competition (e.g. Farrior et al., 2013) but in the simplest case we can ignore this.

The suggestion is that what we need is reductionism: instead of trying to understand processes in n dimensions, often only a couple will do. We will take this as our starting point to develop a more practical concept of the niche.

6.4 Birth and death rates

We begin by conducting a thought experiment into how birth and death rates might be affected by resources or predators. First of all, think about the *per capita* effects of individuals within the population. Every individual will be consuming resources, so whatever the starting level of resources, they will tend to reduce them further (Fig. 6.3a). When

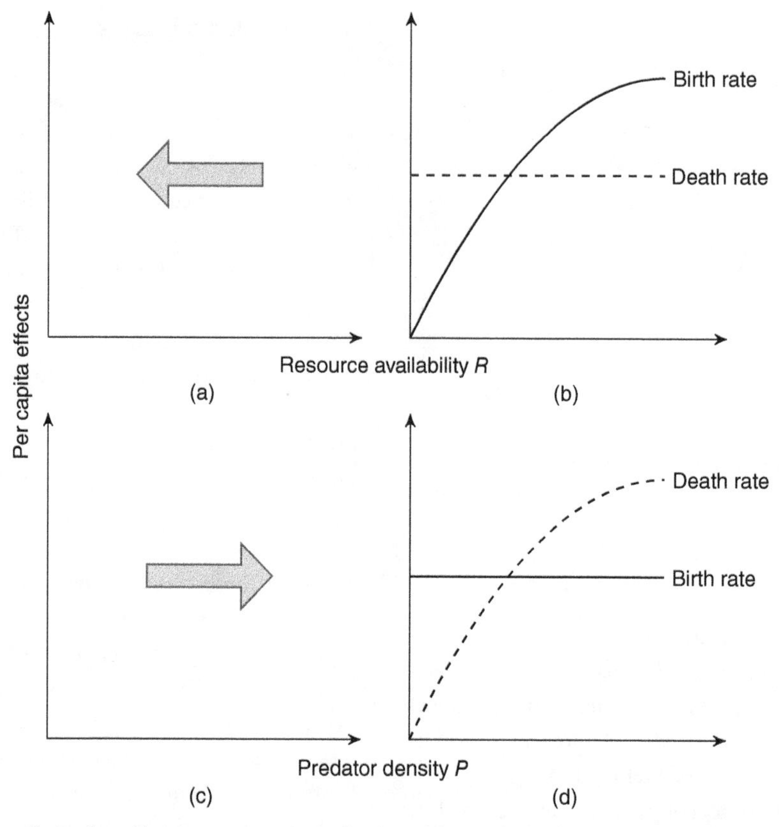

Figure 6.3 Per capita effects of populations on (a) resource levels and hence (b) birth and death rates and on (c) predator density and hence (d) birth and death rates. (*Source*: Adapted from Chase and Leibold (2003).)

resource levels R are high, there is plenty to go round, and the birth rate will exceed the death rate—a growing population (Fig. 6.3b). As resource levels fall, the birth rate will drop as well. We assume here for the sake of simplicity that the death rate remains constant. Actually the shapes of these lines are immaterial; they could take any number of forms. The important principle is that eventually the birth rate will drop to a level where it exactly balances the death rate. The population will then remain stable. Should R fall any further, the population will decline, which will ease the pressure on resources (Fig. 6.3a) and allow R to recover.

Predation can be viewed in a similar way, only here the pattern is in reverse. The *per capita* effect of each individual prey in a population is to increase the abundance of their predators, P (Fig. 6.3c). If you imagine a population of prey living with a low density of predators P, their birth rate will exceed the death rate, which may be very low since relatively few of them are being eaten. As the population grows, the tendency is to increase predator density, and therefore the death rate, until it meets the birth rate, and the population once again balances at a stable level where the lines cross (here we assume no effect of predators on birth rates, and a saturating level of predation, which is common in nature when predators are territorial). Too much predation and the population will go into decline, which will in turn cause predators to leave or die and levels of P will fall again (Fig. 6.3d).

The point at which birth and death rates balance is important. When referring to resources it is known as R^* (Tilman, 1982, Fig. 6.4). Imagine now that you

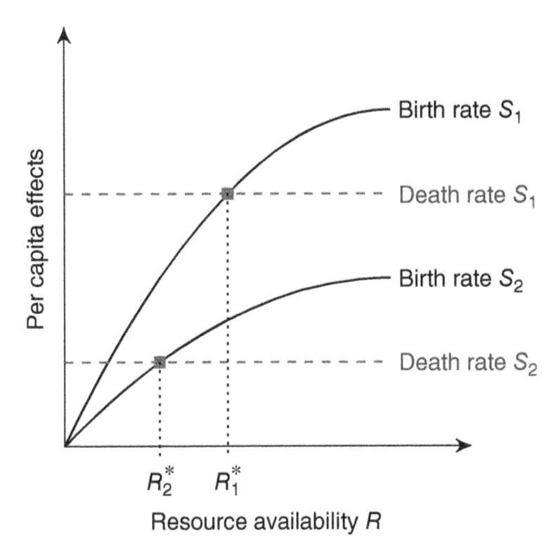

Figure 6.4 Illustration of R^* with two species competing for a common resource. (*Source*: Adapted from Chase and Leibold (2003).)

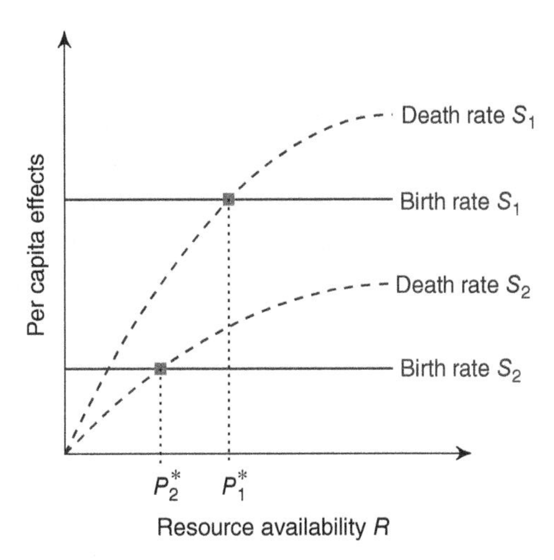

Figure 6.5 Illustration of P^* with two species sharing a common predator. (*Source*: Adapted from Chase and Leibold (2003).)

have two species, S_1 and S_2, competing for the same resource. Resource levels R are initially high, so both populations grow, causing R to decline. Eventually one species reaches the point at which its population is in balance, marked as R_1^*. At this level of resources, however, the other species still has positive population growth rates (even though its birth rate is lower), so it will continue to drag levels of R further down. By the time it is itself in balance (R_2^*), the other species is in decline as its death rate exceeds its growth rate.

In this case species S_2 wins in competition because it has the lowest value of R^*. If you understand this, then everything else should make sense; if not, read through the previous text again carefully and make sure you can work it out, because this is the most important idea to grasp. Technically R^* is defined as the minimum level of resources at which a population can (barely) survive, where its birth rate and death rate are exactly equal. In the mathematics of population biology, we would say $dN/dt = 0$, which means that the rate of change of population numbers N over time t is zero. In competition among species for any given resource, the species with the lowest R^* wins because it reduces levels below that which the other species can handle. This is really a restatement of Gause's principle.

The power of this approach comes through the realisation that it works equally well for other ecological forces (Chase and Leibold, 2003). When two species share a common predator, by analogy, the species with the highest P^* will persist at the expense of the other (Holt et al., 1994, Fig. 6.5). Both prey species tend to increase abundance of the predator, but when one is able to tolerate higher densities of predators than the other, P will continue to rise until only the species with the highest P^* is able to maintain a population. In this case S_1 will be the winner in a struggle that is known as **apparent competition**. The outcome looks like competition but is in fact mediated through the actions of a predator.

It is also possible to work stress gradients into the same system and come up with a similar measure called S^*—the maximum level of stress a population of a given species can tolerate without growth rates becoming negative. Since organisms cannot influence the amount of environmental stress they are subjected to, this works a little differently, as will be seen shortly. One could also add positive effects such as mutualists or ecosystem engineers (Chase and Leibold, 2003), though we will not have space to consider them here.

6.5 The ZNGI

Let's take a small step up in complexity and think about what might happen when two niche dimensions influence a species. In theory there could be a large number, though in practice two is usually sufficient, provided they are selected carefully. Fig. 6.6a illustrates the habitable niche space for a species with two limiting resources. In this case the resources are substitutable: so long as the species has enough of one, it doesn't need any of the other. This line could in theory take almost any shape, but for now a straight line is simplest to demonstrate the basic principles. If you prefer then you can think of this in terms of a particular species—perhaps a population of squirrels that needs a certain amount of either hazelnuts (R_A) or worms (R_B) in order to persist. When resources are above the line, the species persists (population growth rate is positive), whereas below the line the growth rate is negative and the population cannot survive. The line between them represents a population growth rate of zero and where it meets the axes is the R^* value for each resource.

We can play a similar game with two predators (Fig. 6.6b); perhaps here the squirrels are eaten by both foxes (P_A) and owls (P_B). So long as predator density remains below a certain level, the population

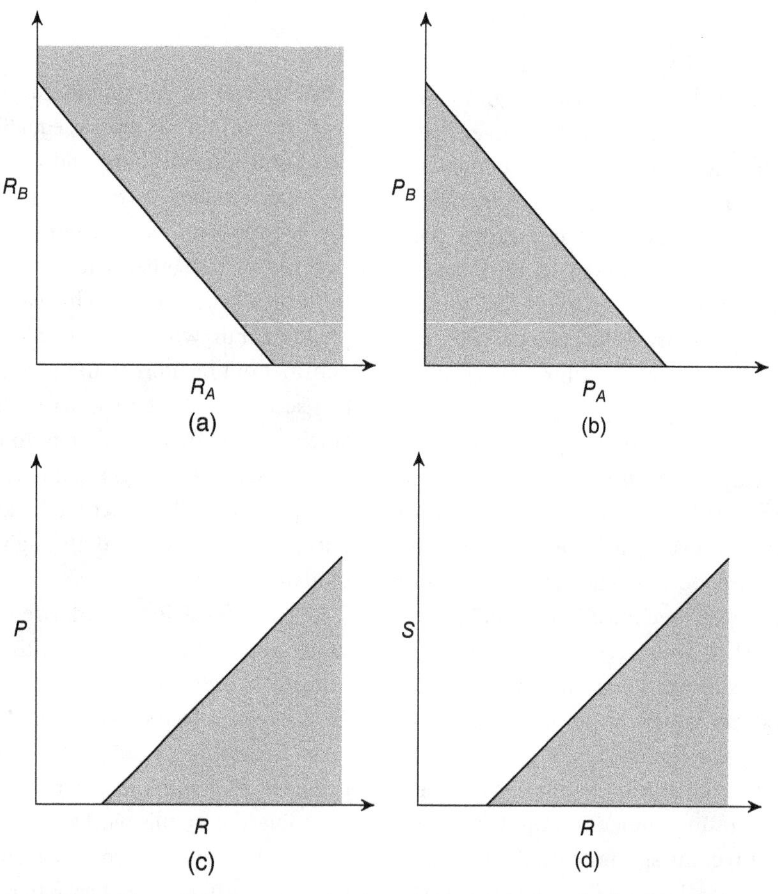

Figure 6.6 Niche space within which a population can maintain positive growth (grey shading) against axes composed of (a) two resources, (b) two predators, (c) a resource and a predator, (d) a resource and a gradient of environmental stress. (*Source*: Adapted from Chase and Leibold (2003).)

of squirrels manages to grow but eventually reaches the point when they are effectively controlled by the predators at a stable level. Once again this line meets the axes at the P^* level for each predator. By extension, we can plot a similar graph for a resource R and a predator P (Fig. 6.6c) or even for a resource and a gradient of environmental stress S (Fig. 6.6d)—perhaps squirrels can cope with severe winter cold provided that there are plenty of worms, but as resource levels fall their ability to tolerate this stress declines. In practice the choice of the two most important axes is up to the investigator and will be informed by an understanding of the natural history and physiology of the species concerned.

The line in niche space that corresponds to the conditions at which the population growth rate is zero is known as the Zero Net Growth Isocline or ZNGI—pronounced 'zingy'—for convenience (Fig. 6.7). It describes an organism's response to the environment and is therefore equivalent in meaning to Hutchinson's 1957 niche concept. It can be obtained by finding out the R^* (or similar) values for the species.

This isn't quite the end of the story though; we still have to incorporate the effects of the organism on the environment, as well as the supply rate of resources and predators.

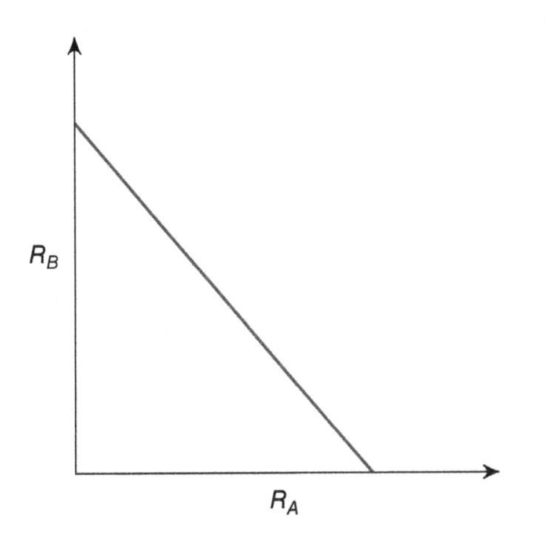

Figure 6.7 The Zero Net Growth Isocline (ZNGI) of a species for two substitutable resources.

6.6 Impact vectors

The impact on the environment of any single individual in a population is to consume some of the available resources. The rate at which it does so describes its effect on niche space. When there are two resources (Fig. 6.8a), the combined effect on both can be turned into a vector that describes how we expect that species to alter levels of the resources in its vicinity.

By extension, each individual of a prey species might provide a certain amount of food for its predators, such that for every prey item consumed their populations will increase by a small amount (Fig. 6.8b). We can even combine impacts of different types into a single vector, as with a resource and a predator (Fig. 6.8c).

The only slight difference comes when incorporating an axis of environmental stress (Fig. 6.8d). Organisms typically do not influence the harshness of the environment, at least not on a local scale: having more squirrels will not stop it getting cold in the winter, nor will more cacti make it rain in the desert. As a result, the impact vector in Fig. 6.8d describes only the effect on resources. You might be able to imagine some special cases where this is not entirely true, but in general it's a safe assumption.

6.7 Supply points

The final element in the scheme is the supply point of the environment. This is the level to which resources would rise in the absence of anything consuming them (Fig. 6.9a). For our imaginary example, this would be the abundance of hazelnuts and worms if there were no squirrels to eat them. With two predators, the picture is very simple—the supply point will be at zero (Fig. 6.9b). This is obvious as if there are no squirrels, then all the foxes and owls will either die or leave for somewhere else, unless they have another food source (which would have implications for the squirrels that we will return to later). Combining a resource and a predator is similarly intuitive (Fig. 6.9c). Finally we can predict

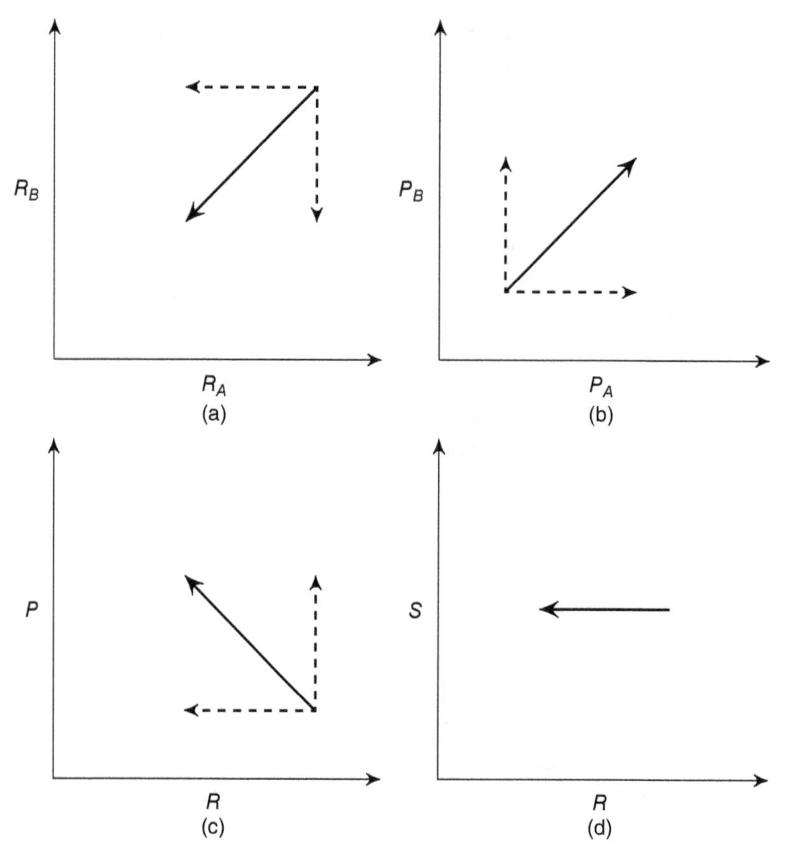

Figure 6.8 Impact vectors describing the per capita effects in niche space of a species on (a) two resources, (b) two predators, (c) a resource and a predator, (d) a resource and a gradient of environmental stress. (*Source*: Adapted from Chase and Leibold (2003).)

that the level of environmental stress will always remain the same whether or not there are any squirrels around to suffer from the cold (Fig. 6.9d).

6.8 Coexistence

We can now pull all these elements together: the ZNGI, impact vector and supply point, into a single model that can predict the conditions under which any two species are able to coexist (Fig. 6.10). Imagine, if you prefer, that our squirrels now have to share the woodland with rats, which also eat both hazelnuts and worms, but at different rates. Provided that the supply point of resources falls somewhere in the middle of potential niche space, they will be able to coexist.

Let's look at this figure more carefully. There are three conditions for coexistence to occur:

1. Their ZNGIs must intersect. If not, then one species is a better competitor for both resources and always wins. If they cross, it means that each has an R^* advantage over the other for one resource. Perhaps squirrels are better at reducing hazelnut levels, while rats deplete worms.
2. Each impact vector must be proportional to the ZNGI; in other words, each species has the greatest impact on the resource it finds most limiting. This is plausible if species forage optimally, as they will expend more effort in harvesting the resource they most require.
3. The supply point must fall at an intermediate level. Resource levels will be inexorably dragged down to the crossing point of the ZNGIs, which

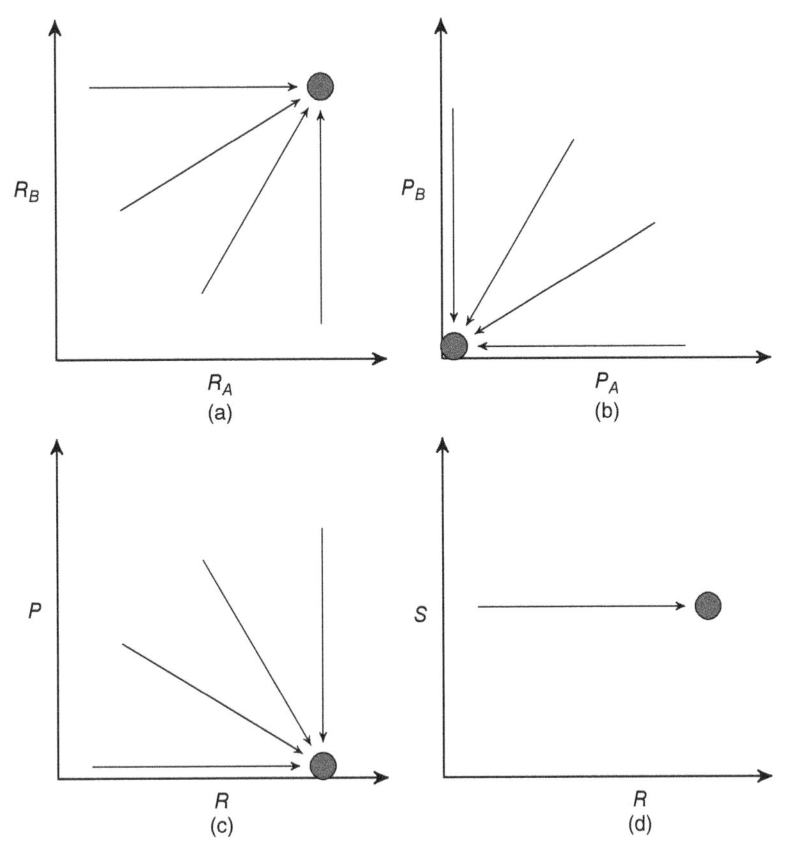

Figure 6.9 Supply points of the environment for (a) two resources, (b) two predators, (c) a resource and a predator, (d) a resource and an environmental stress gradient. In the absence of the focal organism levels will tend to return to the supply point. (*Source:* Adapted from Chase and Leibold (2003).)

in mathematical terminology is a stable attractor. At this point intraspecific competition exceeds interspecific, meaning that each species limits itself more than the other. This is one of the conditions for stable coexistence to be possible (Chesson, 2000). A surfeit of either resource gives one species an advantage and it will then drag resource levels down to the R^* where its ZNGI meets the axis.

When the second condition is violated, and impact vectors are not proportional to their ZNGIs, the crossing point is unstable and the species cannot coexist (Fig. 6.11). In this case, at intermediate resource levels, chance fluctuations dictate which species ends up as the sole winner.

We can employ the same logic to look at two species coexisting when they share two predators (Fig. 6.12). Perhaps squirrels and rats (the two ZNGIs) are both hunted by foxes and owls (P_A and P_B). Identical rules apply, only in this case, the correlation between the ZNGI and the impact vector means that each species must have a stronger impact on the predator to which it is least vulnerable in order for coexistence to occur (Leibold, 1998). Once again we can imagine this working in nature. Foxes might prefer to eat squirrels but have a hard time catching them. While each squirrel makes a great meal for a fox, squirrels can nevertheless tolerate higher fox populations (i.e. they have a higher P^* for foxes). On the ground, owls might prefer to feed on rats but have a higher chance of

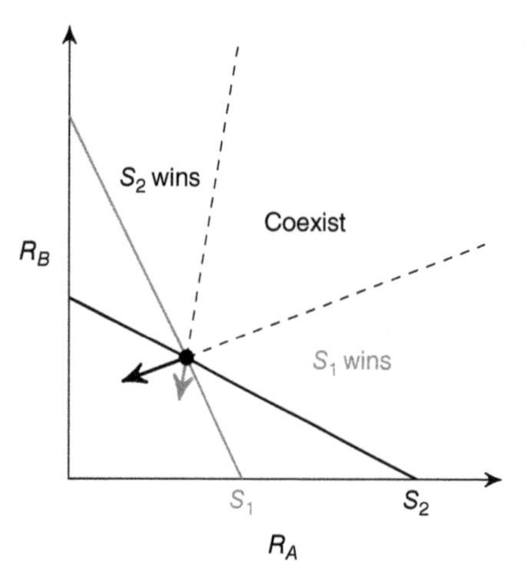

Figure 6.10 Predicting coexistence criteria for two species (grey: S_1, black: S_2) with two substitutable resources (R_A, R_B). Solid lines show ZNGIs which cross at the point marked with a circle. Impact vectors of each species on resources are shown as arrows. Dashed lines trace impact vectors into resource space and determine the outcome of resource competition given the position of the supply point. (*Source*: Adapted from Chase and Leibold (2003).)

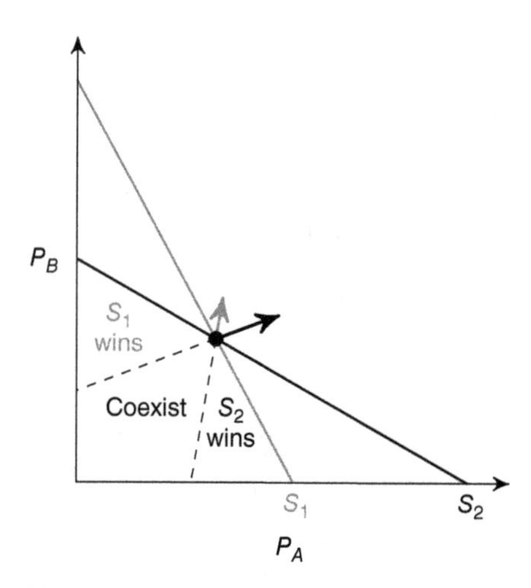

Figure 6.12 Coexistence criteria for two species with two shared predators (P_A, P_B). See Fig. 6.10 for terms. (*Source*: Adapted from Chase and Leibold (2003).)

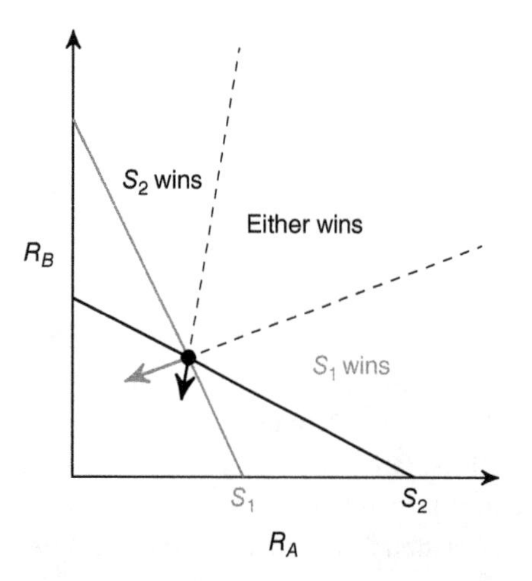

Figure 6.11 Coexistence is unstable when impact vectors are not in proportion to their ZNGIs i.e. each species has a greater impact on the resource it is least limited by. See Fig. 6.10 for terms. (*Source*: Adapted from Chase and Leibold (2003).)

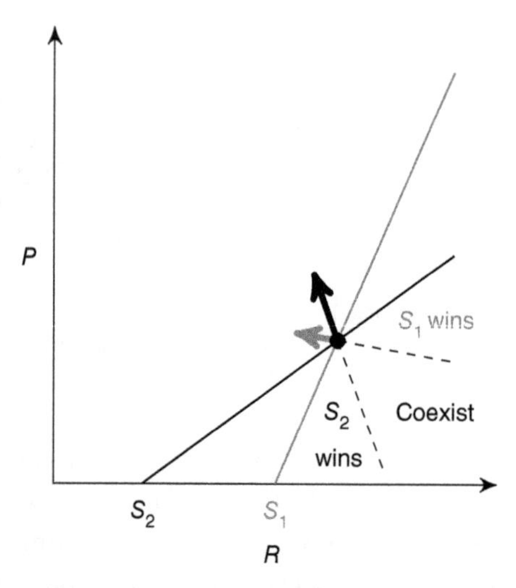

Figure 6.13 Coexistence criteria with both a shared resource R and a shared predator P. See Fig. 6.10 for terms. (*Source*: Adapted from Chase and Leibold (2003).)

catching a squirrel (this may not be true but it works for our analogy). So long as these conditions are met, the prey species can coexist.

The real power of this world view is in combining all possible types of niche dimension. For example, with a shared resource and a shared predator (Fig. 6.13), coexistence can occur so long as the better defended species (higher P^*) is a weaker resource competitor (higher R^*), with impact vectors to match. Imagine an island populated by tortoises (S_1), rabbits (S_2) and eagles (the predator P). Tortoises are better defended against the eagles by their shell but are poor competitors for resources since they are so slow. Yet so long as the eagles prefer to eat rabbits, and the balance of predators and resources is right,

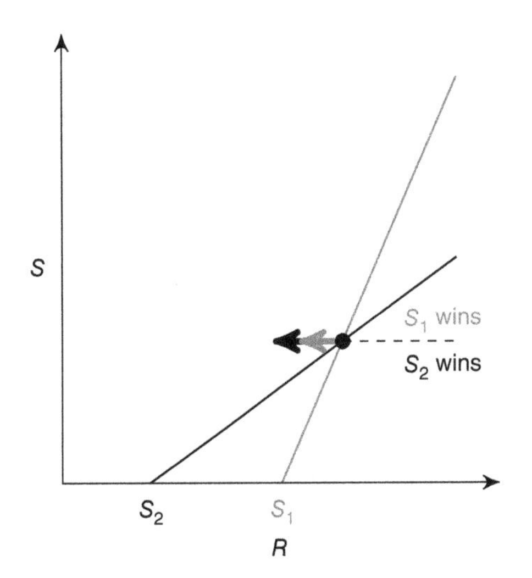

Figure 6.14 Outcomes for two species with a shared resource *R* along a gradient of environmental stress *S*. See Fig. 6.10 for terms. (*Source*: Adapted from Chase and Leibold (2003).)

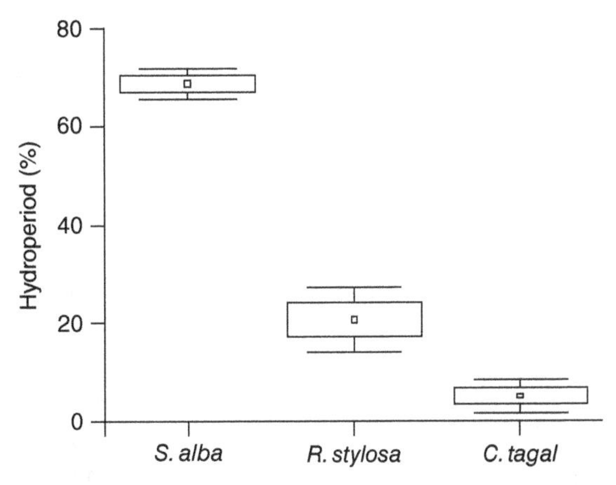

Figure 6.15 Mean, standard error (boxes) and 95% confidence intervals (whiskers) for hydroperiod, which measures the proportion of time sites are inundated by tidal water. Sites dominated by three species of mango tree (*Sonneratia alba, Rhizophora stylosa* and *Ceriops tagal*) in Darwin Harbour, northern Australia. (*Source*: Crase et al. (2013). Reproduced with permission from John Wiley and Sons Ltd.)

both rabbits and tortoises will be able to coexist on the island.

Finally we can think about what might happen when some form of environmental stress limits population growth (Fig. 6.14). This is a little different because species have no impact upon the stress gradient. Here coexistence is simply not possible or at least not stable in the long term. When levels of environmental stress are low, the best resource competitor wins (S_2). However, if the ZNGIs cross, then there will be a level of stress at which the advantage switches, and the species that can tolerate harsher conditions will be the only one to survive (S_1). This pattern is often seen along gradients in nature. For example, on a rocky shore, the best place to be for resources is closer to the sea, where organisms spend more time submerged beneath the water. Higher on the shore conditions become increasingly stressful as a result of desiccation when exposed to the air. Consequently a series of bands is commonly observed with different species dominating at each level. A similar pattern is found in mangrove swamps, where bands form parallel to the shoreline, each dominated by a different tree species. In this case the driving stress factor appears to be the proportion of time the stems are inundated with tidal water (Fig. 6.15). Note the lack of overlap in the range of conditions each species tolerates.

6.9 The evidence

The framework described here is known as resource ratio theory. At this point, having explored it from an entirely theoretical perspective, you should be asking—where's the evidence? The best part of the theory is that it makes some exclusive predictions that we can go out and test on real organisms. These can be summarised as follows:

- The species with the lowest R^* will be the best competitor for that resource.
- The dominant species at any site will vary according to the ratio of two resources.
- The number of species in any single location will be no greater than the number of limiting resources.
- Competitive outcomes depend on both the resource supply point and the impact vectors of the species.
- Coexistence is possible along a gradient of resources through trade-offs among species.
- Species richness should peak when there is an intermediate ratio of resources.

Though many of these ideas have been circulating for a number of years, it is only recently that they have been drawn together into a coherent body of theory, and therefore there have to date been relatively few comprehensive tests (perhaps this is a gap that you might like to fill?). Since the leading figure in this area, David Tilman, is a plant ecologist, the majority of studies have been conducted on plants, as well as microbes. Zoologists have taken longer to catch up, and though this has begun to change, there is plenty of scope for more work. This is largely the result of tradition and inertia; plant and animal ecologists tend to read and publish in different journals, and the problems that dominate research in each field are frequently different.

A good test can still be put together of the first prediction related to R^*. Miller et al. (2005) compiled studies that had assessed whether the species with the lowest R^* always won in competition. Around three quarters of studies were supportive. Another similar attempt to collate evidence found 41 tests, of which 39 backed R^* theory, one was equivocal and only one rejected it (Wilson et al., 2007). On balance the evidence seems to be behind R^*, at least among plants and other producers, but we know relatively little about consumers and detritivores, so as for its wider applicability the jury remains out. There have been fewer direct tests of P^*, and though they have been broadly supportive (e.g. Bohannan and Lenski, 2000), more work is required.

One of the first complete tests of the full model was by Tilman (1977) and examined two photosynthetic diatoms, *Cyclotella* and *Asterionella*. These both required two resources in solution, silica (SiO_2) and phosphate (PO_4). These were both essential resources rather than substitutable because a minimum amount of each was necessary for each species to persist. This meant that ZNGIs were elbow shaped, which is likely to be a common feature in nature (Fig. 6.16). The system is highly amenable to testing and also formed the basis for Gause's initial experiments from which he declared that no two species could share the same niche, at least by his definition (Gause, 1936). Tilman's work challenged this. By creating microcosms (sealed jars), each

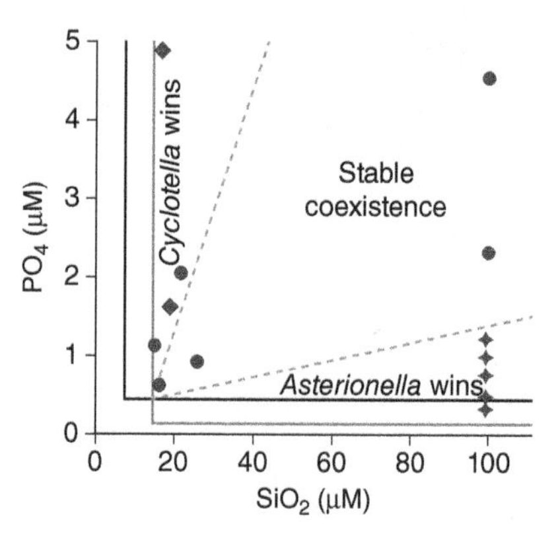

Figure 6.16 Testing coexistence criteria for two diatom species (*Cyclotella* and *Asterionella*) in microcosms with two essential resources (silica and phosphate). ZNGIs as solid lines; dashed lines trace impact vectors into resource space. Symbols are positioned at the experimental supply points of resources. Diamonds indicate supply points for which *Cyclotella* wins in competition, stars where *Asterionella* wins, circles denote coexistence. (*Source*: Redrawn from Tilman (1982). Reproduced with permission from Princeton University Press.)

species could be studied in isolation to find out how quickly it depleted resource levels (its impact vector) and the minimum they could persist at (their R^*). Then, combining the two species, he manipulated the chemistry of the solution to provide different supply points. The outcome was exactly as predicted—intermediate starting levels of the two resources led to stable coexistence of the species. The same patterns can be found using herbivorous zooplankton, indicating that the same processes can enable coexistence among consumers as well (Rothhaupt, 1988).

The theory appears to work well for microorganisms in glass jars, but the real measure of its value has to be in the field. In practice one of the most taxing elements of any such study is to find the impact vectors of the species. Behmer and Joern (2008) managed this for a set of coexisting grasshopper species, whose diets in the laboratory could be seen to trade off between intake of protein and carbohydrate, with each species requiring a slightly different balance (Fig. 6.17). Here there is good evidence that even when there are only two major resource axes, trade-offs occur among species that may permit them to coexist on a shared diet of the same plants.

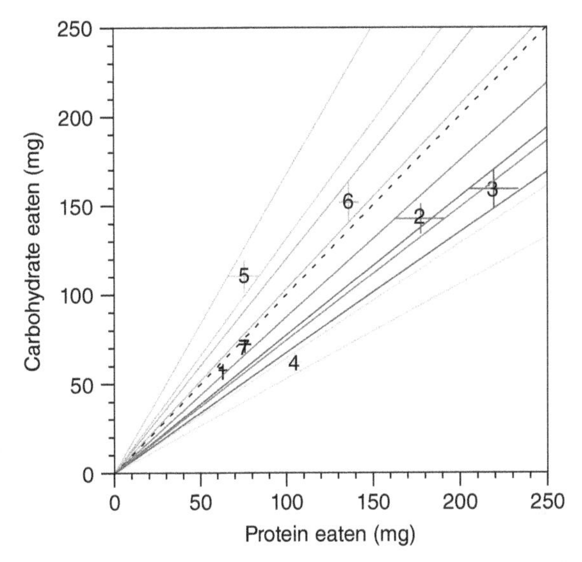

Figure 6.17 Intake targets (means ± SE) of carbohydrates and protein for each of seven coexisting grasshopper species. Dashed line represents 1:1 ratio. All except species 1 and 7 have significantly different targets. Solid lines define the nutrient space occupied by each species; only 1 and 7 overlap appreciably. (*Source*: Behmer and Joern (2008). Reproduced with permission from the National Academy of Sciences, USA.)

A slightly more complicated example comes from browsing mammals in the Serengeti. In this case it is possible to estimate impact vectors based on the consumption rates of zoo animals, which can then be applied to understand what might be happening out on the plains. These browsers all eat the same grass species but in different ways. Grass can be viewed as made up of two resources: the soft, nutritious shoots and the tough stems below (Fig. 6.18). Both are essential resources for the animals; this is indicated by the ZNGIs being almost parallel to the axes, although more leaf is required when the sward is particularly high, suggesting an inhibitive effect of large quantities of stem on digestion or foraging efficiency. Both impact vectors indicate a strong preference by the browsers for leaf material, though the narrower-mouthed topi are slightly more selective in their feeding, whereas wildebeest are relatively unselective and consume the whole sward. Depending on the area the balance of stem and leaf varies, perhaps in response to the amount of water available to the plants.

Unfortunately, due to their indiscriminate munching, wildebeest exert greater pressure on the leaves than the stems, meaning that although the ZNGIs cross, the equilibrium is unstable. When the ratio of stem is low, the browsing of wildebeest consumes the whole sward, and topi cannot forage alongside them (region I in Fig. 6.18). When there is a large amount of stem in the sward, the feeding method of wildebeest is inefficient in harvesting resources, so only topi persist as they are more selective (region II). Only unstable coexistence occurs in region III. Matching this expectation, wildebeest and topi are seldom found together in the same areas.

The approach can be extended across groups of browsers in the Serengeti (Fig. 6.19). Plant biomass can be split into cell walls and contents, with the ratio between them dictating the overall quality of foraging material (the former is effectively a measure of productivity). When cell wall biomass dominates, large browsers (500 kg) occupy the habitats, but when the forage is of higher quality there is a transition to smaller species, though in this case there are stable attractors allowing coexistence of some species. Notice also that the species most similar in size will coexist, rather than the extremes of the range; we will return to this point in Chapter 13 when thinking about community structure.

6.10 Implications

At this stage we can begin to assess what these ideas mean for natural systems. One outcome is that the maximum number of species in a given place should be set by the number of limiting resources (or predators). The more potential limiting factors exist, the more opportunities there are for species to trade off against one another. Another point worth mentioning is that changes in the standing levels of resources observed in the field do not necessarily reflect resource supply but instead depend upon the R^* values of the species present.

The theory only applies to local coexistence, that is, for species interacting at a single point. On a slightly larger scale, many species can coexist so long

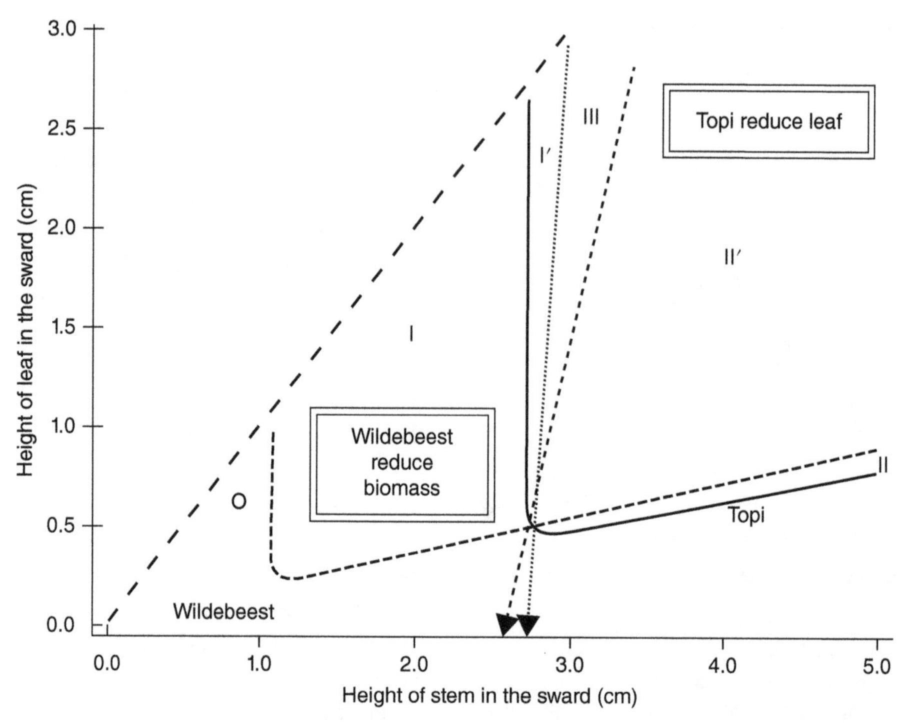

Figure 6.18 Unstable coexistence between topi and wildebeest based on relative height of grass stem and leaf material in a homogeneous grassland. ZNGIs (topi, solid line; wildebeest, closely dashed line), impact vectors (topi, dashed line; wildebeest, dotted line). ZNGIs are constrained to below a 1:1 ratio as it is not possible for leaf height to exceed stem height. (*Source*: Murray and Baird (2008). Reproduced with permission from the Ecological Society of America.)

as there is some habitat heterogeneity, either in space or time (also known as the **storage effect**; Tilman (1994, 2004). Provided that each species has a negative feedback between its resource requirements and its impacts on those resources (i.e. its ZNGI and impact vector), species will separate out along resource gradients (Fig. 6.20). These can be quite fine-scaled, such that with small shifts in the supply point, different species become dominant. Exactly the same can be true of multiple species coexisting alongside two shared predators (Fig. 6.21). We will meet specific examples of these when we consider communities in Part III.

What happens when you put multiple species together? In perhaps the most comprehensive test of R^* theory to date, Dybzinski and Tilman (2007) took a set of prairie plant species and spent over a decade determining their R^* levels for nitrogen and light. They were then grown in various combinations in carefully controlled experiments. The good news was that competition played out as expected—the two species with the lowest ZNGIs drove others extinct, then only those whose ZNGIs crossed were able to coexist (Fig. 6.22).

This seems, on the face of it, like a convincing demonstration. But once you step away from the experiments, which had well-mixed soils, abundant water and no herbivores and were protected from any disturbance, and look at real prairies, it's clear that the real world is more complex. Natural communities nearby contained 15–30 species per square metre and over 100 per hectare; moreover, the species with the lowest ZNGIs did not dominate. Often the species that arrives first (the incumbent) can hold on to a site and prevent a stronger competitor from establishing and dominating; these types of interactions which change through time will be considered in Chapter 12. Just because we can predict

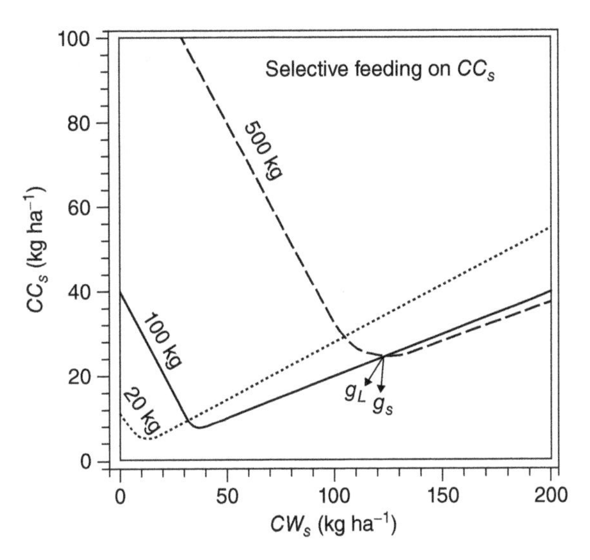

Figure 6.19 Browsers of different size partition habitats along a resource gradient formed by the ratio between the plant biomass per hectare composed of cell wall (CW_S) and cell contents (CC_S). ZNGIs for three nominal browsers are plotted, large (500 kg, dashed line), medium (100 kg, solid line) and small (20 kg, dotted line). The impact vectors from the stable coexistence point between large and medium browsers are shown as arrows. (*Source*: Murray and Baird (2008). Reproduced with permission from the Ecological Society of America.)

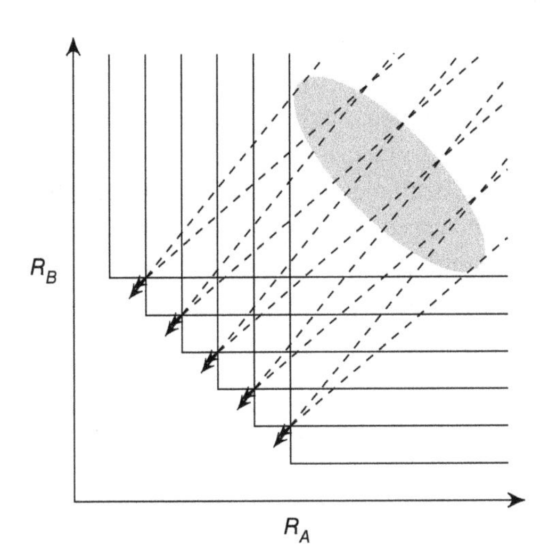

Figure 6.20 Multiple species with overlapping ZNGIs (solid lines) and two essential resources (R_A, R_B) can coexist at larger scales when there is habitat heterogeneity in the position of the supply point (grey shading). Impact vectors are shown as arrows and dashed lines project these into resource space to indicate the range of supply points leading to coexistence. (*Source*: Adapted from Chase and Leibold (2003).)

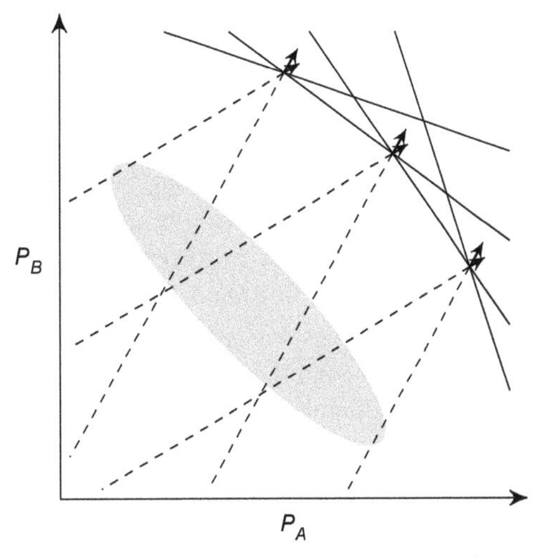

Figure 6.21 Multiple species with overlapping ZNGIs (solid lines) and two shared predators (P_A, P_B) can coexist at larger scales when predator supply varies (grey shading). Other elements as in Fig. 6.20. (*Source*: Adapted from Chase and Leibold (2003).)

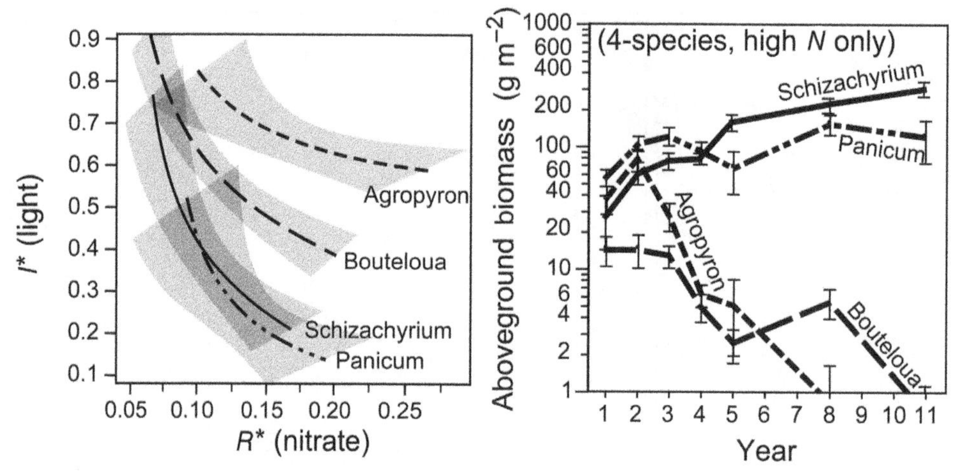

Figure 6.22 (a) ZNGIs for nitrate (R^*) and light (I^*) from four prairie plant species. Grey areas show 95% confidence intervals. (b) Aboveground biomass over time. Bars represent standard errors among replicates. (*Source*: Dybzinski and Tilman (2007). Reproduced with permission from the University of Chicago Press.)

the outcome of competition between two species in an experimental trial doesn't mean we have entirely solved the problem of species coexistence, which will be the theme of the next chapter.

It is worth briefly noting two underlying assumptions that have run throughout the story so far. The first is that individuals of the same species behave in the same way, or at the very least that variation within species is lower than that between. It is strange that this problem is so often overlooked in theory concerning species coexistence, particularly as differences among individuals lie at the core of natural selection, so no ecologist would dispute that they occur. The assumption has been challenged in recent years, with emerging evidence that variation within species can play a vital role in mediating the interactions among species, blurring the differences between them and often making coexistence more likely (Violle et al., 2012). Clark (2010) argues that instead of thinking about species we should look at individuals and how they respond to their immediate neighbours. His work on forest trees shows that even if species overlap considerably in their resource requirements, coexistence is still possible so long as individuals arrange themselves to minimise competition.

The second core assumption is that species do not change, or rather that evolution happens on longer timescales than ecological interactions. An increasing body of evidence suggests that this might not be the case, and even that dynamic evolution can provide novel means for species to coexist (Lankau, 2011), perhaps through altering the interactions among them (e.g. Lawrence et al., 2012). Species are not fixed and unchanging entities; evolution can be both rapid and continuous. These speculative ideas reside at the limits of our present understanding of natural systems but are likely to be a productive avenue for research in the coming years.

6.11 Conclusions

We have now built up a new theory of the niche which solves many of the problems with earlier definitions of the idea, increases its generality and provides predictions that can be tested in natural systems. More data are required to test the theory rigorously, especially in animal communities, but it is a promising line of enquiry, and we can cautiously say that, at present, it is the leading contender for understanding the coexistence of multiple species using the same resources. Following Chase and Leibold (2003) the niche can be broadly defined as

> The joint description of the environmental conditions that allow a local population to persist and the per capita effects on the environment.

More specifically, using the factors identified previously, the niche comprises

> The ZNGI of an organism, combined with the impact vectors on the ZNGI in the multivariate space defined by the environmental factors.

This chapter has deliberately omitted the mathematical proofs that underpin all these graphs, though you can find out more in Chase and Leibold (2003). Opinions remain divided as to whether this should be the definitive niche concept, but the majority of ecologists now acknowledge that Hutchinson's simplistic ideas about the fundamental and realised niche were inadequate. Decide for yourself, and remember that the simplest theories aren't always the best!

6.11.1 Recommended reading

Chase, J. M. and M. A. Leibold, 2003. *Ecological Niches: Linking Classical and Contemporary Approaches*. University of Chicago Press.

Dybzinski, R. and D. Tilman, 2007. Resource use patterns predict long-term outcomes of plant competition for nutrients and light. *American Naturalist* 170:305–318.

Hutchinson, G. E., 1957. Concluding remarks. *Cold Spring Harbor Symposia on Quantitative Biology* 22:415–427.

6.11.2 Questions for the future

- Does R^* work as well for animals as it does with plants and diatoms?
- Is there a consistent way to identify the appropriate axes for trade-offs?
- How might rapid coevolution of species influence coexistence?

References

Behmer, S. T. and A. Joern, 2008. Coexisting generalist herbivores occupy unique nutritional feeding roles. *Proceedings of the National Academy of Sciences of the United States of America* 105:1977–1982.

Blonder, B., C. Lamanna, C. Violle, and B. J. Enquist, 2014. The *n*-dimensional hypervolume. *Global Ecology and Biogeography* 23:595–609.

Bohannan, J. M. and R. E. Lenski, 2000. The relative importance of competition and predation varies with productivity in a model community. *American Naturalist* 156:329–340.

Chase, J. M. and M. A. Leibold, 2003. *Ecological Niches: Linking Classical and Contemporary Approaches*. Chicago University Press.

Chesson, P., 2000. Mechanisms of maintenance of species diversity. *Annual Review of Ecology, Evolution, and Systematics* 31:343–366.

Clark, J. S., 2010. Individuals and the variation needed for high species diversity in forest trees. *Science* 327:1129–1132.

Crase, B., A. Liedloff, P. A. Vesk, M. A. Burgmann, and B. A. Wintle, 2013. Hydroperiod is the main driver of the spatial pattern of dominance in mangrove communities. *Global Ecology and Biogeography* 22:806–817.

Dybzinski, R. and D. Tilman, 2007. Resource use patterns predict long-term outcomes of plant competition for nutrients and light. *American Naturalist* 170:305–318.

Elton, C., 1927. *Animal Ecology*. Sidgwick and Jackson.

Farrior, C. E., D. Tilman, R. Dybzinski, P. B. Reich, S. A. Levin, and S. W. Pacala, 2013. Resource limitation in a competitive context determines complex plant responses to experimental resource additions. *Ecology* 94:2505–2517.

Gause, G. F., 1936. *The Struggle for Existence*. Williams and Wilkins.

Grinnell, J., 1917. The niche-relationships of the California thrasher. *Auk* 34:427–433.

Holt, R. D., J. P. Grover, and D. Tilman, 1994. Simple rules for interspecific dominance in systems with expoitative and apparent competition. *American Naturalist* 144:741–771.

Hubbell, S. P., 2001. *The Unified Neutral Theory of Biodiversity and Biogeography*. Princeton University Press.

Hutchinson, G. E., 1957. Concluding remarks. *Cold Springs Harbor Symposia on Quantitative Biology* 22:415–427.

Lankau, R. A., 2011. Rapid evolutionary change and the coexistence of species. *Annual Review of Ecology, Evolution, and Systematics* 42:335–354.

Lawrence, D., F. Fiegna, V. Behrends, J. G. Bundy, A. B. Phillimore, T. Bell, and T. G. Barraclough, 2012. Species interactions alter evolutionary responses to a novel environment. *PLoS Biology* 10:e1001330.

Leibold, M. A., 1998. Similarity and local co-existence of species in regional biotas. *Evolutionary Ecology* 12:95–110.

von Liebig, J., 1840. *Chemistry and its Application to Agriculture and Physiology*. Taylor and Walton.

Lotka, A. J., 1924. *Elements of Physical Biology*. Williams and Wilkins.

MacArthur, R. H., 1972. *Geographical Ecology: Patterns in the Distribution of Species*. Harper and Row.

MacArthur, R. H. and R. Levins, 1967. The limiting similarity, convergence, and divergence of coexisting species. *American Naturalist* 101:377–385.

May, R. M. and R. H. MacArthur, 1972. Niche overlap as a function of environmental variability. *Proceedings of the National Academy of Sciences of the United States of America* 69:1109–1113.

Miller, T. E., J. H. Burns, P. Munguia, E. L. Walters, J. M. Kneitel, P. M. Richards, N. Mouquet, and H. L. Buckley, 2005. A critical review of twenty years' use of the Resource-Ratio Theory. *American Naturalist* 165:339–448.

Murray, M. G. and D. R. Baird, 2008. Resource-ratio theory applied to large herbivores. *Ecology* 89:1445–1456

Real, L. A. and S. A. Levin, 1991. Theoretical advances: the role of theory in the rise of modern ecology. In L. A. Real and J. H. Brown, editors, *Foundations of Ecology: Classic Papers With Commentaries*, pages 177–197. University of Chicago Press.

Rothhaupt, K. O., 1988. Mechanistic resource competition theory applied to laboratory experiments with zooplankton. *Nature* 333:660–662.

Tilman, D., 1977. Resource competition between planktonic algae: an experimental and theoretical approach. *Ecology* 58:338–348.

Tilman, D., 1982. *Resource Competition and Community Structure*. Princeton University Press.

Tilman, D., 1994. Competition and biodiversity in spatially structured habitats. *Ecology* 75:2–16.

Tilman, D., 2004. Niche tradeoffs, neutrality, and community structure: a stochastic theory of resource competition, invasion, and community assembly. *Proceedings of the National Academy of Sciences of the United States of America* 101:10854–10861.

Turnbull, L. A., 2014. Ecology's dark matter: the elusive and enigmatic niche. *Basic and Applied Ecology* 15:93–100.

Violle, C., B. J. Enquist, B. J. McGill, L. Jiang, C. H. Albert, C. Hulshof, V. Jung, and J. Messier, 2012. The return of the variance: intraspecific variability in community ecology. *Trends in Ecology & Evolution* 27:244–252.

Volterra, V., 1926. Fluctuations in the abundance of a species considered mathematically. *Nature* 118:558–560.

Wilson, J. B., E. Spijkerman, and J. Huisman, 2007. Is there really insufficient support for Tilman's R^* concept? A comment on Miller et al. *American Naturalist* 169:700–706.

CHAPTER 7

Patterns in species richness

7.1 The big question

The number of species found from place to place varies immensely. Some correlates of species richness can be found, though it is not always obvious what the underlying processes causing them are. If you were to visit patches of wet forest across Queensland, Australia, and estimate the number of reptile species in each, you would find all sorts of differences—with forest size, elevation, age, distance from the sea and no doubt many others. Are there consistent patterns to this variation throughout the world and for all groups of life? Can we understand what causes them? Note that variation in species richness occurs at all scales, from the local to the global, though this chapter is restricted to the former, leaving global patterns until the biogeography section (see Chapter 15).

7.2 Area

The first key difference among habitat patches is their size. One of the oldest and most fundamental laws of ecology was first noticed by Olof Arrhenius, son of the more famous Nobel laureate chemist Svante Arrhenius, the man who postulated the greenhouse effect. In 1921 the younger Arrhenius published a paper which forms the starting point for understanding patterns of species richness in nature (Arrhenius, 1921). He recognised that the number of species S in an area A follows a curved relationship of the form

$$S = cA^z$$

This is a saturating curve where the rate of discovery of new species decreases as the area surveyed increases (Fig. 7.1a). When plotted on logarithmic scales, this equation transforms into that of a straight line (Fig. 7.1b), with a slope of z and crossing the y-axis at log c:

$$\log S = \log c + z \log A$$

This is known as the **species–area relationship**, often abbreviated as SAR. Typically area alone accounts for more than 50% of the variation in species richness among samples: larger sites contain more species. The gradient of this relationship, z, determines the sensitivity of S to A. The parameter c is known as the biotic richness of the region and gives the number of species which would be present in a single unit of area.

The relationship holds over the full range of scales in nature, from small experimental microcosms in the laboratory up to whole regions of the earth, as shown in Fig. 7.2 (Smith et al., 2005). There is no shortage of comparable cases. Some debate remains over the precise form of the relationship and how it changes with the scales of either measurement or study, but the basic form is remarkably robust.

The relationship has important connotations throughout ecology. In conservation biology it is

Natural Systems: The organisation of life, First Edition. Markus P. Eichhorn.

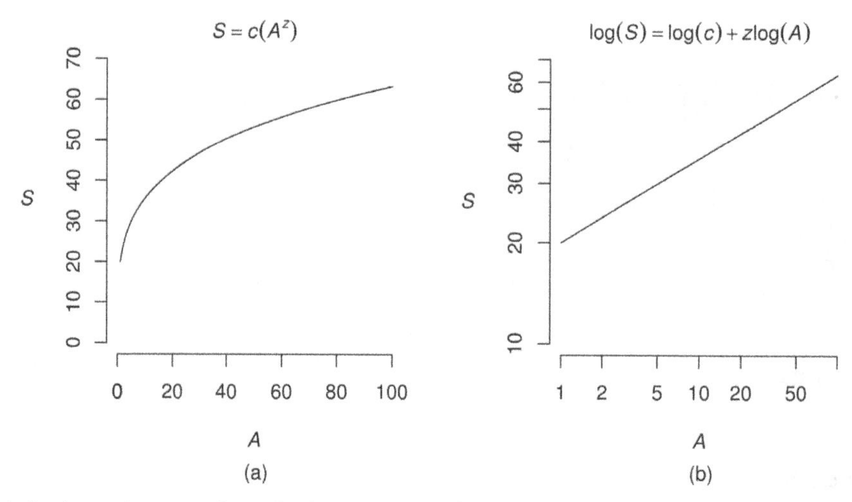

Figure 7.1 The Arrhenius species–area relationship between species richness S and area surveyed A displayed on (a) normal and (b) log-transformed scales.

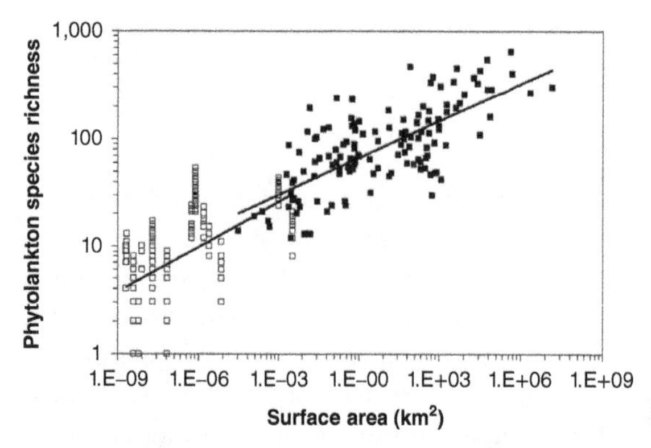

Figure 7.2 Relationship between phytoplankton species richness and ecosystem surface area (km^2) in natural (filled dots) and experimental (clear) aquatic ecosystems. (*Source*: Smith et al. (2005). Reproduced with permission from the National Academy of Sciences, USA.)

often used to anticipate the number of species that will go extinct when habitat is lost. Once the area of habitat shrinks, these species are committed to extinction, even if they persist temporarily, a syndrome known as **extinction debt**. Losses of species are seldom as catastrophic as the SAR predicts though, especially if patches remain connected to one another through dispersal of individuals. The reverse application, projecting the number of species in a larger (or unknown) area, is fraught with many errors and not recommended. Non-parametric estimation techniques are preferable (see Chapter 5).

Understanding why the pattern occurs is less straightforward, and it is worth noting that there are two major kinds of SAR (even further subdivision is possible; Scheiner, 2003). The first starts with a small sample and increases its size, spreading over a larger area. In this case more species might be collected simply by chance—the more individuals captured, the more opportunities there are to find each, until eventually no new ones are there to be found. Individuals of the same species are often clustered together in space and larger areas are necessary to find everything. This was the basis of the species accumulation curves presented in Chapter 5. An increased area also encompasses more heterogeneity in the environment, and subtle shifts in habitats can allow multiple species to coexist, as seen in Chapter 6. In this conventional SAR the value of the slope z usually falls between 0.25 and 0.30 (Preston, 1962; Drakare et al., 2006).

The second type of SAR occurs when a series of islands or patches vary in size and is known as the island species–area relationship or ISAR. These behave rather differently as a whole suite of additional processes come into play, including the probability of species being able to reach them, maintaining populations large enough to persist and interacting with those species already present. The value of z is greater, giving the line a steeper slope. This makes islands a special case, and they will be considered separately in Chapter 19.

7.3 Local and regional species richness

Despite the general rule that patches of different size tend to vary in species richness, there is still great variation between patches even of the same area. For example, a single hectare of forest from northern Russia contains only two or three species of tree. Repeating the same survey in northern China would yield 20–25 species, more than doubling in northern Laos, while in tropical Malaysia there could be several hundred tree species in the same area, despite containing roughly the same density of individual trees.

It turns out that there is a strong relationship between local (patch level, S_L) and regional species richness (S_R). The more species there are in a country, the more species are found in any standard sample. But what form does this relationship take? Three theoretical possibilities should be considered (Fig. 7.3). There might be a 1:1 relationship, where all species in the region are found in every location. This is impossible and defies the logic of what can be seen by looking out of the window—not every bird species in the country lives in your garden. We therefore have to consider the other two. It might be that S_R and S_L are correlated, such that a fixed proportion of the species from a region is found in any given area A, known as a Type I response. Alternatively, there could be a limit to local species richness S_L so that as regional species richness S_R increases, an asymptote is reached, after which no more species can fit into a single patch (Type II). Which is true?

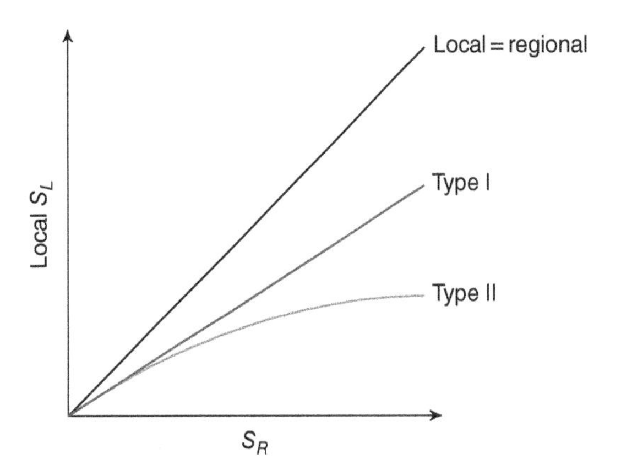

Figure 7.3 Possible relationships between regional and local species richness (S_R, S_L). Either all regional species are present in all localities ($S_R = S_L$), a fixed proportion of S_R occurs in each locality (Type I) or there is a limit to the number of species S_L in any locality (Type II). (*Source*: Adapted from Gaston and Spicer (2004).)

Before a satisfactory answer can be reached, first we need to consider what the evidence would be for species richness having reached a limit. We might speculate, following Cornell (1999), that:

- The relationship between S_R and S_L would be curved in real data from natural systems due to density competition among species.
- Assemblages with more species would resist invasion by new species.
- Communities that are similar in size and character would converge on the same S_L.

We will examine each of these in turn. Firstly, what does the relationship between regional and local species richness look like? It usually appears close to a straight line. There is much debate over this, and often a slight curvature can be detected, but at much larger scales than apply to local communities, and in practice even these show no sign of reaching an asymptote within the range of values seen in nature (e.g. Fig. 7.4). It more closely matches the Type I pattern of Fig. 7.3.

The implication is that local species richness depends on regional levels, and in fact around

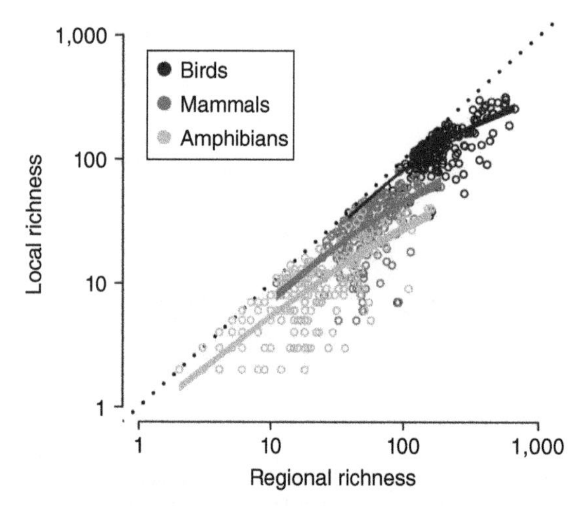

Figure 7.4 Relationship between local and regional richness for bird, mammal and amphibian assemblages. Solid curves are best-fit predictions ± SE; dotted line represents richness equivalence. Regional richness was defined as from a minimum 100 km radius (31,400 km^2), while average local richness was from areas of 11.5 km radius (411 km^2). (*Source*: Belmaker and Jetz (2012). Reproduced with permission from the University of Chicago Press.)

75% of variation in S_L can be accounted for by this relationship alone (keeping sample areas constant). There is limited evidence of density competition among species, though we can't be entirely certain, as it is always possible that insufficient values have been sampled to see a curve properly. Maybe saturation would occur, though only at much greater levels of species richness than are normally seen in nature. Note also that at the very smallest scales, such as within a single square metre, there is an obvious limit set by the number of individuals that can fit inside—this is trivial and not relevant at the community scale. Nevertheless the pattern extends to surprisingly small scales down to the level of 10 m transect lines of corals on reefs (Fig. 7.5). In this case, the lines were perhaps very slightly curved but would not reach a limit within any realistic level of regional species richness.

Not everyone accepts the ubiquity of linear relationships. If both S_R and S_L are determined by the same climatic variables, then any correlation will simply be an artifact, though this can be controlled

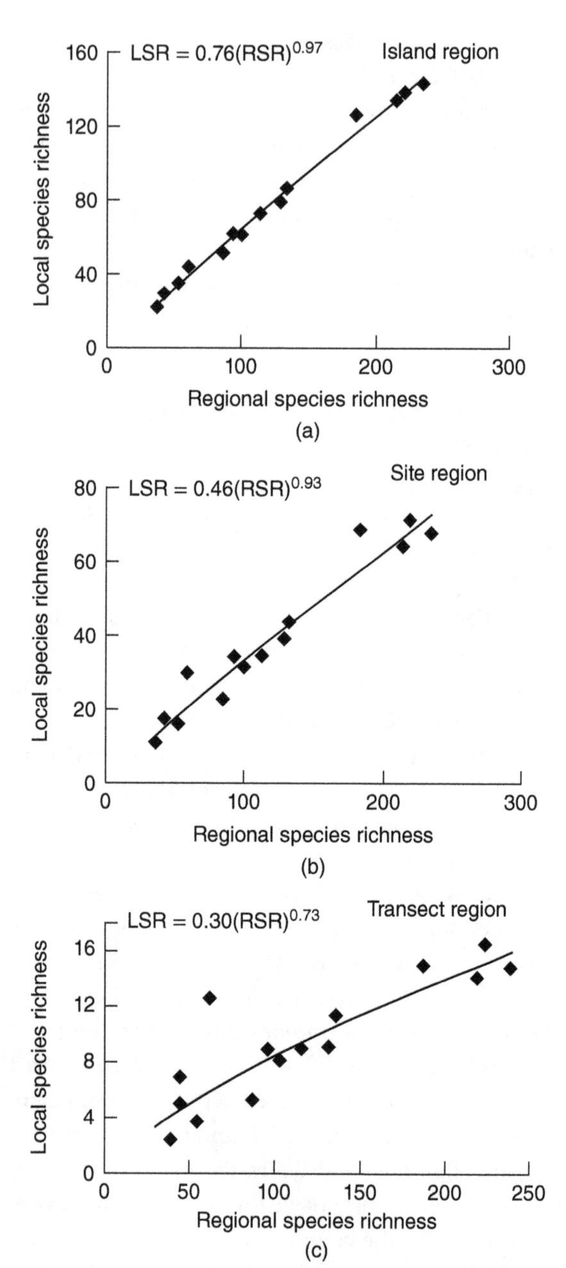

Figure 7.5 Local to regional species richness relationships for coral assemblages. Regional richness was the total number of species found in three habitat types in each of five regions in the west–central Pacific. Local richness for each habitat was determined at the scales of (a) three whole islands per region (40 × 10 m transect lines (b) four sites on each island (10 × 10 m transect lines and (c) single 10 m transect lines within sites, averaged within regions (Cornell et al., 2008; Cornell and Harrison, 2013).

for in analyses (Harrison and Cornell, 2008). In some cases inappropriate statistical tests have been used which might make local saturation harder to detect (Gonçalves-Souza et al., 2013), and once these have been accounted for, the predominance of strictly linear patterns disappears (Szava-Kovats et al., 2013). Most importantly, it is impossible to attribute causation to this relationship, whatever its shape, and it should be taken as an indication rather than definitive evidence of regional drivers.

We should therefore test the other two postulates. Many people have suggested that diverse communities should resist invasion by new species, either because the chance of including a species that will block the invader increases with total richness (identity effects) or because complementarity in resource use means that fewer opportunities are available for new species to enter. Evidence for these processes has often been found in experimental systems (e.g. Fargione et al., 2003).

On the other hand, there are many cases where no evidence of saturation occurs. Lee and Bruno (2009) added individuals of numerous grazer species to **microcosms** of seabed communities. Far from seeing any sign of resistance to new species, the more they added, the more species were able to coexist, with no sign of levelling off (Fig. 7.6). The same was found for grassland plants by Tilman (1997), though in the latter case the highest diversity plots showed some signs of resisting further invasion.

In an unintentional natural experiment, we have released many invasive species which are spreading throughout the world, but their net effect is often to increase local species richness, certainly in the case of plants and fish, and more so on islands than continental areas (Fig. 7.7). For birds this doesn't seem to be the case, and instead there may be a balance in the number of bird species, with each invasive simply replacing one of the incumbent species, though this is more often the result of changes in habitats or predation than competition (see Chapter 20.7).

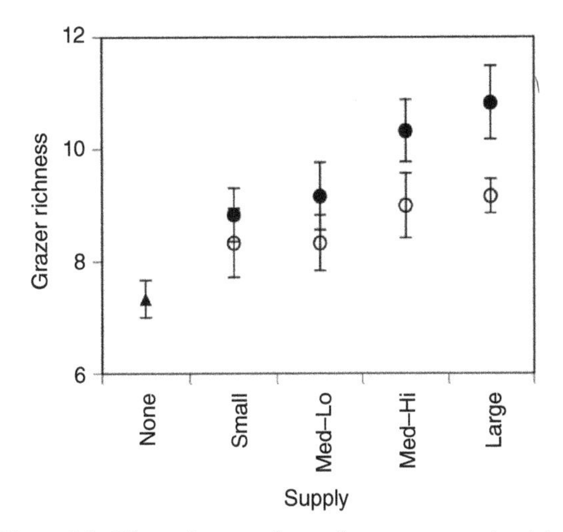

Figure 7.6 Effects of propagule supply on grazer species richness, single (hollow dots) or multiple (solid) additions, means ± SE. (*Source*: Lee and Bruno (2009). Reproduced with permission from the National Academy of Sciences, USA.)

Overall communities do not consistently resist invasion. In fact, counter-intuitively, communities with greater species richness accept more invasives.

This means that, around the world, introductions of species by humans are tending to maintain or increase S_L even as global species richness goes down. How can this be the case? The answer is the process of biotic homogenisation, whereby we are introducing the same species everywhere and allowing endemic species and local specialties to go extinct (McKinney and Lockwood, 1999). Recall from Chapter 5 that α-scale richness is not necessarily correlated with γ-scale, and what is really happening is a reduction in β-diversity.

Further evidence of the impacts of invasions can be drawn from historical patterns of biogeography, where new combinations of groups have performed a natural experiment on diversity patterns. The Great American Biotic Interchange around 15 Mya occurred when the two continents of North and South America began to drift closer together and

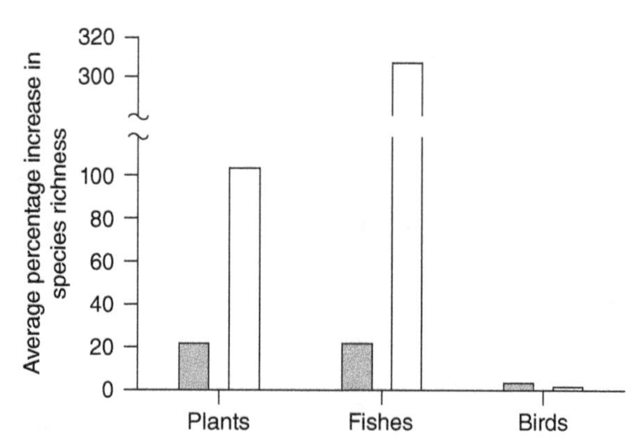

Figure 7.7 Change in species richness of plants, fish and birds in mainland sites (filled columns) and on islands (clear) due to introduction of invasive species. (*Source*: Sax and Gaines (2003). Reproduced with permission from Elsevier.)

organisms were able to disperse between them. Groups of species which had never encountered one another were suddenly combined. These included the terraranan frogs about which Pinto-Sánchez et al. (2014) made a simple prediction. If communities were saturated, then places which contained a combination of clades from both North and South America would have the same species richness as those with frogs from only one region. Alternatively, if there were no saturation, combining frog species from the two continents would increase species richness. They discovered the latter pattern, indicating that the more frog species can colonise an area, the more species are found.

The final postulate made previously was that similar communities should have consistent values of species richness. This was tested by a group of ecologists who may be reckoned among the luckiest of scientists based on their study system. Karlson et al. (2004) set up a large-scale project in which they visited coral reefs across the South Pacific, starting in the Indo-Pacific off the coast of Sulawesi and moving eastwards past New Guinea and into the Solomon and Society Islands. At every reef they carried out a standard survey of habitats, including the reef slopes, crests and flats, each of which contains different coral species as well as other animal groups. The core question was what determined patterns of local species richness—was the region or the habitat type most important?

The results are shown in Fig. 7.8. In every area the ranking of species richness between habitats was the same: slopes always contained more species per unit area than crests, with flats having the fewest. Yet in different regions these habitats varied greatly in species richness; there wasn't a set number of species expected on a reef crest.

The next, and most important step, was to add up all the species found in each habitat within the different regions and ask how S_R and S_L were related (Fig. 7.9). What leaps out is that once again a strong correlation emerges between regional and local species richness but with a separate line for each habitat. The variation among regions is much greater than that between habitats, and we can safely say that similar communities do not converge in species richness on a common limit, otherwise the three lines would all be flat.

We can therefore conclude that, as a general rule, the relationship with regional species richness accounts for around 75% of variation in local species richness. At this point you may be wondering what factors determine regional species richness levels; this is something we will return to in Chapter 15. For the remainder of this chapter we will focus on trying to understand the remaining 25% that can be accounted for by processes at the local scale.

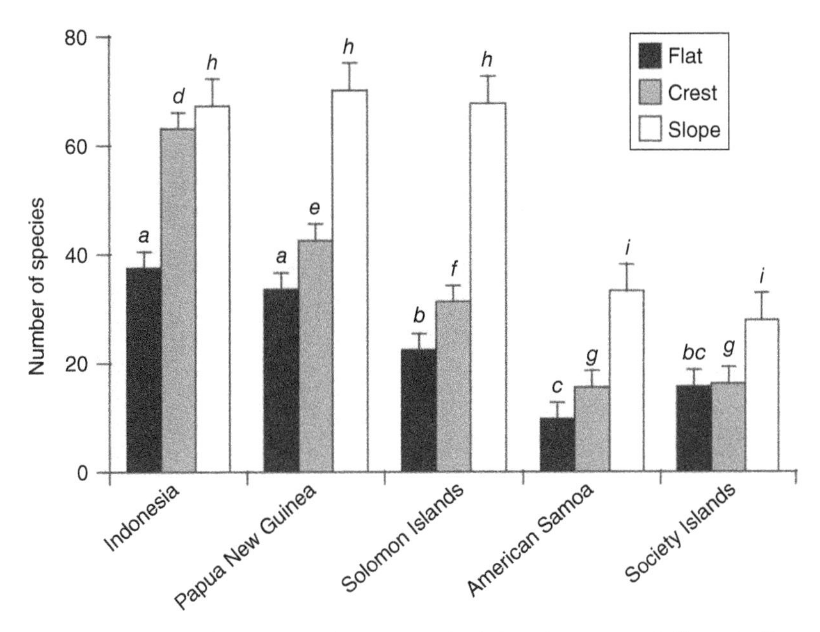

Figure 7.8 Coral species richness on reef flats, crests and slopes in each of six regions. Letters indicate significant differences among values. (*Source*: Karlson et al. (2004). Reproduced with permission from Nature Publishing Group.)

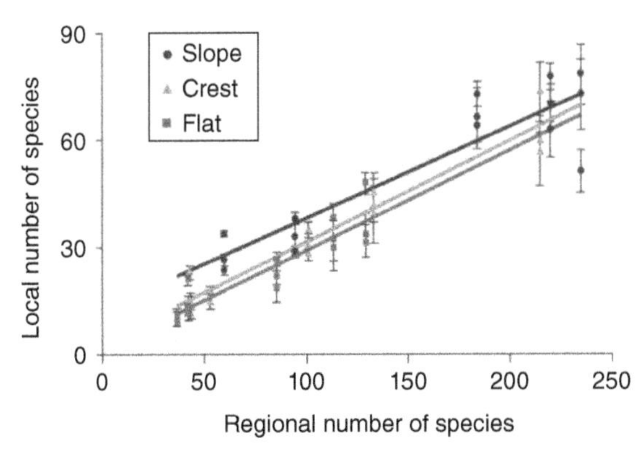

Figure 7.9 Local versus regional coral species richness of reef flats, crests and slopes. (*Source*: Karlson et al. (2004). Reproduced with permission from Nature Publishing Group.)

7.4 Local patterns in species richness

Even within a single region, and when considering a single type of habitat (e.g. alpine meadows), different patches of the same area contain varying numbers of species. There are other, more local influences on species richness *S*. These include trends with elevation on land and depth in the ocean, along with impacts of local geography and patch isolation.

7.4.1 Elevation

Ascending a mountain the number of species found usually decreases, as shown in Fig. 7.10a–c. There are confounding effects since it could simply be the result of a decline in area. If a mountain were divided into a series of bands representing different height zones, those at the top would necessarily be smaller than those lower down (unless the mountain is cylindrical!). Also, mountain peaks are relatively isolated from one another while lowlands are connected. Nevertheless, neither of these confounding effects fully accounts for the decline in species richness with elevation.

While recognising the common rule, there are also numerous exceptions, and note that the clines in Fig. 7.10a–c do not all start at sea level. Often, when corrected for area and once the full range is observed, *S* peaks at mid-elevations (Rahbek, 1995). In the Italian Alps, fern species richness

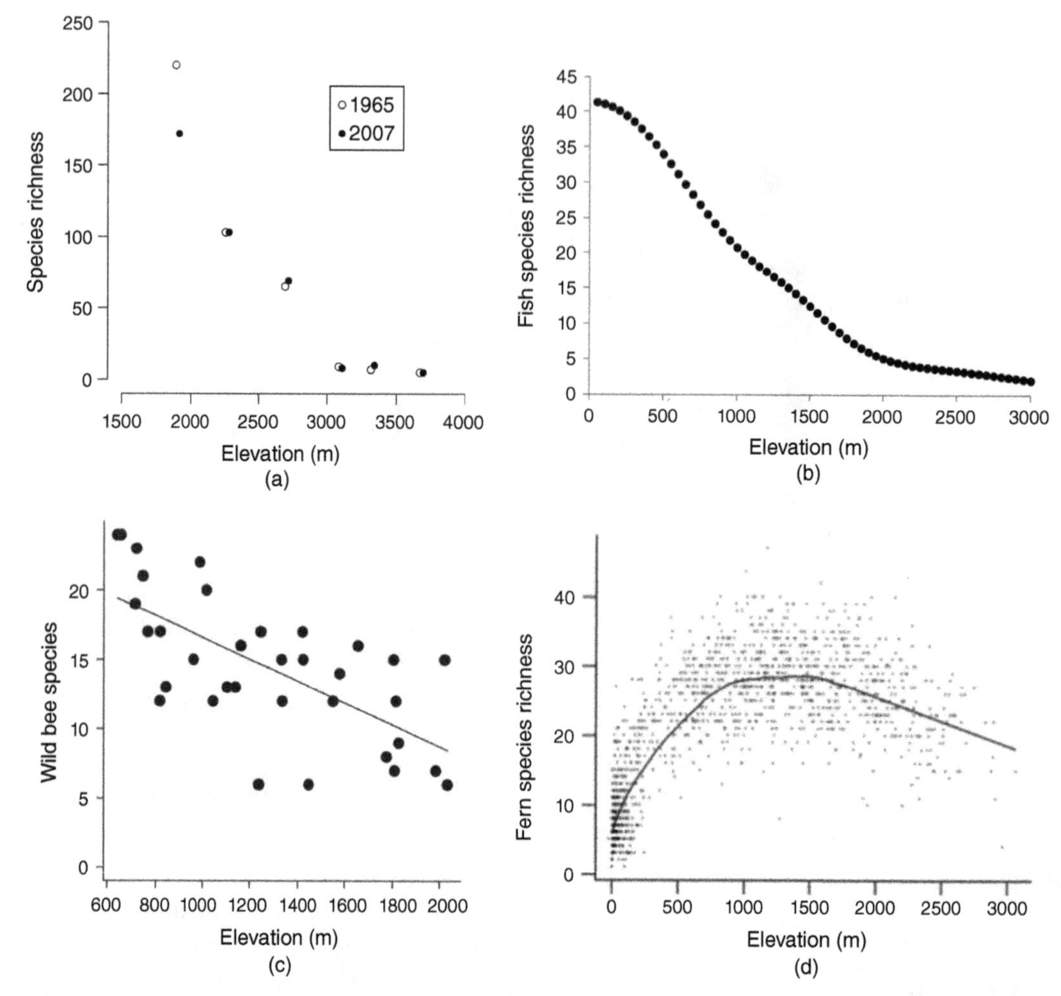

Figure 7.10 (a) Species richness of geometrid moths with elevation on Mount Kinabalu; surveys were repeated in two years, and datapoints are horizontally offset for clarity. (*Source*: Adapted from Chen et al. (2009).) (b) Species richness of freshwater river fish along a Himalayan elevation gradient. (*Source*: From Bhatt et al. (2012). Used under CC-BY-SA 2.5 http://creativecommons.org/licenses/by-sa/2.5/.) (c) Species richness of wild bees with elevation in the German Alps. (*Source*: Hoiss et al. (2012). Reproduced with permission from the Royal Society.) (d) Fern species richness in the Italian alps with elevation in 35.7 km^2 grid cells with fitted smoother. (*Source*: Marini et al. (2011). Reproduced with permission from John Wiley & Sons Ltd.)

is highest from 800 to 1,500 m. This is due to a combination of human disturbance in the lowlands, higher rainfall and environmental heterogeneity on the slopes, then declining temperature towards the peaks (Fig. 7.10d).

The explanations for elevational trends are usually related to the climate—a drop in average temperatures typically accompanies an increase in altitude, restricting the number of species that can survive there. In terms of temperature, ascending a mountain is much the same as moving away from the equator, and similar trends occur in both species richness and the types of habitat found. Climbing a mountain in the tropics can lead you from tropical rain forest at its base through what look ever more like temperate broadleaved then coniferous forests, followed by

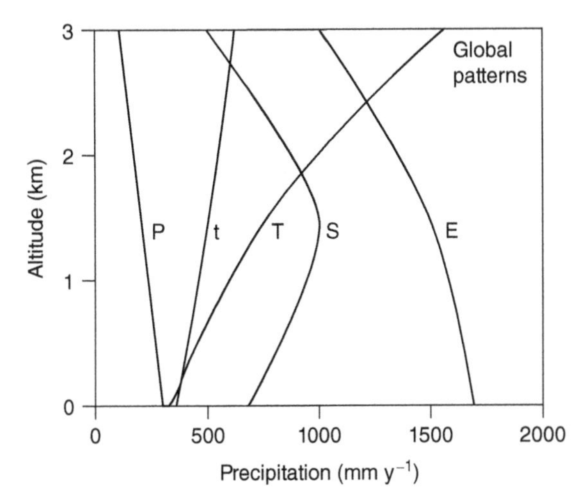

Figure 7.11 Trends in precipitation with elevation in equatorial (E, 0–10°), subtropical (S, 10–30°), transitional (t, 30–40°), temperate (T, 40–60°) and polar (P, Greenland) latitudes. (*Source*: Körner (2007). Reproduced with permission from Elsevier.)

vegetation corresponding to tundra before reaching snow-capped peaks. Elevational trends are therefore a condensed equivalent of latitudinal trends, a pattern that has been recognised for over 200 years (von Humboldt, 1808 see Chapter 14). Note that the decline in S with elevation has nothing to do with reduced atmospheric pressure or lack of oxygen.

It's not always so simple though, as elevation is a much more complex environmental parameter than the number of metres above sea level might suggest (Körner, 2007). The relationship between elevation and precipitation varies considerably in different parts of the world (Fig. 7.11), making it difficult to determine the root cause of trends in S_L. Even on the same mountain different taxa can vary in their response. In the Rocky Mountains, Colorado, bacteria show a simple decline in richness with elevation, while plants display a humped relationship (Bryant et al., 2008). In some cases linear declines exist only because greater human impacts in lowlands have obscured the lower part of a humped shape (Nogués-Bravo et al., 2008). There are even examples of inverted trends, such as is seen with ant

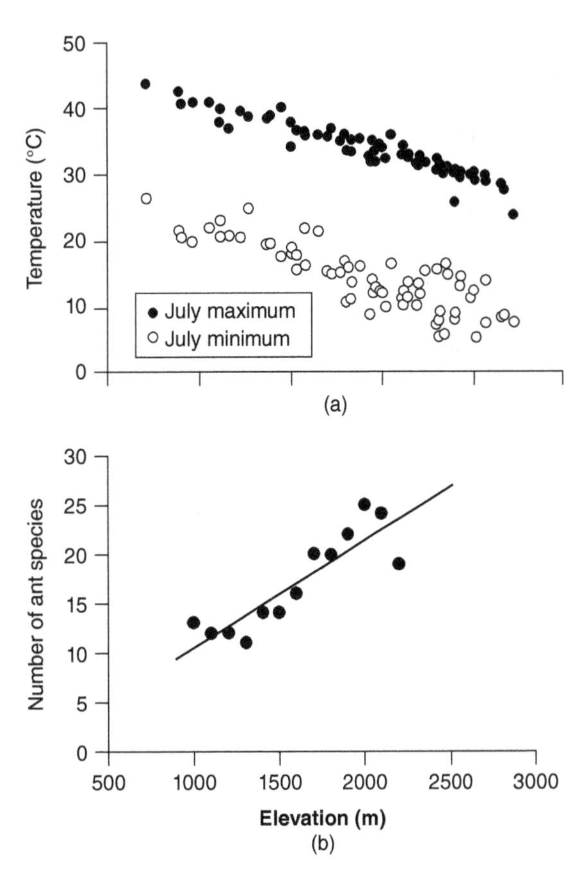

Figure 7.12 (a) July temperature range and (b) species richness of ants with elevation in the Nevada desert. (*Source*: Smith and Brown (2002). Reproduced with permission from John Wiley & Sons Ltd.)

species richness in the Nevada desert (Fig. 7.12). Here there are more species at higher elevations, a peculiar observation caused by the local climate which results in extremely hot, dry conditions in the valleys. Oppressive temperatures and low water availability restrict the species that are able to live at lower elevations.

Therefore, although there are frequently trends with elevation, they often conflict among taxa even on the same mountain. They are associated with changes in environmental conditions, resource availability, species interactions, community composition, ecosystem processes and many other factors (Sundqvist et al., 2013). Making sense of this requires a realisation that altitude itself is not the real driving force.

7.4.2 Depth

Declines in species richness also occur with water depth. Similar patterns may occur below ground, though little is known. Highly specialised chemosynthetic bacteria can be recovered from up to 4,000 m below the land surface, but whether it is meaningful to talk of any diversity trends is debatable.

In the ocean, however, the average depth is 3.8 km below the surface, falling to abyssal depths of over 10 km in a few trenches, and life is known to continue all the way down. In general benthic (seabed) species richness declines with depth, while patterns of pelagic (free-swimming) organisms are more complex. In the northeast Pacific there is a decline in species richness of fish with depth, albeit with a peak around 200 m below the surface (Fig. 7.13). A concurrent increase in the depth range inhabited by fish complicates observed patterns.

As with elevation, not all trends are linear, and some can be humped (Fig. 7.14). The causes of patterns are complicated and can vary among groups. There is a fall in temperature with depth, combined with a decrease in its variability, making the environment more stable and predictable. Barely any light penetrates below the top few metres of the ocean, which means this has little influence except right at the surface. Nutrient availability does decline with depth though. Crucially for understanding the patterns of filter-feeding animals on the seabeds of continental shelves, the diversity of particle sizes changes, and peaks in diversity of benthic fauna correspond to the places with the greatest variety of food sizes. In the deep ocean and abyssal regions, there is hardly any input from above and resources are extremely scarce, hence low species richness.

7.4.3 Peninsulas and Bays

Species richness tends to decline along peninsulas moving from the mainland towards the tip. As with elevation, this effect is difficult to separate from the decline in area, as well as increasing isolation from the mainland, and is also likely to be at least in part due to the change in environment that comes with being a spit of land protruding into the sea. Comparable patterns are also seen in estuaries, where changes in water salinity are influential.

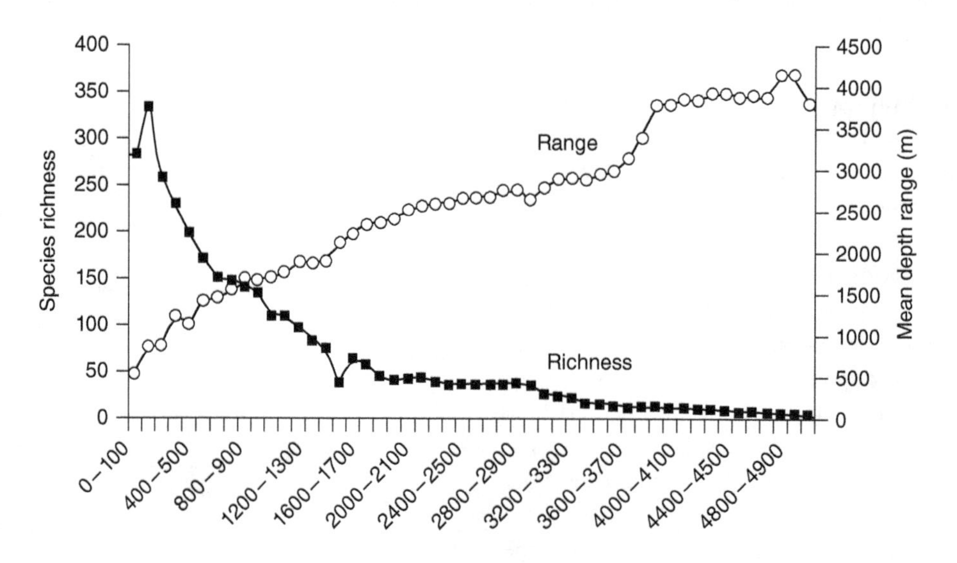

Figure 7.13 Species richness and mean depth range plotted as a function of depth for pelagic marine fishes along a gradient of depth in the northeast Pacific. (*Source*: Smith and Brown (2002). Reproduced with permission from John Wiley & Sons Ltd.)

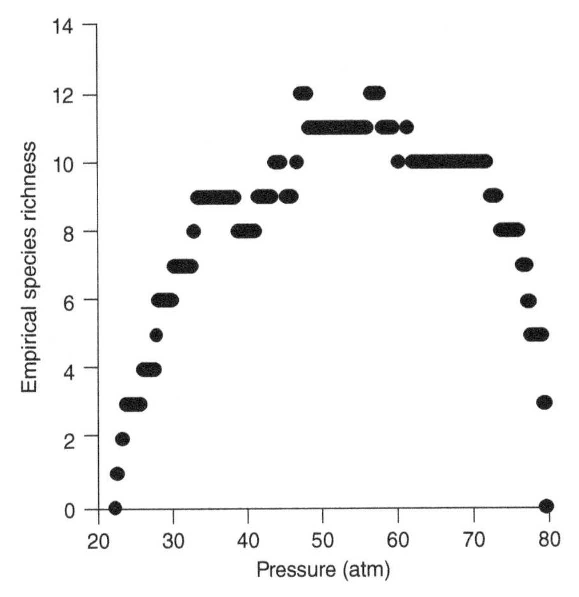

Figure 7.14 Species richness of demersal (swimming close to the seabed) fish between 200 and 800 m depth (measured as pressure in atmospheres) on the Chatham Rise, New Zealand. (*Source*: Geagne et al. (2012). Reproduced with permission from John Wiley & Sons Ltd.)

7.4.4 Isolation

More isolated patches tend to contain fewer species because the probability of any species reaching them is reduced and limited by their dispersal ability. This observation, combined with the decline in species richness with area, was at the heart of the Equilibrium Model of Island Biogeography (MacArthur and Wilson, 1967) which will be considered in more detail in Chapter 19. For now it is enough to know that it can be a powerful influence on species richness. Its importance can be demonstrated experimentally. Examining the diversity of invertebrates attached to rock faces in the intertidal zones of both Alaska and Maine, Palardy and Witman (2014) discovered that increasing the flow rate of water boosted the number of individuals recruiting into the communities and thereby increased their species richness (see also Fig. 7.6). This is a good indication of how isolation doesn't simply mean distance, but rather how likely it is that an individual will arrive there and be able to colonise.

7.4.5 Mid-Domain Effects

Imagine a fixed area of habitat with clear boundaries—an island, perhaps. Then take a selection of species, each with its own range, and randomly scatter them across the area. By chance alone, there will be more overlaps in the middle. This is the fundamental basis of the mid-domain effect, a controversial idea that suggests more species will be found in the centre of a region (Colwell and Lees, 2000). While some evidence exists in support, even the classic examples have been disputed, and its importance in nature remains in doubt (Currie and Kerr, 2008). Among its problems are that it assumes that the species range is an innate property, rather than the outcome of interactions with the environment, and there is no reason to assume that any given species would have the same range size if you were to move it to another place. Nevertheless the pattern is found in many places and it is often used as a null model.

7.5 Congruence

All these discussions of trends might give the impression that many taxa will follow similar patterns. If this were the case, it would be a great aid in conservation, as it would allow us to sample only those groups that were easy to observe, collect and identify (e.g. birds or butterflies) and use them to infer patterns in those groups about which less is known. This is therefore an important property to test: how much correspondence is there in species richness trends among groups?

In some cases the approach works remarkably well. The species richness of a wide variety of insect groups in Panamanian rain forest correlate strongly with plant species richness, even for predatory or soil-dwelling insects with no direct connection to plants (Basset et al., 2012). In this example it is likely that plant richness acts as an index both of the heterogeneity in food resources and habitats (abiotic and biotic). The finding has been reproduced across a wide variety of animal groups and habitats

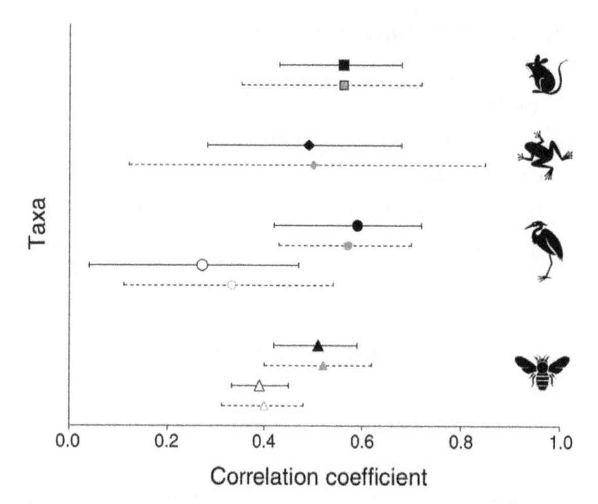

Figure 7.15 Mean correlation coefficients r (with 95% confidence intervals) between plant and animal species richness for α (open symbols) and γ (solid) diversity studies on mammals (squares), amphibians and reptiles (diamonds), birds (circles) and arthropods (triangles) for all data (solid lines) and for data averaged at the taxon level within publications (dashed). (*Source*: Castagneyrol and Jactel (2012). Reproduced with permission from the Ecological Society of America.)

throughout the world (Fig. 7.15). Even so, on average plant diversity only accounts for around 20% of animal diversity, which leaves a lot to be desired when making predictions for unsurveyed areas.

Among other groups the unfortunate truth is that there are often mismatches, and the correlations observed are typically weak and unreliable, both at local and regional scales (though become much stronger when endemics are considered separately; Kier et al., 2009). This is a shame as it makes life more difficult, and also a problem as the existence of correlations among taxa is so often assumed rather than tested. When assessing a wide range of studies of congruence in species richness, Westgate et al. (2014) found that it was usually weak (around 35%), but giving an average figure is misleading because the results span such a wide range and depend at least as much on study design as the actual biology of the species concerned.

Even when congruence does occur, it can have a number of causes. Several groups might share the same drivers, perhaps a common response to an environmental gradient. They can be related trophically, such that diversity of predators follows their prey, or by other interactions such as mutualisms. Diversity of groups might however be driven by entirely different mechanisms which happen to covary in the region of study—correlation doesn't always imply causation. This can lead to mistakes if the relationship is sought or expected elsewhere.

As a further cautionary note, it is also likely that patterns are confounded by sampling efforts as areas that have been well characterised for one group of species tend to have good records for others. For example, there is typically an apparent peak in species richness around scientific field stations where many visitors have taken records. Another non-biological cause of apparent congruence is that simply by sampling a greater number of species, by chance alone there will be more in each group.

These problems are especially acute in conservation biology where there is great interest in finding indicator taxa that could be used as proxies for other groups of species. Lawton et al. (1998) attempted this by assembling a team of expert taxonomists in Yaoundé Forest, Cameroon. Sadly, of the taxonomic groups they sampled, none was a good predictor of the others, with typical levels of covariance only around 11%, and in some cases negative correlations. This cautions against using favourite groups such as birds or butterflies as substitutes for sampling unknown taxa solely because we have the data or enjoy surveying them.

7.6 Assembling a model

We can start now to build a model of the forces that determine species richness at the local scale (Fig. 7.16). Regional species richness S_R is the major force, which on average accounts for approximately 75% of the variation observed in local species richness S_L. There are also differences due to the local environment and isolation. While correlations are also found with elevation, depth and geography, these are often mediated through the concurrent

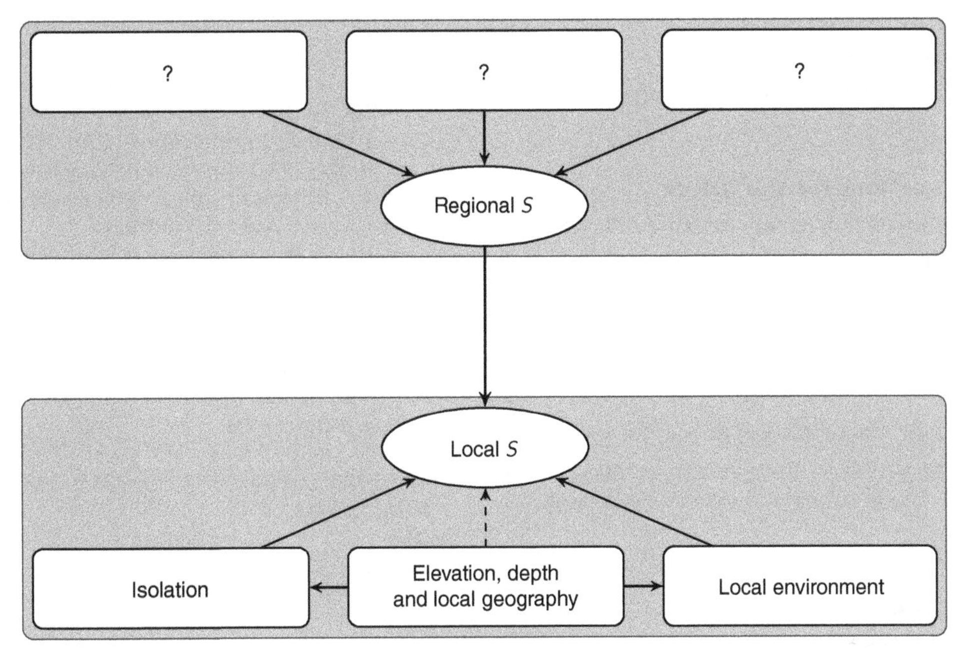

Figure 7.16 Drivers of local-scale species richness S within a given area. Direct links with solid lines, indirect with dashed lines. Determinants of regional species richness will be resolved in later chapters.

impacts of environment and isolation. We will continue to build on this model in the coming chapters to better understand how processes at both local and regional scales fit together.

7.7 Conclusions

The Species–Area Relationship discovered by Arrhenius, $S = cA^z$, is so widespread as to be an ecological law (Schoener, 1976). Among patches of a given area, however, there is still great variation in species richness. The majority of this variation—around three quarters on average—correlates with regional richness, the drivers of which remain unresolved at this point in the story. Communities are not full, and a greater input of species at the regional level increases the number found in any single location. Nevertheless, there are still patterns within the remaining variation, which can include gradients with elevation, depth, local geography and isolation. These are not always linear and can often be humped. Separating the causes when there are multiple factors at play is difficult, and it is likely that taxa respond in different ways to each as congruence between diversity trends is notoriously poor and unreliable. The next chapter will consider what the underlying causes of these local-scale patterns might be.

7.7.1 Recommended reading

Cornell H.V. & Harrison S.P. (2013). Regional effects as important determinants of local diversity in both marine and terrestrial systems. *Oikos* 122, 288–297.

Gardner T.A., Barlow J., Araujo I.S., Avila-Pires T.C., Bonaldo A.B., Costa J.E., Esposito M.C., Ferreira L.V., Hawes J., Hernandez M.I.M., Hoogmoed M.S., Leite R.N., Lo-Man-Hung N.F., Malcolm J.R., Martins M.B., Mestre L.A.M., Miranda-Santos R., Overal W.L., Parry L., Peters S.L., Ribeiro-Junior M.A., da Silva M.N.F., Motta C.D.S. & Peres C.A. (2008). The cost-effectiveness of biodiversity surveys in tropical forests. *Ecology Letters* 11, 139–150.

Gaston K.J. & Spicer J.I. (2004). *Biodiversity: An Introduction*. Blackwell Science Ltd.

7.7.2 Questions for the future

- Why do taxonomic groups respond differently to environmental gradients?
- Most evidence so far for the '75% rule' of regional species richness determining local comes from marine systems—do the same patterns occur on land?
- If no single taxonomic group is a good indicator of overall species richness patterns, could we use another feature of natural systems as a predictor instead?

References

Arrhenius, O., 1921. Species and area. *Journal of Ecology* 9:95–99.

Basset, Y., L. Cizek, P. Cuénoud, R. K. Didham, F. Guilhaumon, O. Missa, V. Novotny, F. Ødegaard, T. Roslin, J. Schmidl, A. K. Tishechkin, N. N. Winchester, D. W. Roubik, H.-P. Aberlenc, J. Bail, H. Barrios, J. R. Bridle, G. Casta no Meneses, B. Corbara, G. Curletti, W. Duarte da Rocha, D. De Bakker, J. H. C. Delabie, A. Dejean, L. L. Fagan, A. Floren, R. L. Kitching, E. Medianero, S. E. Miller, E. Gama de Oliveira, J. Orivel, M. Pollet, M. Rapp, S. P. Ribeiro, Y. Roisin, J. B. Schmidt, L. Sørensen, and M. Leponce, 2012. Arthropod diversity in a tropical forest. *Science* 338:1481–1484.

Belmaker, J. and W. Jetz, 2012. Regional pools and environmental controls of vertebrate richness. *American Naturalist* 179:512–523.

Bhatt, J. P., K. Manish, and M. K. Pandit, 2012. Elevational gradients in fish diversity in the Himalaya: water discharge is the key driver of distribution patterns. *PLoS ONE* 7:e46237.

Bryant, J. A., C. Lamanna, H. Morlon, A. J. Kerkhoff, B. J. Enquist, and J. L. Green, 2008. Microbes on mountainsides: contrasting elevational patterns of bacterial and plant diversity. *Proceedings of the National Academy of Sciences of the United States of America* 105:11505–11511.

Castagneyrol, B. and H. Jactel, 2012. Unravelling plant-animal diversity relationships: a meta-regression analysis. *Ecology* 93:2115–2124.

Chen, I.-C., H.-J. Shiu, S. Benedick, J. D. Holloway, V. K. Chey, H. S. Barlow, J. K. Hill, and C. D. Thomas, 2009. Elevation increases in moth assemblages over 42 years on a tropical mountain. *Proceedings of the National Academy of Sciences of the United States of America* 106:1479–1483.

Colwell, R. K. and D. C. Lees, 2000. The mid-domain effect: geometric constraints on the geography of species richness. *Trends in Ecology & Evolution* 15:70–76.

Cornell, H. V., 1999. Unsaturation and regional influences on species richness in ecological communities: a review of the evidence. *Ecoscience* 6:303–315.

Cornell, H. V. and S. P. Harrison, 2013. Regional effects as important determinants of local diversity in both marine and terrestrial systems. *Oikos* 122:288–297.

Cornell, H. V., R. H. Karlson, and T. P. Hughes, 2008. Local-regional species richness relationships are linear at very small to large scales in west-central Pacific corals. *Coral Reefs* 27:145–151.

Currie, D. J. and J. T. Kerr, 2008. Tests of the mid-domain hypothesis: a review of the evidence. *Ecological Monographs* 78:3–18.

Drakare, S., J. J. Lennon, and H. Hillebrand, 2006. The imprint of the geographical, evolutionary and ecological context on species–area relationships. *Ecology Letters* 9:215–227.

Fargione, J., C. S. Brown, and D. Tilman, 2003. Community assembly and invasion: an experimental test of neutral versus niche processes. *Proceedings of the National Academy of Sciences of the United States of America* 100:8916–8920.

Gaston, K. J. and J. I. Spicer, 2004. *Biodiversity: An Introduction*. Blackwell Science Ltd., second edition.

Geange, S. W., A. W. Connell, P. J. Lester, M. R. Dunn, and K. C. Burns, 2012. Fish distributions along depth gradients of a sea mountain range conform to the mid-domain effect. *Ecography* 35:557–565.

Gonçalves-Souza, T., G. Q. Romero, and K. Cottenie, 2013. A critical analysis of the ubiquity of linear local-regional richness relationships. *Oikos* 122:961–966.

Harrison, S. and H. Cornell, 2008. Toward a better understanding of the regional causes of local community richness. *Ecology Letters* 11:969–979.

Hoiss, B., J. Krauss, S. G. Potts, S. Roberts, and I. Steffan-Dewenter, 2012. Altitude acts as an environmental filter on phylogenetic composition, traits and diversity in bee communities. *Proceedings of the Royal Society Series B* 279:4447–4456.

von Humboldt, A., 1808. *Ansichten de Natur mit Wissenschaflichen Erlauterungen*. J. G. Cotta.

Karlson, R. H., H. V. Cornell, and T. P. Hughes, 2004. Coral communities are regionally enriched along an oceanic biodiversity gradient. *Nature* 429:867–870.

Kier, G., H. Kreft, T. M. Lee, W. Jetz, P. L. Ibisch, C. Nowicki, J. Mutke, and W. Barthlott, 2009. A global assessment of endemism and species richness across island and mainland regions. *Proceedings of the National Academy of Sciences of the United States of America* 106:9322–9327.

Körner, C., 2007. The use of 'altitude' in ecological research. *Trends in Ecology & Evolution* 22:569–574.

Lawton, J. H., D. E. Bignell, B. Bolton, G. F. Bloemers, P. Eggleton, P. M. Hammond, M. Hodda, R. D. Holt, T. B. Larsen, N. A. Mawdsley, N. E. Stork, D. S. Srivastava, and A. D. Watt, 1998. Biodiversity inventories, indicator taxa and effects of habitat modification in tropical forest. *Nature* 391:72–76.

Lee, S. C. and J. F. Bruno, 2009. Propagule supply controls grazer community structure and primary production in a benthic marine ecosystem. *Proceedings of the National Academy of Sciences of the United States of America* 106:7052–7057.

MacArthur, R. H. and E. O. Wilson, 1967. *The Theory of Island Biogeography*. Princeton University Press.

Marini, L., E. Bona, W. E. Kunin, and K. J. Gaston, 2011. Exploring anthropogenic and natural processes shaping fern species richness along elevational gradients. *Journal of Biogeography* 38:78–88.

McKinney, M. L. and J. L. Lockwood, 1999. Biotic homogenization: a few winners replacing many losers in the next mass extinction. *Trends in Ecology & Evolution* 14:450–453.

Nogués-Bravo, D., M. B. Araújo, T. Romdal, and D. Rahbek, 2008. Scale effects and human impacts on the elevational species richness gradient. *Nature* 453:216–219.

Palardy, J. E. and J. D. Witman, 2014. Flow, recruitment limitation, and the maintenance of diversity in marine benthic communities. *Ecology* 95:286–297.

Pinto-Sánchez, N. R., A. J. Crawford, and J. J. Wiens, 2014. Using historical biogeography to test for community saturation. *Ecology Letters* 17:1077–1085.

Preston, F. W., 1962. The canonical distribution of commonness and rarity, part II. *Ecology* 43:410–432.

Rahbek, C., 1995. The elevational gradient of species richness—a uniform pattern. *Ecography* 18:200–205.

Sanders, N. J., J. Moss, and D. Wagner, 2003. Patterns of ant species richness along elevational gradients in an arid ecosystem. *Global Ecology and Biogeography* 12:93–102.

Sax, D. F. and S. D. Gaines, 2003. Species diversity: from global decreases to local increases. *Trends in Ecology & Evolution* 18:561–566.

Scheiner, S. M., 2003. Six types of species-area curves. *Global Ecology and Biogeography* 12:441–447.

Schoener, T. W., 1976. The species–area relation within archipelagos: models and evidence from island land birds. *Proceedings of the International Ornithological Congress* 16:628–642.

Smith, K. F. and J. H. Brown, 2002. Patterns of diversity, depth range and body size among pelagic fishes along a gradient of depth. *Global Ecology and Biogeography* 11:313–322.

Smith, V. H., B. L. Foster, J. P. Grover, R. D. Holt, M. A. Leibold, and F. J. deNoyelles, 2005. Phytoplankton species richness scales consistently from laboratory microcosms to the world's oceans. *Proceedings of the National Academy of Sciences of the United States of America* 102:4393–4396.

Sundqvist, M. K., N. J. Sanders, and D. A. Wardle, 2013. Community and ecosystem responses to elevational gradients: processes, mechanisms, and insights for global change. *Annual Review of Ecology, Evolution, and Systematics* 44:261–280.

Szava-Kovats, R., A. Ronk, and M. Pärtel, 2013. Pattern without bias: local–regional richness relationship revisited. *Ecology* 94:1986–1992.

Tilman, D., 1997. Community invasibility, recruitment limitation, and grassland biodiversity. *Ecology* 78:81–92.

Westgate, M. J., P. S. Barton, P. W. Lane, and D. B. Lindenmayer, 2014. Global meta-analysis reveals low consistency of biodiversity congruence relationships. *Nature Communications* 5:3899.

CHAPTER 8

Drivers of diversity

8.1 The big question

Species richness varies among sites, as revealed in the last chapter, with several common patterns. This has however revealed few clues as to what the underlying processes giving rise to them are. The enquiry leads into one of the most fundamental questions in ecology: how can so many species coexist in the same location? The issue has been a long-standing one in ecology and is often referred to as the "paradox of the plankton" after a paper by Hutchinson (1961). Ever since the earliest days of microscopy, investigators had peered into droplets of water and found many minute forms of life swimming around. How could it be that a single drop of water contained so many different species? Surely there isn't much variation in the environment within these droplets, so why didn't one dominant species take over? Hutchinson's paper provided no definitive answers, but it set the tone for a debate that has continued for over 50 years.

8.2 Coexistence or co-occurrence?

Before beginning, we should clear up an important semantic issue, which is the definition of coexistence. There are many species that happen to live in the same place but have no direct interactions; perhaps they share no resources or occupy contrasting microhabitats (e.g. above or below ground). This is simple co-occurrence and requires no special theory to understand. True coexistence, on the other hand, implies that species live together despite substantial overlap in their resources or other conflicts. Two families of mechanisms can maintain this (Chesson, 2000). These are either *stabilising*, when a species is able to recover should it become rare or is controlled if its abundance becomes excessive, or *equalising*, when differences among species are evened out, such that all end up with the same average fitness. Only stabilising mechanisms can lead to long-term coexistence and therefore these are the ones we should seek to identify (Siepielski and McPeek, 2010). Equalising forces may stave off competitive exclusion almost indefinitely but will not prevent species from going locally extinct through random demographic fluctuations. When assessing candidate theories for species coexistence, always ask, what is the advantage to being rare or the cost to being common?

8.3 Energy and resources

One of the most obvious early lines of enquiry was the input of energy and resources into natural systems. Where more are available, is it possible for greater numbers of species to coexist? Superficially this seems straightforward, but actually there are two possible outcomes. If the resources entering a system increase, will this increase the number of species or

Natural Systems: The organisation of life, First Edition. Markus P. Eichhorn.

simply increase the number of individuals of those species already present? To clarify our thinking, let's conduct a thought experiment as to what might happen as energy input or resource supply increases.

On land, more energy from sunlight would increase net primary productivity (NPP) as plants would be able to produce more carbohydrates via photosynthesis. This would also increase the amount of food available to their herbivores, making the gains for plants uncertain, as they might be eaten as fast as they grow. Meanwhile, for ectotherms (organisms relying on external conditions to regulate their body temperature), their activity would increase, enhancing their ability to capture resources and consequently improving their fitness. For endotherms (organisms which generate their own body heat), less energy would be required to maintain body temperatures at optimal levels, so long as it didn't get too hot, leaving more for reproduction.

Put together, this means that the fitness of both plants and animals should be enhanced as they acquire greater resources for investing in growth, survival and reproduction. But the end result will be that their populations rise to new, higher levels until resources once again become limiting, and competition as fierce as ever. The only potential difference is that, with a larger density of individuals, there might be the potential for some species to split into more specialised taxa while still maintaining stable populations.

On the global scale there certainly seems to be a link between energy and species richness as many more are found in the warm tropics (see Chapter 15). This isn't the full story though. In some systems even abundant energy and resources fail to obviously increase species richness. Kelp forests occur in warm ocean waters where a continuous supply of nutrients, usually derived from deep ocean upwelling, foster the growth of gigantic algae. These systems have the greatest NPP in the oceans and a profusion of life. Yet there are only 100–120 species of kelp worldwide (Santelices, 2007). In contrast, coral reefs typically occur in nutrient-poor waters and cover a much smaller global area, but in total there are around 10 times as many species of true corals.

Much speculation can be found in the literature, one traditional view being that there ought to be a humped relationship between productivity and species richness—few species when resources are scarce, a peak at intermediate levels, and then a small number of dominants taking over through competitive exclusion when resources are abundant (Grime, 1973a,b). This narrative was accepted for many years, but a vast study by Adler et al. (2011) has cast doubt upon it by conducting replicated surveys of the link between plant biomass (a surrogate of primary productivity) and plant species richness across the entire world (Fig. 8.1). In the majority of sites, there was no relationship at all; some found linear increases or decreases, and while a few were curved, these where either concave or convex. Likewise reviews of multiple studies have failed to find a consistent single pattern (Waide et al., 1999). As for animals, the most common pattern is for an increase in species richness with plant productivity, though this is by no means universal, and it is uncertain what the underlying cause might be (Cusens et al., 2012).

The link between energy, resources and species richness in nature is therefore not as straightforward as might have first been supposed. Can any more be learnt from experimental approaches?

The longest-running experiment in ecology, the Park Grass Experiment at Rothamsted research station, began in 1856 (Crawley et al., 2005). While the design is not as rigorous as many modern experiments, its duration allows some powerful lessons to be drawn. For over 150 years researchers have been adding fertilisers in different combinations to grassland communities and monitoring the number of plant species present. Over time the trend has been for fertilised plots to lose many of their species. Even more striking is that the more types of nutrients are provided—nitrogen, phosphorus, potassium—the greater the loss of species (Fig. 8.2).

A nice experiment was set up which sheds light on the processes occurring. Hautier et al. (2009)

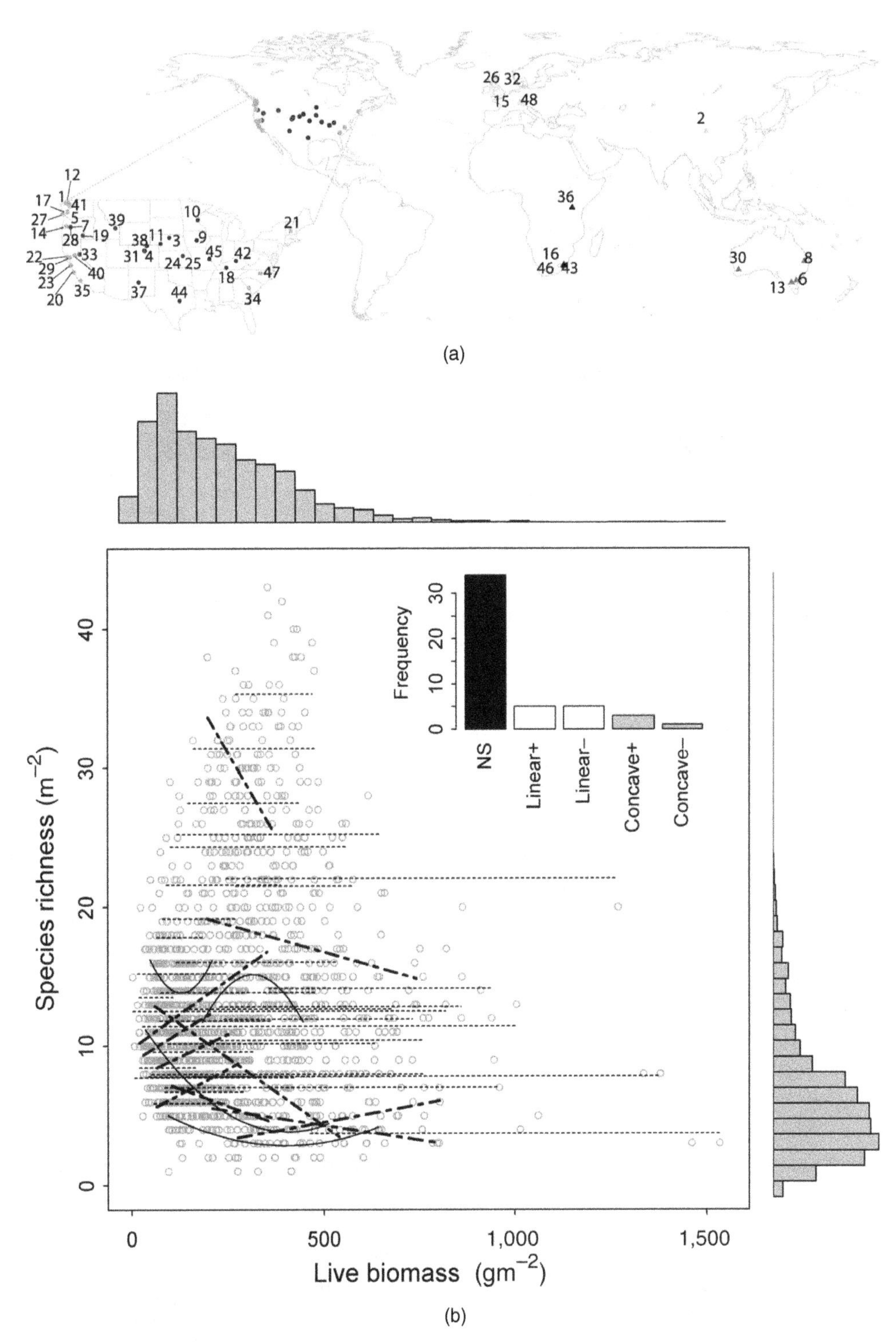

Figure 8.1 (a) Locations of 48 study sites and (b) relationships between productivity and plant species richness. Plots were 1 m^2 in size. Productivity was measured as dry weight of peak live biomass. Inset shows number of relationships that were either non-significant (thin dashed lines), linear (thick dashed lines) or curved (solid lines). Histograms on each axis represent the frequency of species richness and peak live biomass levels across all sites. (*Source*: Alroy et al. (2008). Reproduced with permission from the American Association for the Advancement of Science.)

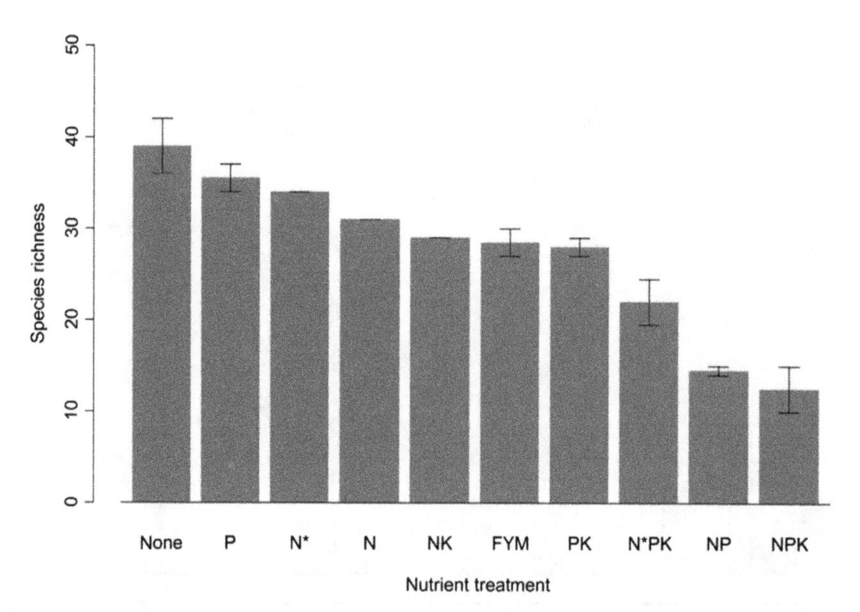

Figure 8.2 Species richness in 25×50 cm samples taken from plots with differing nutrient treatments across the Park Grassland Experiment. Plots varied in size (0.013–0.05 ha) and were treated with either no added nutrients or a combination of phosphorus (P), nitrogen applied as either sodium nitrate (N*) or ammonium sulphate (N), potassium (K) or farmyard manure (FYM). Means ± SE except for those treatments with no replication (N, N*, NK). (*Source*: Crawley et al., (2005). Reproduced with permission from the University of Chicago Press.)

created microcosms which contained a random combination of six grassland species (Fig. 8.3). These then received a combination of two factorial treatments. Half were fertilised and half provided with additional light at ground level. (The design was very clever—fluorescent tubes were inserted with gaps that either pointed downwards, so only ground plants received the light, or entirely enclosed, to act as controls and ensure that it wasn't an effect of heat from the tubes.)

As expected, total productivity in the microcosms went up when fertilised. Additional light on its own had little effect on productivity, perhaps because it was only available at ground level. Combining both light and fertiliser, however, led to massively greater growth overall (Fig. 8.4a).

What about species richness? In fertilised plots, species richness fell as a few dominant competitors took over. Yet when provided with additional light, this loss of species was prevented (Fig. 8.4b). What had taken place is very revealing for understanding the effects of resources on diversity. With additional nutrients, the strongest competitors had grown tall, reducing the amount of light reaching ground level from 13% to 5% of photosynthetically active radiation (PAR; light at wavelengths that plants can use). This meant that smaller or slower-growing plants were shaded out. Adding extra light above the level of the tallest plants would have made little difference, but by putting the tubes below the dominant species, they were able to rescue species that would otherwise have been crowded out. The taller species were limited by nutrients, whereas the shorter species were limited by light. The number of species is not simply responding to overall resource availability but to the number of *limiting resources*.

Harpole and Tilman (2007) tried a different approach in which they added varying numbers of resources to grassland patches. As in the previous study, the productivity of the system, measured as biomass produced, went up with more types

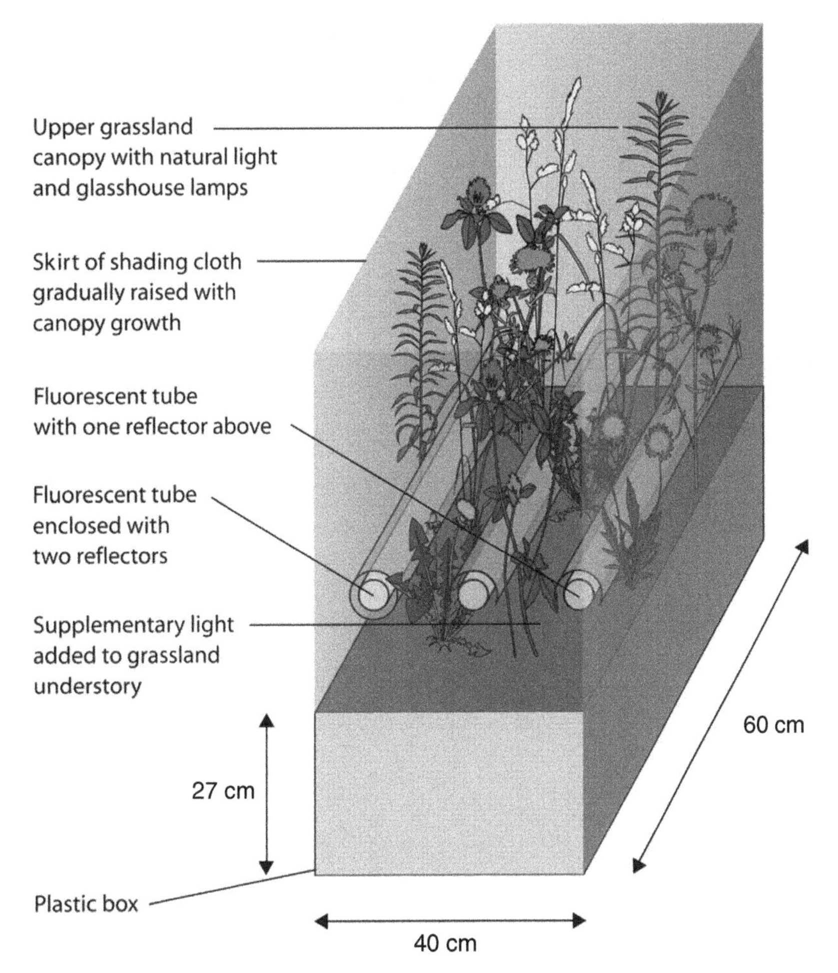

Figure 8.3 Microcosm from the experiments of Hautier et al. (2009). (*Source*: Hautier and Hector (2009). Reproduced with permission from the American Association for the Advancement of Science.)

of added resources (Fig. 8.5a). At the same time, increased productivity correlated with reduced diversity (measured here as e^H; see Chapter 5) both in observational and experimental plots (Fig. 8.5b). Each added resource led to the loss of, on average, 3.3 species. This confirms that the link between productivity and species richness cannot be assumed to be positive.

Having started with the relatively simple belief that levels of resources might tell us something about species richness, we've found something more complicated. The more types of resources are added, the fewer species are present (though note that this only applies to studies of small-scale plots). Recall from Chapter 6 that this is exactly in line with the predictions of R^* theory: the number of species in any given site is set by the number of limiting resources, with species richness peaking at intermediate ratios. By adding resources, they cease to be limiting, and opportunities for species to trade off against each other are removed. That we often don't find a correlation between measured resource

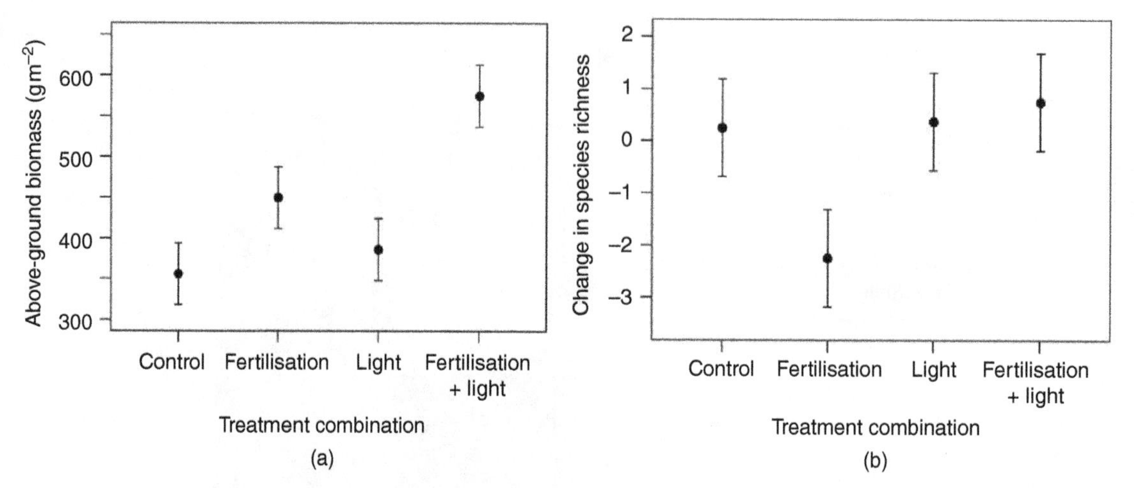

Figure 8.4 (a) Above-ground biomass and (b) species richness changes in experimental treatments. (*Source*: Hautier and Hector (2009). Reproduced with permission from the American Association for the Advancement of Science.)

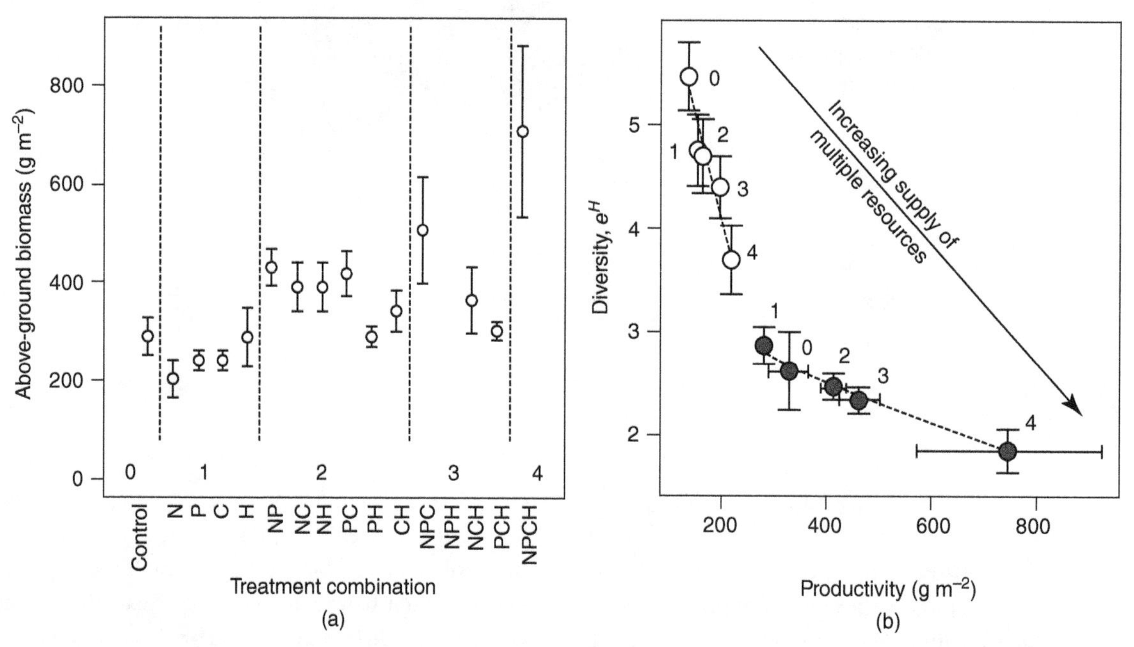

Figure 8.5 (a) Above-ground biomass in resource addition treatments (N, nitrogen; P, phosphorus; C, cations; H, water) and (b) change in diversity in observational (open circles) and experimental (closed circles) plots with numbers of resources added. (*Source*: Harpole and Tilman (2007). Reproduced with permission from Nature Publications Group.)

levels and diversity is also unsurprising: remember that R^* theory tells us that observed resource levels are the result of what's left after competition, not the supply rate.

In short, the search for a causative link between energy or resources and species richness has turned into something more nuanced; we will return to see how species richness influences productivity in the next chapter. In the meantime we should look for other possible drivers of diversity.

8.4 Diversity begets diversity

8.4.1 Heterogeneity in space

There are many ways in which the environment can vary, allowing species to coexist because each can be specialised on a particular set of conditions. As a general rule, more heterogeneous environments support a greater number of species (Stein et al., 2014). There are frequent correlations between structural diversity of habitat patches and the species richness of a wide variety of animal groups (Tews et al., 2004). The most obvious pattern is for different locations to vary slightly in their balance of resources, thereby shifting which species persist (we saw the theory behind this in Fig. 6.20). A gradient of environmental stress would achieve the same outcome. Heterogeneity need not be solely abiotic though—there can also be trends in the biotic environment, such as in the distribution of predators (Fig. 6.21).

Patterns of increased species richness with environmental heterogeneity are widespread. For example, looking across the Iberian–Balearic region, the richness of ferns correlates strongly with the span of elevation in each area, which captures a range of environmental conditions (Fig. 8.6).

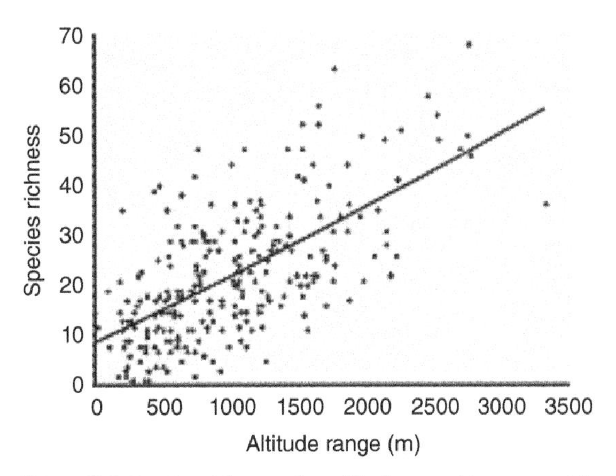

Figure 8.6 Species richness of pteridophytes with topographic range (m) in 50 × 50 km grid squares. Moreno Saiz and Lobo (2008). Reproduced with permission from Springer Business and Print Media.

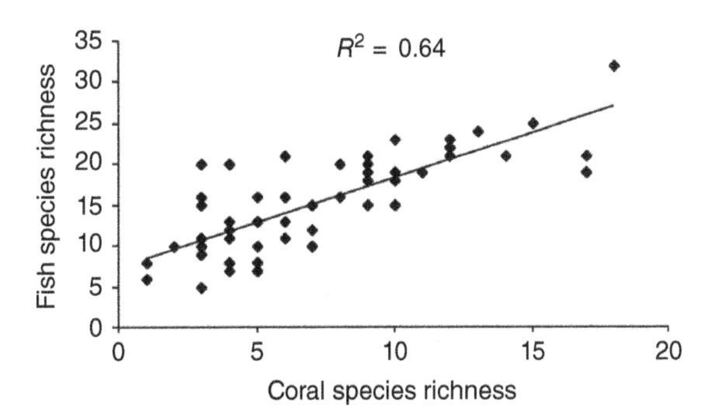

Figure 8.7 Relationships between fish and coral species richness in the lagoon of Lizard Island, Great Barrier Reef (Komyakova et al., 2013). (*Source*: From http://creativecommons.org/licenses/by/4.0/.)

Gradients in heterogeneity may lie at the root of some of the positive trends found between species richness and productivity. As more biomass is produced, it creates a greater range of microenvironments and resource gradients, along with novel resources made by the primary producers. Increased diversity among those organisms responsible for creating the structure of the habitat, such as forest trees or reef corals, often increases the complexity of habitats and thereby increases the diversity of other groups. For example, Komyakova et al. (2013) found that coral species richness was a powerful correlate of the species richness of associated fish (Fig. 8.7). Likewise structural complexity is at least one reason why there are many more animal species in tropical rain forests as compared to the simpler boreal forests, despite high-latitude forests tending to form on more fertile soils compared to the ancient, leached soils of the tropics. Nutrients are scarce in tropical rain forests, leading to intense competition and specialisation among species.

The effect of heterogeneity in habitat structure on patterns of diversity holds even at very small scales. A truly remarkable pattern was discovered in a study of the diversity of twig-nesting ants in a Mexican coffee plantation. Armbrecht et al (2004) placed mesh bags on the floor of the plantation as habitats for the ants. These contained bundles of twigs of the

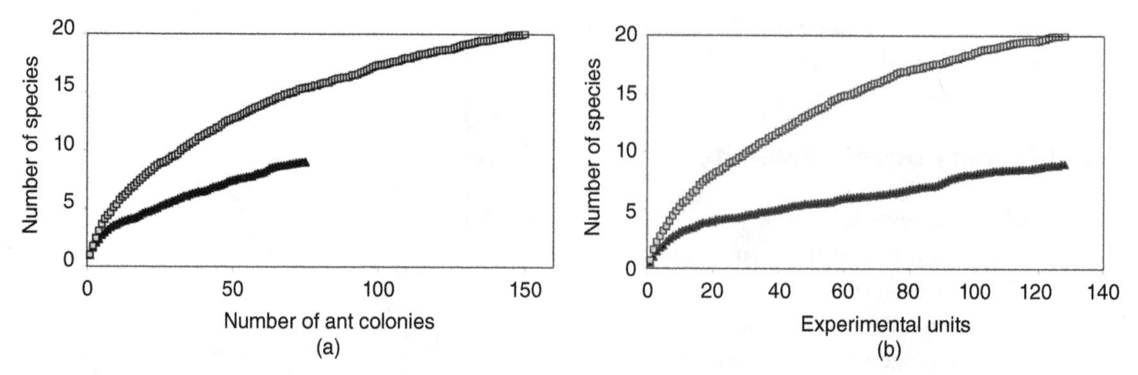

Figure 8.8 Species accumulation curves for twig-nesting ants assessed by (a) number of colonies or (b) number of mesh bags sampled. Clear squares, mixed-species bags of twigs; filled triangles, single-species bags of twigs. (*Source*: Armbrecht et al. (2004). Reproduced with permission from American Association for the Advancement of Science.)

same weight which either all came from the same tree species or from a variety of different trees. Once collected, bags with mixed twigs contained a greater variety of ant species, whether measured in terms of number of colonies or number of bags (Fig. 8.8). It didn't matter which tree species the twigs came from, and there was no evidence that the ants were specialised to one twig species or another. Even on this tiny scale, increasing heterogeneity was enough to promote greater species richness.

The link between climate and habitat characteristics goes some way towards explaining the apparent effect of local environment on species richness. Ferger et al. (2014) found that the species richness of birds on Mount Kilimanjaro correlated positively with temperature. Rather than being caused by the thermal tolerance of bird species, in fact it was due to indirect effects on the vegetation. Greater fruit availability increased the species richness of frugivorous birds, while greater invertebrate abundance and vegetation heterogeneity increased the species richness of insectivorous birds. The correlation between bird species richness and temperature was only part of the story, and biotic heterogeneity provided the real mechanism.

In terrestrial systems this positive link between plant diversity and that of animal groups is widespread, even when the animals concerned are not directly dependent on the plants for nutrition (c.f. Fig. 7.15). Moreover, the phylogenetic diversity of plants can be an even stronger predictor, which suggests that more unrelated plants support a higher diversity of other species (Dinnage et al., 2012). As a result of increased biotic complexity among the producers, other species follow suit. In the Mojave desert, species richness of rodents was found to correlate with that of perennial plants (Fig. 8.9). What makes this such a striking result is that for these largely seed-eating rodents the effect of resource heterogeneity was much greater than that of the total amount of resources. Variety matters more than quantity.

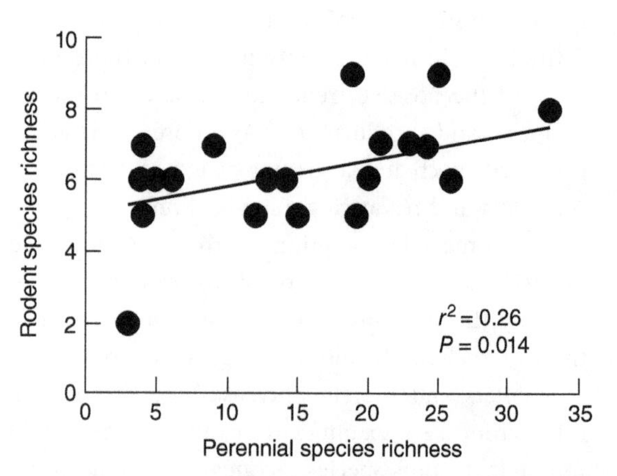

Figure 8.9 Relationship between rodent and perennial plant species richness in 31 Mojave desert communities. (*Source*: Stevens et al. (2012). Reproduced with permission from John Wiley & Sons Ltd.)

Increased spatial heterogeneity doesn't always mean more species though; there can be a confounding effect of area. Too much variation in a confined space would make individual patches too small to maintain stable populations, reducing overall species richness, depending on the tolerances of the species present (Allouche et al., 2012). The relationship between heterogeneity and diversity therefore depends upon both the scale of the study and the species concerned and in theory could take almost any form. The greatest species richness for a given area might be found when there is an intermediate degree of heterogeneity; too much and only generalist species will be able to persist, and too little and there will be insufficient variety to maintain specialists. This is analagous to temporal heterogeneity, which we turn to next.

8.4.2 Heterogeneity in time

Heterogeneity doesn't just occur in space; it can also occur through time as environments change. If the abiotic environment fluctuates, whether predictably or randomly, species will vary in their fitness and recruitment into the community. Variability in climate is at least as important in generating patterns of species richness as average conditions (e.g. for plants in south-east Australia, Letten et al., 2013). Average measures of reproductive output or competitive ability may be misleading as in the long term there can be no best species for the habitat. In fact, even if one species is clearly better adapted than the others, inferior competitors can still persist if they receive just enough chances to keep them in the game. Their populations can be buffered against extinction if occasional good spells provide sufficient opportunities to survive the intervals between them.

This can be difficult to imagine so we will illustrate it with an example. In Oneida Lake in New York State, two species of zooplanktonic water fleas were found by Cáceres (1997) to coexist, with negative density correlation—when one did well, the other tended to perform poorly. As with most small species, there were massive fluctuations in populations from year to year, but one species had by far the upper hand. Why did the other not go extinct? The reason comes from a peculiar feature of their life history. In the autumn, as temperatures in the lake cool, the water fleas lay diapausing eggs. These sink to the bottom of the lake and can emerge the following spring or not—in fact, they remain viable for at least 125 years.

As in many lakes, sediment tends to accumulate through time, which provided an unusual opportunity. By digging down through the mud at the bottom of the lake with a sediment corer, Cáceres was able to reconstruct populations of the water fleas based on the number of eggs they had deposited each year (Fig. 8.10). One species, *Daphnia pulicaria*, was a strong competitor and was able to recruit enough new individuals every year to maintain a population. In some years it produced vast numbers of eggs. Its partner species, *Daphnia galeata mendotae*, performed less well. Even in its best years, it produced less than a tenth as many eggs as *D. pulicaria*, and in many years there were no eggs produced at all.

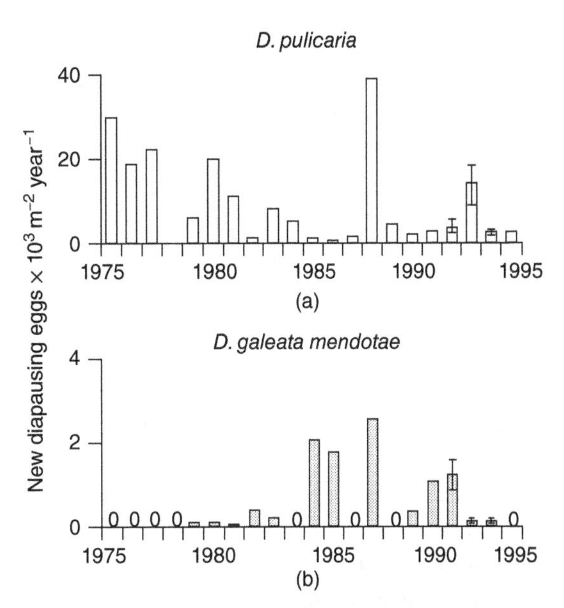

Figure 8.10 Diapausing egg production per year by two species of water flea in a lake. Note differing scales on each graph. (*Source*: Cáceres (1997). Reproduced with permission from National Academy of Sciences, USA.)

The weaker species survived in this system as a result of the **storage effect** (Warner and Chesson, 1985). By having long-lived eggs and intermittently producing just enough offspring, it could remain in a lake dominated by a more abundant competitor. The major requirement of the storage effect is some form of buffering mechanism. The diapausing eggs achieve this by ensuring that populations can return even after bad years.

In larger species it often takes the form of long-lived adults who survive through lean periods and capitalise on occasional opportunities to reproduce. Oceanic fish are a prime example, as many are surprisingly long lived (Secor, 2007). Species commonly found at the fishmonger might survive for as long as a century in the water. Their reproduction is highly unpredictable, both through space and time. The adult population in any given year tends to split itself between different spawning grounds. There are good years and poor years, but also successful sites and unsuccessful, which might vary between breeding seasons. Maintenance of the population depends on some spawning grounds, in some years, being able to produce sufficient recruits, while long lifespans increase the number of chances any single fish has. This also indicates one of the major problems for sustainable fisheries: as spawning grounds become depleted, there are fewer chances for populations to buffer themselves, while long-lived species take a considerable time to recover from harvesting.

The storage effect can also work across seasons, allowing a succession of species to dominate at different times of year. From studying the composition of fish in the Bristol Channel over more than 30 years, Shimadzu et al. (2013) recognised that there were four different groups of species, each of which dominated for a season. Within these seasonal groupings, species foraged in different places—there was evidence of niche partitioning. The overall abundance of fish in the Bristol Channel remained relatively stable throughout the year, while each seasonal group came and went.

Put together, heterogeneity, both abiotic and biotic, through both space and time, can have important effects in modulating patterns of species richness. There are further ways in which the environment can influence species richness though.

8.5 Disturbance

It is sometimes naïvely assumed that disrupting natural systems reduces the number of species they contain. This turns out to not be the case. The oldest known organisms on earth are bristlecone pines, *Pinus aristata* Engelm. They can live for many thousands of years in communities that are hardly ever disturbed, yet the forests contain remarkably few species. A tropical rain forest, on the other hand, undergoes a continuous cycle of tree growth and death; even the largest trees cannot compete for age with the bristlecone pine. A common joke among habitat managers is that the quickest way to increase the species richness of a woodland is to cut half of it down, which is entirely true, though it is seldom seen as a reasonable conservation strategy.

Following disturbance of a patch of habitat, initially colonists arrive and increase its species richness. Local biotic heterogeneity increases and leads to a set of ever more diverse interactions. With time, however, dominant species arrive, outcompeting and replacing the first colonists. In theory this results in a peak of species richness occurring around the mid-point of succession. Many interacting forces are involved in driving this pattern, including dispersal, establishment, facilitation and competition. This prompted the *intermediate disturbance hypothesis* (Connell, 1978), which proposes that species richness in communities should peak when disturbance occurs occasionally but not too often (or too seldom). Constant disturbance would result in communities permanently filled with weeds and pioneers, whereas a lack of disturbance would see dominant species taking over.

The theory soon garnered support from a study by Sousa (1979a, b) who studied the algae found on boulders in the intertidal zone in California. Wave action moves boulders around, disrupting the dominant algae and preventing them from taking over. Sousa assumed that the size of the boulder reflected the force required to move it, and therefore the likely rate of disturbance, since small boulders would be frequently shifted, whereas only a major storm would move the largest. This was reflected in the diversity of algae, which was greatest on medium-sized boulders. It's not enough to simply observe this though; the best studies are those which include an experiment, which in this case involved artificially stabilising boulders. When small boulders were prevented from moving, their algal communities became more similar to those of the largest boulders.

A more recent study by Biswas and Mallik (2010) determined the effects of human disturbance on streamside plants in Canadian boreal forests. Their study was particularly insightful as they also collected information on functional traits of the species, including measures of productivity, competitive ability, reproduction, disturbance tolerance, life history and tolerance to habitat instability. Both measures of diversity, species and functional richness, peaked with moderate disturbance (Fig. 8.11).

The intermediate disturbance hypothesis has become a popular explanation for diversity patterns and is widely cited, perhaps because the predictions of Connell (1978) seem intuitive, so it is disappointing to find that in the majority of studies the expected humped pattern is not found (Mackey and Currie, 2001). There can also be feedback between species richness and disturbance, such that each influences the other (Hughes et al., 2007). The safest thing to say is that disturbance can influence diversity, but it depends on the species concerned and their interactions, and the intermediate disturbance hypothesis is no substitute for a more detailed understanding of the system (Fox, 2013).

8.6 Top-down control

How can one dominant species be prevented from crowding out all others? The answer is to make sure that it receives some form of handicap, reducing its competitive impact on the species alongside it. This might come from specialised herbivores, predators, diseases or parasites that keep its growth in check. There can also be abiotic controls, particularly from disturbance.

These biotic processes are all referred to as top-down mechanisms for controlling diversity as they originate from higher levels in the food chain.

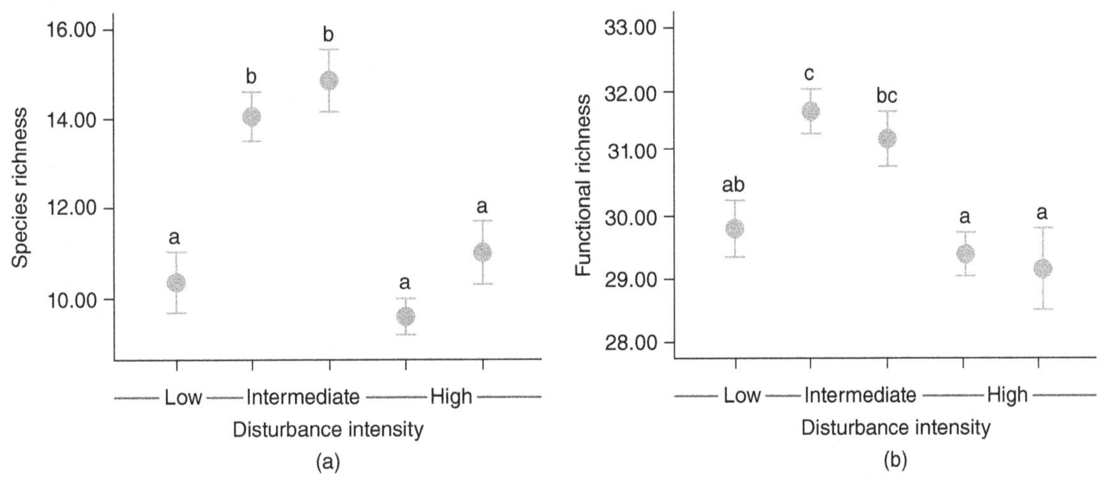

Figure 8.11 (a) Plant species and (b) functional richness responses to disturbance treatments in riparian habitats, means ± SE; points which do not share letters differ significantly. (*Source*: Biswas and Mallik (2010). Reproduced with permission from Ecological Society of America.)

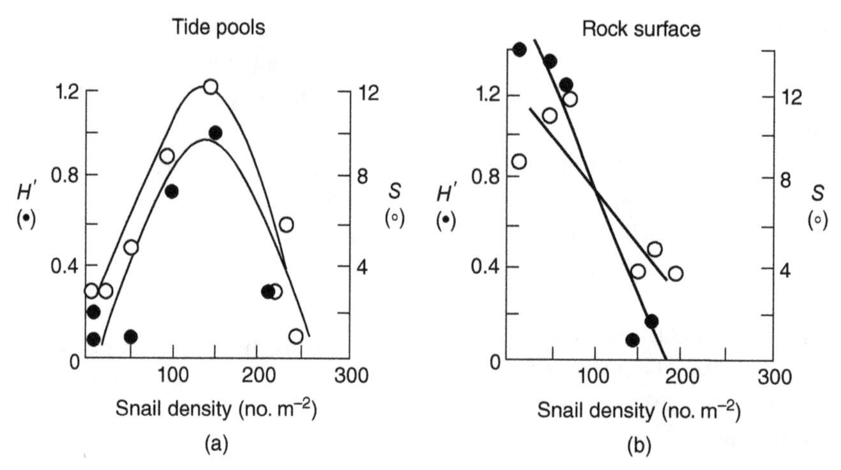

Figure 8.12 Relationship between herbivorous snail density and algal diversity (measured as species richness S and Shannon's diversity index H') (a) in tidal pools and (b) on rock surfaces. (*Source*: Lubchenco (1978). Reproduced with permission from University of Chicago Press.)

They can be stabilising, in the sense that they can allow many species to coexist, but only when their action is **density dependent**. In other words, whenever a dominant species starts to take over, it has to be impeded, while species at low densities are favoured and tend to increase. If these conditions are satisfied, then competitive exclusion can be permanently prevented.

The potential for such outcomes was noticed by Lubchenco (1978) when working in rock pools on the coast of New England. These were occupied by the herbivorous snail *Littorina littorea* which browsed the algae that grew on the rocks. The snails also varied greatly in density. Where there were few snails, one dominant species of green alga, *Enteromorpha intestinalis,* swiftly took over. This was, however, the alga that snails preferred to feed on in cafeteria trials, in which snails were offered a range of algae to choose from.

Lubchenco then set up an experiment in which she manipulated density of snails in the pools. She found that at low snail densities, the dominant alga took over, and both species richness and diversity (measured as Shannon's H'; see Chapter 5) were low. Once snail densities reached high levels, species richness also fell, as they began to eat not only their favourite alga but also everything else until eventually there was none left (Fig. 8.12). Species richness and diversity peaked at intermediate levels of herbivory.

This pattern was clear in the tidal pools, but something else happened on the rock surfaces which were exposed to wave action. Here increased snail density simply reduced species richness, so there was no beneficial effect. Why the contradiction? The dominant alga *Enteromorpha* is very sensitive to wave disturbance, so it is naturally absent from the rock surfaces. As a result, there is no advantage offered to the other algae by having their prime competitor removed, and any further herbivore addition is bad for them.

Further work on plant communities has found that top-down control by herbivores can rescue species that would go extinct after nutrient addition. In a series of experiments related to that by Hautier et al. (2009) described earlier, grasslands around the world were either fertilised, grazed or both (Borer et al., 2014). Fertilised plots lost species as the dominant plants outgrew and shaded their competitors, but when herbivores were present, the diversity of plant communities was maintained because the dominants were grazed back.

Top-down control doesn't always increase diversity. In some cases predator populations can build up beyond the ability of local prey to sustain them, leading to consumer fronts in which waves of predators sweep through a habitat, consuming everything in reach (Silliman et al., 2013). Examples include the crown-of-thorns seastar on Indo-Pacific coral reefs or snails in salt marshes. Predator outbreaks are unusual events but confirm Lubchenco's observation that a hungry predator will eat almost anything.

The indirect effects of top-down control can be surprising. Schmitz (2008) studied grassland communities in which the top predators were spiders. These preyed on grasshoppers, which in turn fed on the plants. There were two species of spider, though present in different areas. When an active hunter *Phidippus rimator* was present, it went out searching for grasshoppers and was more effective at reducing their populations. As a result, the ability of grasshoppers to control the dominant plant species *Solidago rugosa* was impaired, leading to a 168% increase in its biomass. When the spiders were sit-and-wait predators, *Pisaurina mira*, they tended to hide among the grasses, which were the grasshoppers' favoured food, causing them to focus their foraging on the safer *Solidago* (Fig. 8.13). This meant that there were indirect positive effects on the grasses and other forbs.

These subtly different hunting strategies caused cascading effects on not only the species richness of the grassland but also on its ecosystem processes such as NPP, soil organic matter and nitrogen cycling. When actively hunting spiders were present, the biomass of *Solidago* was increased at the expense of other herbaceous plants (Fig 8.14a). Sit-and-wait predators had the effect of not only increasing the diversity of the plant community (measured as evenness; Fig. 8.14b) but also its total productivity. Top-down control can therefore not only affect the layer immediately below predators but can cascade through food chains and affect the processes of entire ecosystems.

An important class of top-down mechanisms known as **Janzen–Connell effects** are named after the two great ecologists, Dan Janzen and Joseph Connell, who came up with the same idea almost simultaneously (Janzen, 1970; Connell, 1971). Their theory, otherwise known as the escape hypothesis, proposes that organisms are spatially restricted in their ability to disperse offspring. For example, a tree might produce a large crop of seeds, but most of these will tend to fall close to the parent, with relatively few being spread further away. Adult trees are also hosts to specialist predators and diseases. This means that for a seedling with the misfortune of germinating close to its parent (or another member of the same species), its natural enemies will find it quickly. Only those at a greater distance have a chance of escaping.

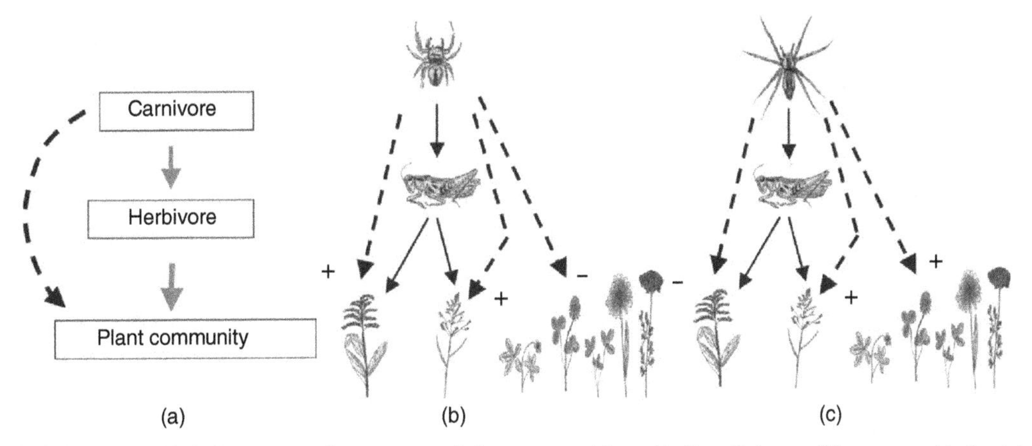

Figure 8.13 (a) Potential links between predators, prey and plant communities, with direct links as solid arrows and indirect dashed; (b) impacts of an active predator (*Solidago* at bottom left); (c) impacts of a sit-and-wait predator. (*Source*: Schmitz (2008). Reproduced with permission from American Association for the Advancement of Science.)

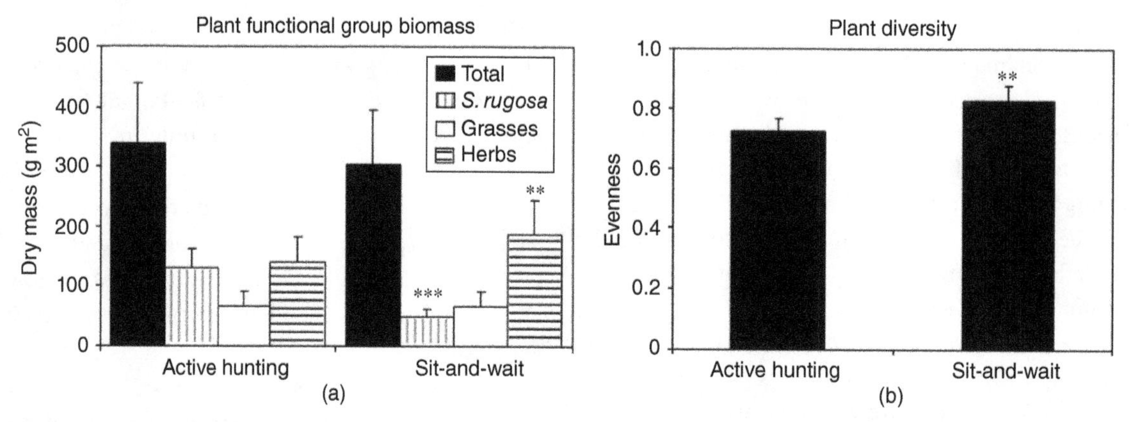

Figure 8.14 Plant (a) functional group biomass and (b) diversity in response to two different predators. Asterisks indicate significant differences between bars of the same type ** $P < 0.01$. *** $P < 0.001$. (*Source*: Schmitz (2008). Reproduced with permission from American Association for the Advancement of Science.)

This should have the effect of making sure that individuals of the same species are spread out in space, preventing them from taking over entirely. It was first put forward to account for the astonishing diversity of tree species present in tropical forests, where often no two neighbouring trees are alike. Simulations suggest that if both offspring and natural enemies can only disperse short distances, plant species richness could be more than doubled (Adler and Muller-Landau, 2005).

Evidence for these effects occurring in rain forests has been accumulating rapidly in the last few years, but one of the best studies, by Packer and Clay (2000), came from a more temperate clime. In the United States, the cherry *Prunus serotina* has classically limited dispersal as most fruit falls directly beneath the parent tree, with a smaller proportion being eaten by animals and hence defecated further away. Most animals simply chew the fruit and spit the seed out (as do we). Nevertheless, in native woodlands, cherry trees tend to be spread out in space. Could this be an example of Janzen–Connell processes?

To investigate, Packer and Clay monitored seedlings growing at different distances from adult trees. Initially there were many more seedlings close to adults than further away. By 4 months those further away had a greater probability of survival,

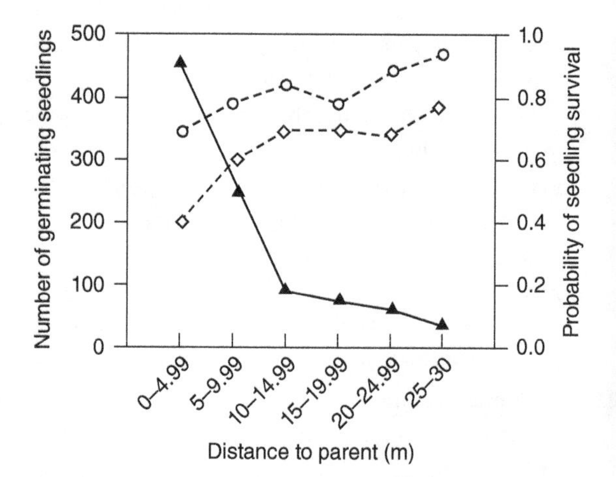

Figure 8.15 Number of germinating seedlings (solid line) and probability of survival at 4 and 16 months (upper and lower dashed lines, respectively) for cherry seedlings at increasing distance from adults. (*Source*: Packer and Clay (2000). Reproduced with permission from Nature Publishing Group.)

and by 16 months the trend had become even stronger (Fig. 8.15).

To tease out the mechanism, Packer and Clay planted cherry seeds in soil collected either from directly beneath an adult cherry tree or from a certain distance away. To control for potential differences in soil quality, they sterilised the soil for half the seedlings in each batch. Overall, seedlings survived well in soil that came from further away from adults. Seedlings also had no problems in soil from beneath

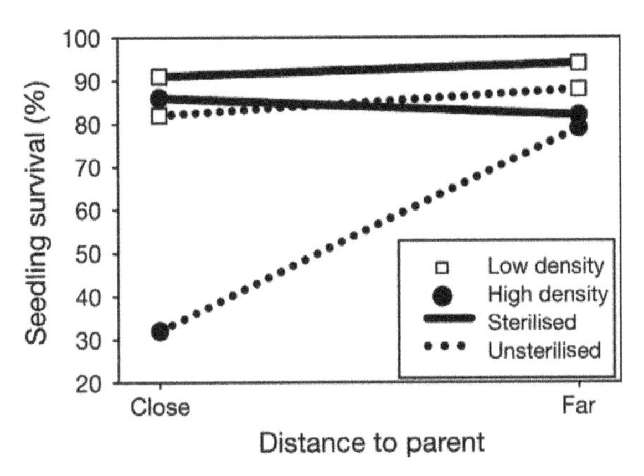

Figure 8.16 Seedling survival in soils collected from close to or far from adult trees, which were sterilised or unsterilised, at two levels of density. (*Source*: Packer and Clay (2000). Reproduced with permission from Nature Publishing Group.)

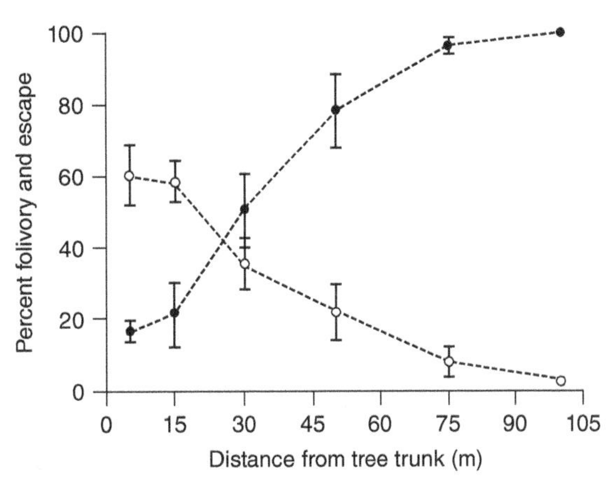

Figure 8.17 Distance from mature fruiting trees and herbivory of *Swietenia macrophylla* seedlings in Pará, Brazil. Proportion of leaf area damaged (open circles), and the percentage of seedlings escaping attack by moth caterpillars (closed circles). (*Source*: Norghauer et al. (2010). Reproduced with permission from Springer Business and Print Media.)

adults that had been sterilised. Unsterilised soil, however, resulted in high levels of mortality when seedlings were planted at high density (Fig. 8.16). They suspected a soil-borne pathogen might be the cause of this pattern and later were able to culture a species of the disease-bearing fungus *Pythium* from the roots of cherry trees. This could not spread far in the soil beyond the adult trees, hence the restricted range of its effects. A Janzen–Connell process was confirmed.

It's not only pathogens that can cause this effect—natural enemies such as insect herbivores can have similar impacts, as was found in the Brazilian rain forest for the mahogany *Swietenia macrophylla* (Fig. 8.17). Caterpillars of a particular species of moth, *Steniscadia poliophaea*, caused greater amounts of leaf damage and reduced survival of tree seedlings that grew near mature trees. A tiny insect controlled the distribution of a gigantic tree.

Janzen–Connell effects may not be restricted within species; some evidence suggests that there is also a phylogenetic pattern, with closely related species influencing one another (Liu et al., 2012). This should not be surprising; we have already seen how insect herbivores frequently feed on several related host plants (Fig. 5.11), and diseases are more likely to transfer between sister species. The net result should be an increase in the phylogenetic diversity of plant communities. It's not just about plants either; Marhaver et al. (2013) have demonstrated that the same effects apply to reef-dwelling corals.

8.7 Expanding our model

We can now look back on the outline that was begun in the last chapter (Fig. 7.16) and expand it further to include numbers of limiting resources, disturbance and top-down control, which can all act to modulate species richness (Fig. 8.18). Effects of the local environment depend upon its heterogeneity, both abiotic and biotic, and through space and time. The environment also influences NPP, which in turn sometimes correlates with species richness, though this relationship can be positive, negative or humped. Productivity almost always increases with greater availability of resources.

These are more satisfying drivers of variation in species richness than trends in elevation, depth or local geography, which reflect other processes rather than anything inherently special about the

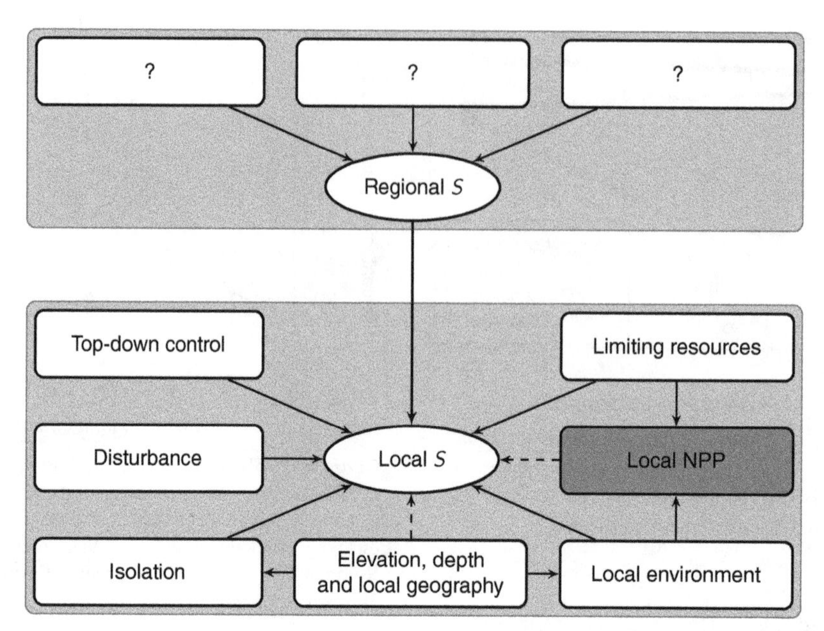

Figure 8.18 Drivers of local-scale species richness *S* and Net Primary Productivity (NPP) within a given area. Direct links in bold, indirect with dashed lines. Determinants of regional species richness will be resolved in later chapters.

number of metres above or below sea level. We therefore treat them as correlates rather than causes.

8.8 Conclusions

Why do some areas have more species than others? We began by considering the obvious candidates, inputs of energy and resources, but found that though they reliably predict patterns in NPP, their correlation with diversity can be positive, negative or humped. When multiple resources are limited, however, species richness increases in line with resource ratio theory. Another major cause of patterns in species richness is heterogeneity, whether abiotic or biotic, spatial or temporal. Diversity begets diversity—the more species there are, the greater the complexity of the habitat and the more can be supported. This does little to explain where the original diversity comes from (the mystery 75% from regional causes) but helps to understand how massively diverse habitats such as tropical rain forests or coral reefs hold so many species. Finally, internal top-down forces can permit stable coexistence by controlling the populations of dominant competitors which would otherwise take over. These processes can include predation, herbivory, parasites and diseases. In the special case when these are spatially restricted in their effects, they are known as Janzen–Connell effects. Abiotic factors such as disturbance can also play a part in enabling coexistence so long as they satisfy the rule of disproportionately affecting the most abundant or competitively dominant species and therefore have a stabilising effect.

8.8.1 Recommended reading

Armbrecht, I., I. Perfecto, and J. Vandermeer, 2004. Enigmatic biodiversity correlations: ant diversity responds to diverse resources. *Science* 304:284–286.

Chesson, P., 2000. Mechanisms of maintenance of species diversity. *Annual Review of Ecology, Evolution, and Systematics* 31:343–366.

Fox, J., 2013. The intermediate disturbance hypothesis should be abandoned. *Trends in Ecology & Evolution* 28:86–92.

8.8.2 Questions for the future

- Can we form a unified theory to predict when species richness will increase or decrease with environmental change?
- How much heterogeneity is too much?
- If rare species are at an advantage then why are the majority of species rare?

References

Adler, F. R. and H. C. Muller-Landau, 2005. When do localized natural enemies increase species richness? *Ecology Letters* 8:438–447.

Adler, P. B., E. W. Seabloom, E. T. Borer, H. Hillebrand, Y. Hautier, A. Hector, W. S. Harpole, L. R. O'Halloran, J. B. Grace, T. M. Anderson, J. D. Bakker, L. A. Biederman, C. S. Brown, Y. M. Buckley, L. B. Calabrese, C.-J. Chu, E. E. Cleland, S. L. Collins, K. L. Cottingham, M. J. Crawley, E. I. Damschen, K. F. Davies, N. M. DeCrappeo, P. A. Fay, J. Firn, P. Frater, E. I. Gasarch, D. S. Gruner, N. Hagenah, J. H. R. Lambers, H. Humphries, V. L. Jin, A. D. Kay, K. P. Kirkman, J. A. Klein, J. M. H. Knops, K. J. L. Pierre, J. G. Lambrinos, W. Li, A. S. MacDougall, R. L. McCulley, B. A. Melbourne, C. E. Mitchell, J. L. Moore, J. W. Morgan, B. Mortensen, J. L. Orrock, S. M. Prober, D. A. Pyke, A. C. Risch, M. Schuetz, M. D. Smith, C. J. Stevens, L. L. Sullivan, G. Wang, P. D. Wragg, J. P. Wright, and L. H. Yang, 2011. Productivity is a poor predictor of plant species richness. *Science* 333:1750–1753.

Allouche, O., M. Kalyuzhny, G. Moreno-Rueda, M. Pizarro, and R. Kadmon, 2012. Area-heterogeneity tradeoff and the diversity of ecological communities. *Proceedings of the National Academy of Sciences of the United States of America* 109:17495–17500.

Armbrecht, I., I. Perfecto, and J. Vandermeer, 2004. Enigmatic biodiversity correlations: ant diversity responds to diverse resources. *Science* 304:284–286.

Biswas, S. R. and A. U. Mallik, 2010. Disturbance effects on species diversity and functional diversity in riparian and upland plant communities. *Ecology* 91:28–35.

Borer, E. T., E. W. Seabloom, D. S. Gruner, W. S. Harpole, H. Hillebrand, E. M. Lind, P. B. Adler, J. Alberti, T. M. Anderson, J. D. Bakker, L. Biederman, D. Blumenthal, C. S. Brown, L. A. Brudvig, Y. M. Buckley, M. Cadotte, C. Chu, E. E. Cleland, M. J. Crawley, P. Daleo, E. I. Damschen, K. F. Davies, N. M. DeCrappeo, G. Du, J. Firn, Y. Hautier, R. W. Heckman, A. Hector, J. HilleRisLambers, O. Iribarne, J. A. Klein, J. M. H. Knops, K. J. La Pierre, A. D. B. Leakey, W. Li, A. S. MacDougall, R. L. McCulley, B. A. Melbourne, C. E. Mitchell, J. L. Moore, B. Mortensen, L. R. O'Halloran, J. L. Orrock, J. Pascual, S. M. Prober, D. A. Pyke, A. C. Risch, M. Schuetz, M. D. Smith, C. J. Stevens, L. L. Sullivan, R. J. Williams, P. D. Wragg, J. P. Wright, and L. H. Yang, 2014. Herbivores and nutrients control grassland plant diversity via light limitation. *Nature* 508:517–520.

Cáceres, C. E., 1997. Temporal variation, dormancy and coexistence: a field test of the storage effect. *Proceedings of the National Academy of Sciences of the United States of America* 94:9171–9175.

Chesson, P., 2000. Mechanisms of maintenance of species diversity. *Annual Review of Ecology, Evolution, and Systematics* 31:343–366.

Connell, J. H., 1971. On the role of natural enemies in preventing competitive exclusion in some marine animals and in rain forest trees. In P. J. den Boor and G. R. Gradwell, editors, *Dynamics of Populations, Proceedings of the Advanced Study Institute in Dynamics of Numbers in Populations, Oosterbeck*, pages 298–310. Centre for Agricultural Publishing and Documentation.

Connell, J. H., 1978. Diversity in tropical rainforests and coral reefs. *Science* 199:1302–1310.

Crawley, M. J., A. E. Johnston, J. Silvertown, M. Dodd, C. de Mazancourt, M. S. Heard, D. F. Henman, and G. R. Edwards, 2005. Determinants of species richness in the Park Grass Experiment. *American Naturalist* 165:179–192.

Cusens, J., S. D. Wright, P. D. McBride, and L. N. Gillman, 2012. What is the form of the productivity—animal-species-richness relationship? A critical review and meta-analysis. *Ecology* 93:2241–2252.

Dinnage, R., M. W. Cadotte, N. M. Haddad, G. M. Crutsinger, and D. Tilman, 2012. Diversity of plant evolutionary lineages promotes arthropod diversity. *Ecology Letters* 15:1308–1317.

Ferger, S. W., M. Schleuning, A. Hemp, K. M. Howell, and K. Böhning-Gaese, 2014. Food resources and vegetation structure mediate climatic effects on species richness of birds. *Global Ecology and Biogeography* 23:541–549.

Fox, J. W., 2013. The intermediate disturbance hypothesis should be abandoned. *Trends in Ecology & Evolution* 28:86–92.

Grime, J. P., 1973a. Competitive exclusion in herbaceous vegetation. *Nature* 242:344–347.

Grime, J. P., 1973b. Control of species density on herbaceous vegetation. *Journal of Environmental Management* 1:151–167.

Harpole, W. S. and D. Tilman, 2007. Grassland species loss resulting from reduced niche dimension. *Nature* 446:791–793.

Hautier, Y., P. A. Niklaus, and A. Hector, 2009. Competition for light causes plant biodiversity loss after eutrophication. *Science* 324:636–638.

Hughes, A. R., J. E. Byrnes, D. L. Kimbro, and J. J. Stachowicz, 2007. Reciprocal relationships and potential feedbacks between biodiversity and disturbance. *Ecology Letters* 10:849–864.

Hutchinson, G. E., 1961. The paradox of the plankton. *American Naturalist* 95:137–145.

Janzen, D. H., 1970. Herbivores and the number of tree species in tropical forests. *American Naturalist* 104:501–528.

Komyakova, V., P. L. Munday, and G. P. Jones, 2013. Relative importance of coral cover, habitat complexity and diversity in determining the structure of reef fish communities. *PLoS ONE* 8:e83178.

Letten, A. D., M. B. Ashcroft, D. A. Keith, J. R. Gollan, and D. Ramp, 2013. The importance of temporal climate variability for spatial patterns in plant diversity. *Ecography* 36:1341–1349.

Liu, X., M. Liang, R. S. Etienne, Y. Wang, C. Staehelin, and S. Yu, 2012. Experimental evidence for a phylogenetic Janzen-Connell effect in a subtropical forest. *Ecology Letters* 15:111–118.

Lubchenco, J., 1978. Plant species diversity in a marine intertidal community: importance of herbivore food preference and algal competitive abilities. *American Naturalist* 112:23–39.

Mackey, R. L. and D. J. Currie, 2001. The diversity-disturbance relationship: is it generally strong and peaked? *Ecology* 82:3479–3492.

Marhaver, K. L., M. J. A. Vermeij, F. Rohwer, and S. A. Sandin, 2013. Janzen-Connell effects in a broadcast-spawning Caribbean coral: distance-dependent survival of larvae and settlers. *Ecology* 94:146–160.

Moreno Saiz, J. C. and J. M. Lobo, 2008. Iberian–Balearic fern regions and their explanatory variables. *Plant Ecology* 198:149–167.

Norghauer, J. M., J. Grogan, J. R. Malcolm, and J. M. Felfili, 2010. Long-distance dispersal helps germinating mahogany seedlings escape defoliation by a specialist caterpillar. *Oecologia* 162:405–412.

Packer, A. and K. Clay, 2000. Soil pathogens and spatial patterns of seedling mortality in a temperate tree. *Nature* 404:278–281.

Santelices, B., 2007. The discovery of kelp forests in deep-water habitats of tropical regions. *Proceedings of the National Academy of Sciences of the United States of America* 104:19163–19164.

Schmitz, O. J., 2008. Effects of predator hunting mode on grassland ecosystem function. *Science* 319:952–954.

Secor, D. H., 2007. The year-class phenomenon and the storage effect in marine fishes. *Journal of Sea Research* 57:91–103.

Shimadzu, H., M. Dornelas, P. A. Henderson, and A. E. Magurran, 2013. Diversity is maintained by seasonal variation in species abundance. *BMC Biology* 11:98.

Siepielski, A. M. and M. A. McPeek, 2010. On the evidence for species coexistence: a critique of the coexistence program. *Ecology* 91:3153–3164.

Silliman, B. R., M. W. McCoy, C. Angelini, R. D. Holt, J. N. Griffin, and J. van de Koppel, 2013. Consumer fronts, global change, and runaway collapse in ecosystems. *Annual Review of Ecology, Evolution, and Systematics* 44:503–538.

Sousa, M. E., 1979a. Disturbance in marine intertidal boulder fields: the nonequilibrium maintenance of species diversity. *Ecology* 60:1225–1239.

Sousa, M. E., 1979b. Experimental investigation of disturbance and ecological succession in a rocky intertidal algal community. *Ecological Monographs* 49:227–254.

Stein, M., K. Gerstner, and H. Kreft, 2014. Environmental heterogeneity as a universal driver of species richness across taxa, biomes and spatial scales. *Ecology Letters* 17:866–880.

Stevens, R. D., M. M. Gavilanez, J. S. Tello, and D. A. Ray, 2012. Phylogenetic structure illuminates the mechanistic role of environmental heterogeneity in community organization. *Journal of Animal Ecology* 81:455–462.

Tews, J., U. Brose, V. Grimm, K. Tielbörger, M. C. Wichmann, M. Schwager, and F. Jeltsch, 2004. Animal species diversity driven by habitat heterogeneity/diversity: the importance of keystone structures. *Journal of Biogeography* 31:79–92.

Waide, R. B., M. R. Willig, C. F. Steiner, G. Mittelbach, L. Gough, S. I. Dodson, G. P. Juday, and R. Parmenter, 1999. The relationship between productivity and species richness. *Annual Review of Ecology, Evolution, and Systematics* 30:257–300.

Warner, R. R. and P. L. Chesson, 1985. Coexistence mediated by recruitment fluctuations—a field guide to the storage effect. *American Naturalist* 125:769–787.

CHAPTER 9

Does diversity matter?

9.1 The big question

Why should we care about species? The question is one which might baffle the average environmentally conscious person, and faced with a global extinction crisis, the understandable reaction is to worry about the effects this might have on natural systems. Of course extinctions matter! But most species are not big cats or appealing primates, rather obscure insects and other creepy-crawlies that will disappear before they have even been named. The passing of tigers would be mourned by the world, but should the loss of a nondescript weevil inspire the same passion? In seeking to protect these species, we are likely to invoke justifications such as their potential commercial value in the production of new drugs, crops or biological control agents or their contribution to ecosystem functions and services. However, do we really understand what role the number of species plays in effective ecosystem functioning? Does having all the species matter?

With the foundation in January 2013 of the Intergovernmental Platform on Biodiversity and Ecosystem Services (IPBES), the world is beginning to focus greater attention on the changes occurring in natural systems across the globe. Making the case for biological diversity is more necessary than ever, and a firm understanding of the scientific basis behind concerns is essential. While it would make sense to try to save everything, the precautionary principle is not the most compelling argument when faced with the competing demands of economic development and human livelihoods. Firm evidence is required that all these species are necessary.

9.2 Ecosystems

Species come together to form communities with a composition often determined by the abiotic environment. These communities ultimately begin to influence the environment itself, at which point we refer to them as ecosystems. Their effects can include changes to the regional climate, perhaps caused by clouds formed through transpiration or differences in solar reflectance by vegetation. Water that falls as rain can be retained in the ground rather than draining swiftly into rivers. Roots break up rocks and help form soil, and the essential chemistry of life leads to cycling of carbon, nitrogen and other elements.

There is insufficient space in this book to discuss the manifold ways in which life both responds to and shapes both climate and geochemistry; it forms an entire discipline, environmental science, and there are other excellent introductions to this (e.g. Chapin et al., 2012). We should however be aware of two sets of processes, known as **ecosystem functions** and **services**. The former refers to the movement or transformation of abiotic materials. The latter are those processes which have direct implications for supporting and sustaining human life; some are outlined in Fig. 9.1.

Natural Systems: The organisation of life, First Edition. Markus P. Eichhorn.

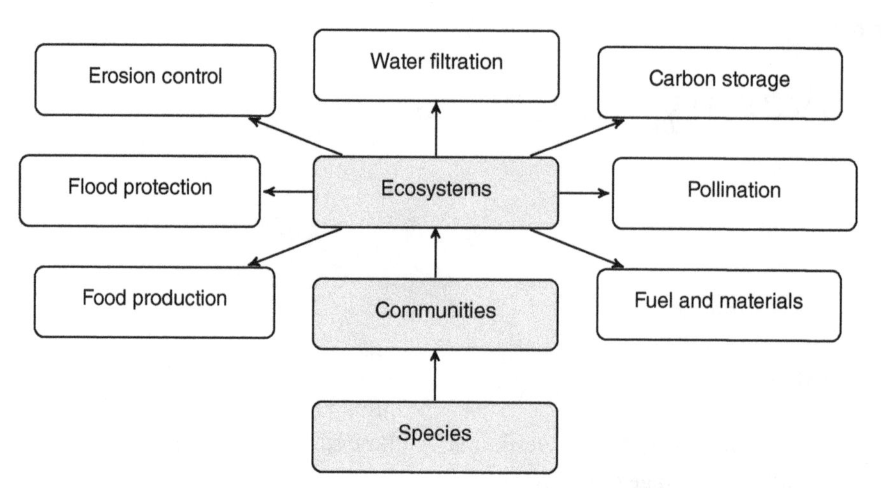

Figure 9.1 Examples of ecosystem services. Components in the hierarchy of nature are highlighted in grey; the ecosystem scale is the one at which services are measured.

Ecosystems therefore play crucial roles. We often depend on the structures of natural systems to protect us, as with mangrove forests shielding coastal communities from storms and tsunamis; their conversion to shrimp farms was a contributor to deaths and damage in Southeast Asia following the tsunami of 2004. Vegetation also serves to clean and filter water and protect against flooding. China is currently reforesting large parts of its mountains in response to the finding that denuded slopes lead to greater run-off and tragedy further downstream. In the United States, removal of the oyster beds of Chesapeake Bay destroyed a natural means of water filtration that would perhaps negate the need for expensive sewage treatment. Due to the role played by deforestation in carbon emissions and climate change, international agreements are now seeking to place a price on the amount of carbon in forest trees and soils.

Set against a background of local and regional extinctions, we have to ask—what will this mean for these ecosystem functions and services? Four possible responses are commonly assumed (Fig. 9.2). It could be that all species are equal in value, and extinctions will lead to a linear decline in ecosystem functioning. This seems unlikely, however, as species are not identical. We might therefore see cases where some species are redundant and there is little effect on ecosystems until extinctions reach a certain level. More troubling still is the catastrophic scenario: extinctions may have no discernable impact up to a certain point, following which a dramatic collapse takes place, perhaps with little prior warning, much like a game of Jenga®. Finally, if each species makes a unique but variable contribution, there could be an idiosyncratic response as links and dependencies are broken down.

You should look upon these models with scepticism. These dire scenarios are all informed by the founding assumption that loss of biodiversity is A Bad Thing. Imagine each of these trends inverted—why shouldn't it be the case that ecosystem functions improve with reduced species richness? After all, the history of agriculture has been a gradual process of removing ever more species, of pests and weeds, to focus on highly productive monocultures that optimise the ecosystem service of food production. You seldom see farmers argue that they need more species in their fields to boost yields, so why do we expect the same of natural processes? There are even cases in nature where functions are maximised at low species richness (Bracken and Williams, 2013). We need to challenge our preconceptions and look closely at the evidence.

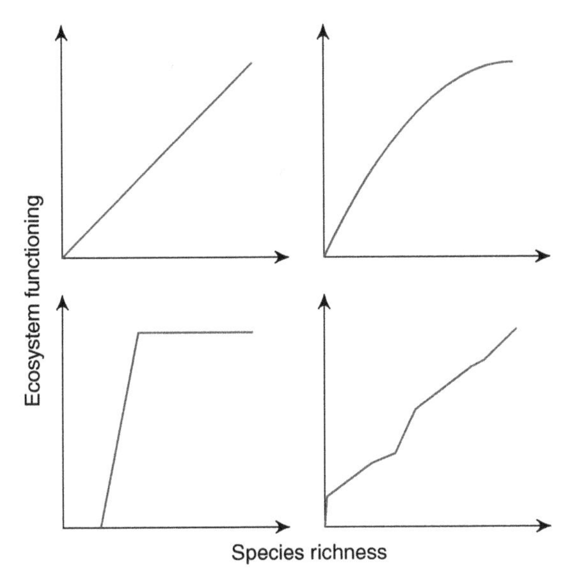

Figure 9.2 Possible relationships between species richness and ecosystem function.

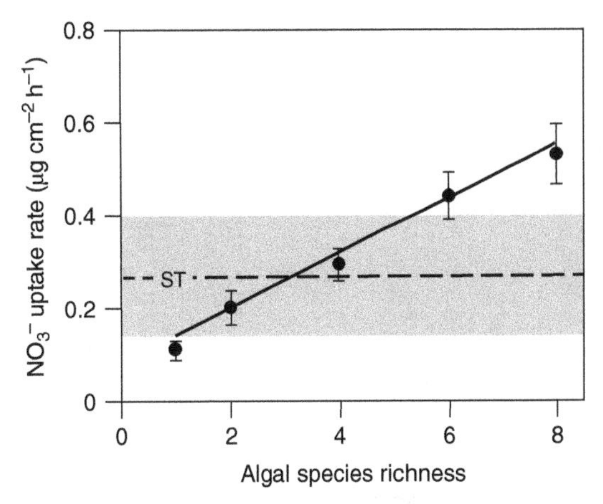

Figure 9.3 Algal species richness effects on nitrate uptake in streams with heterogeneous environments; mean±SE; the horizontal line and grey shaded area show mean±SE for the best-performing species in monoculture. (*Source*: Cardinale (2011). Reproduced with permission from Nature Publishing Group.)

9.3 What shape is the relationship?

One of many ecosystem functions that also provides a service to people is control of nutrient levels in waterways. Nitrogen pollution has become a major problem in many freshwater systems, but this nitrogen can be taken up by algae. When more species of algae are present, they can be more efficient at stripping the excess nutrients from the water. This was the basis of a study by Cardinale (2011), who set up model streams and ran them through varying numbers of algal species, monitoring how much nitrate they absorbed. More species increased the rate of this process (Fig. 9.3), though only when the streams had some built-in environmental heterogeneity, as each species depended on slightly different rates of water flow and disturbance. In fact, the most diverse systems had greater rates of uptake than even the best-performing species grown on its own, implying that the effect of combining species is more than additive.

We should be wary of inferring too much from this though; it's easy to make a line from only five points. What about when there are literally thousands of species? This brings us back to the plankton and specifically the photosynthesising phytoplankton, which are responsible for 50% of global net primary productivity (NPP). Plankton are hyperdiverse, with hundreds of species present in a few millilitres of sea water. There are also many different groups of species with slightly different roles and tolerances. Are they all necessary?

Ptacnik et al. (2008) sampled phytoplankton communities from oceans in Northern Europe and looked at their resource use efficiency (RUE), specifically the concentration of chlorophyll in the water (from all species combined) per unit phosphorus, which is the main limiting nutrient in these systems. Not only did the presence of more species increase RUE, but there was also less variability (Fig. 9.4). This finding has potential importance in applications such as pollution control. Rather than developing a single species of microbe to help clean up oil or chemical spills, it might be more efficient to provide a diversity of microbes that can work together.

These field observations are supported by controlled laboratory studies. McGrady-Steed et al. (1997) placed wheat seeds into aquatic microcosms, sealed flasks with a predetermined number of bacteria species present, and monitored the loss of seed

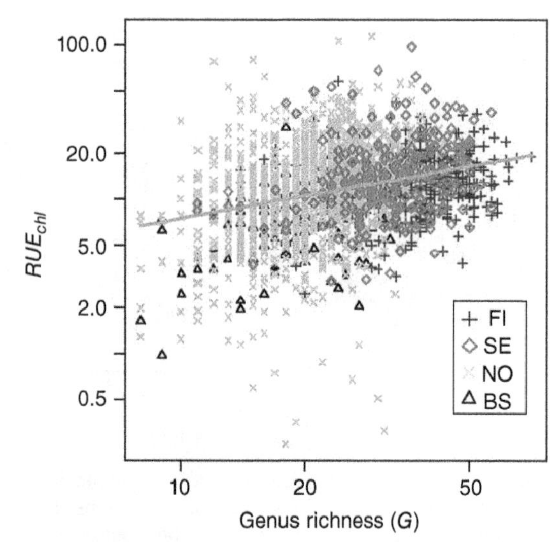

Figure 9.4 Resource use efficiency (measured as chlorophyll-*a* per unit phosphorus) by generic richness of phytoplankton in four regions, FI, Finland; SE, Sweden; NO, Norway; and BS, Baltic Sea. (*Source*: Ptacnik et al. (2008). Reproduced with permission from National Academy of Sciences, USA.)

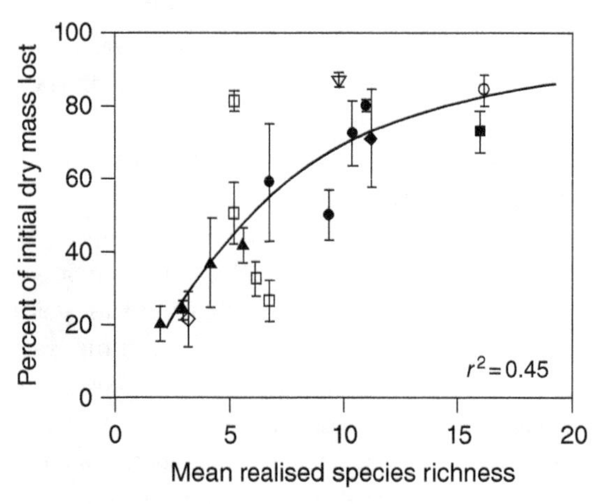

Figure 9.5 Mass loss of wheat seeds in bacterial microcosms with variable species richness. Communities of differing species composition but equal initial species richness plotted using the same symbol. (*Source*: McGrady-Steed et al. (1997). Reproduced with permission from Nature Publishing Group.)

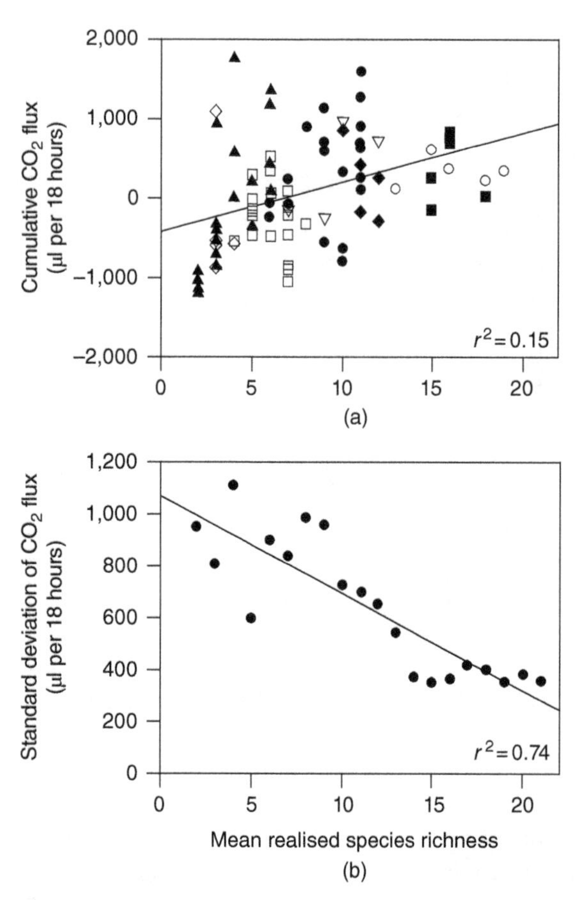

Figure 9.6 Carbon dioxide flux with realised species richness of bacterial microcosms after 6 weeks represented as (a) raw values, with communities of differing species composition but equal initial species richness plotted using the same symbol and (b) standard deviation of values. (*Source*: McGrady-Steed et al. (1997). Reproduced with permission from Nature Publishing Group.)

mass after 6 weeks. The selection of bacterial species was entirely random and included multiple trophic levels. Microcosms with more species managed to break down a greater proportion of the wheat seed (Fig. 9.5). This was a saturating relationship, with limited further increase in decomposition rates when there were more than 10 species.

The study also monitored the flux of carbon dioxide, a measure of the respiration of the entire bacterial community. This declined as species richness fell, but more importantly, the variation in values increased. At low numbers of species, the response became unpredictable, whereas with more than 13 species, it remained roughly steady (Fig. 9.6). This leads to an interesting possibility when it comes to understanding the link between species richness and ecosystem processes. It might not simply be the case that rates of processes go up or down when species are removed—they may simply become more unpredictable.

Why might this occur? Imagine the population of a single species through time (Fig. 9.7). If this species occurs on its own, then as its population rises and falls, the ecosystem function being measured will track it closely. If all species respond in the same way over time, perhaps to a changing environment, then their combined effect will simply be the sum across species and equally volatile. But if two species have slightly different responses, the mismatches in their behaviour will tend to smooth out the ecosystem function and with a large number of species rates will become more stable, even if the individual species act entirely randomly and independently of one another.

9.4 Field experiments

This is an important idea, and one worth testing in an experimental system where more information is available about the constituent species than for randomly selected bacteria. There have been two major international projects undertaking such research, both of which have focussed on plants, since these are the primary producers on land and their diversity tends to influence all the organisms that live on and around them.

At Cedar Creek, Minnesota, a large grassland was divided up into 207 plots across four fields into which various numbers of species were added as seed. For any single species, the variability in their biomass actually increased with total species richness, though the relationship was rather weak, explaining only around 2% of variation in the data. The biomass of

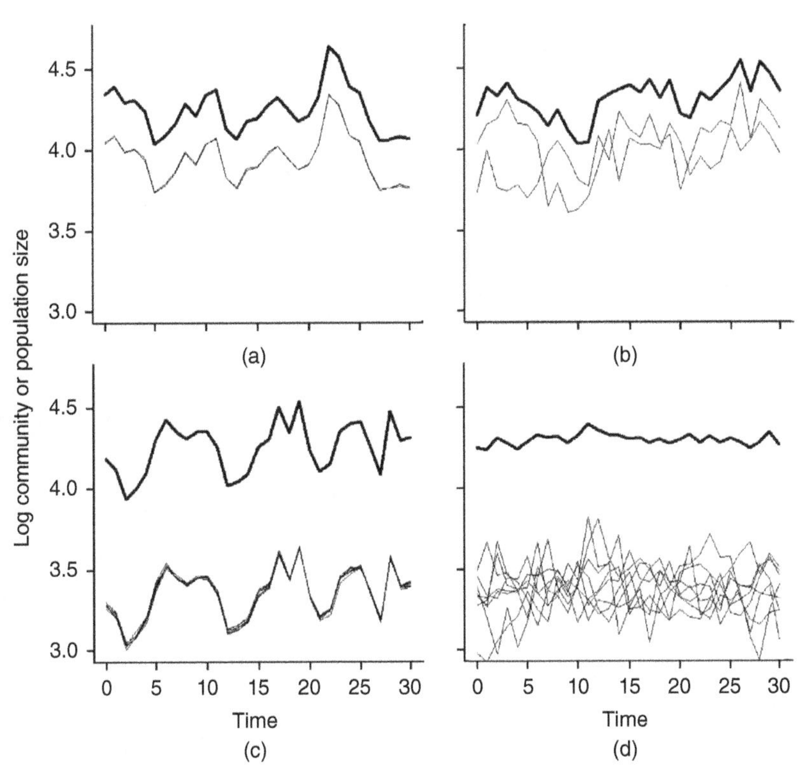

Figure 9.7 Effects on ecosystem processes of combining species with differing functional responses. Thin lines are levels provided by each individual species; dark lines are the total for the entire assemblage. Two species with (a) identical or (b) random responses or many species with (c) identical or (d) random responses. (*Source*: (Loreau, 2010) https://creativecommons.org/licenses/by/4.0/.)

individual species does not become more predictable when greater numbers of other species are present around them. But what about the biomass of the community as a whole? Here the opposite pattern was observed—more species made the system more predictable, a trend found in all four experimental fields, although not significantly so in one of them (Fig. 9.8). The individual components might behave with greater variability, but the community as a whole was more stable. This backs up the idea that more species tend to even out ecosystem-level processes.

A rare change in the weather in one year revealed an unexpected finding. Many scientists would have been appalled when a severe drought took hold, causing the grassland to dry out and die back. The team kept following their plots though and were able to look at the resilience of the communities to this dramatic event. Examining the ratio of biomass before and after the drought, they discovered that plots with more species recovered better than those with fewer (Fig. 9.9). Not only were plots with more species more stable, they were also more resilient.

The effects of diversity extended beyond the plants. Examining nitrogen levels in plant tissues, Zak et al. (2003) found that in plots with more plant species, the foliage had a greater concentration of nitrogen. This was exciting because plants themselves are unable to fix nitrogen from the air; they depend on symbiotic microbes in the soil to create soluble forms

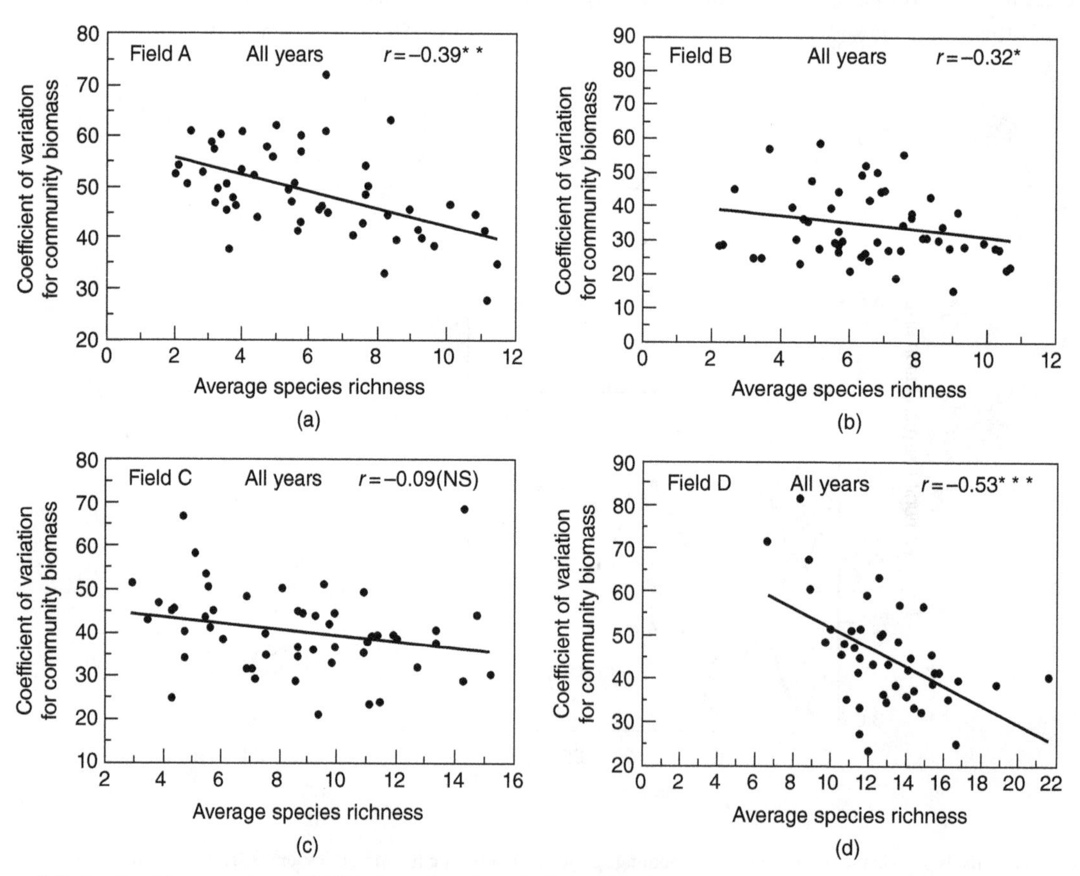

Figure 9.8 Species richness against variability in total community biomass across four grassland sites in Cedar Creek, MN. (*Source:* Tilman (1996). Reproduced with permission from Ecological Society of America.)

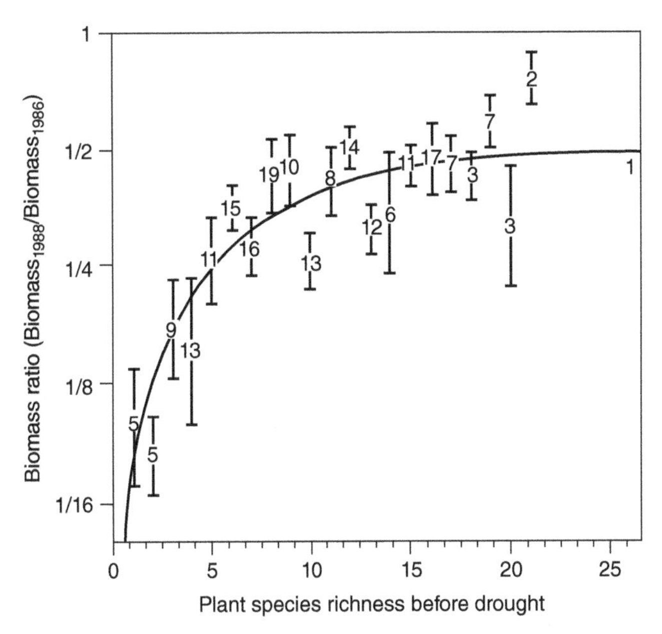

Figure 9.9 Biomass ratio before and after a drought (1986 vs. 1988) by number of species in the 1986 grassland community in Cedar Creek, MN (Tilman, 1996). Means ± SE, with numbers indicating the number of plots at each level of species richness. (*Source*: Tilman (1996). Reproduced with permission from Ecological Society of America.)

that they can use. Greater plant species richness was apparently having effects on other organisms below the ground.

Similar results were being uncovered in another study in Europe, the BIODEPTH[1] project. In this case there were eight sites across the continent, from Greece to Sweden, in which multiple plots had been sown with between 1 and 32 randomly selected species. The researchers then followed 11 different variables, including not merely biomass production but also resource usage (in terms of space, light and nitrogen) and decomposition rates (Spehn et al., 2005). This breadth of study was critical for the insights obtained.

The first finding was that total biomass production fell when fewer species were present (Fig. 9.10) and by roughly the same degree in all eight sites, despite differences in the overall amount of growth between them. There are two possible explanations for this pattern, which might be acting independently or in concert. One is that there is complementarity among species. By having niches that differ from one another, species are able to increase their net efficiency of resource use. The other is that facilitation could be taking place, with positive mutualistic interactions among plant species (and their microbial partners). Either of these might lead to overyielding, where the total community biomass exceeds that expected based on the sum of its parts. If you grew each species separately and then predicted the biomass of a mixture, it turned out to be an underestimate. This appeared at first sight to be a strong argument for preserving greater species richness in communities. Not everyone, however, was in agreement.

Some ecologists, on seeing the BIODEPTH results, were less than convinced. They pointed out that many of the plots contained legumes, plants from the family Fabaceae, which includes peas. Legumes differ from most other plant species in that they possess microbial symbionts that are able to fix nitrogen, which not only enhances their own growth

[1] Bio-Diversity and Ecological Processes in Terrestrial Herbaceous ecosystems.

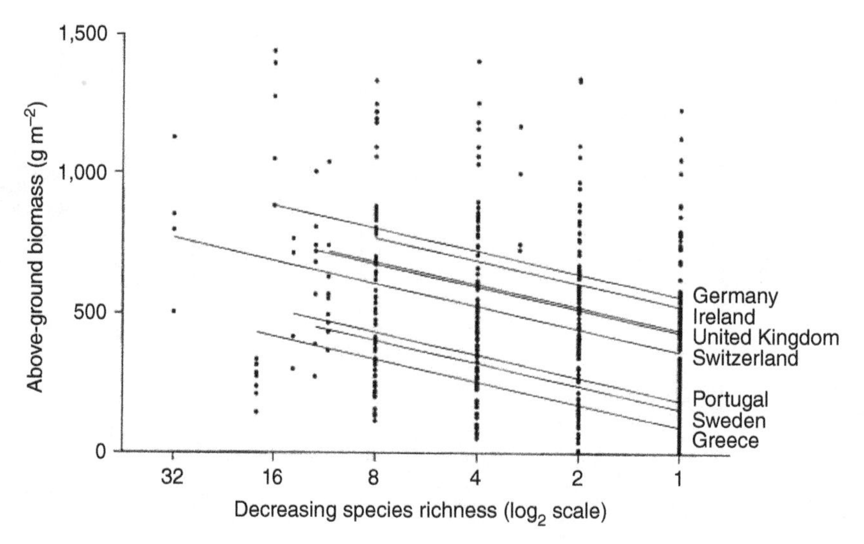

Figure 9.10 Relationship between above-ground biomass and plant species richness in grassland plots within the BIODEPTH project. (*Source*: Hector et al. (1999). Reproduced with permission from American Association for the Advancement of Science.)

but boosts the concentration of nitrogen in the soil and fertilises the entire community. This has been well known for many years; agricultural rotations in previous centuries recognised the beneficial effects of growing clover in a field for a fallow year, thereby increasing harvests in subsequent years, long before the role of microbes was even suspected. There is good evidence that overyielding is driven by interactions with organisms in the soil—including pathogens—rather than by the plants themselves (Hendriks et al., 2013).

Perhaps, in the BIODEPTH project, larger numbers of species simply meant that plots were more likely to contain a legume—a sampling artefact, which would mean that the observed overyielding with more species was the result of chance alone. This may sound like an inconsequential discussion, but lined up on each side were some of the greatest names in modern ecology. In support of overyielding stood notable scientists including Andy Hector, Michel Loreau, John Lawton and David Tilman, while set against them and pointing at the peas were figures such as Lonnie Aarssen, Phil Grime and John Vandermeer. These are all names you are likely to encounter again in your wider reading, and the dispute is ongoing.

Overyielding is not restricted to grassland plots with legumes though. As seen in Fig. 9.3 it happens when combining algal species in streams (Cardinale, 2011), and the same effect can be found in forests across Europe—wood production is 24% higher in mixed forests than when one tree species is grown alone (Vilà et al., 2013). The phenomenon of apparent complementarity requires a better explanation than pure chance alone. A species by itself is on average less productive than a mixture. Note however that transgressive overyielding, where the mixture performs better than the best single species alone, is extremely rare (Turnbull et al., 2013). In the majority of systems the fastest-growing species always outperforms any mixture when grown alone.

How large is the impact of species richness on productivity? This is an important question, as there are many factors that can shape plant growth, including addition of water or CO_2, stress such as drought or removal of biomass by fire or herbivores. Based on the specific treatments applied within the Cedar Creek plots, Tilman et al. (2012) demonstrated that species richness was as important as any of

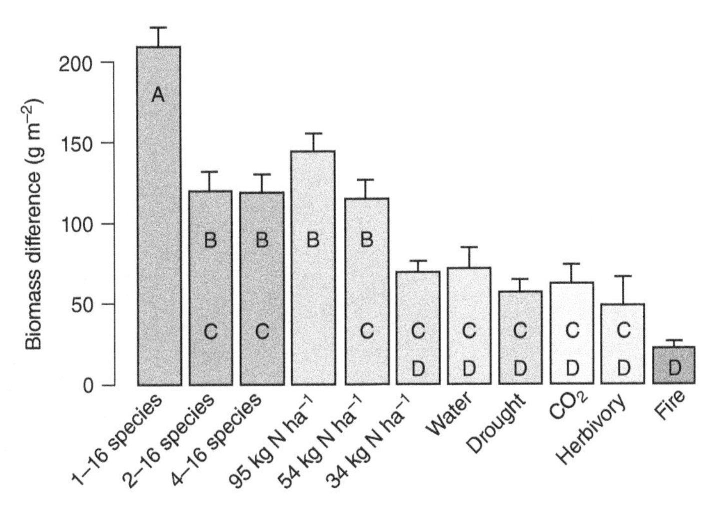

Figure 9.11 Relative influences of biotic and abiotic factors on grassland plant productivity in experimental plots, shown as biomass differences. Species richness contrasts are given as the difference between two levels of diversity. Bars with the same letter do not significantly differ; means ± SE. (*Source*: Tilman et al. (2012). Reproduced with permission from National Academy of Sciences, USA.)

these in driving greater productivity, if not more so (Fig. 9.11). At least within this highly controlled experimental system, and at the relatively small scales studied, species richness was one of the main determinants of productivity. There is uncertainty as to how far this might be extended though, as the effect only occurs at very small scales (0.04 ha) in forests, with environmental factors dominating in larger plots (Chisholm et al., 2013).

9.5 Other measures of diversity

So far this chapter has focussed exclusively on species richness, the most commonly used and widely understood measure of diversity. Chapter 5 showed however that there is more to diversity than just counts of species, and this can have implications for ecosystem functioning. Perhaps natural systems are organised with different units? One possibility is **functional groups**, sets of species with roughly the same roles in the community. This might offer a way of resolving the dispute over peas.

Hector et al. (1999) checked for this in the BIODEPTH project by splitting their plant species into three groups: grasses, nitrogen-fixing legumes and other forbs (herbaceous plants). They found that total community biomass closely correlated with the number of functional groups present and much more so than with the total number of species (Fig. 9.12). This fits with the assertion that legumes have a disproportionate influence but also suggests that it's not only about them. More recent studies have found that both species richness and functional group richness make separate contributions to the total yield (Marquand et al., 2009). These are complex and sophisticated analyses, however, and it is difficult to make a definitive judgement about which factor is more important.

Another approach has suggested something more subtle: it may be that the evolutionary relatedness of the species is crucial. This is measured as phylogenetic diversity, the distance among species based on their relative positions on the evolutionary tree (a more refined version of Δ^+ from Section 5.7). Greater phylogenetic diversity correlates with total community biomass and stability much better than species richness alone and even better than the number of functional groups (Cadotte et al., 2008, 2012). Could this be evidence that species have evolved to be complementary when grown together? This is an intriguing possibility, and it does seem that the

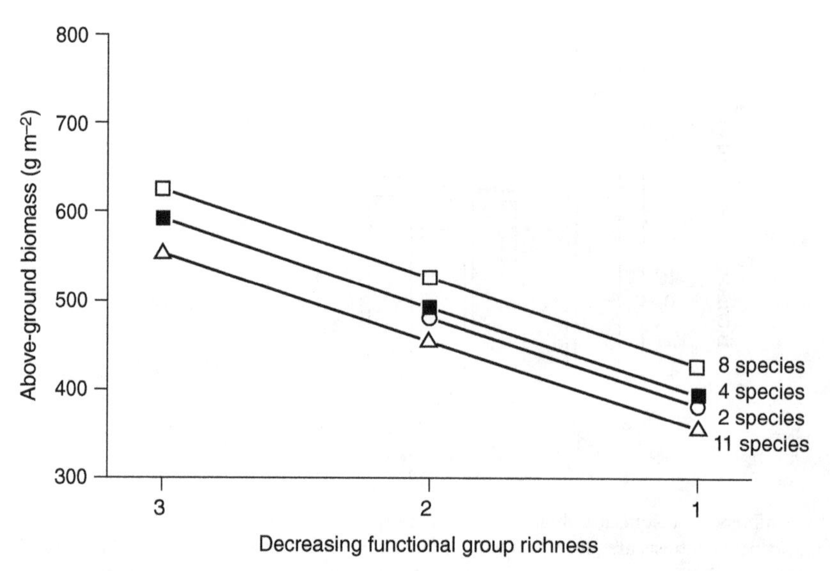

Figure 9.12 Relationship between above-ground biomass and plant functional group richness in the BIODEPTH project. (*Source:* Hector et al. (1999). Reproduced with permission from American Association for the Advancement of Science.)

more unrelated a set of species is, the greater the enhancement in productivity (Cadotte, 2013). The mechanism behind this must involve species traits, though it remains uncertain which ones.

Remember also that the abundances of species can matter at least as much as their total richness. For instance, on organic potato farms, greater evenness among the natural enemies of crop pests leads to higher yields, an effect that bears no relation to the species richness or identities of the species (Crowder et al., 2010). Whether such effects are widespread deserves further investigation.

9.6 Multifunctionality

Reviewing the link between diversity and ecosystem processes, Balvanera et al. (2006) compiled a vast number of studies and demonstrated that in almost all cases, the effects were positive (Fig. 9.13). This included measures of productivity, nutrient cycling, regulation of diversity and stability. The most important beneficial effects of greater diversity may not be shown by studies of single processes though but instead from seeing how communities respond across a wide range of measures. This was one of the major findings of the BIODEPTH project, which allowed for many different ecosystem processes to be studied simultaneously (Fig. 9.14). For any single process there was a saturating curve; in other words, some species were redundant but loss of greater numbers caused severe impacts. When summing the species involved in multiple processes, it turned out that the species contributing to each tended to overlap by only 20–50% (Hector and Bagchi, 2010). The value of diversity is in multifunctionality rather than in supporting any single process. A similar pattern was found for tree species richness in production forests by Gamfeldt et al. (2013), who strike a more cautionary note, as some ecosystem services may trade off against one another, so we should not expect greater species richness to increase all of them simultaneously.

A recognition of the multiple functions of diversity assumes particular importance in a fluctuating environment when individual species respond differently but their aggregate response has a stabilising effect. In this case few species are truly redundant, as all contribute to some processes at least some of the time.

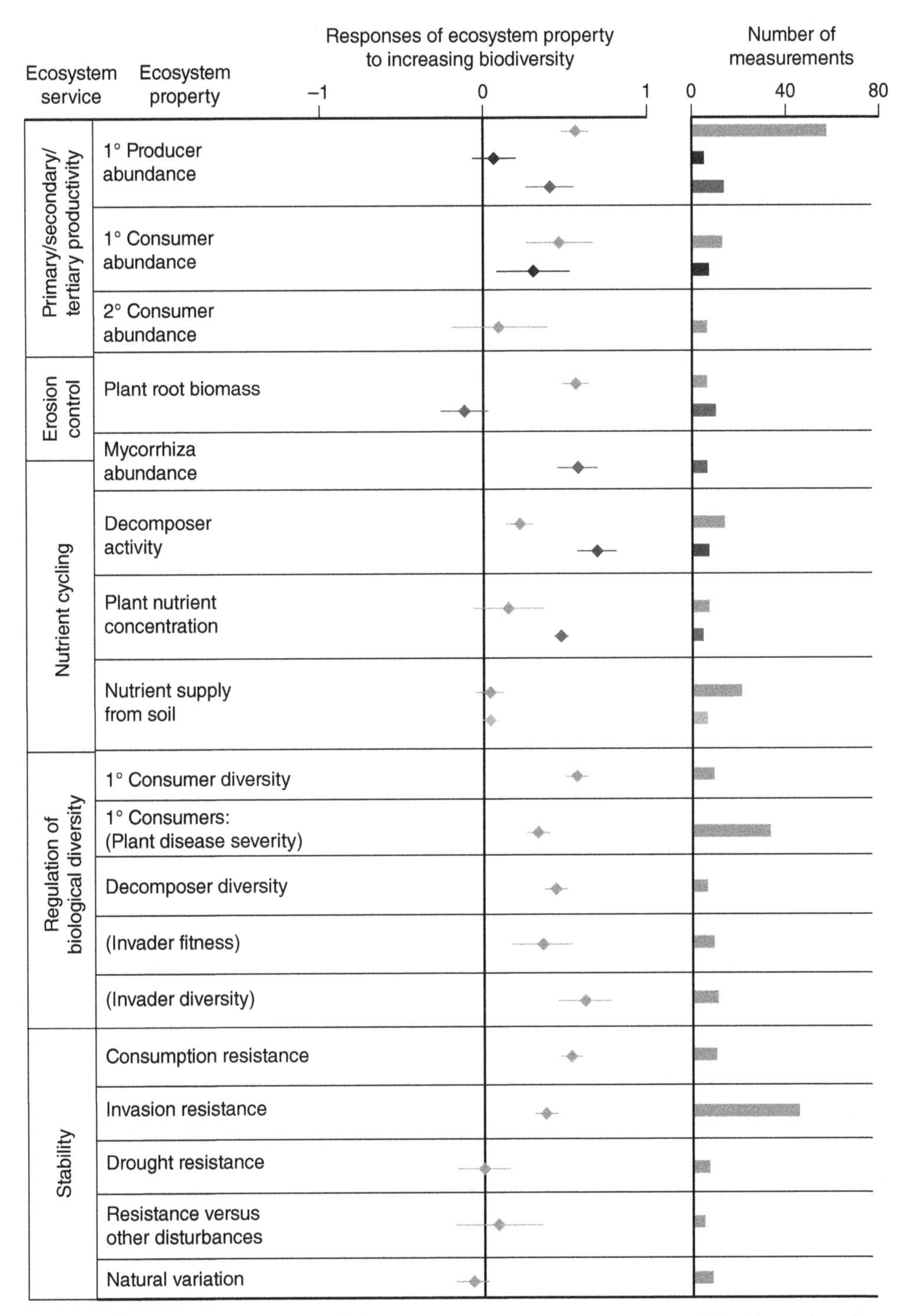

Figure 9.13 Magnitude and direction of effects of diversity on ecosystem processes (mean ± SE) with number of measurements available. Ecosystem properties shown in parentheses were considered of negative value for human well-being, and thus inverse effect sizes are shown. Coloured bars show trophic level manipulated: green, primary producers; blue, primary consumers; pink, mycorrhizae; brown, decomposers; grey, multiple levels. (*Source*: Balvanera et al. (2006). Reproduced with permission from John Wiley & Sons Ltd.) (*See colour plate section for the colour representation of this figure.*)

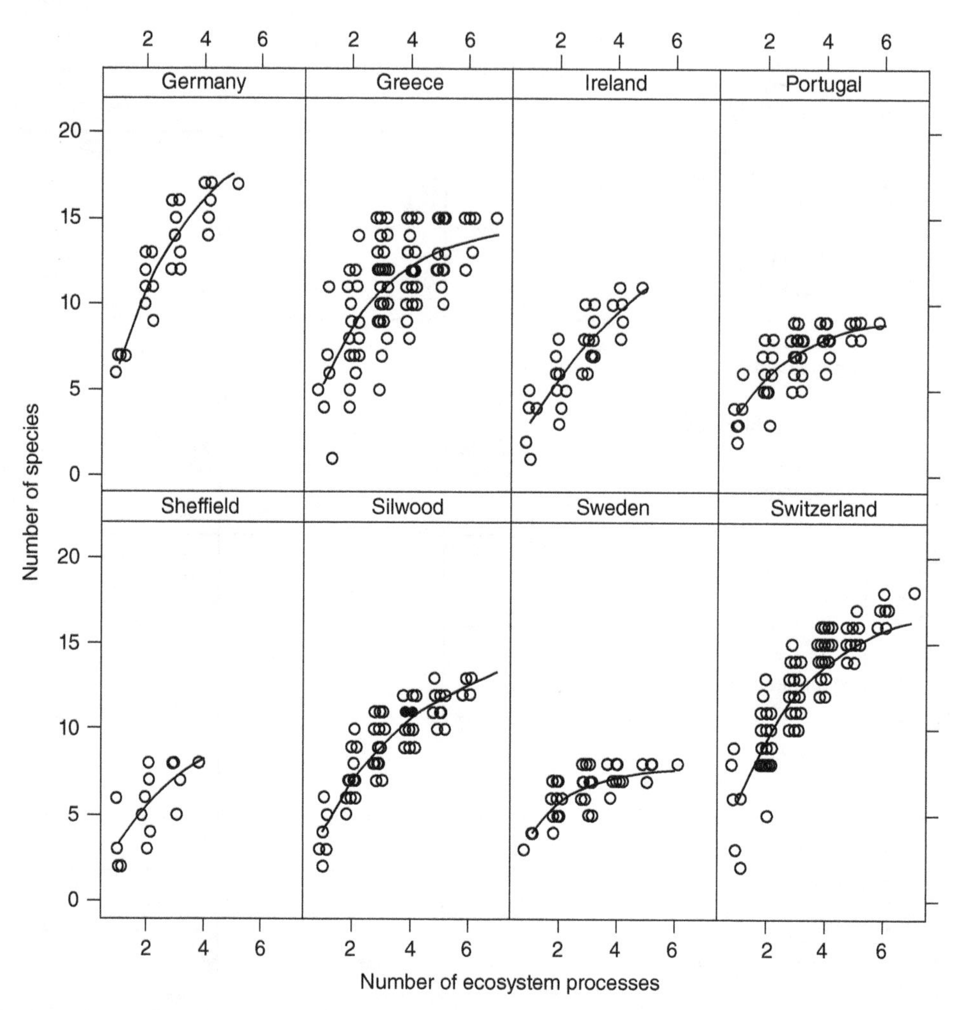

Figure 9.14 Increasing numbers of species are identified as influential when greater numbers of ecosystem processes are considered across eight European grassland sites in the BIODEPTH project. (*Source*: Hector and Bagchi (2007). Reproduced with permission from Nature Publishing Group.)

Over time the saturating effect of high species richness becomes less evident, causing the benefit of high species richness to increase the longer a study lasts (e.g. Reich et al., 2012).

Combining information from multiple experiments on grassland species suggests that, in fact, most species are important for maintaining ecosystem functions (Isbell et al., 2011). In 17 different experiments, covering 147 species, 84% were found to influence ecosystem functioning at least once, while the remainder may have done so if extra processes were considered. At any one time about 27% were involved in regulating any single process, regardless of the total number of species present. When combining through time, or space, or different functions, or in a fluctuating environment, it turns out that ever more species are involved in regulating ecosystem functions. This is a powerful argument for the principle that we should maintain high species richness even when some species might initially appear to be redundant.

9.7 The real world

Up to this point we have largely considered evidence from experimental systems that have been manipulated. These generate intriguing results, but in the real world the species richness and composition of communities are determined by natural processes rather than human design. What happens once we leave the laboratory or the field station and look at real assemblages of species?

A study by Bracken et al. (2008) points to the possible problems with relying on random groupings of species. On shores with up to seven species of seaweed, they examined rates of ammonium uptake from the water. The species varied considerably in their individual rates, and in randomly assembled communities there was little variation in the overall response. When compared to real communities, however, the researchers found that assemblages with fewer species took up less ammonium, as the species that were present in depauperate systems tended to be those with lower individual rates (Fig. 9.15).

This is concerning as the identity of species that go extinct is non-random. If experiments randomly choose which species to pull out of communities, they may underestimate the likely impacts of changes in species richness in the real world.

Bioturbation (mixing) of marine sediments provides another case study (Solan et al., 2004). This is a crucial process as it determines the concentration of oxygen at the ocean floor and hence the amount of biomass and productivity that can be supported. Within sea bed communities, and natural systems more generally, some species are always more likely to go extinct than others. Those at risk include species that are already rare or low in number and those with larger than average body size. Modelling the effects of extinctions based on a random sequence, or following body size or rarity, revealed some interesting patterns (Fig. 9.16). Random extinctions had little effect on average mixing depths, but the variability among assemblages increased—a pattern which should be familiar by now. Taking out species

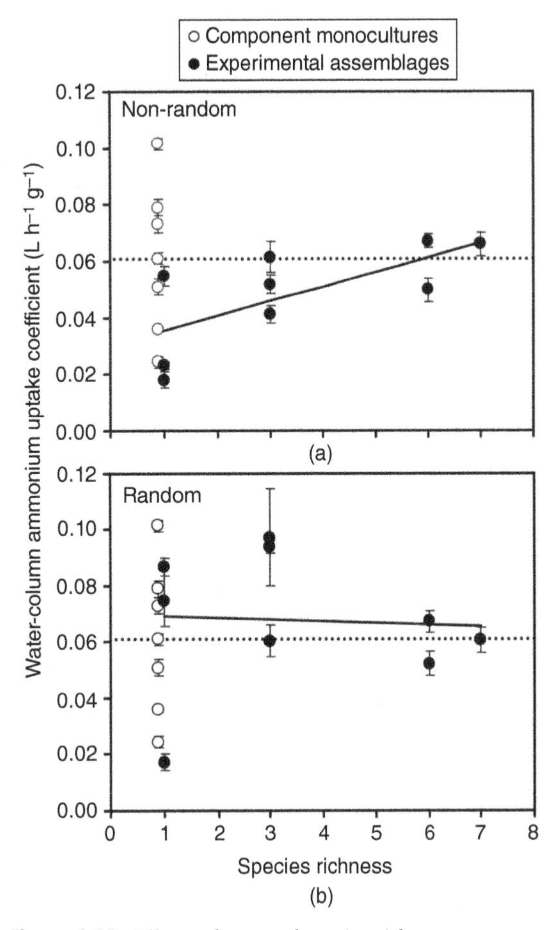

Figure 9.15 Effects of seaweed species richness on ammonium uptake with (a) non-random variation in species composition, as observed in nature versus (b) random composition. (*Source*: Bracken et al. (2008). Reproduced with permission from National Academy of Sciences, USA.))

on the basis of rarity had hardly any effect at all. Rare species had little impact on ecosystem processes and that it was only when the major players were removed that things began to change.

The most important finding was that removal by body size—a common pattern for extinctions—led to a steady decline in the process of biogenic mixing of sediments. The step change in the graph (also seen in the random patterns) occurs whenever the burrowing brittlestar, *Amphiura filiformis*, is lost from the community. This points back to something seen earlier: it's not simply the number of species

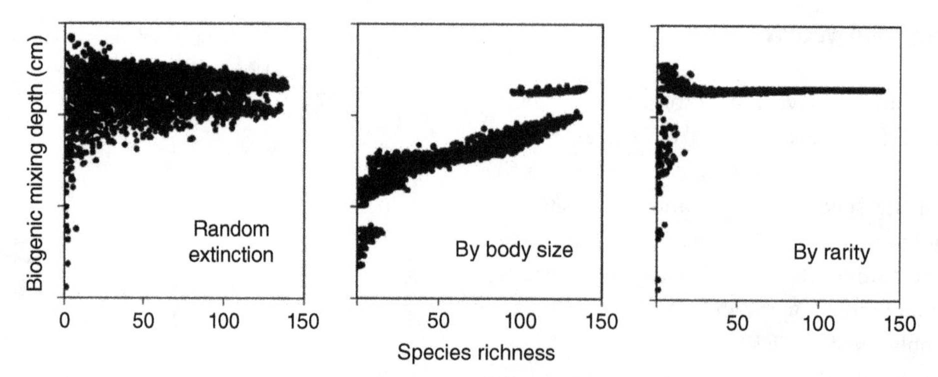

Figure 9.16 Simulated effects on biogenic mixing depth of removal of species based on differing criteria. (*Source*: Solan et al. (2004). Reproduced with permission from American Association for the Advancement of Science.)

that matters but the identity of those species, and some are more important for any single ecosystem function than others. The same was found in the BIODEPTH experiments, where clover had a disproportionate effect on total yields (Hector et al., 1999). This issue will be discussed further in the next chapter in the context of keystone species.

These studies and others suggest an unnerving conclusion. Natural systems are often composed of species that have evolved ways to coexist stably through mechanisms that might include niche complementarity and facilitation. Compared to experimental systems with random combinations of species, does this mean that extinctions are likely to have an even greater impact in the real world? Some authors certainly think so—Flombaum and Sala (2008) suggest that the effect of real changes in species richness on primary productivity in their study of plants on the Patagonian Steppe was greater in magnitude than in previous experiments. There is much more work to be done before we can be sure of this.

It's worth ending on a more sceptical note though. Vellend et al. (2013) collected information on the species richness of plants in 16,000 plots across the world that had been monitored over 5–261 years and found that local species richness was not going down—if anything, it had increased slightly on average. This does mask profound local changes, as no net change implies that there were as many places where species richness went down as up, but it does call into question the assumption that loss of species globally is the same as locally. Their analysis did not, however, include the impacts of land-use change. Where natural habitats are converted to anthropogenic ones, species richness is almost certain to have fallen considerably.

9.8 species richness and productivity

We can now further modify our model of local species richness patterns to include one more relationship (Fig. 9.17). Having previously decided in Chapter 7 that NPP doesn't have a consistent effect on species richness, the opposite has emerged: local species richness can be a driver of NPP. If this seems paradoxical then consider the ways in which the two possible directions of effects were assessed. Comparing multiple sites with different NPP didn't show a common relationship with species richness, but manipulative experiments within a single community demonstrate that in any single place species richness is a determinant of NPP.

There is scope for debate here as the majority of the evidence for species richness influencing productivity comes from primary producers in controlled experiments at relatively small spatial scales. It is also not intuitively obvious whether the same would apply to

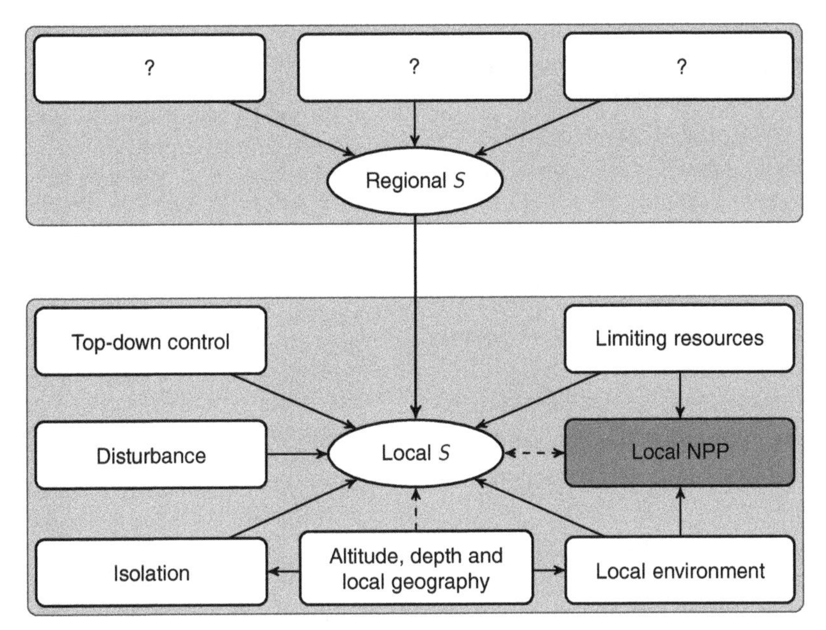

Figure 9.17 Determinants of local species richness *S* and Net Primary Productivity NPP within a given area. Direct links with solid lines, indirect with dashed lines. Determinants of regional species richness will be resolved in later chapters.

animals—would a greater species richness of herbivores or carnivores have any impact on productivity? For this reason the link remains tentative.

9.9 Conclusions

Ecosystems perform valuable services, and a consensus has formed regarding the importance of species richness in maintaining these (Cardinale et al., 2012). Based on the evidence to date, we can say that:

- Reduced species richness causes declines in ecosystem function
- More species increase stability in these functions
- The link between species richness and ecosystem functioning is non-linear and saturating so that extinctions have an ever-increasing impact
- Diverse systems are more likely to contain key species or a range of functional traits which support functions
- Evidence from phylogenetic diversity indicates that there are evolved complementarities among species

The impacts of diversity loss on ecosystem functions are at least as great as other drivers of environmental stress, such as climate change, nitrogen pollution or biotic interactions (Hooper et al., 2012; Tilman et al., 2012). Moreover, the impacts of species loss increase through time, and to maintain multiple services in the long term requires many more species than any single process at any given moment. When it comes to predicting the likely effects of extinctions, in the real world their impact could be even greater than simple experiments suggest. The significance of these debates for the future of the natural world accounts for why this has become one of the hottest topics in ecology.

9.9.1 Recommended reading

Cardinale, B. J., J. E. Duffy, A. Gonzalez, D. U. Hooper, C. Perrings, P. Venail, A. Narwani, G. M. Mace, D. Tilman, D. A. Wardle, A. P. Kinzig, G. C. Daily, M. Loreau, J. B. Grace, A. Larigauderie, D. S. Srivastava, and S. Naeem, 2012. Biodiversity loss and its impact on humanity. *Nature* 486:59–67.

Duffy, J. E. (2009). Why biodiversity is important to the functioning of real-world ecosystems. *Frontiers in Ecology and the Environment* 7:437–444.

Naeem, S., J. E. Duffy, and E. Zavaleta, 2012. The functions of biological diversity in an age of extinction. *Science* 336:1401–1406.

9.9.2 Questions for the future

- Do manipulative experiments over- or underestimate the effects of species richness on ecosystem processes?
- Should agricultural systems be designed to maximise production or multifunctionality, and what implications might this have for the services we require of them?
- What matters more—species richness, identities or their relative abundance?

References

Balvanera, P., A. B. Pfisterer, N. Buchmann, J.-S. He, T. Nakashizuka, D. Raffaelli, and B. Schmid, 2006. Quantifying the evidence for biodiversity effects on ecosystem functioning and services. *Ecology Letters* 9:1146–1156.

Bracken, M. E. S., S. E. Friberg, C. A. Gonzalez-Dorantes, and S. L. Williams, 2008. Functional consequences of realistic biodiversity changes in a marine ecosystem. *Proceedings of the National Academy of Sciences of the United States of America* 105:924–928.

Bracken, M. E. S. and S. L. Williams, 2013. Realistic changes in seaweed biodiversity affect multiple ecosystem functions on a rocky shore. *Ecology* 94:1944–1954.

Cadotte, M. W., 2013. Experimental evidence that evolutionarily diverse assemblages result in higher productivity. *Proceedings of the National Academy of Sciences of the United States of America* 110:8996–9000.

Cadotte, M. W., B. J. Cardinale, and T. H. Oakley, 2008. Evolutionary history and the effect of biodiversity on plant productivity. *Proceedings of the National Academy of Sciences of the United States of America* 105:17012–17017.

Cadotte, M. W., R. Dinnage, and D. Tilman, 2012. Phylogenetic diversity promotes ecosystem stability. *Ecology* 93:S223–S233.

Cardinale, B. J., 2011. Biodiversity improves water quality through niche partitioning. *Nature* 472:86–89.

Cardinale, B. J., J. E. Duffy, A. Gonzalez, D. U. Hooper, C. Perrings, P. Venail, A. Narwani, G. M. Mace, D. Tilman, D. A. Wardle, A. Kinzig, G. C. Daily, M. Loreau, J. B. Grace, A. Larigauderie, D. S. Srivastava, and S. Naeem, 2012. Biodiversity loss and its impact on humanity. *Nature* 486:59–67.

Chapin, F. S. III, P. A. Matson, and P. M. Vitousek, 2012. *Principles of terrestrial ecosystem ecology*. Springer, second edition.

Chisholm, R. A., H. C. Muller-Landau, K. Abdul Rahman, D. P. Bebber, Y. Bin, S. A. Bohlman, N. A. Bourg, J. Brinks, S. Bunyavejchewin, N. Butt, H. Cao, M. Cao, D. Cárdenas, L.-W. Chang, J.-M. Chiang, G. Chuyong, R. Condit, H. S. Dattaraja, S. Davies, A. Duque, C. Fletcher, N. Gunatilleke, S. Gunatilleke, Z. Hao, R. D. Harrison, R. Howe, C.-F. Hsieh, S. P. Hubbell, A. Itoh, D. Kenfack, S. Kiratiprayoon, A. J. Larson, J. Lian, D. Lin, H. Liu, J. A. Lutz, K. Ma, Y. Malhi, S. McMahon, W. McShea, M. Meegaskumbura, S. Mohd. Razman, M. D. Morecroft, C. J. Nytch, A. Oliveira, G. G. Parker, S. Pulla, R. Punchi-Manage, H. Romero-Saltos, W. Sang, J. Schurman, S.-H. Su, R. Sukumar, I.-F. Sun, H. S. Suresh, S. Tan, D. Thomas, S. Thomas, J. Thompson, R. Valencia, A. Wolf, S. Yap, W. Ye, Z. Yuan, and J. K. Zimmerman, 2013. Scale-dependent relationships between tree species richness and ecosystem function in forests. *Journal of Ecology* 101:1214–1224.

Crowder, D. W., T. D. Northfield, M. R. Strand, and W. E. Snyder, 2010. Organic agriculture promotes evenness and natural pest control. *Nature* 466:109–112.

Flombaum, P. and O. E. Sala, 2008. Higher effect of plant species diversity on productivity in natural than artificial systems. *Proceedings of the National Academy of Sciences of the United States of America* 105:6087–6090.

Gamfeldt, L., T. Snäll, R. Bagchi, M. Jonsson, L. Gustafsson, P. Kjellander, M. Ruiz-Jaen, M. Fröberg, J. Stendahl, C. D. Philipson, G. Mikusiński, E. Andersson, B. Westerlund, H. Andrén, F. Moberg, J. Moen, and J. Bengtsson, 2013. Higher levels of multiple ecosystem services are found in forests with more tree species. *Nature Communications* 4:1340.

Hector, A. and R. Bagchi, 2010. Biodiversity and ecosystem multifunctionality. *Nature* 448:188–190.

Hector, A., B. Schmid, C. Beierkuhnlein, M. C. Caldeira, M. Diemer, P. G. Dimitrakopoulos, J. A. Finn, H. Freitas, P. S. Giller, J. Good, R. Harris, P. Högberg, K. Huss-Danell, J. Joshi, A. Jumpponen, C. Körner, P. W. Leadley, M. Loreau, A. Minns, C. P. H. Mulder, G. O'Donovan, S. J. Otway, J. S. Pereira, A. Prinz, D. J. Read, M. Scherer-Lorenzen, E.-D. Schulze, A.-S. D. Siamantziouras, E. M. Spehn, A. C. Terry, A. Y. Troumbis, F. I. Woodward, S. Yachi, and J. H. Lawton, 1999. Plant diversity and productivity experiments in European grasslands. *Science* 286:1123–1127.

Hendriks, M., L. Mommer, H. de Caluwe, A. E. Smit-Tiekstra, W. H. van der Putten, and H. de Kroon, 2013. Independent variations of plant and soil mixtures reveal soil feedback effects on plant community overyielding. *Journal of Ecology* 101:287–297.

Hooper, D. U., E. C. Adair, B. J. Cardinale, J. E. K. Byrnes, B. A. Hungate, K. L. Matulich, A. Gonzalez, J. E. Duffy, L. Gamfeldt, and M. I. O'Connor, 2012. A global synthesis reveals biodiversity loss as a major driver of ecosystem change. *Nature* 486:105–108.

Isbell, F., V. Calcagno, A. Hector, J. Connolly, W. S. Harpole, P. B. Reich, M. Scherzer-Lorenzen, B. Schmid, D. Tilman, J. van Ruijven, A. Weigelt, B. J. Wilsey, E. S. Zavaleta, and M. Loreau, 2011. High plant diversity is needed to maintain ecosystem services. *Nature* 477:199–202.

Loreau, M., 2010. Linking biodiversity and ecosystems: towards a unifying ecological theory. *Philosophical Transactions of the Royal Society Series B* 365:49–60.

Marquand, E., A. Weigelt, V. M. Temperton, C. Roscher, J. Schumacher, N. Buchmann, M. Fischer, W. W. Weisser, and B. Schmid, 2009. Plant species richness and functional composition drive overyielding in a six-year grassland experiment. *Ecology* 90:3290–3302.

McGrady-Steed, J., P. M. Harris, and P. J. Morin, 1997. Biodiversity regulates ecosystem predictability. *Nature* 390:162–165.

Ptacnik, R., A. G. Solimini, T. Andersen, T. Tamminen, P. Brettum, L. Lepistö, E. Willén, and S. Rekolainen, 2008. Diversity predicts stability and resource use efficiency in natural phytoplankton communities. *Proceedings of the National Academy of Sciences of the United States of America* 105:5134–5138.

Reich, P. B., D. Tilman, F. Isbell, K. Mueller, S. E. Hobbie, D. F. B. Flynn, and N. Eisenhauer, 2012. Impacts of biodiversity loss escalate through time as redundancy fades. *Nature* 336:589–592.

Solan, M., B. J. Cardinale, A. L. Downing, K. A. M. Engelhardt, J. L. Ruesink, and D. S. Srivastava, 2004. Extinction and ecosystem function in the marine benthos. *Science* 306:1177–1180.

Spehn, E. M., A. Hector, J. Joshi, M. Scherer-Lorenzen, B. Schmid, E. Bazeley-White, C. Beierkuhnlein, M. C. Caldeira, M. Diemer, P. G. Dimitrakopoulos, J. A. Finn, H. Freitas, P. S. Giller, J. Good, R. Harris, P. Hogberg, K. Huss-Danell, A. Jumpponen, J. Koricheva, P. W. Leadley, M. Loreau, A. Minns, C. P. H. Mulder, G. O'Donovan, S. J. Otway, C. Palmborg, J. S. Pereira, A. B. Pfisterer, A. Prinz, D. J. Read, E. D. Schulze, A. S. D. Siamantziouras, A. C. Terry, A. Y. Troumbis, F. I. Woodward, S. Yachi, and J. H. Lawton, 2005. Ecosystem effects of biodiversity manipulations in European grasslands. *Ecological Monographs* 75:37–63.

Tilman, D., 1996. Biodiversity: Population versus ecosystem stability. *Ecology* 77:350–363.

Tilman, D., P. B. Reich, and F. Isbell, 2012. Biodiversity impacts ecosystem productivity as much as resources, disturbance, or herbivory. *Proceedings of the National Academy of Sciences of the United States of America* 109: 10394–10397.

Turnbull, L. A., J. M. Levine, M. Loreau, and A. Hector, 2013. Coexistence, niches and biodiversity effects on ecosystem functioning. *Ecology Letters* 16:s116–s127.

Vellend, M., L. Baeten, I. H. Myers-Smith, S. C. Elmendorf, R. Beauséjour, C. D. Brown, P. De Frenne, K. Verheyen, and S. Wipf, 2013. Global meta-analysis reveals no net change in local-scale plant biodiversity over time. *Proceedings of the National Academy of Sciences of the United States of America* 110:19456–19459.

Vilà, M., A. Carrillo-Gavilàn, J. Vayreda, H. Bugmann, J. Fridman, W. Grodzki, J. Haase, G. Kunstler, M. Schelhaas, and A. Trasobares, 2013. Disentangling biodiversity and climatic determinants of wood production. *PLoS ONE* 8:e53530.

Zak, D. R., W. E. Holmes, D. C. White, A. D. Peacock, and D. Tilman, 2003. Plant diversity, soil microbial communities, and ecosystem function: are there any links? *Ecology* 84:2042–2050.

PART III
Communities

CHAPTER 10

Organisation at the community scale

10.1 The big question

An entire subfield of ecology is devoted to the study of communities, taking up substantial sections of major scientific journals and with dedicated specialist textbooks. You might therefore be surprised to learn that there are some distinguished ecologists who have disputed whether communities exist at all (e.g. Ricklefs, 2008), though these comments have provoked a strong backlash (Brooker et al., 2009). It seems like a strange source of controversy, but stems from a common problem within ecology, which is that once again we are faced with a term that is broadly used but often only vaguely defined.

Are communities real biological entities or merely study units created by ecologists as a matter of convenience? To confirm that they exist, evidence is required that they can be demarcated accurately and consistently in the field. This necessitates a rigorous quantitative definition. Having formed one, we can begin to investigate the internal structure of communities, their components, and how they are connected into food chains and webs, and ask what these features mean for ecosystem processes or temporal stability. What can investigations at the community scale reveal that other scales cannot?

10.2 Definitions

Community ecology suffers from a surfeit of ambiguous terminology, despite previous attempts to provide standardised definitions (e.g. Fauth et al., 1996). The conventional ecological definition of a community is a set of interacting species, occurring together in the same location, which is stable through space and time. The concept has been extended in recent years to include **metacommunities**, which are networks of communities linked by the dispersal of species among them (Holt et al., 2005), and will be considered in Chapter 13. The definition is distinct from that of an **ecosystem**, which includes interactions with the abiotic environment. The community scale is solely concerned with the biotic components and the relationships among them.

Another common term which has already appeared in previous chapters is **assemblage**. Usually this denotes a group of species that are found together but where there is insufficient evidence to categorically state that they form a community. If you take a sample of frogs from a peat swamp, you can say nothing more than that they were found together at the same time; whether they are interacting or in any sense stable in composition remains unknown. Often the literature is not as rigorous as it could be, and it is commonly assumed that any biological sample represents a community. This is less

Natural Systems: The organisation of life, First Edition. Markus P. Eichhorn.

of a problem for plants and other sessile organisms, where comprehensive surveys are possible, than for animals which move around continuously.

The community concept has a contested history in ecological theory, much like that of the niche, with two rival schools of thought. This dispute has now largely been resolved though the echoes still persist in folk perceptions of nature. The original idea put forward by Clements (1916) was that of a set of closely knit and interdependent species working together as a superorganism. The community, whether it be a stream bed or alpine grassland, was believed to be a synthetic unit composed of co-evolved elements that required one another to survive. This idea is equivalent to belief in the correct balance and unity of nature: that a natural system is a finely tuned engine whose components must all be present in order for it to function.

This was challenged by Gleason (1926), another plant ecologist, who had a much more individualistic view of species. Each has its own set of resource requirements and environmental tolerances (what we would call its niche), which determine where it is able to occur. Communities in Gleason's world were therefore contingent assemblies of whatever species happened to be able to survive in that location. For sure there would be interactions among species once they established, but these were not crucial for the community to form in the first place.

To cut a long story short, the latter individualistic view has come to dominate opinions among modern ecologists, though most would admit to accepting a mix of both views, with the precise blend depending on the system under study. In fact, while much has been made of this debate, the two ancestral figures differed in emphasis rather than being implacably opposed. In both cases the community exists as a level of organisation in nature rather than a fixed unit in space or time.

10.3 Communities in the field

Outside the detached scope of academic disputation, a more practical definition of communities tends to be used on the ground by field ecologists. This is largely driven by convenience or 'common sense' rather than a strict application of theories espoused in the literature. Here is where real confusion begins to creep into ecological thinking.

Consider the landscape represented in Fig. 10.1, where shallow pothole lakes have been left scattered across the tundra in Siberia following the retreat of glaciers. Each one of these lakes is separate from the others, forming distinct units. To the ecologist on the ground, surrounded by a swarm of mosquitoes, it seems obvious to treat each lake as a single community. Yet even the cloud of biting insects around her unsettle that simple assumption. Mosquito larvae develop as filter-feeders on the surface of water bodies before emerging as airborne adults to breed and, in this case, feed on land mammals. They will then lay their eggs in water again. What community do mosquitoes belong to?

The perspective of any single species alters our view of where the community begins and ends. The eagle flying overhead perceives the whole landscape as its province, freely moving between lakes in search of fish or over land for other prey. A bank vole will have a more restricted range but will still forage in the waters of one lake while living on the adjacent land. Perhaps the lake is a true community when experienced by a stickleback, but for a tiny diatom floating in the water it is in fact an immensely heterogeneous environment. Very different species of microbiota occur in the surface layers, the water column or the anoxic depths; the lake could therefore contain any number of planktonic communities.

The working definition of a community is therefore usually formed by the scale of any particular study. This might be the species of host plant (e.g. communities of lichen on oak trees), the dominant organism (e.g. kelp forests) or features of the landscape (e.g. limestone cliffs). These are entirely arbitrary and pay no regard to the actual movements or interactions of species, which may overlap communities in surprising ways. Salmon are most familiar to us through their appearance in freshwater rivers, though this is only a transient and terminal stage of their life

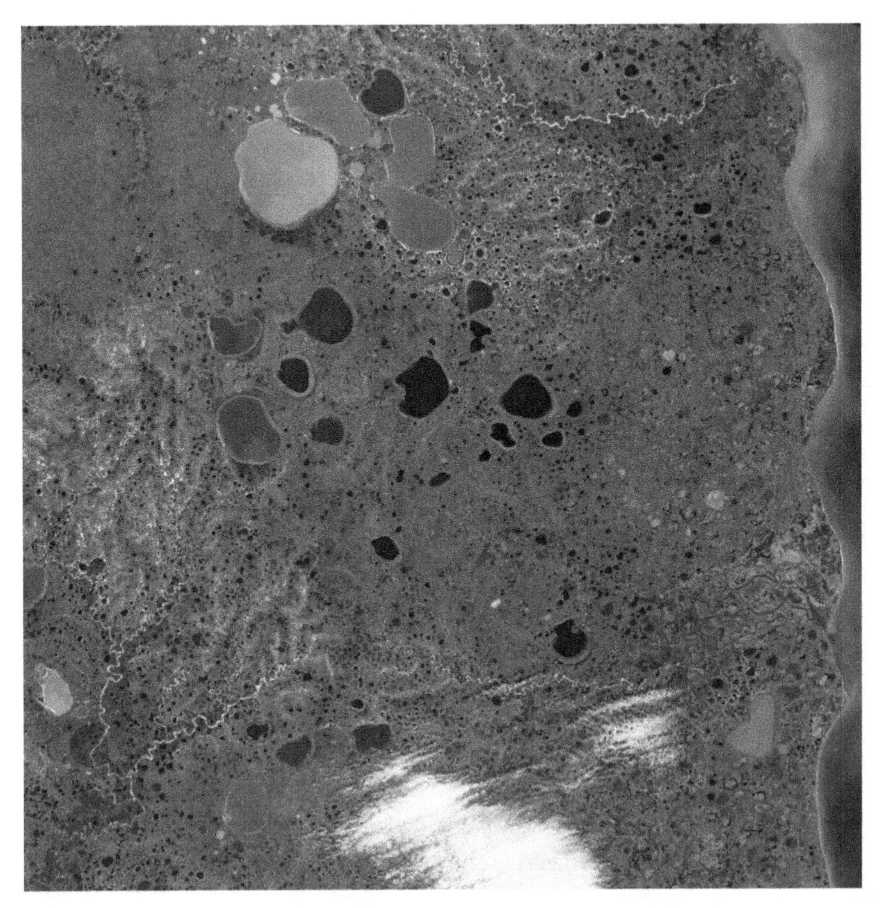

Figure 10.1 A tundra landscape of Siberian pothole lakes. (*Source*: From NASA Earth Observatory.)

cycle, just before their reproduction and death. For most of their lives salmon are marine species, only briefly becoming an important element of freshwater communities and in the process becoming vital food sources for land mammals. As a result, the nutrients contained within their bodies, which originate far out in the ocean, end up distributed across the land and influence the composition of plant communities (Morris and Stanford, 2011). Our definition of communities needs to acknowledge this flexibility.

10.4 Quantitative approaches

As in most areas of science, the only way to ensure that our assumptions are genuinely defensible and not merely a matter of opinion is to employ statistical techniques to make the definition of communities quantitative. These tend to revolve around multivariate analyses that allow the presence (or absence) and abundance of many species to be compared simultaneously. Chapter 5 gave a simple example in dissimilarity measures such as Sørensen's index. More sophisticated ordination techniques allow comparison of multiple samples at once, plotting them in virtual space to visualise the variation. The approach reduces the bias present in human judgements, even those of experts. The question is whether consistent combinations of species are found; the alternative would be a lack of any pattern, indicating that species vary independently of one another.

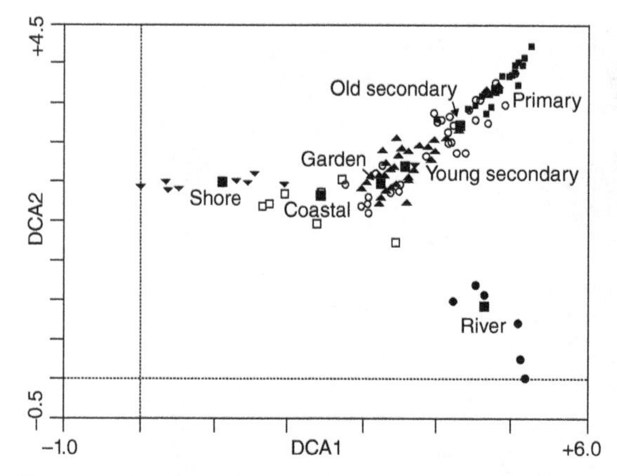

Figure 10.2 Plant community samples from Papua New Guinea represented using the two major axes obtained through detrended correspondence analysis (DCA). (*Source*: Lepš et al. (2001). Reproduced with permission from John Wiley & Sons Ltd.)

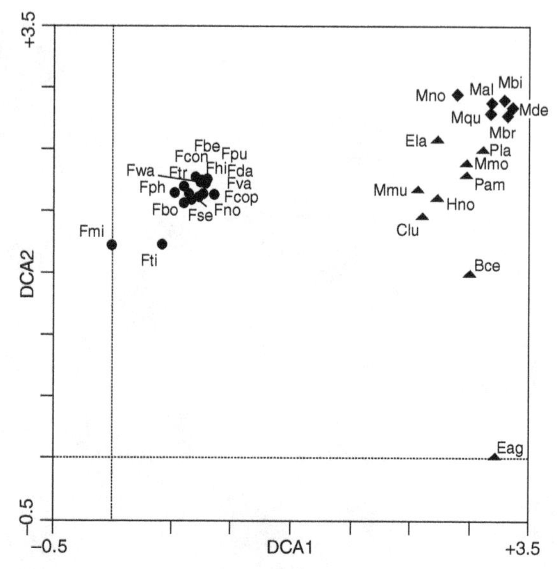

Figure 10.3 Insect herbivore community samples from Papua New Guinea represented using the two major axes obtained through detrended correspondence analysis (DCA). Host plants were either *Ficus* spp. (circles), *Macaranga* spp. (diamonds) or Euphorbiaceae spp. (triangles). (*Source*: Lepš et al. (2001). Reproduced with permission from John Wiley & Sons Ltd.)

In Papua New Guinea, Lepš et al. (2001) demonstrated the existence of discrete communities on two scales. First they assessed whether plant assemblages were distinct from one another (Fig. 10.2). In this graph a technique known as detrended correspondence analysis (DCA) has been used to place samples along two axes which reflect variation in the species present as well as their relative abundance. Points that are closer to one another represent samples with more similar composition. Many similar techniques exist (e.g. principal component analysis (PCA), NMDS, ANOSIM), all of which can be interpreted in the same way.

Samples taken from vegetation along shores, coasts or rivers form clear clusters, suggesting that each has a distinct composition. They even appear to fall along a single axis of separation. Even more striking is the relationship between the plant assemblages found in agricultural gardens tended by native villagers, regrowing forest ('young secondary', abandoned following agriculture) and older secondary and primary forests. These form a natural sequence of forest recovery following clearance. Not only is there evidence that they form discrete and identifiable stages, but there is overlap among them, and the trajectory is well represented by this graph. This is instructive because it allows us to not simply classify communities but also recognise the linkages among them and where the boundaries are blurred.

Lepš et al. (2001) also investigated the insect herbivores present on certain plants and found another set of communities (Fig. 10.3). Figs (*Ficus* spp.) have very similar insect herbivore assemblages, forming a tight and distinct cluster. The same is true of the fast-growing pioneers in the genus *Macaranga*. There is more of a spread among plant species from the Euphorbiaceae; this may reflect the fact that samples were drawn from across an entire plant family rather than a single genus. Once again it seems sensible to treat the insects found on different plant taxa as making up communities.

Do similar communities always form in comparable locations? A study of marine fish by González-Cabello and Bellwood (2009) compared the fish present in habitats formed by coral heads, rubble, vertical or horizontal boulders in two areas, one in the Great Barrier Reef and the other in the Gulf of California. The first of these lies in the peak

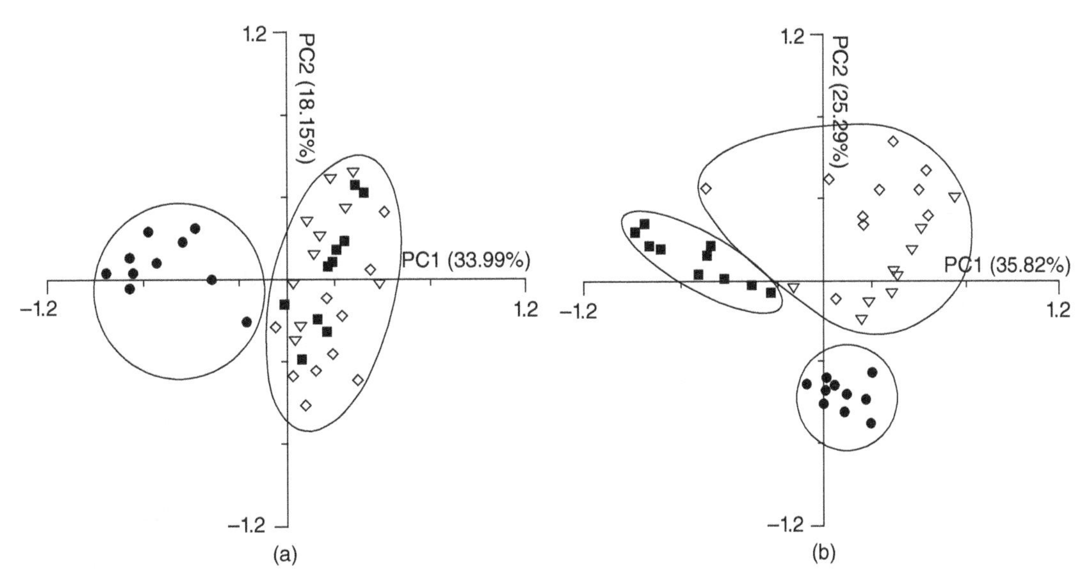

Figure 10.4 Assemblages of fish present on coral heads (black circles), rubble/rock (open triangles), horizontal boulders (black squares) and vertical boulders (open diamonds). (a) Lizard Island, Great Barrier Reef; (b) Loreto Bay, Gulf of California. (*Source*: Gonzalez-Cabello and Bellwood (2009). Reproduced with permission from John Wiley & Sons Ltd.)

of species richness of oceanic fish, whereas the latter contained relatively few species. Would this alter the existence of distinct communities?

Their results are shown in Fig. 10.4, which uses PCA, another method of achieving separation of samples along a few common axes of variation. In the Great Barrier Reef site there is a well-defined community of fish on coral heads, though other habitats share roughly the same species composition. Perhaps surprisingly, given the reduced number of species, in the Bay of California there is a further separation, with a distinctive fish community forming above horizontal boulders, while those on vertical boulders and rubble remain indistinguishable. This implies that the partitioning of natural systems into component communities can vary between regions, and one should be wary of carrying assumptions from one location to another.

Further refinements to these quantitative methods can be made by the use of cluster analysis to reliably identify groups of related samples, removing entirely the personal judgement involved in identifying recognisable subsets. These are not considered here as in practice an ecological survey seldom produces sufficient samples to satisfy the demands of complex statistical approaches.

10.5 Community structure

Having determined that communities are real, consistent groupings of species, we can begin to study them in more depth. The first stage is usually to produce some descriptive plots and statistics using the methods in Chapter 5, revealing patterns of species richness and relative abundance. This may not always be the most appropriate approach; for example, patterns of biomass distribution among species tend to correlate better with their resource usage. The roles and identities of species are often of crucial importance too.

Communities are therefore often broken up into functional groups or guilds (*sensu* Root, 1967). These are sets of species with shared resource requirements or similar lifestyles (e.g. leaf-gleaning insectivores or filter-feeders). The problem of defining their niches and how they can coexist on the same resources was dealt with in Chapter 6. There is no difficulty understanding how species from different functional groups coexist; there is no reason why an owl and an oak tree can't both live in the same forest. The separation of niches is only necessary when you look within these groups. Sometimes partitioning can be subtle. In the Southern Ocean four species of petrel often coexist and breed on the same islands, feeding on the same zooplankton. They achieve this through foraging at different times, in different places and at varying depths in the water (Navarro et al., 2013). All however belong to the same guild.

As with the uneven distribution of species among taxonomic groups (see Section 3.4), functional groups are also not equal in size. Mouillot et al. (2014) divided 6,316 fish species present on tropical reefs around the world into a set of 646 functional groups. On any given reef, at least a third of the species were in groups with multiple members, while another third were the sole representative of their group. This implies that even in highly species-rich communities, the insurance value provided depends strongly on whether extinctions happen in diverse or species-poor functional groups.

Note that functional groups are usually descriptive and based on opinion and judgement. As ever it is better to apply a quantitative metric. One which has been recently developed is to assign species to *archetypes* which group species according to their environmental responses rather than the locations in which they are found, their taxonomy or their trophic level (Dunstan et al., 2011). Applying this method to marine fish and macro-invertebrate species off southern Australia, Leaper et al. (2014) found that at any one site the community was built from a selection of species drawn from those archetypes which were able to survive there, rather than having a consistent and predictable membership. This view is closer to Gleason's perspective.

The *dominant* species or group is usually the one that is most abundant, either numerically or (more commonly) in terms of biomass. Sometimes this is self-evident and the clue lies in the name of the habitat; kelp forests, coral reefs and grasslands are all named after the taxon that defines them. In other cases a little more knowledge about the system is required before dominants can be identified.

The *basal* species or group is the foundation of the rest of the food web. In many cases this is a primary producer (**autotroph**), perhaps photosynthetic plants or algae, or chemosynthetic bacteria on deep-sea vents. The basal group is not always the same as the dominant group; paradoxically in some communities it is possible for greater amounts of biomass to be held in higher levels of the food chain, as in coral reefs, where algae are relatively scarce yet support a whole network of herbivorous fish and their predators. The basal species is also not always an autotroph. In benthic marine environments, the main resource input is from dead material floating down through the water column, and therefore decomposers or filter-feeders (**heterotrophs**) lie at the base of the food web.

Reference is often made to **keystone species**, whose impact on the structure and functioning of a community greatly outweighs their biomass or abundance. Examples include the snails studied by Lubchenco (1978) or the spiders that were found by Schmitz (2008) to control not only herbivores but also the structure of plant communities and their ecosystem processes (see Section 8.6). Herbivores and predators are often keystone species when their activities have ramifications for the composition and diversity of other functional groups. Keystones can have more subtle effects; **mutualists** often perform services such as pollinating flowers or fixing atmospheric nitrogen that benefit a wide range of species.

Many keystone species are also **ecosystem engineers**, species whose activities create or modify their habitat, often providing opportunities for those species that live alongside them. Classic examples include beavers, which fell trees to create the pools within which they promote and feed on populations of fish that would otherwise not be present. Trees create the forest structure that enables the most complex of all natural systems to form, but given their large biomass, they would not be regarded as

Figure 10.5 (a) The giant kangaroo rat, *Dipodomys ingens*; (b) a burrow mound; (c) mounds in the Carrizo Plain National Monument, CA; (d) a kangaroo rat exclosure experiment (left side of image). (*Source*: Prugh and Brashares (2012). Reproduced with permission from John Wiley & Sons Ltd.) (*See colour plate section for the colour representation of this figure.*)

keystone species. Humans are both keystone species and ecosystem engineers; think how few farmers it takes to transform entire landscapes from wild nature to agricultural production.

Kangaroo rats are a less immediately obvious example of ecosystem engineers. They change soil characteristics by digging burrows, grazing the vegetation and leaving behind abandoned tunnels which become a resource for other species. Prugh and Brashares (2012) studied the significance of the giant kangaroo rat *Dipodomys ingens* in California (Fig. 10.5). Alongside its impacts on plants through seed predation and competitive interactions with other rodents, the rat produced burrows which influenced the diversity of plants and invertebrates, along with the density of squirrels, and hence had many other linked effects.

There is only one sure way to identify a keystone species, which is to remove it and measure the consequences. Only when they are withdrawn does their importance become apparent. Watson and Herring (2012) managed to remove mistletoe from entire woodland blocks in Australia. Though mistletoes make up only a tiny proportion of the total biomass of the woodland, they have a large number of interactions with other species and form an important resource. Woodlands without mistletoe lost 20% of their bird species, with over a third of resident woodland-dependent species vanishing. The cause was not the absence of berries or nesting habitat, but something more subtle and indirect. Mistletoes drop large numbers of nutrient-enriched leaves which can double or even triple the supply of leaves to the forest floor. This in turn boosts populations of invertebrates and the species that feed upon them and hence shapes the entire community.

The keystone species concept has received much criticism on the grounds that it is rather vague in its formulation, often applied indiscriminately, and nearly impossible to test definitively (e.g. Cottee-Jones and Whittaker, 2012). How large an effect is required for a species to be a keystone? Is it enough to remove just that species, or do you have to remove every species separately and compare their effects? What does it mean to say the effect is large relative to its abundance (or biomass)—how should that be determined? In practice, while a useful idea in thinking about community structure, it is hard to apply rigorously.

A final group of species, often neglected when considering community structure, are **migrants**. These can have special effects due to their sudden arrival and the way in which they connect communities together (for an excellent review see Bauer and Hoye, 2014). They often transport nutrients, as mentioned earlier for salmon, or transport seeds, propagules, diseases and parasites. Around 20% of bird species are migratory, making them critical links across landscapes. Migrants can even be keystone species; migratory long-nosed bats are only fleetingly present in Mexico yet are responsible for pollinating cacti as they pass through.

10.6 Food chains

Species—and whole functional groups—are linked together through trophic interactions into food chains. The basal group supplies one or more levels of consumers which are dependent upon them. Chains tend to have a restricted length, most falling within the range of three to four levels, seldom as low as two or greater than five. Even where longer chains occur, only a tiny amount of energy gets beyond the fifth level, and high-level consumers are subsidised by feeding on organisms from multiple levels (Baird and Ulanowicz, 1989). Humans, for example, may feed on a wide variety of species, but their average trophic level is only 2.21—about the level of the anchoveta (Bonhommeau et al., 2013). Typically chains do not include parasites or decomposers, not because they are unimportant, but because their effects on the overall transfer of energy and resources are either negligible (parasites) or form the basis for entirely new food chains (decomposers).

The stereotypical food chains that we might sketch out as starting points (diatoms → zooplankton → shrimp → fish → birds) are usually too simplistic to capture the complexity of natural systems. In fact chains are very seldom linear and are more often nested within food webs, which improve their stability.

Why are food chains usually so short? There are many potential reasons for this. One is a simple issue of design constraints on species at the upper levels. Once a top predator has entered (or evolved), why would any species bother trying to feed on it? There would be no point trying to be a predator of lions since they are large, dangerous and difficult to capture. A predator might as well feed on wildebeest, which are more abundant and of similar size. There are exceptions; king cobra (*Ophiophagus hannah*) are predators of other snakes. Nevertheless, the difficulty of evolving to be a super-predator seems to have made this an infrequent occurrence. Moreover, predators tend to exist at low densities, so an 'uber-predator' would need to not only be fiercer but also cover a much larger territory, and there may simply not be enough habitat space to maintain a stable and interconnected population.

A second possibility is that dynamical instability prevents food chains becoming too stacked with levels. Any fluctuation in the abundance of the basal group leads to amplified effects on the species feeding upon them, whose body size precludes their populations from responding as quickly. Imagine a house of cards, where the merest tremble at the base is enough to shake the whole structure.

Thirdly, it might simply be that resources run out at higher levels. The average efficiency of transfer between trophic groups in terrestrial communities is around 10% (Pimm, 1982; Post, 2002), so there may not be enough left to maintain populations in further levels at the top. In marine environments the efficiency of transfer is higher since most species are ectotherms (with reduced energetic demands) and the costs of movement in water are lower due to buoyancy. This might be part of the reason why oceanic food chains are the longest in nature, with up to five levels, though it is also likely that the larger overall size of the system provides a greater productivity base.

Even within apparently similar communities, there can be pronounced variation in food chain length, for example, when comparing among freshwater lakes. What drives this? There are three hypotheses (Post et al., 2000; Post, 2002). Perhaps overall productivity determines length, such that systems with the highest rate of carbon fixation support more levels. Alternatively, it could simply be a consequence of total size; larger systems support more levels because there is greater space available for consumer species to maintain sustainable populations. Finally, there could be some combination of the two.

The results of comparing 25 lakes in North America, in which productivity was limited by phosphorus availability, can be seen in Fig. 10.6. The size of the system was of primary importance, with larger lakes supporting longer food chains. This explains 80% of the variation—a remarkably strong relationship. The productivity of those lakes had no discernable effect. In other words, it may not simply be about the amount of energy going in at the base; there can be absolute effects of area (or volume in this case). Perhaps consumers and predators require territories of a particular size in order to enter a system. Likewise, among islands off Finland, smaller islands have shorter food chains (Roslin et al., 2014).

These outcomes may be unusual though. Comparing across a large range of studies, Takimoto and Post

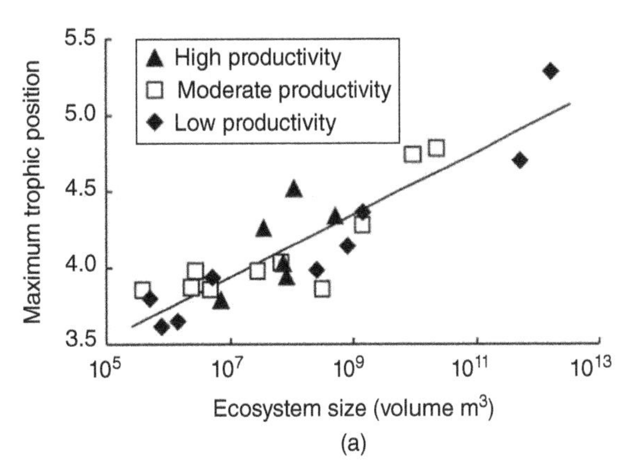

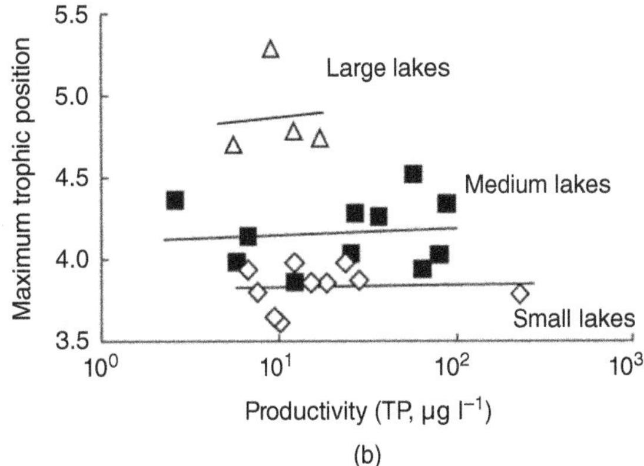

Figure 10.6 Observed variation in lake food chain length (maximum trophic position) with either (a) size or (b) productivity measured using total phosphorus levels as a proxy. (*Source*: Post et al. (2000). Reproduced with permission from Nature Publishing Group.)

(2013) found that ecosystem size and productivity were both important in determining food chain length, with variation between different types of community. Although highly disturbed sites might be expected to have shorter chains, no consistent effect of disturbance was found. It remains to be determined in what circumstances each factor is most important.

There are some examples of very short food chains, with as few as two levels, which is obviously the minimum while still being a chain. Cases include very young communities, where perhaps higher trophic levels have yet to arrive, or isolated patches that might not have been reached. There can however also be a role for productivity, which perhaps conflicts with the previous conclusions but is likely to apply only to those systems with extremely low net primary productivity (NPP) (i.e. $< 10\,g\,C\,m^{-2}\,y^{-1}$). Deserts and caves fall into this category. Bear in mind though that caves are also small and isolated, causing confounding effects.

10.7 Food webs

Food chains do not exist in isolation, but are instead nested and cross-linked within food webs, complex networks of feeding relationships that can encompass hundreds or even thousands of species. Often they are presented with taxa merged into particular functional groups (e.g. fungivores), though even within webs there can be a mix of categories, such that lower levels are often grouped (e.g. 'phytoplankton'), while the top predators are split into species (e.g. killer whales).

An example of a food web from the Serengeti National Park, Tanzania, is shown in Fig. 10.7. Note that although there are many species, a few are highly interconnected, while others lie on the margins.

Single chains are dynamically unstable, but once connected to others the web becomes increasingly robust. Most of the linkages are weak and represent relatively minor flows of energy or resources, but they serve to correct imbalances and damp down the fluctuations that might spread if one element were to change dramatically, for example by going extinct.

The size and complexity of food webs can be staggering. Novotný et al. (2010) constructed a food web based on a single hectare of forest in lowland Papua New Guinea. It contained 224 plant species and 1,490 insect herbivore species across 11 feeding guilds, making up a total of 6,818 individual feeding links. They estimated that less than 20% of the true web was reflected in their sample, which is likely to extend to around 50,000 feeding interactions—only considering plants and their insect herbivores. It is unlikely that communities on this scale will ever be fully described, which means the identification of feeding guilds is a necessary simplification in order to understand their dynamics.

Most species within a single community are tightly interconnected. This can be measured using

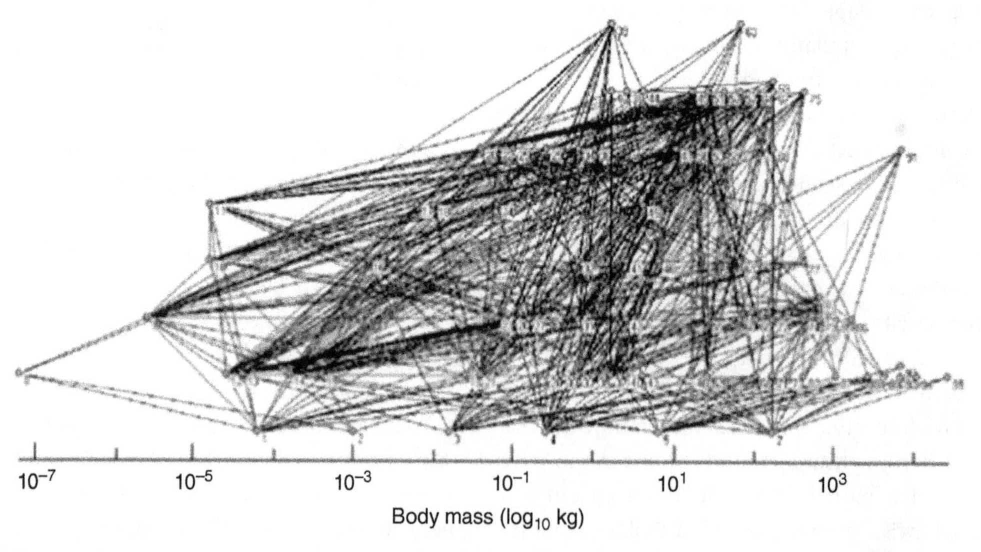

Figure 10.7 Food web representing invertebrates, mammals, birds, reptiles and amphibians in Serengeti National Park, Tanzania, scaled by body mass ($\log_{10}$ kg). Each node represents a group of species with similar feeding. Nodes 1–7 at the base are resources which are not ordered by body size. (1: decaying material, 2: plant juices, 3: fruits and nectar, 4: grains and seeds, 5: grasses and herbs, 6: crops and 7: trees and shrubs). Nodes 8–95 are consumers; for details of species see de Visser et al. (2011). (*Source*: de Visser et al. (2011). Reproduced with permission from John Wiley & Sons Ltd.)

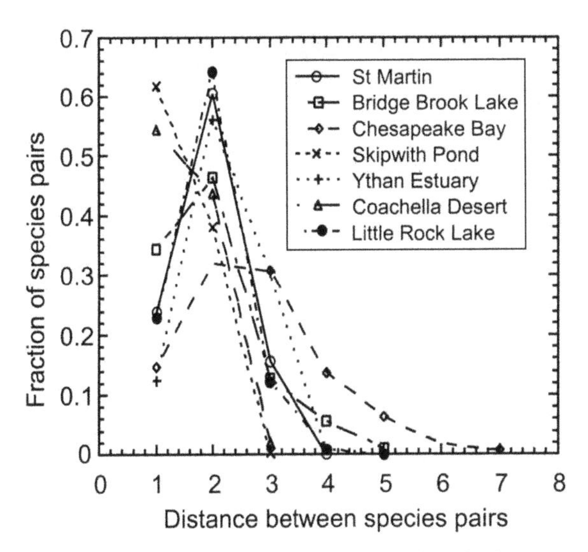

Figure 10.8 Number of links separating species pairs in seven food webs. (*Source*: Williams et al. (2002). Copyright (2002) National Academy of Sciences, USA.)

a property known as **connectance** C, which documents the ratio between the actual number of links and the maximum possible:

$$C = \frac{L}{S(S-1)}$$

where L is the number of links and S is the number of species.

Connectance of food webs provides another means of demarcating the boundaries between communities. Though there are often links that span habitats, these tend to be less numerous and influential than those within. Webs are frequently compartmented by habitat, but highly connected within. The extent to which this is true is striking. Comparing a set of seven food webs, Williams et al. (2002) found that 80% of species were within two links of one another and 97% within three (Fig. 10.8). In comparison, on Facebook the average number of links between users is 4.74 (data from November 2011), and for Twitter users the figure is similar[1]. Food webs appear to be particularly closely interconnected relative to these other complex networks.

[1] See http://en.wikipedia.org/wiki/Six_degrees_of_separation for other examples.

Species are also more connected than random chance would suggest. If you took a natural food web and redistributed the same number of links at random, the average distance between species would be much greater. This is because most communities contain a core of generalist species that have a large number of connections right across the web.

An important implication of this finding is that specialism tends to be one sided. It is extremely rare in nature to find two species that are mutually interdependent. Even where one species is a specialist (e.g. a herbivore or parasite), this is not reciprocated by the species it depends upon. Specialists tend to interact with generalists, maintaining their linkage to the wider network. For example, Fig. 10.9 shows an interaction network of flowers and their pollinators. Some flowers are specialised and will only accept one pollinator species, whereas the pollinator they interact with is a generalist visiting many species of flower. Similarly, there are specialist pollinators that will only go to a few species of flower, but these flowers are able to be pollinated by a range of insects. Note that there are no points in the bottom right corner. Interactions between guilds show little evidence of mutual specialisation.

This leads to the concept of *nestedness* in networks. If a core set of generalist species all interact with each other while specialists only interact with generalists, then the community will be perfectly nested—there are no modules or groups of species which are separate from the others. Figure 10.9 illustrates this using a line; a perfectly nested network would have all points above this. Nestedness is clearly far from being absolute, but the density of points is much higher above than below the line.

The existence of compartments presents another potential way of identifying distinct communities. There are seldom enough data to evaluate this, but a rare example comes from a study by Pocock et al. (2012), in which the entire interaction web of a farm in England was characterised (Fig. 10.10).

As well as the trophic connections among species, that is, what eats what, there is also a multitude

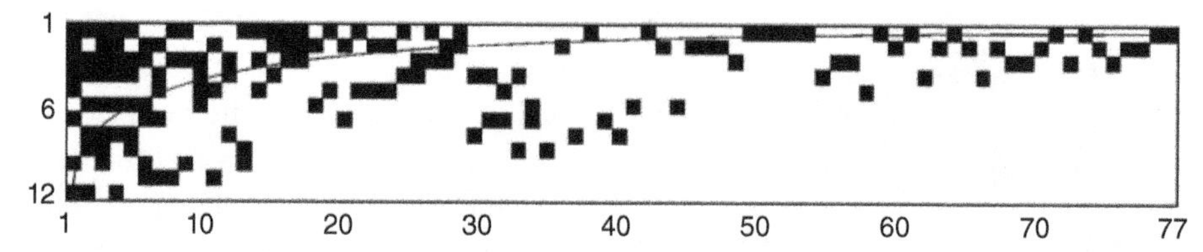

Figure 10.9 Interaction matrix between plants (horizontal) and pollinators (vertical) in a boreal forest in south-east Norway. Line represents the isocline of perfect nestedness; a completely nested community would have all points on or above this. (*Source*: Nielsen and Bascompte (2007). Reproduced with permission from John Wiley & Sons Ltd.)

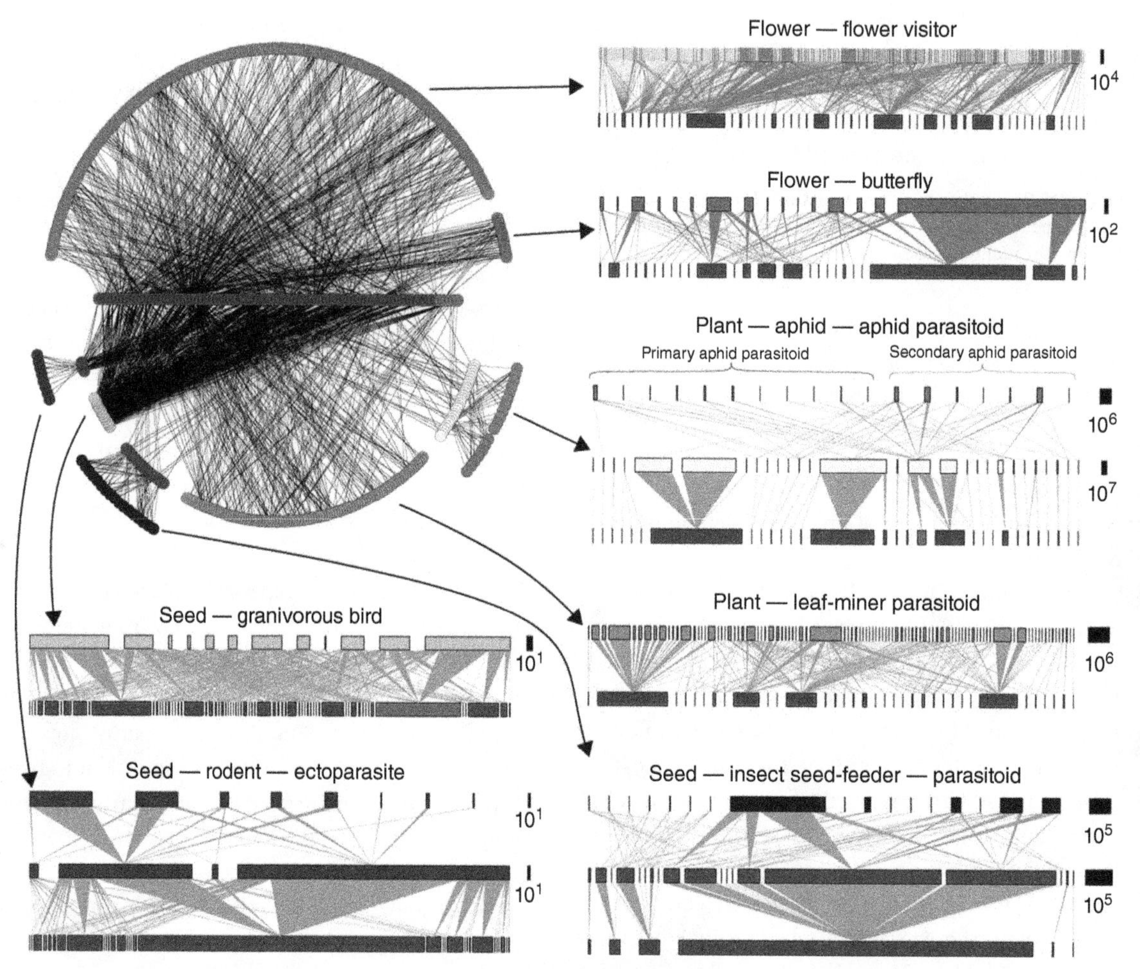

Figure 10.10 Species interaction networks for plants and insects from Norwood Farm, Somerset, United Kingdom. The full interaction web at the top left (dots represent single species) with plants in the centre is decomposed into quantitative networks within which each block represents a species and their width indicates relative abundance (scale bars offset), except for plants on the bottom rows, which are scaled by the number of interactions with animals recorded. (*Source*: Pocock et al. (2012). Reproduced with permission from the American Association for the Advancement of Science.)

of non-trophic interactions. These are often summarised by examining the relative benefits on each side and include mutualism (+,+), competition (−,−), antagonism (+,−), commensalism (+,0) and amensalism (−,0). Building these into an understanding of the structure and organisation of communities presents an opportunity for future research (Kéfi et al., 2012). For example, Karban et al. (2012) studied two herbivores that share the host plant *Lupinus arboreus*. Bushes with high abundance of tiger moth caterpillars supported fewer tussock moth caterpillars later in the season, showing clear evidence of competition. Yet the following year, bushes with more tussock moth caterpillars had many more tiger moth caterpillars, which implied facilitation. This occurred due to the accumulation of litter beneath the bushes, a non-trophic interaction whereby feeding by one species supported the other. The connection between these three species is therefore more complex than first appears from direct feeding links, and it is likely that similar cases occur in many natural systems.

In one of the most important ecological findings of recent years, Eklöf et al. (2013) demonstrate that despite their apparent complexity, the number of dimensions required to explain all the interactions within communities is remarkably low (refer back to Chapter 6 to see how this also applies to niches). Most communities require measurements of only two or three species traits to account for the pattern of connections among them. In food webs body size is invariably involved, but the same principle applies to mutualistic (e.g. pollination) or host-parasite webs.

The two properties of networks described earlier—their high connectance and nested structure—seem to make them more resilient when disrupted and explain how communities can remain stable over long periods of time.

10.8 Complexity and stability

The relationship between the structure of food webs and their stability is one of the most rapidly developing topics in ecology, with new papers emerging continuously, often with contradictory findings. There isn't space here to delve into the finer details, especially as they tend to involve sophisticated computer models, but it's worth taking a moment to distil the essence of the question at their heart. Will a complex web be more or less stable?

The simplest expectation is that more species increase the number of links, making a few extinctions tolerable and increasing the number of pathways through which energy can flow (MacArthur, 1955; Elton, 1958). These naïve speculations were thrown into disarray by May (1972), who created a simple set of artificial food webs which revealed the opposite outcome: as species richness increased, stability actually fell. Much of the literature since has been concerned with discovering what aspects of food webs allow them to escape the apparent trap of instability, since the patterns evident in nature cannot be readily replicated *in silico*.

One line of evidence comes from taking real food webs and simulating what happens when species are removed. This suggests that webs with higher connectance are more stable and suffer fewer extinction cascades (Dunne et al., 2002). Removing the most connected species causes a more rapid collapse. This suggests that high connectance improves stability, though some simulations have suggested that the most connected species are more likely to go extinct, which would make complex webs unlikely to persist (Saavedra et al., 2011). This was challenged by de Visser et al. (2011), who found for their Serengeti food web that the species most threatened with extinction in the real world were actually the least connected and had little effect on stability. Poorly connected species were lost first. A gap therefore remains between theory and practice that will need to be bridged if we are to predict the impacts of human-derived extinctions on food web stability.

Others have argued that nestedness improves stability, by ensuring the close linkage of all species, or that compartmentation is better, as it prevents perturbations from spreading and destabilising the

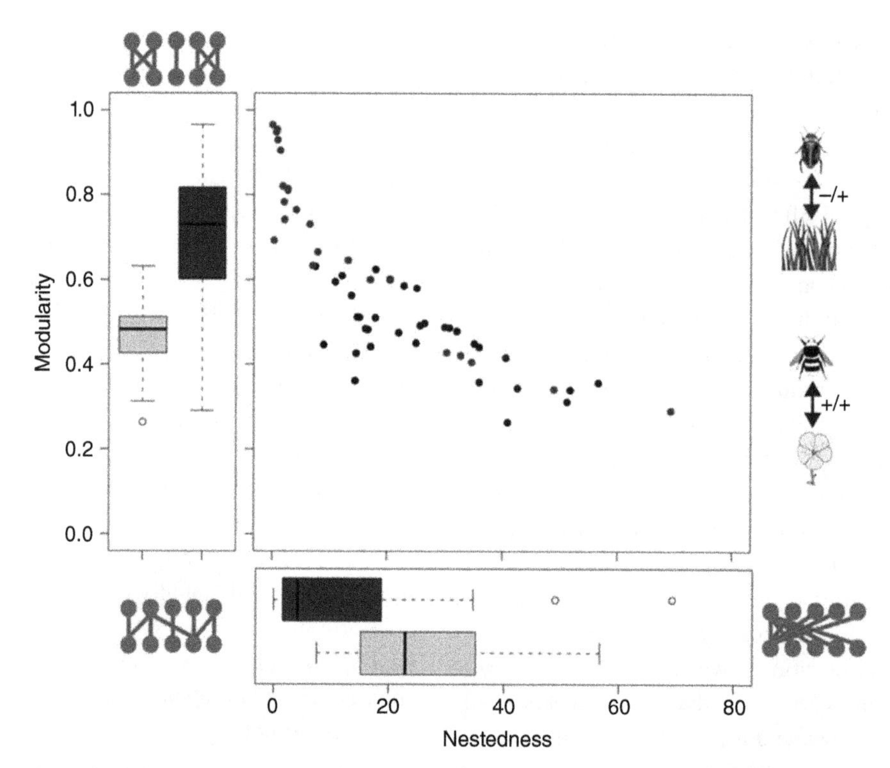

Figure 10.11 Relationship between network nestedness and modularity, with box plots of nestedness (below) and a measure of network modularity (left). Higher values are more nested or modular, respectively. Each dot represents an empirical network, either pollination (black) or herbivory (red), with corresponding box plots showing median line and interquartile range with bars ±1.5 interquartile range. (*Source*: Thébault and Fontaine (2010). Reproduced with permission from the American Association for the Advancement of Science.) (*See colour plate section for the colour representation of this figure.*)

whole web. Evidence from nature suggests that there are fundamental differences between the networks formed by various types of interaction. Thébault and Fontaine (2010) found that mutualistic networks tend to be nested, whereas trophic networks (those based on feeding relationships) are modular (i.e. compartmented), in each case serving to increase their stability (Fig. 10.11). Perhaps a diversity of interaction types can maintain community stability, with a portfolio of predators, competitors and mutualists (Mougi and Kondoh, 2012).

An increasing amount of attention has been paid to the strength of interactions within food webs, since not all links are equal. Some occur more frequently, or represent larger flows of energy, and are therefore likely to be more important. In mutualistic networks the strongest interactions are among generalists (those with the greatest number of connections), making them keystone species as they are the spine upon which the rest of the food web depends (Gilarranz et al., 2012). Nevertheless, weak links can still be crucial, much like stabilisers on a child's bike, which are only small wheels that make little corrections but are sufficient to prevent the whole from falling over.

Even though food webs are a mesh of links, the majority of these are weak, and only a few carry most of the energy through the system. This means that even when there are hundreds of links among thousands of species, most food webs contain a compact core of around a dozen nodes or 40 links with only three to four trophic levels (Ulanowicz et al., 2014), and increasing their complexity beyond this makes little difference to their overall behaviour. Perhaps nature is not as complicated as it first appears.

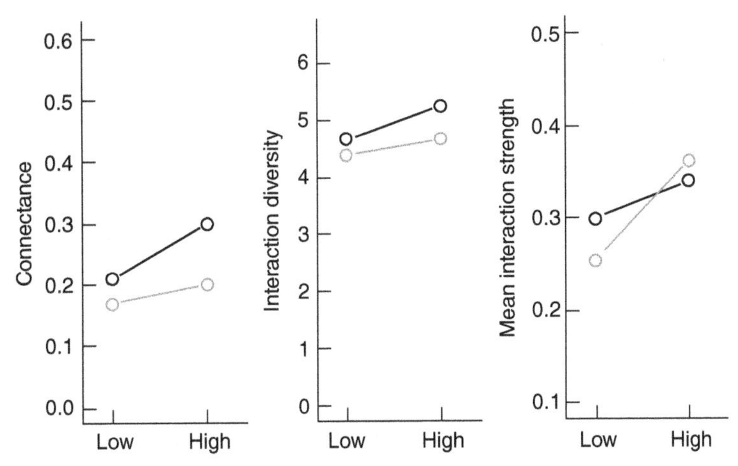

Figure 10.12 Network metrics for interaction webs of plants, insects and spiders in experimental plots with low and high plant species richness (4 vs. 16 species) in 2003 (grey) and 2005 (black). (*Source*: Rzanny and Voigt (2012). Reproduced with permission from John Wiley & Sons Ltd.)

Finally, there is an apparent link between diversity and food web complexity which has become an important direction of study. Rzanny and Voigt (2012) found that grasslands with lower plant diversity had lower connectance, interaction diversity and mean interaction strength (Fig. 10.12). Even with the same spread of functional groups, webs with more species at the base are more connected and complex, showing the importance of producer richness both for maintaining other species (recall from Section 8.4 how 'diversity begets diversity') and the stability of ecosystem services. Whether complex food webs support high species richness, or *vice versa*, is a question to which we can as yet provide no definitive answer.

10.9 Trophic cascades

Section 8.6 showed how top-down control by predators, pests or parasites can influence the species richness of their prey by suppressing dominant species. The effects of higher trophic levels can be profound, not merely changing the behaviour and population dynamics of the level below them, but with cascading effects on species several links removed in the food web.

Horsehair worms infect crickets, manipulating their behaviour to force them to enter streams where the parasites are able to reproduce. The crickets are then eaten by fish. Sato et al. (2012) found that when fish can eat crickets, they eat fewer invertebrates from the stream bed, which means these in turn consume more algae and increase the breakdown rate of leaves (Fig. 10.13). All this was discovered by manipulating the input of crickets into streams. What makes this such a remarkable study is that it is the influence of parasites, which make up a trivially small fraction of the total community biomass, that leads to dramatic repercussions throughout the food web and ecosystem.

Similar effects can occur over large scales. Cod populations in the Baltic Sea occasionally spill over into the Gulf of Riga, where their arrival prompts a series of changes in the behaviour of the entire ecosystem (Casini et al., 2012). An influx of cod causes decreased abundance of herring, which are also adversely affected by fishing and lower temperatures (herring size also falls when their population levels rise, an indication of density-dependent competition). Herring in turn feed on zooplankton, which feed on phytoplankton (Fig. 10.14). A chain

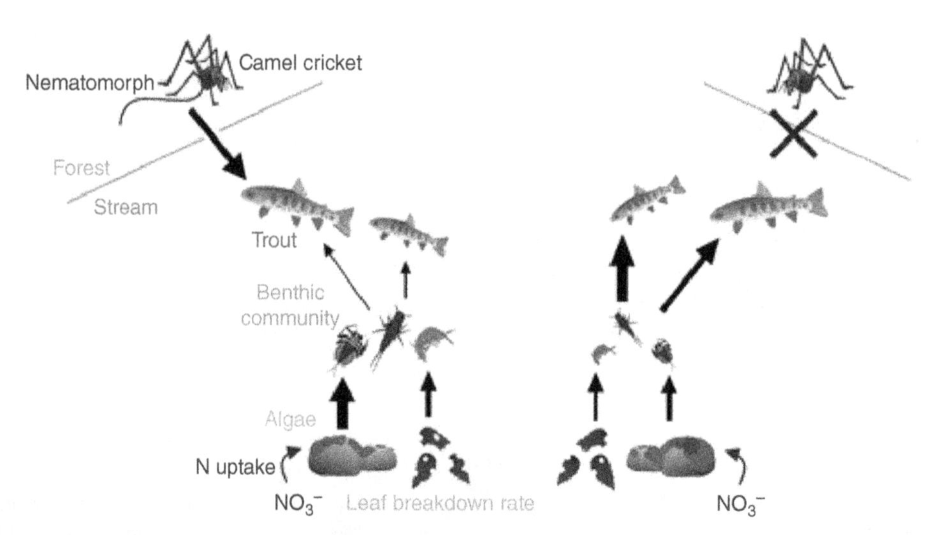

Figure 10.13 Trophic cascade in a stream modified by input of camel crickets, which occurs when they become infected with a nematomorph parasite (left); arrows denote consumption, with thickness indicating interaction strength. (*Source*: Sato et al. (2012). Reproduced with permission from John Wiley & Sons Ltd.)

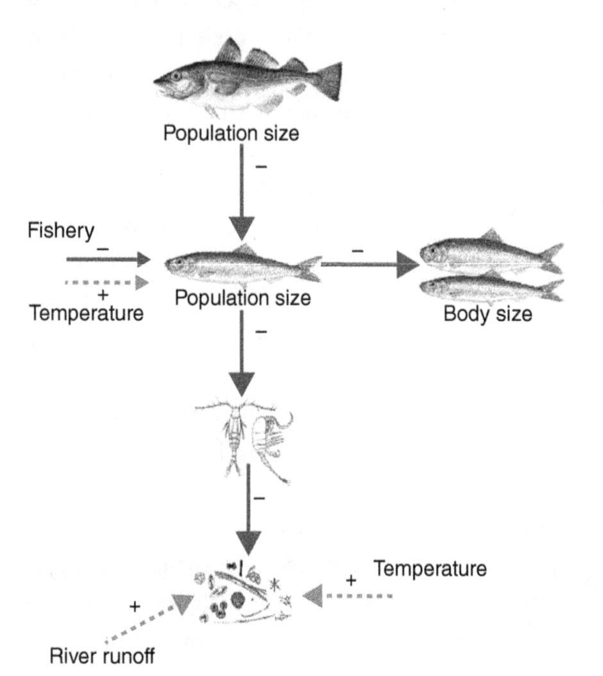

Figure 10.14 Schematic illustrating the effects of cod influx into the Gulf of Riga throughout the food web. Connections between cod (top), herring, zooplankton and phytoplankton (bottom) are indicated, with direction and strength indicated by the size of the arrow. Both biotic (solid) and abiotic (dashed) effects are included. (*Source*: From Casini et al. (2012).)

of interactions means that the arrival of cod indirectly causes a decrease in the abundance of primary producers, leading to greater transparency of the water. Given this, it is worth contemplating what effects the industrial extraction of cod from the world's oceans might have had, right down to the level of phytoplankton and hence overall productivity.

10.10 SAD again

Thus far little attention has been paid to one of the major differences among species within communities, which is that they vary greatly in abundance. This can have major implications for community structure and the manner in which species interact with one another. Recall from Section 5.5 that abundance distributions tend to exhibit a lognormal distribution when presented as a Preston plot. This is not always the case however; under-sampled communities might show a veil line, while other times no obvious pattern is clear. These observations challenge the generality of the rule, which means we need to query why the lognormal appears so often and what might cause departures from it.

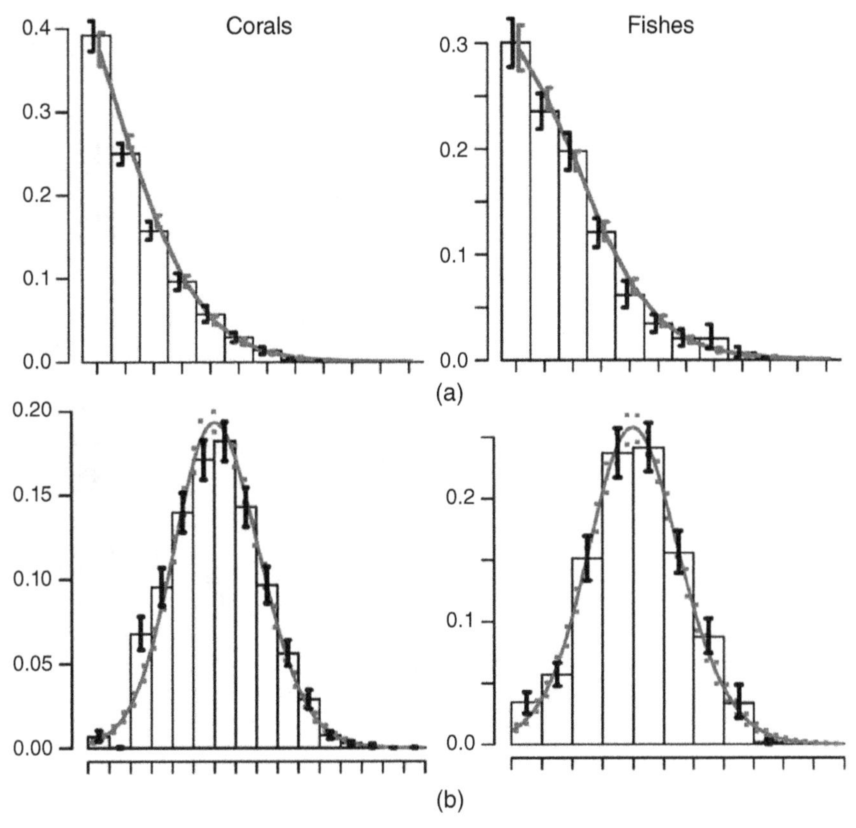

Figure 10.15 Preston plots of coral and fish ranked (a) by abundance or by (b) cover and biomass. (*Source*: Connolly et al. (2005). Reproduced with permission from the American Association for the Advancement of Science.)

Whether the lognormal appears at all depends on the data chosen. To illustrate this we return to the great coral reef transect first described in Section 7.3. Connolly et al. (2005) recorded the abundances of coral and fish species on a wide range of reefs, as well as estimating the percentage cover of each coral species and the total biomass for each fish species. When presenting their data in terms of numerical abundance, it seemed as though there was no lognormal distribution. However, once they transformed their data into cover and biomass, the lognormal became absolutely clear (Fig. 10.15). In both cases, cover and biomass convey something meaningful about resource usage by the species involved, far more so than abundance alone. The lognormal reflects something important about how communities are structured.

Yet some have noticed a peculiar feature in the shape of the curve. Volkov et al. (2007) examined the SAD for trees in the 50 ha forest plot on Barro Colorado Island (BCI) in Panama. As they took ever greater samples from the total dataset, the lognormal emerged from the veil line. This wasn't entirely true though; there were consistently more rare species than the lognormal predicted. In other words, the curve has a shoulder on the left-hand side. This is referred to as log-left skew.

Is this a common phenomenon? Based on Volkov et al. (2007) it is widespread in tropical tree communities, as six plots spread across the tropics show the same pattern (Fig. 10.16). In fact, in many other communities with very high species richness (measured at the local, α scale), log-left skew is a common

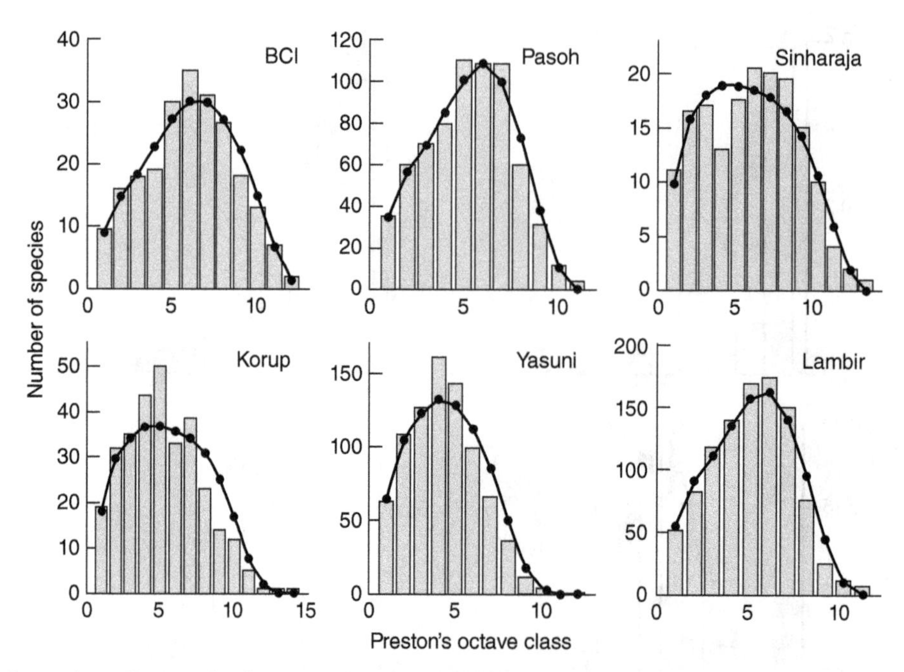

Figure 10.16 Comparison of Preston plots for tree species communities from BCI, Panama (50 ha); Pasoh, Malaysia (50 ha); Sinharaja, Sri Lanka (25 ha); Korup, Cameroon (50 ha); Yasuni, Ecuador (50 ha); and Lambir Hills, Sarawak, Malaysia (52 ha). (*Source*: Volkov et al. (2007). Reproduced with permission from Nature Publishing Group.)

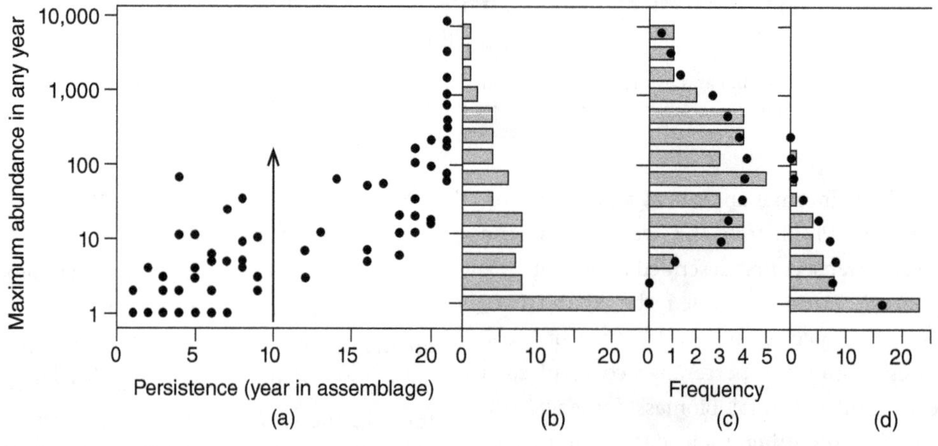

Figure 10.17 Abundance of fish species in the Bristol Channel based on a 21-year time series. (a) Species ranked by number of years in which they were present; (b) Preston plot for full assemblage, followed by subsets with species present for (c) more than 10 years or (d) under 10 years. (*Source*: Magurran and Henderson (2003). Reproduced with permission from Nature Publishing Group.)

feature. Does this invalidate the generality of the log-normal distribution?

One explanation for the skewed pattern was revealed by Magurran and Henderson (2003) and only became clear thanks to a monumental sampling effort. Monthly collections of fish were taken from the Bristol Channel over 21 years. Combining all the samples into a single Preston plot left a rather messy distribution with no clear trend and a mass of apparently rare species. This was resolved by splitting the sample into two groups (Fig. 10.17). Species that were resident in the Bristol Channel were be found in most years, whereas a large number of species were found in fewer than 10 years. Once these groups were partitioned, it was revealed that the resident species were lognormally distributed in their abundance. The occasional species were probably rare migrants which had entered the channel by chance every so often. They therefore represented an incomplete sample of the larger pool of species present in the Atlantic Ocean. Their SAD showed a log-series pattern, typical of small samples. The failure to see an overall lognormal was a consequence of adding these groups together, and only a substantial amount of natural historical information could resolve the problem. Similar results have been found elsewhere, with the most abundant species in communities being those which are widespread or persistent (e.g. Hercos et al., 2013).

10.11 Complex systems

The SAD discovered by Frank Preston was an important insight, but academics in other fields were beginning to notice comparable patterns in their own data. Social scientists described the 'distribution of wealth' curve, which was eerily familiar, while in economics the Pareto distribution characterised personal incomes, and in physics the Boltzmann curve was fit to the distribution of molecular kinetic energies in gases. Wherever people looked, the same pattern kept popping up, even in bizarre examples such as the lifespan of glass tumblers or the age at first marriage of American women. More are shown in Fig. 10.18 (Nekola and Brown, 2007).

The reason for this common pattern across not merely ecological samples but scientific fields was that Preston had stumbled across a general feature of what we now recognise as *complex systems*. This is a term that arose in mathematics and applies to systems that satisfy a number of conditions (it means more than looking complicated):

1 Many differing components
2 Interactions on multiple scales
3 Contain non-linear dynamics
4 A combination of stochastic (i.e. unpredictable) and deterministic processes
5 Both positive and negative feedbacks
6 Open systems (require external input)
7 Historically contingent (dependent on starting conditions)
8 Often nested

You should be able to quickly spot how each of these conditions is satisfied by natural systems based on the evidence so far.

Where does this leave us? The implication is that the lognormal SAD is something that would appear anyway and has no special place in ecological theory. Many pages of academic journals have been spent coming up with elegant explanations based on biological processes and sophisticated models that purport to lead to similar SADs. If any complex system would behave in the same way, then these are all unnecessary. Have ecologists been wasting their time?

Another troubling implication comes when considering whether there will ever be a fully predictive theory of community ecology. The laws governing the behaviour of complex systems imply that we will not be able to predict the future abundance of any single species with certainty, given that it is embedded in a network of interactions. For a conservation biologist who might like to anticipate whether their favourite salamander species will have a reasonable population in a century's time, this is a worrying thought. It is possible to predict what the overall distribution of abundance of elements across whole systems will be, but not that of any individual component.

In fact this problem has been recognised in other fields, and they have already made their peace with it. An economist, looking at a cohort of university graduates, would be able to state with some certainty that their salaries in 20 years time would follow the Pareto curve. For any single graduate, she could collect a wealth of information on their grades, intended career, interests and aptitudes, and yet her success rate in predicting their salary would be poor. Similarly for climate predictions, we can make confident

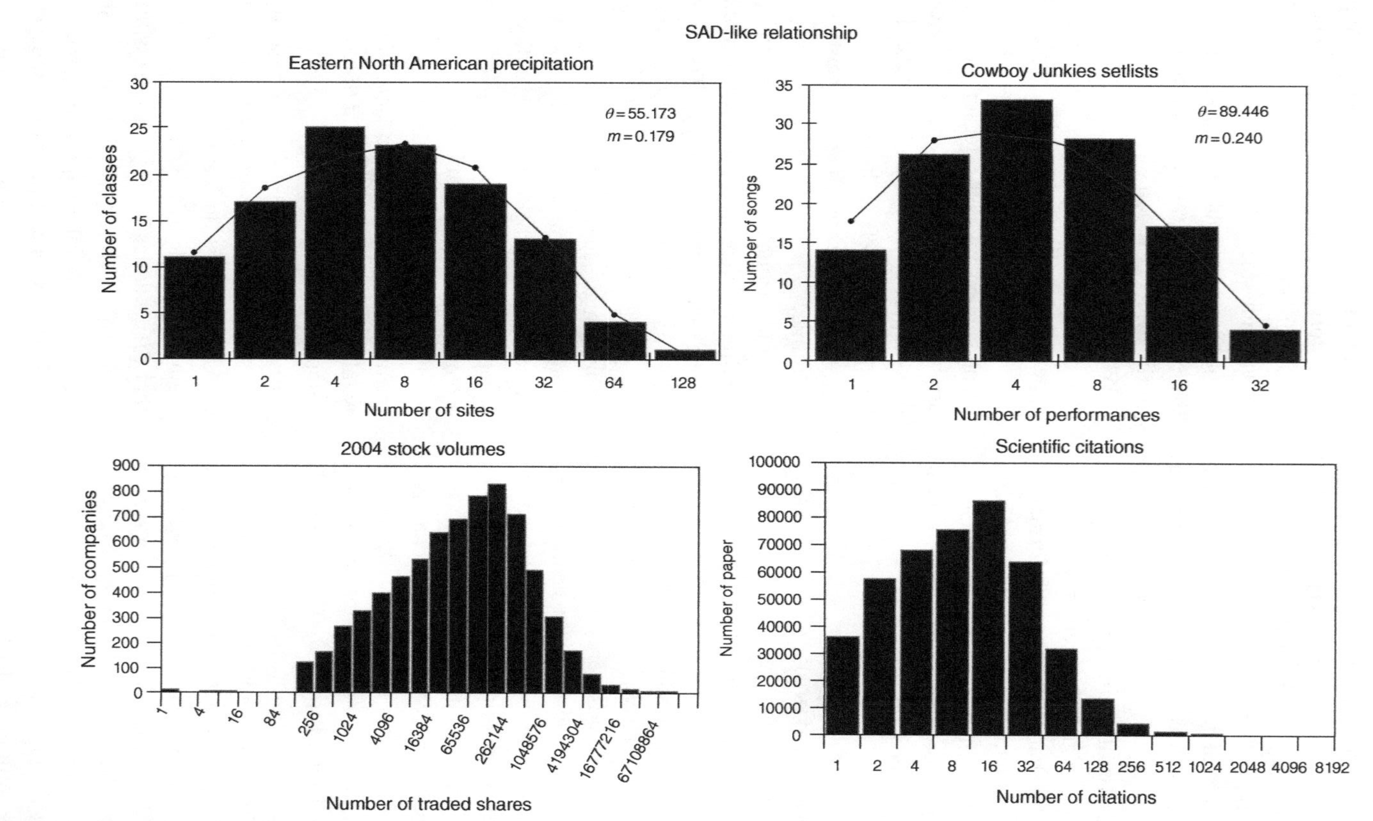

Figure 10.18 Preston plots of North American precipitation classes by site, stock volumes for publicly traded US corporations, song performances in Cowboy Junkies setlists and citation frequencies of scientific papers. (*Source*: Nekola and Brown (2007). Reproduced with permission from Nature Publishing Group.)

assessments of the trajectory of increasing global temperatures, but there is no knowing what the weather will be like in Crewe on 23 June 2073.

Each scientific field that has encountered this issue has needed to revise its expectations of what can and can't be predicted about the behaviour of complex systems. In all cases, progress has been made by seeking the broad patterns and dropping the fine details, no matter how critical they may seem to those at the coalface. In physics, the ideal gas equation succeeds in predicting the behaviour of gases despite completely excluding interactions among molecules, which are known to be important. The equation is therefore in a sense fundamentally wrong—but it works.

Uncomfortable though it may be, ecologists might need to throw out their carefully assembled data on species interactions when they move up to study the bigger picture. This bold statement was the rationale behind the most controversial idea in modern ecology: neutral theory.

10.12 Unified neutral theory

In 2001, Steve Hubbell published a book that shook the foundations of ecology: 'The Unified Neutral Theory of Biodiversity and Biogeography' (Hubbell, 2001). Hubbell had been working for decades in the tropical rain forests of Panama. These are hyperdiverse systems containing an astonishing level of local species richness. How could literally hundreds of species of trees coexist when they require the same basic resources? These species certainly looked different, and much research effort had been expended on divining the ecological variation among them.

But what if, despite all the apparent differences, these species actually behaved identically from a demographic perspective? In other words, what would happen if you assumed that any individual tree had exactly the same chances of survival and reproduction as any other, regardless of its identity? This was Hubbell's starting point.

He assumed that within the region there was a metacommunity of M species. This was divided into arbitrary patches, within which species richness was proportional to the area. He then assigned each individual a small probability v of evolving into a new species. There were several key assumptions. The first was dispersal limitation: no species was able to spread everywhere. This is clearly true of tropical trees, whose large seeds are vital to allow survival of young plants on the dark forest floor, but which preclude them travelling long distances. He also assumed that forests were saturated: every available space for an individual was taken, as with the closed canopy of a rain forest, which only allows for a certain number of adult trees. Finally he added the crucial rule of equality *per capita*: each individual would behave the same way in terms of its demographic parameters, regardless of its species.

Through a process of mathematical logic which we will bypass *ad rem*, Hubbell arrived at a constant θ which he dubbed the 'biodiversity number' and could be derived as

$$\theta = M\frac{v}{1-v}$$

Once estimated, θ could be used to predict the relative abundances of species in a community according to a pattern known as the zero sum multinomial (ZSM). If v were to increase while M remained constant, more rare species would evolve. This model could be fit to observed data and corresponded closely to the lognormal (Fig. 10.19).

Neutral theory has served to reshape the debate over the causes of patterns in natural systems. When it works, it implies that highly detailed, complex models of species interactions are unnecessary. When it fails, it provokes ecologists to ask what processes need to be added. The fact that neutral theory appears to match real data is intriguing, though not necessarily evidence for its veracity, because many models can be shown to do the same thing. Fierce debates have ensued over what information is (and isn't) required to capture the same patterns.

While neutral theory performs well for static patterns, it has been less successful at predicting the dynamics of systems, such as changes in the

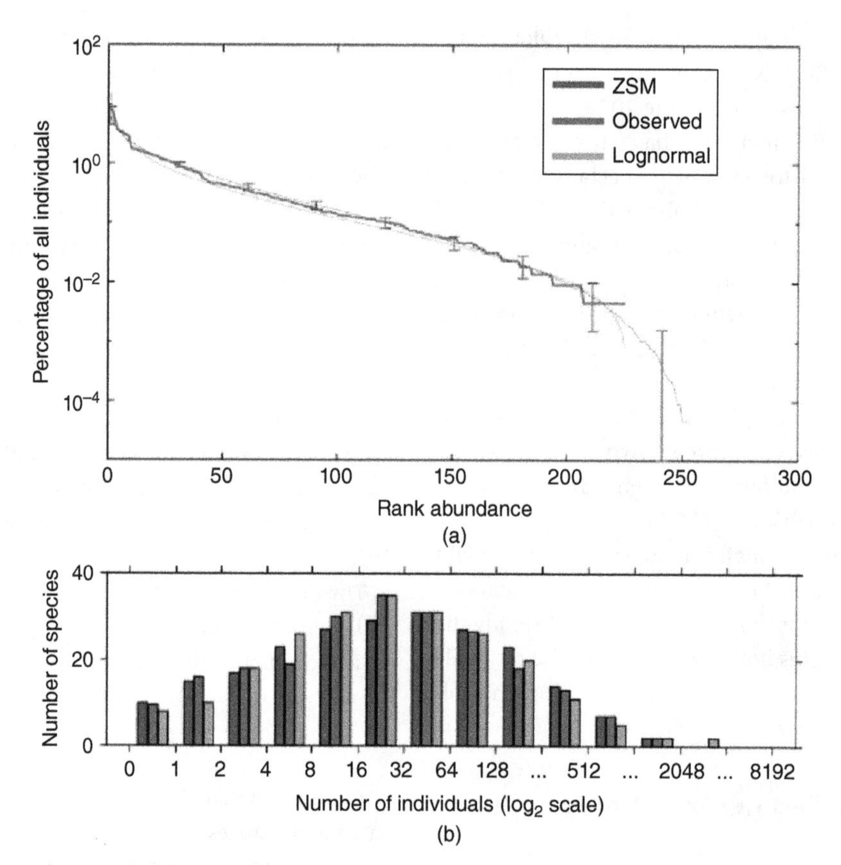

Figure 10.19 Observed species abundance distributions for the Barro Colorado Island tree dataset compared with predictions based on the lognormal and neutral theory (zero sum multinomial (ZSM)). (*Source*: McGill (2003). Reproduced with permission from Nature Publishing Group.) (*See colour plate section for the colour representation of this figure.*)

abundances of particular species or the pattern of species ages (Nee, 2005). It is particularly poor at predicting the dynamics of the most abundant species, which in nature remain stable through time and are often the most crucial for driving the behaviour of ecosystems (Connolly et al., 2014). For example, in the Bristol Channel, four core species make up 98% of total fish biomass, and their abundance remains stable through time due to density-dependent regulation (Henderson and Magurran, 2014). Meanwhile it is the rarer species that are constantly entering and leaving, but perhaps having little impact on overall ecosystem processes.

Neutral theory remains interesting because, despite containing patently incorrect assumptions, it often works. We know that there are differences among organisms; while these might cancel each other out in fitness terms, this is so implausible as to be effectively impossible (Purves and Turnbull, 2010). Its most savage critics point out, with good cause, that the theory provides no mechanism and merely substitutes a stochastic probability distribution for actual knowledge of the underlying processes (Clark, 2012). The performance of US college basketball teams is a highly structured set of interactions, based on a clear competitive hierarchy, yet the number of wins across teams follows exactly the same SAD-like form (Warren et al., 2011). If we don't need these interactions to create the SAD, that doesn't mean they're not happening, more that the SAD isn't providing any information about them. Perhaps ecologists have been measuring the wrong outcome, and we should instead be looking for new ways to describe communities.

Another interesting possibility is that some communities might behave as if they're neutral, while others behave as if niche based, with a transition between them much like that which occurs in water as it moves between solid, liquid and gas phases (Fisher and Mehta, 2014). Niche-structured communities might occur in constant environments with large populations of each species, while neutral communities are found in unpredictable environments where many species occur in low numbers. Perhaps both occur simultaneously but apply to different elements of the community. There are exciting developments in the fundamental theory of ecology which will surely lead to major new insights in the coming years.

Though neutral theory is unlikely to be the fundamental basis behind SADs, some common statistical process must be giving rise to their distinctive shape. White et al. (2012) took over 15,000 SADs from across taxa and the whole globe and showed that the majority of the variation (83–93%) could be accounted for with a simple model that included only the species richness S and total abundance N. In principle any model producing log-series distributions would achieve similar results; what is interesting is how little information is required to predict the SAD. Once we know the emergent properties of complex systems, their overall patterns are readily predictable. The enticing prospect remains that a unified theory of biodiversity is within reach and these two simple measures, S and N, are likely to lie at its core, along with three simple assumptions (McGill, 2010):

- Individuals are clustered within species.
- Abundance among species is unequal.
- No interactions occur among species.

These simple rules are sufficient to explain a large number of observed patterns in nature; the question therefore is why, and how much else do we need to know? What determines S and N?

10.13 Metabolic theory of ecology

The best guess for a foundational theory that can explain variation in the relative abundance of organisms is rooted in chemistry and the transfer of energy within food webs. It is known as the metabolic theory of ecology. It was first set out in detail by Brown et al. (2004), whose research was concerned with the long-standing observation that larger-bodied species occur at lower abundances than smaller (e.g. Colinvaux, 1979). Why is this?

The amount of resources required to maintain a large body scales as M^{α}, where M is body mass and α is a positive scaling parameter. If you assume that all species, large and small, have equal access to resources, then their abundance N should follow the relationship

$$N = iM^{-\alpha}$$

where i is a normalisation constant. It's not quite so simple though, because an increase in temperature causes each individual to require more resources. The basic formula therefore needs rescaling by an additional parameter D which represents temperature dependence:

$$N = iM^{-\alpha}D^{-1}$$

The term D can be expected to follow normal thermodynamic principles, which are beyond the scope of this book but should be familiar to anyone with a basic training in chemistry. A little algebra leads to a 'temperature-corrected' abundance $N_{temp} = ND$, and by log transformation this converts to

$$\log N_{temp} = \log i - \alpha \log M$$

Whether or not you follow the logic, the end result is simple—this is the equation of a straight line. In nature, metabolic rates scale as $M^{0.75}$, which means the abundance of organisms should scale as $M^{-0.75}$. This is the basic theory, though it often requires inclusion of some additional processes in order to apply to the real world. Hechinger et al. (2011) took three estuarine communities in California containing species spanning 11 orders of magnitude

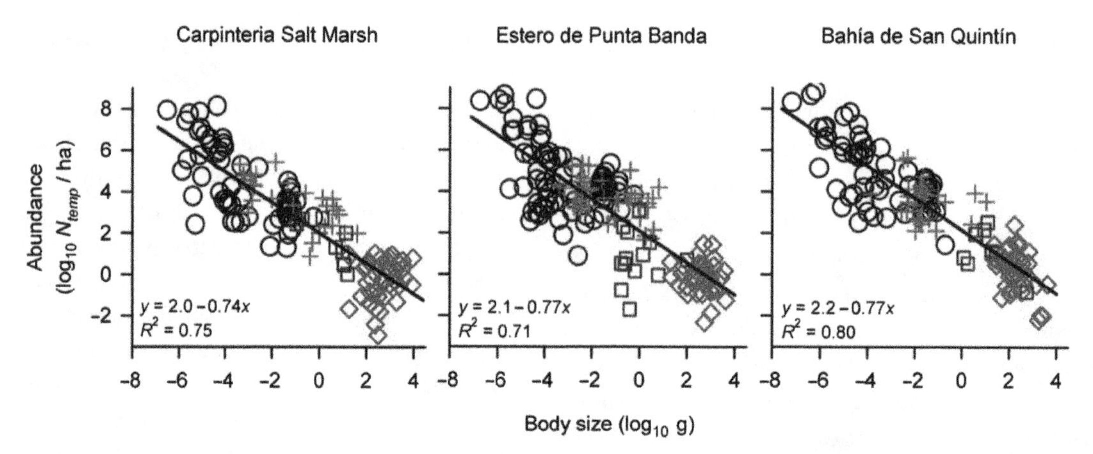

Figure 10.20 Temperature-corrected abundance as a function of body size for species in three estuaries after controlling for trophic level; circles, parasites; crosses, invertebrates; squares, fish; diamonds, birds. (*Source*: Hechinger et al. (2011). Reproduced with permission from the American Association for the Advancement of Science.)

in both body size and abundance. After correcting for the efficiency of energy transfer between trophic levels, they were able to fit a single relationship to all species—from the very smallest to the largest, ectotherms and endotherms, vertebrates and invertebrates, even including parasites (Fig. 10.20). Best of all, the fitted values for α were all close to 0.75, in agreement with the basic model.

Extensions of the theory exist which attempt to account for the number of species in a system (S) and their range of sizes, which enables prediction of the total number of individuals N. Metabolic theory has its detractors; some are uncomfortable with reducing the complexity of natural systems to simple transfers of energy, while others dispute whether its predictions are actually met in real systems. It is probable that the underlying logic is correct; the remaining issues surround how much detail is needed to capture the form and dynamics of natural systems. There will no doubt for be further enhancements and applications of the theory over the coming years.

10.14 Conclusions

Communities are sets of interacting species that occur together over space and time. Deciding what constitutes a community has often been arbitrary but increasingly relies on statistical techniques that support their existence as genuine entities. Their parts can be broken down into functional groups that are linked within food chains, the lengths of which are driven by multiple factors, though area seems to be particularly influential, along with a lesser role for productivity. These chains are combined within food webs Certain structural features of webs confer stability on communities, in particular their degree of connectance, which exceeds random expectations. The most connected species are also the most important for holding the community together.

The distribution of abundance among species tends to follow a predictable lognormal pattern. Even when it does not appear on first inspection, close attention to the structure of the community often reveals the lognormal pattern. Many attempts have been made to explain this, but a wider appreciation of the behaviour of complex systems suggests that we might not need to look for detailed bottom-up models. Important lessons can be learnt from physics and other fields about the usefulness of simple models which can be expanded as required. These include neutral and metabolic theories.

If ecology is to answer truly big questions, it might need to break free from an obsession with the fine details of natural history. To say that not everyone agrees with this would be an understatement, but it is hard to see how general rules can be established without a reductive approach.

10.14.1 Recommended reading

Brown, J. H., J. F. Gillooly, A. P. Allen, V. M. Savage, and G. B. West, 2004. Toward a metabolic theory of ecology. *Ecology* 85:1771–1789.

Chesson, P., 2013. Species competition and predation. pp. 223–256 in Leemans, R. (ed) *Ecological Systems*. Springer-Verlag.

Hubbell, S. P., 2001. *The Unified Neutral Theory of Biodiversity and Biogeography*. Princeton University Press.

10.14.2 Questions for the future

- What features of food webs maintain their stability?
- How much do we need to know about the differences among species to predict community-level properties?
- Can the structure of natural systems be reduced to the transfer of energy and materials?

References

Baird, D. and R. E. Ulanowicz, 1989. The seasonal dynamics of the Chesapeake Bay ecosystem. *Ecological Monographs* 59:329–364.

Bauer, S. and B. J. Hoye, 2014. Migratory animals couple biodiversity and ecosystem functioning worldwide. *Science* 344:1242552.

Bonhommeau, S., L. Dubroca, O. Le Pape, J. Barde, D. M. Kaplan, E. Chassot, and A.-E. Nieblas, 2013. Eating up the world's food web and the human trophic level. *Proceedings of the National Academy of Sciences of the United States of America* 110:20617–20620.

Brooker, R. W., R. M. Callaway, L. A. Cavieres, Z. Kikvidze, C. J. Lortie, R. Michalet, F. I. Pugnaire, A. Valiente-Banuet, and T. G. Whitman, 2009. Don't diss integration: a comment on Ricklef's disintegrating communities. *American Naturalist* 174:919–927.

Brown, J. H., J. F. Gillooly, A. P. Allen, V. M. Savage, and G. B. West, 2004. Toward a metabolic theory of ecology. *Ecology* 85:1771–1789.

Casini, M., T. Blenckner, C. Möllmann, A. Gårdmak, M. Lindegren, M. Llope, G. Kornilovs, M. Plikshs, and N. C. Stenseth, 2012. Predator transitory spillover induces trophic cascades in ecological sinks. *Proceedings of the National Academy of Sciences of the United States of America* 109:8185–8189.

Clark, J. S., 2012. The coherence problem with the Unified Neutral Theory of Biodiversity. *Trends in Ecology & Evolution* 27:198–202.

Clements, F. E., 1916. *Plant Succession: Analysis of the Development of Vegetation*. Carnegie Institute of Washington.

Colinvaux, P. A., 1979. *Why Big Fierce Animals Are Rare: An Ecologist's Perspective*. Princeton University Press.

Connolly, S. R., T. P. Hughes, D. R. Bellwood, and R. H. Karlson, 2005. Community structure of corals and reef fishes at multiple scales. *Science* 309:1363–1365.

Connolly, S. R., M. A. MacNeil, J. Caley, N. Knowlton, E. Cripps, M. Hisano, L. M. Thibaut, B. D. Bhattacharya, L. Benedetti-Cecchi, R. E. Brainard, A. Brandt, F. Bulleri, K. E. Ellingsen, S. Kaiser, I. Kroencke, K. Linse, E. Maggi, T. D. O'Hara, L. Plaisance, G. C. B. Poore, S. K. Sarkar, K. K. Satpathy, U. Schueckel, A. Williams, and R. S. Wilson, 2014. Commonness and rarity in the marine biosphere. *Proceedings of the National Academy of Sciences of the United States of America* 111:8524–8529.

Cottee-Jones, H. E. W. and R. J. Whittaker, 2012. The keystone species concept: a critical appraisal. *Frontiers of Biogeography* 4:117–127.

Dunne, J. A., R. J. Williams, and N. D. Martinez, 2002. Network structure and biodiversity loss in food webs: robustness increases with connectance. *Ecology Letters* 5:558–567.

Dunstan, P. K., S. D. Foster, and R. Darnell, 2011. Model based grouping of species across environmental gradients. *Ecological Modelling* 222:955–963.

Eklöf, A., U. Jacob, J. Kopp, J. Bosch, R. Castro-Urgal, N. P. Chacoff, B. Dalsgaard, C. de Sassi, M. Galetti, P. R. Guimarães, S. B. Lomáscolo, A. M. Martín González, M. A. Pizo, R. Rader, A. Rodrigo, J. M. Tylianakis, D. P. Vázquez, and S. Allesina, 2013. The dimensionality of ecological networks. *Ecology Letters* 16:577–583.

Elton, C. S., 1958. *The Ecology of Invasions by Animals and Plants*. Methuen.

Fauth, J. E., J. Bernardo, M. Camara, W. J. J. Resetarits, J. Van Buskirk, and S. A. McCollum, 1996. Simplifying the jargon of community ecology: a conceptual approach. *American Naturalist* 147:282–286.

Fisher, C. K. and P. Mehta, 2014. The transition between the niche and neutral regimes in ecology. *Proceedings of the National Academy of Sciences of the United States of America* 111:13111–13116.

Gilarranz, L. J., J. M. Pastor, and J. Galeano, 2012. The architecture of weighted mutualistic networks. *Oikos* 121:1154–1162.

Gleason, H. A., 1926. The individualistic concept of the plant association. *Torrey Botanical Club Bulletin* 53:7–26.

González-Cabello, A. and D. R. Bellwood, 2009. Local ecological impacts of regional biodiversity on reef fish assemblages. *Journal of Biogeography* 36:1129–1137.

Hechinger, R. F., K. D. Lafferty, A. P. Dobson, J. H. Brown, and A. M. Kuris, 2011. A common scaling rule for abundance, energetics, and production of parasitic and free-living species. *Science* 333:445–448.

Henderson, P. A. and A. E. Magurran, 2014. Direct evidence that density-dependent regulation underpins the temporal stability of abundant species in a diverse animal community. *Proceedings of the Royal Society Series B* 281:20141336.

Hercos, A. P., M. Sobansky, H. L. Queiroz, and A. E. Magurran, 2013. Local and regional rarity in a diverse tropical fish assemblage. *Proceedings of the Royal Society Series B* 280:20122076.

Holt, M. H., M. A. Leibold, and R. D, editors, 2005. *Metacommunities: Spatial Dynamics and Ecological Communities*. The University of Chicago Press.

Hubbell, S. P., 2001. *The Unified Neutral Theory of Biodiversity and Biogeography*. Princeton University Press.

Karban, R., P. Grof-Tisza, and M. Holyoak, 2012. Facilitation of tiger moths by outbreaking tussock moths that share the same host plants. *Journal of Animal Ecology* 81:1095–1102.

Kéfi, S., E. L. Berlow, E. A. Wieters, S. A. Navarrete, O. L. Petchey, S. A. Wood, A. Boit, L. N. Joppa, K. D. Lafferty, R. J. Williams, N. D. Martinez, B. A. Menge, C. A. Blanchette, A. C. Iles, and U. Brose, 2012. More than a meal ... integrating non-feeding interactions into food webs. *Ecology Letters* 15:291–300.

Leaper, R., P. K. Dunstan, S. D. Foster, N. S. Barrett, and G. J. Edgar, 2014. Do communities exist? Complex patterns of overlapping marine species distributions. *Ecology* 95:2016–2025.

Lepš, J., V. Novotný, and Y. Basset, 2001. Habitat and successional status of plants in relation to the communities of their leaf-chewing herbivores in Papua New Guinea. *Journal of Ecology* 89:186–199.

Lubchenco, J., 1978. Plant species diversity in a marine intertidal community: importance of herbivore food preference and algal competitive abilities. *American Naturalist* 112:23–39.

MacArthur, R. H., 1955. Fluctuations of animal populations and a measure of community stability. *Ecology* 36:533–536.

Magurran, A. E. and P. A. Henderson, 2003. Explaining the excess of rare species in natural species abundance distributions. *Nature* 422:714–716.

May, R. M., 1972. Will a large complex system be stable? *Nature* 238:413–414.

McGill, B. J., 2003. A test of the unified neutral theory of biodiversity. *Nature* 422:881–885.

McGill, B. J., 2010. Towards a unification of unified theories of biodiversity. *Ecology Letters* 13:627–642.

Morris, M. R. and J. A. Stanford, 2011. Floodplain succession and soil nitrogen accumulation on a salmon river in southwestern Kamchatka. *Ecological Monographs* 81:43–61.

Mougi, D. and M. Kondoh, 2012. Diversity of interaction types and ecological community stability. *Science* 337:349–351.

Mouillot, D., S. Villéger, V. Parravicini, M. Kulbicki, J. E. Arias-González, M. Bender, P. Chabanet, S. R. Floeter, A. Friedlander, L. Vigliola, and D. R. Bellwood, 2014. Functional over-redundancy and high functional vulnerability in global fish faunas on tropical reefs. *Proceedings of the National Academy of Sciences of the United States of America* 111:13757–13762.

Navarro, J., S. C. Votier, J. Aguzzi, J. J. Chiesa, M. G. Forero, and R. A. Phillips, 2013. Ecological segregation in space, time and trophic niche of sympatric planktivorous petrels. *PLoS ONE* 8:e62897.

Nee, S., 2005. The neutral theory of biodiversity: do the numbers add up? *Functional Ecology* 19:173–176.

Nekola, J. C. and J. H. Brown, 2007. The wealth of species: ecological communities, complex systems and the legacy of Frank Preston. *Ecology Letters* 10:188–196.

Nielsen, A. and J. Bascompte, 2007. Ecological networks, nestedness and sampling effort. *Journal of Ecology* 95:1134–1141.

Novotný, V., S. E. Miller, L. Baje, S. Balagawi, Y. Basset, L. Cizek, K. J. Craft, F. Dem, R. A. I. Drew, J. Hulcr, J. Lepš, O. T. Lewis, R. Pokon, A. J. A. Stewart, G. A. Samuelson, and G. D. Weiblen, 2010. Guild-specific patterns of species richness and host specialization in plant-herbivore food webs from a tropical forest. *Journal of Animal Ecology* 79:1193–1203.

Pimm, S. L., 1982. *Food Webs*. University of Chicago Press.

Pocock, M. J. O., D. M. Evans, and J. Memmott, 2012. The robustness and restoration of a network of ecological networks. *Science* 335:973–977.

Post, D. M., 2002. The long and short of food-chain length. *Trends in Ecology & Evolution* 17:269–277.

Post, D. M., M. L. Pace, and N. G. Hairston, 2000. Ecosystem size determines food-chain length in lakes. *Nature* 405:1047–1049.

Prugh, L. R. and J. S. Brashares, 2012. Partitioning the effects of an ecosystem engineer: kangaroo rats control community structure via multiple pathways. *Journal of Animal Ecology* 81:667–678.

Purves, D. W. and L. A. Turnbull, 2010. Different but equal: the implausible assumption at the heart of neutral theory. *Journal of Animal Ecology* 79:1215–1225.

Ricklefs, R. E., 2008. Disintegration of the ecological community. *American Naturalist* 172:741–750.

Root, R., 1967. The niche exploitation pattern of the blue-grey gnatcatcher. *Ecological Monographs* 37:317–350.

Roslin, T., G. Várkonki, M. Koponen, V. Vikberg, and M. Nieminen, 2014. Species–area relationships across four trophic levels—decreasing island size truncates food chains. *Ecography* 37:443–453.

Rzanny, M. and W. Voigt, 2012. Complexity of multitrophic interactions in a grassland ecosystem depends on plant species diversity. *Journal of Animal Ecology* 81:614–627.

Saavedra, S., D. B. Stouffer, B. Uzzi, and J. Bascompte, 2011. Strong contributors to network persistence are the most vulnerable to extinction. *Nature* 478:233–235.

Sato, T., T. Egusa, K. Fukushima, T. Oda, N. Ohte, N. Tokuchi, K. Watanabe, M. Kanaiwa, I. Murakami, and K. D. Lafferty, 2012. Nematomorph parasites indirectly alter the food web and ecosystem function of streams through behavioural manipulation of their cricket hosts. *Ecology Letters* 15:786–793.

Schmitz, O. J., 2008. Effects of predator hunting mode on grassland ecosystem function. *Science* 319:952–954.

Takimoto, G. and D. M. Post, 2013. Environmental determinism of food-chain length: a meta-analysis. *Ecological Research* 28:675–681.

Thébault, E. and C. Fontaine, 2010. Stability of ecological communities and the architecture of mutualistic and trophic networks. *Science* 329:853–856.

Ulanowicz, R. E., R. D. Holt, and M. Barfield, 2014. Limits on ecosystem trophic complexity: insights from ecological network analysis. *Ecology Letters* 17:127–136.

de Visser, S. N., B. D. Freymann, and H. Olff, 2011. The Serengeti food web: empirical quantification and analysis of topological changes under increasing human impact. *Journal of Animal Ecology* 80:484–494.

Volkov, I., J. R. Banavar, S. P. Hubbell, and A. Maritan, 2007. Patterns of relative species abundance in rainforests and coral reefs. *Nature* 450:45–49.

Warren, R. J., D. K. Skelly, O. J. Schmitz, and M. A. Bradford, 2011. Universal ecological patterns in college basketball communities. *PLoS ONE* 6:e17342.

Watson, D. M. and M. Herring, 2012. Mistletoe as a keystone resource: an experimental test. *Proceedings of the Royal Society Series B* 279:3853–3860.

White, E. P., K. M. Thibault, and X. Xiao, 2012. Characterizing species abundance distributions across taxa and ecosystems using a simple maximum entropy model. *Ecology* 93:1772–1778.

Williams, R. J., E. L. Berlow, J. A. Dunne, A.-L. Barabási, and N. D. Martinez, 2002. Two degrees of separation in complex food webs. *Proceedings of the National Academy of Sciences of the United States of America* 99: 12913–12916.

CHAPTER 11

Stability

11.1 The big question

The communities observed in nature are only a subset of all possible combinations of species and not a random selection. Particular sets of species form (and reform) repeatedly. In some systems, no matter how many times they are perturbed, they are able to reassemble and continue as before. In other cases, however, disrupting a natural system leads to catastrophic changes that appear to be irreversible. What underlies these differences? Why do some combinations of species fail to persist?

11.2 Stable states

We begin by addressing the theoretical background to community stability. Natural systems can be thought of as possessing stabilising properties which act either *locally* or *globally* (Fig. 11.1). Communities can be locally stable but globally unstable (top right panel, Fig. 11.1), which means they are unlikely to change of their own accord, but once moved past a certain point will collapse and be unable to return. In other cases, local stability might be low but global stability high (lower left panel, Fig. 11.1). In this case the community is unlikely to persist, as any disruption will dislodge it, but it will not alter greatly.

This idea can be extended to more complex virtual landscapes. Two further properties can be envisioned to describe the system itself. Whether it can be shifted at all depends on its *resistance*, which is defined as its innate ability to avoid displacement. Once moved, its *resilience* determines the rate at which it returns to its former state (Fig. 11.2).

Based on these properties we can come to a more nuanced understanding of what makes certain systems either 'fragile' or 'robust'. The resistance of a community can be measured as its tolerance of external pressures such as environmental variation or disturbance. Once its resilience is exceeded, it becomes permanently altered to a new state. Together these properties describe the stability of natural systems.

The concept can be linked to broader ecological theory, for example, the gradient between ***r*-selected** species, which maximise reproduction, and ***K*-selected** species, which are adapted to persist in populations with high levels of intraspecific competition (MacArthur and Wilson, 1967). In an analogous fashion, *r*-selected communities are those which form in variable and unpredictable environments. They tend to have low resistance, but high resilience, allowing them to tolerate frequent disturbance. A grassland is a typical example; regularly cut back to ground level by grazing animals, the system puts up little resistance but recovers rapidly. In contrast, *K*-selected communities dominate in stable, predictable environments and are characterised by high resistance. This however trades off against their resilience and reduces their ability to restore themselves. Consider a mature forest, whose large trees are mechanically strong and require an

Natural Systems: The organisation of life, First Edition. Markus P. Eichhorn.

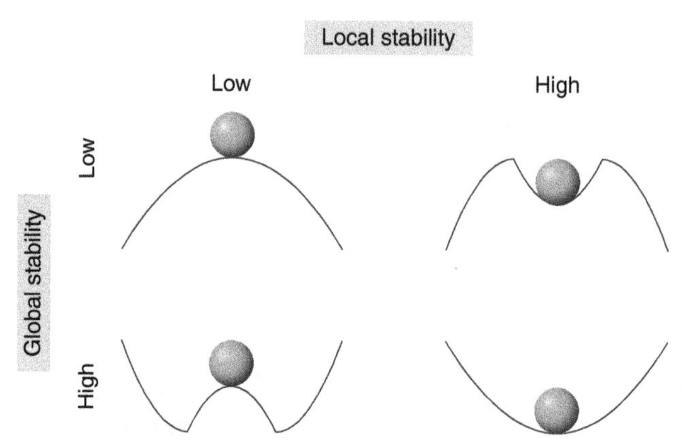

Figure 11.1 Schematic representation of local versus global stability in natural systems.

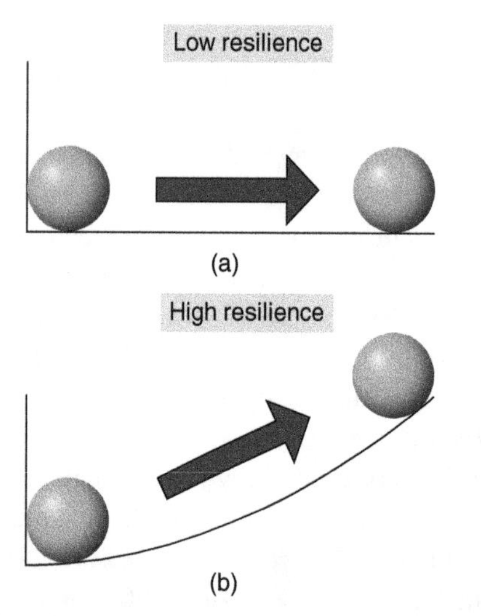

Figure 11.2 Schematic representation of systems with (a) low and (b) high resilience. The force required to move the system determines its resistance.

immense amount of energy to fell, but once removed take centuries to recover (if at all).

When Clements formed his original concept of the community in 1916, he assumed that any site would have a particular community type that was most suited to occur there. Given enough time, this was the set of species that should eventually come to dominate. This ideal is referred to as the 'climax' community, though it is not widely accepted in modern ecology. One reason is that often there are multiple possible communities, none of which can lay claim to being the default state.

Consider two genera of freshwater plants often found in lakes: *Elodea*, which grows upwards from the bottom, and the duckweed *Lemna*, which floats on the surface. Can these two types coexist, or will they be mutually exclusive (Fig. 11.3)? In an influential study, Scheffer et al. (2003) set up 158 experimental lakes combining these two species with varying starting proportions. Depending on very fine differences in the initial abundances of each, lakes quickly diverged into being dominated by one type or the other (Fig. 11.4a). Overall, when measured in terms of the cover of floating plants (i.e. *Lemna*), almost all the communities fell one or other end of the spectrum of near-complete coverage or near-absence (Fig. 11.4b). This is reminiscent of the unstable equilibrium of resource ratio theory from Section 6.8.

Similar cases are widespread (see Scheffer et al., 2001). In the Brazilian Central Plateau, patches of grassy savannah intermingle with closed-canopy woodlands (Dantas et al., 2013). The open areas are filled with grasses which burn in the dry season, preventing sensitive woodland trees from developing. In the woodlands, the shade cast by trees prevents the grasses from building up and thereby protects them from fire. The two communities occur side by side in identical conditions, forming alternative stable states.

Other stable states might be formed as a result of environmental factors. Krummholz vegetation is

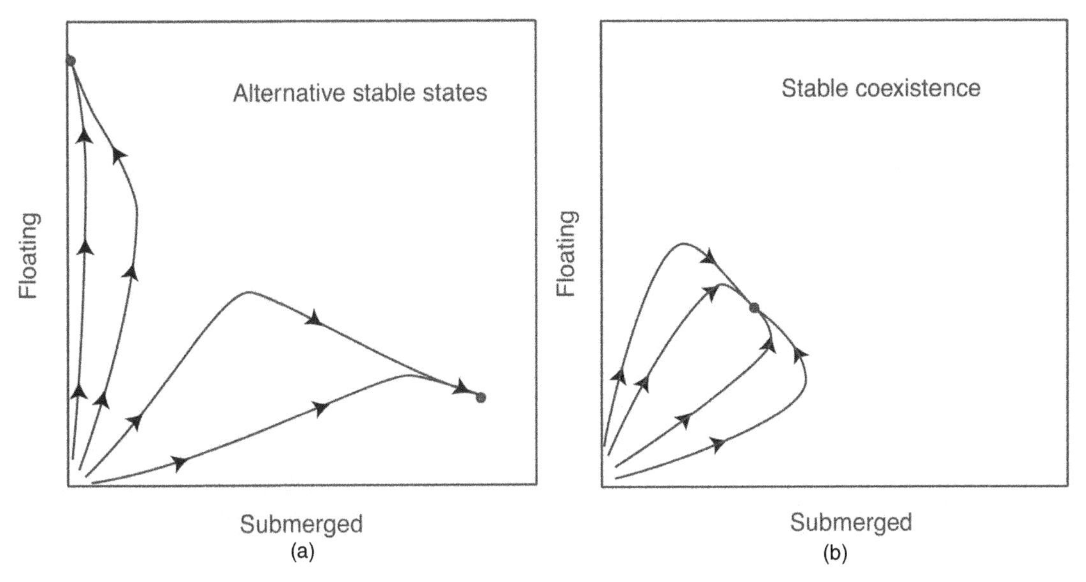

Figure 11.3 Possible outcomes of competition between submerged *Elodea* and floating *Lemna*. Either (a) one will dominate at the expense of the other or (b) both will be able to coexist. (*Source*: Scheffer et al. (2003). Copyright (2003) National Academy of Sciences, USA.)

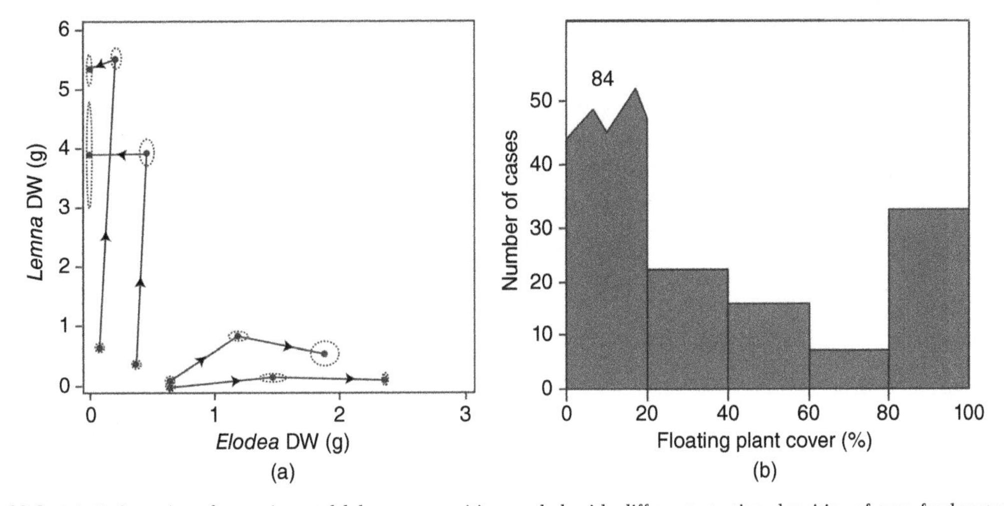

Figure 11.4 (a) Trajectories of experimental lake communities seeded with different starting densities of two freshwater plants measured as dry weight (DW); (b) final community states of 158 experimental lakes measured as the proportion of the lake surface area covered with floating plants. (*Source*: Scheffer et al. (2003). Copyright (2003) National Academy of Sciences, USA.)

common in alpine habitats, where dwarf pines or other shrubs occur in patches or stripes with clear areas in between. Gaps are maintained by winter snow, which swirls as it falls, collecting in the spaces, where it forms a deep covering over the ground that takes much longer to melt in spring. In contrast, the pines shed the snow that falls directly on them. Removal of snow would allow pine seedlings to survive in the open areas, and the total amount of winter snow determines the balance between these two stable states.

Note however that many of these anecdotal examples of alternative stable states remain exactly that. Direct experimental evidence demonstrating their existence is remarkably rare (Schröder et al., 2005). While hard to identify in community assembly, however, it can be seen in experimental evolution of communities. Celiker and Gore (2014) took six bacterial species and allowed them to evolve from the same starting abundances. They neither converged to a single, identical community nor were they randomly organised. Instead a relatively small number of configurations developed and stabilised. The extent to which dynamic evolution and coadaptation shape the communities of larger organisms remains to be fully explored.

11.3 Changing environments

The environment is constantly changing. Whether this occurs on the long timescales of glacial retreat, through the troubling pace of human-induced climatic change, or even over much shorter durations, it can be sufficient to perturb less resistant communities. We might envision a system following a transition between two stable states, but this journey can take a variety of forms, from a smooth and gradual process of change to sudden, dramatic shifts (Fig. 11.5). These might occur unpredictably when a certain point is reached, perhaps without any obvious warning. In some cases these switches can be entirely natural.

In the Pacific Ocean, Hare and Mantua (2000) documented changes in a range of 100 biological and environmental metrics, including plankton abundance, fish catches and temperatures. A surprising outcome is shown in Fig. 11.6. In 1977, for no apparent reason, the system switched rapidly to another state. Further observation saw it switch again in 1989, though it was not a simple reversal. The causes remain mysterious; climatic drivers or human impacts such as fishing cannot be excluded, but it would be difficult to isolate a single cause (and in truth there may not be one). Flipping between states might be a natural occurrence. Dramatic shifts in communities of benthic marine species have also occurred in the fossil record, driven by past climate change (Yasuhara et al., 2008).

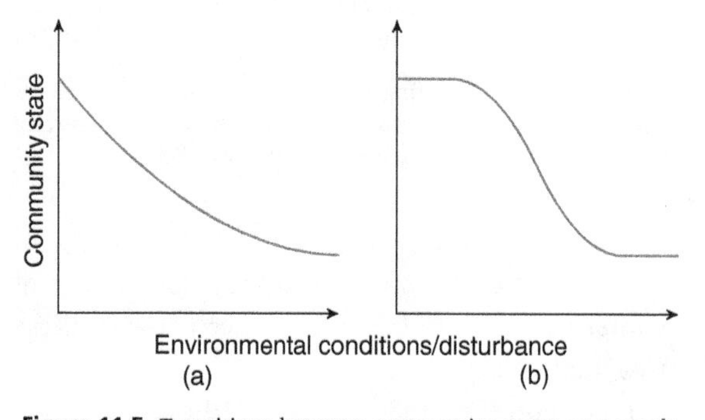

Figure 11.5 Transitions between community states as a result of changes in environmental conditions or disturbance may be (a) continuous or (b) abrupt.

Other systems show a dramatic and unpredictable shift, referred to as *catastrophic bifurcation*. As conditions change incrementally, a sudden switch occurs that is difficult to reverse. This happened in the Sahara which, though implausible to the modern observer, was covered in lush vegetation only 6,000 years ago. As glacial retreat occurred further north, it initiated a protracted period of gradual climatic change and drying in North Africa. At a point around 5,500 years ago, the Saharan vegetation had absorbed as much change as it could and collapsed (Foley et al., 2003). This was despite the relatively smooth trend in summer insolation (i.e. input of solar energy) through time.

To understand why the lush Saharan vegetation vanished, we need to recognise the role played by plants in regulating local climate (Fig. 11.7). Plants generate clouds through transpiration and therefore aid in spreading rainfall across entire regions (Sheil and Murdiyarso, 2009). As total levels of precipitation fell, water availability eventually reached the minimum that would allow vegetation to persist. This caused a dramatic shift to another stable state, one with no vegetation, and hence reduced precipitation, which reinforced the change and made it hard for plants to recover. In order to revegetate the region, water inputs would need to rise again, back to a level that could independently sustain plant communities. This was therefore an ecosystem-level change.

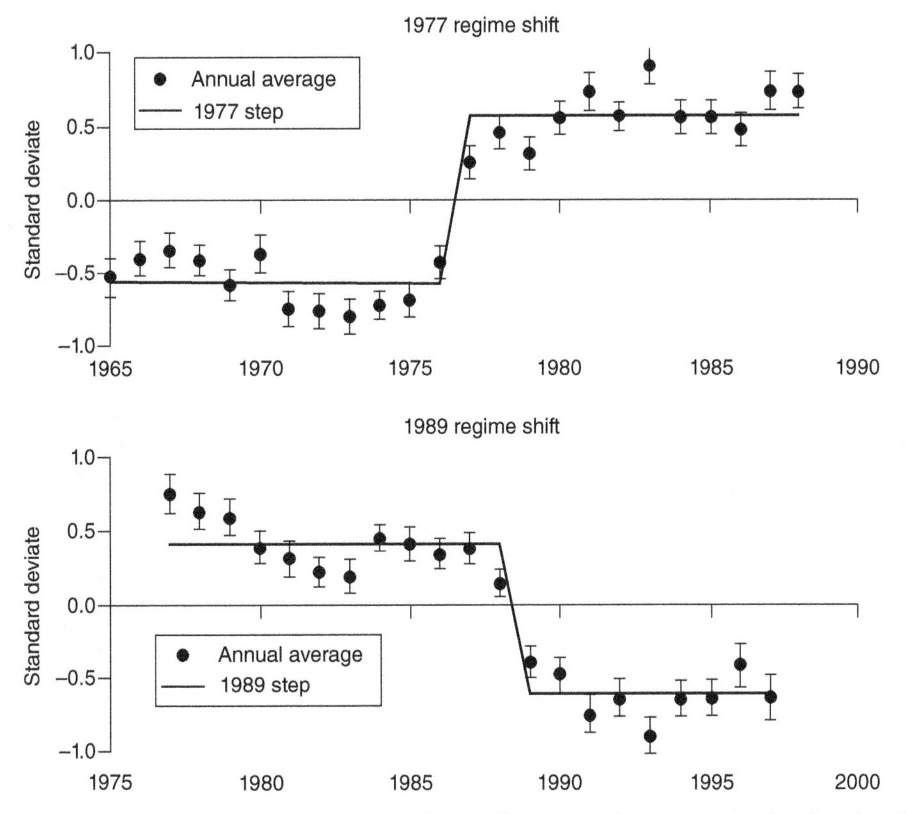

Figure 11.6 Shifts in the state of the Pacific Ocean ecosystem. The overall state is based on a composite of 31 climatic and 69 biological time series, plotted as the mean deviation ± SE of each time series from the average for the whole graph; hence the two panels do not align. (*Source*: Hare and Mantua (2000). Reproduced with permission from Elsevier.)

If this palæoecological example seems remote and inconsequential, consider what effects land-use change is having in many parts of the world. In forested regions their expanse acts as a conveyor belt, moving rainfall across the continent from the coast to the interior (Sheil and Murdiyarso, 2009). If too much of the rain forest is removed, this may lead to a self-reinforcing shift to a drier climate, causing the remainder of the forest to collapse. This is a very real fear, perhaps most of all because we have no means of predicting if or when it might happen. All we know is that it has happened in Africa before, leaving the last remnants of a formerly widespread forest stranded on the tops of volcanoes in the Canary Islands. It may recur in South America if too much of the Amazon, or the Atlantic coastal forest, is lost to development.

11.4 Hysteresis

The process just described, where a shift between stable states is hard to reverse, is known as *hysteresis*. A system can remain stable as the environment changes until it either reaches a critical point or is perturbed just enough to force it to switch to an alternative state. Reversing this process is difficult because conditions have to be brought back much further than the level at which the initial change occurred. When there are two stable states with an unstable equilibrium between them, abrupt shifts can occur, often with no prior warning.

The process can be illustrated using marble plots (Fig. 11.8). As the environment changes, the community might appear to remain in the same state, but its global resilience is impaired up to the point where

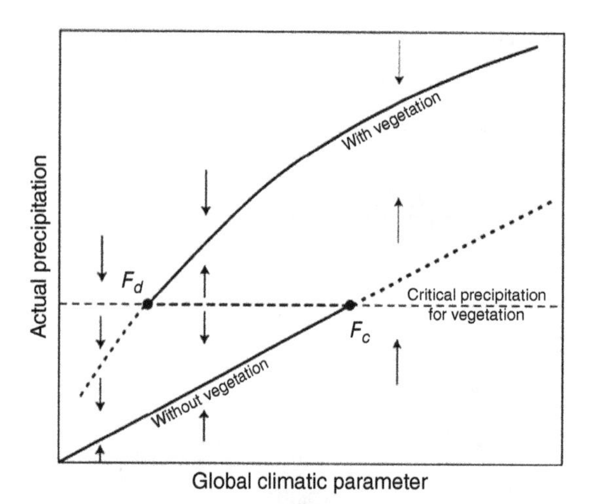

Figure 11.7 Causes of phase shift between vegetated and arid landscapes in the Sahara and Sahel region. Climatic change (moving leftwards) caused a reduction in precipitation, which led to a sudden drop in vegetation cover at point F_d. This further reduced precipitation levels and made the return of vegetation unlikely, unless climatic conditions return to point F_c. (*Source*: Foley et al. (2003). Reproduced with permission from Springer Business and Print Media.)

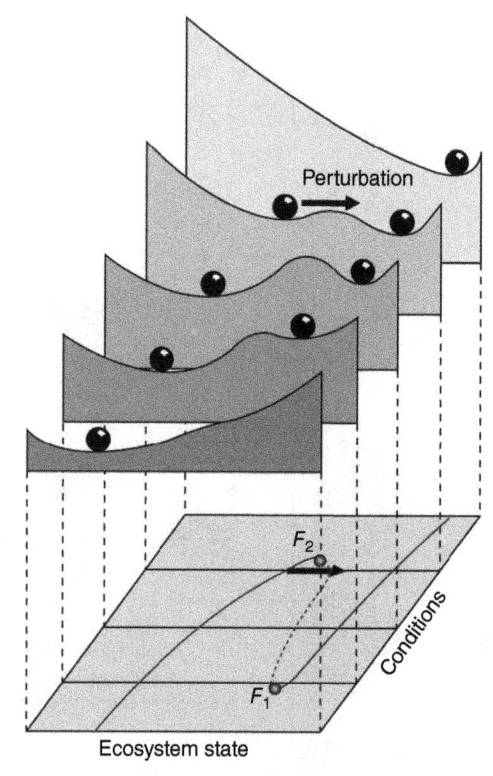

Figure 11.8 Marble plot to illustrate the process of hysteresis. F_1 and F_2 indicate critical tipping points where transition to the alternative state occurs. Solid lines represent stable equilibria, and the dotted line is unstable. (*Source*: Scheffer and Carpenter (2003). Reproduced with permission from Elsevier.)

even a small and innocuous disruption can cause radical changes. Not seeing any obvious change does not imply that a community is stable.

Before allowing ourselves to be won over by pretty diagrams, we should ask whether such patterns are ever witnessed in nature. The answer is a resounding yes. A striking case study is shown in Fig. 11.9. **Eutrophication** (addition of surplus nutrients, often through agricultural run-off or sewage) in lakes in the Netherlands increases phosphorus levels, while the degree of vegetation cover by freshwater plants is a good index of lake health. Nutrient-poor freshwater lakes tend to exist in a stable state with clear water and submerged vegetation. Adding nutrients promotes the growth of suspended algae, which cause the water to become turbid and shade out the plants. As a result, the community of plants on the lakebed collapses, with major implications for the organisms that use them as a habitat, especially fish. Reversing this shift requires major intervention because vegetation recovery depends upon phosphorus levels being brought down far below the level that caused the original change. The algal-dominated state is also stable but in this case undesirable.

Another example, also from an aquatic system, is that of Tasmanian kelp forests (Ling et al., 2009). In this region warming of the coastal waters through climate change is happening at four times the global average, allowing long-spined sea urchin (*Centrostephanus rodgersii*) to expand their ranges. At the same time overfishing has decimated populations of the spiny lobster (*Jasus edwardsii*), an important predator of urchins. Once urchin density passes a critical point, excessive grazing causes a shift from a kelp bed to a new and less species-rich stable state known as a sea urchin barren (Fig. 11.10).

Marine fishing has caused other dramatic shifts between stable states. Overconsumption of cod has facilitated a switch towards systems dominated

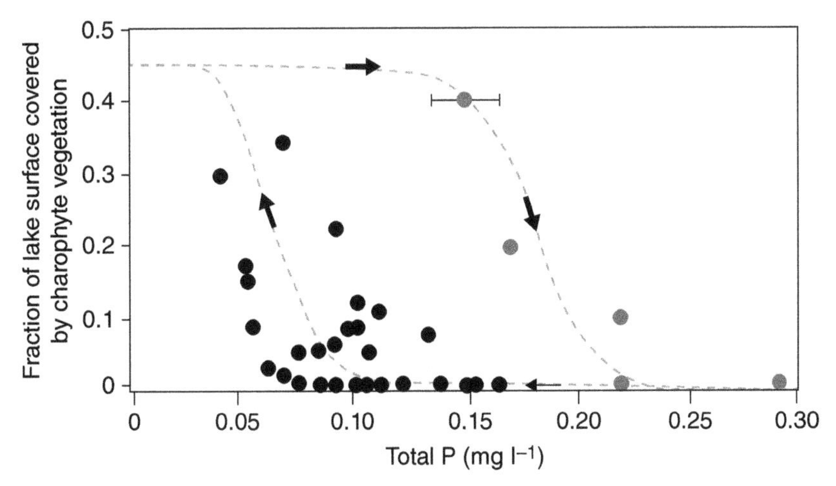

Figure 11.9 Relationship between lake phosphorus levels and cover of freshwater plants (charophytes) in the shallow Lake Veluwe. Phosphorus concentrations were increased in the late 1960s and early 1970s (gray dots) and then gradually reduced in the 1990s until a return transition took place (black dots). (*Source*: Scheffer et al. 2001. Reproduced with permission from Nature Publishing Group.)

instead by herring (Fouchald, 2010). Falling cod numbers allow herring to increase, and their larvae not only compete for the same food, but herring also eat cod eggs and larvae. This could allow herring dominance to form a new stable state, at least until humans deplete them as well. When cod levels are high, they suppress herring populations, which would otherwise impede their recruitment.

The rearrangement of a system into a new state can cause grave problems for conservation. In the Galápagos Islands, invasive plants and insects, introduced by humans, have integrated themselves into pollination networks (Traveset et al., 2013). In the process they have increased the nestedness of the communities (see Section 10.7), ironically making them more stable. This example serves to demonstrate that stability on its own should not be used as a criterion for measuring the health of a system.

11.5 Predicting changes

The theory of alternative stable states is still young and until recently it was not known whether it would be possible to predict when the transition between two states would occur. Some very recent research, borrowing an idea from physics, suggests that in at least some cases it might be. This involves the concept of *critical slowing down*, whereby the recovery of a system after perturbation acts as an indicator of how close it is to reaching the point of transition. Veraart et al. (2012) tested this in microcosms filled with cyanobacteria, which were subjected to gradually increasing light levels (Fig. 11.11). This process caused them to become increasingly stressed, as indicated by reduced efficiency of photosynthesis. When occurring at high density in the water column, cyanobacteria are self-shading, which creates a positive feedback allowing the population to be maintained even at high light levels that would otherwise be damaging. This facilitation outweighs the costs of competition. Microcosms were perturbed by sudden shock dilutions which reduced the total biomass by 3–5%, after which the population recovered. The rate of recovery from this decreased as they approached the point at which the population crashed.

Another potential indicator of an approaching transition is *flickering*, where systems switch rapidly between alternative states, giving an indication of what is to come (Wang et al., 2012, Dakos et al., 2013). This does not mean that a transition can be accurately predicted but that it might be able to

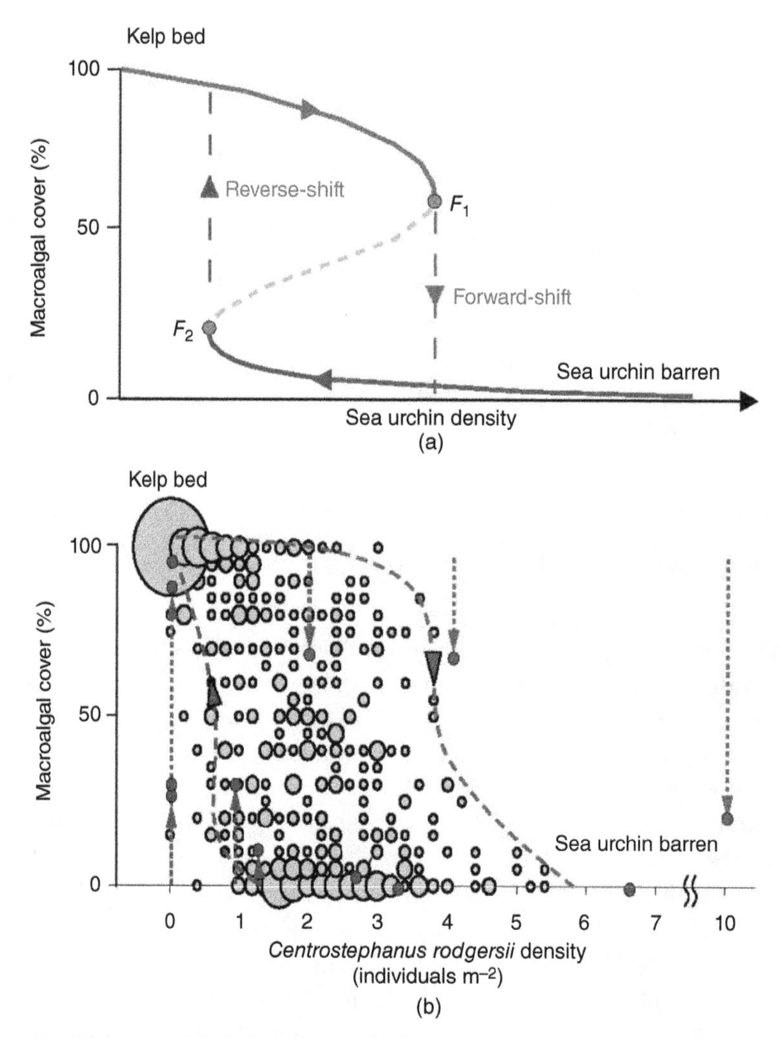

Figure 11.10 Catastrophic shift between kelp beds and sea urchin barrens. (a) Schematic of phase shift between kelp state on the upper line close to the threshold F_1, where a slight increase in sea urchin density induces a switch to the alternative and stable sea urchin barren state. Reverting to the kelp state occurs only if sea urchin density falls below the return threshold at F_2. Dashed line indicates the region of instability between the alternative stable states. (b) Macroalgal cover versus *Centrostephanus rodgersii* (sea urchin) density in eastern Tasmania. Bubble size represents relative frequency of particular urchin density and macroalgal cover combinations measured in 5 m^2 plots. Overlaid arrows indicate the response to experimental manipulations of *C. rodgersii* density. (*Source*: From Ling et al. (2009).)

monitor systems and record whether their resilience is changing through time. It could potentially act as an indicator of where conservation efforts are most needed. That said, incorporating deliberate perturbations into management of natural systems would be very controversial and possibly unwise. In a rare example, Carpenter et al. (2011) added top predators to a lake over 3 years, monitoring its food web, with an adjacent unmodified lake acting as a control. Signs of change were apparent more than a year before the food web completed a reorganisation to a new state. These signals can, however, be seen before non-catastrophic transitions as well, which might limit their usefulness (Kéfi et al., 2013).

Having learnt a great deal about the theory of community structure and stability, a single case study

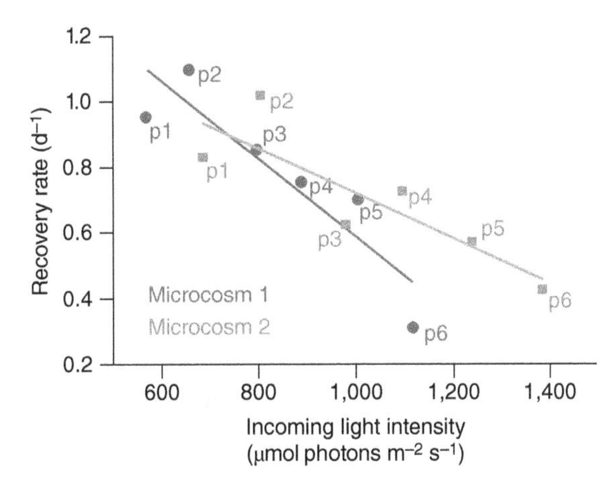

Figure 11.11 Recovery rate of two cyanobacterial microcosms (parameter obtained through model fits; higher values indicate faster recovery) following six perturbations (shock dilutions, p1–p6) along a gradient of light intensity, a measure of stress. (*Source*: Veraart et al. (2012). Reproduced with permission from Nature Publishing Group.)

follows to demonstrate how these principles can lead to a better understanding of natural systems and the appropriate means of managing them to maintain ecosystem services and promote conservation goals.

11.6 Coral reefs

Coral reefs are habitats in which space competition is of paramount importance. Their name implies a preponderance of corals, which actually make up only a small proportion of the living biomass but build the reef structure through calcareous deposits which accumulate through time; they are both keystone species and ecosystem engineers. Nevertheless, corals are not the dominant competitors for space, and instead macroalgae (including seaweeds) swiftly take over when permitted. If dead coral is not removed, or macroalgae are not grazed at a sufficient rate, reefs are always poised to switch to a state where algae become the dominant group.

The stability of coral reefs is maintained by herbivores which browse algae, thereby regulating the dominant space competitor (a form of top-down control; see Section 8.6). On Caribbean reefs where parrotfish are present, they graze 40% of the algae, but recent overfishing has caused declines in their populations, such that present levels of grazing are closer to 5%. Fortunately sea urchins are also present and more effective as grazers. As parrotfish have declined, urchins have taken over their role in grazing, an example of how multiple pathways in a food web can promote overall stability (Mumby et al., 2007). The final element in the story is disturbance by hurricanes, which strike the region every year, but only hit any particular reef every 20 years on average.

We can now examine what has happened to one particular reef in recent years (Fig. 11.12). Chronic overfishing caused a fall in parrotfish numbers, leaving urchins as the main grazers. After damage by Hurricane Allen, recovery of the reef began, but diseases struck the urchin *Diadema* and two important coral species. A decline in the urchins, the last link preserving the reef community, led to a gradual community change. Then, by poor fortune, Hurricane Gilbert caused further damage to the remaining coral. Cover dropped and the macroalgae took over. The reef was lost.

Based on their knowledge of the system and its elements, Mumby et al. (2007) were able to predict whether a reef was likely to collapse given variation

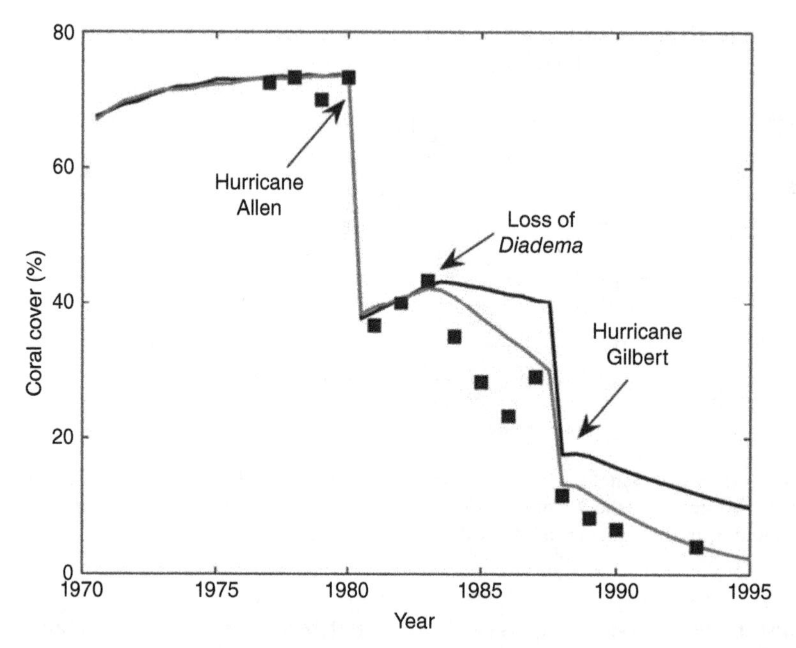

Figure 11.12 Trends in coral cover on a Caribbean reef over 25 years. Dots are real data, while lines represent predictions of two simulation models based on minimum or maximum estimates of the rate of algal overgrowth of coral. (*Source*: Mumby et al. (2007). Reproduced with permission from Nature Publishing Group.)

in grazing (determined by presence of parrotfish and urchins) and initial coral cover. So long as grazing remained above a certain level, and freak hurricanes did not damage the reef too heavily, it would remain stable. Loss of grazers or frequent disturbance would cause it to switch to an algal-dominated state.

Coral reefs elsewhere in the world are suffering similar declines due to a variety of causes (Bellwood et al., 2004). These predominantly involve either heavy fishing pressure or eutrophication. The interactions between these drivers cause transitions between a number of recognisable states (Fig. 11.13). A 'healthy' reef, once placed under stress, loses resilience and can be pushed down a path that leads to eventual collapse.

As usual, however, we must maintain a dispassionate scepticism when presented with this type of narrative. After all, who are we to judge what constitutes a 'healthy' system, and can we be certain that this is its natural state?

11.7 Shifting baselines

It is extremely difficult to determine what community would otherwise be present when impacts of humans have been prolonged and widespread, exceeding living memory. When engaged in conservation or restoration of natural systems, can we be sure that our target is appropriate? The problem is known as shifting baselines. It is a common human tendency to assume that what we became familiar with when we were young is normal, and any departures from this are problematic. Most National Parks in the British Isles maintain wholly unnatural systems. People may adore the landscape of the English Lake District, but what is being preserved is a cultural memory of how society feels it ought to look rather than the dense broadleaf forests that would take over were it to be left alone. In New Zealand, offshore islands are cloaked in forests that are generally assumed to be the natural end point of succession, but palæobotanical evidence reveals that

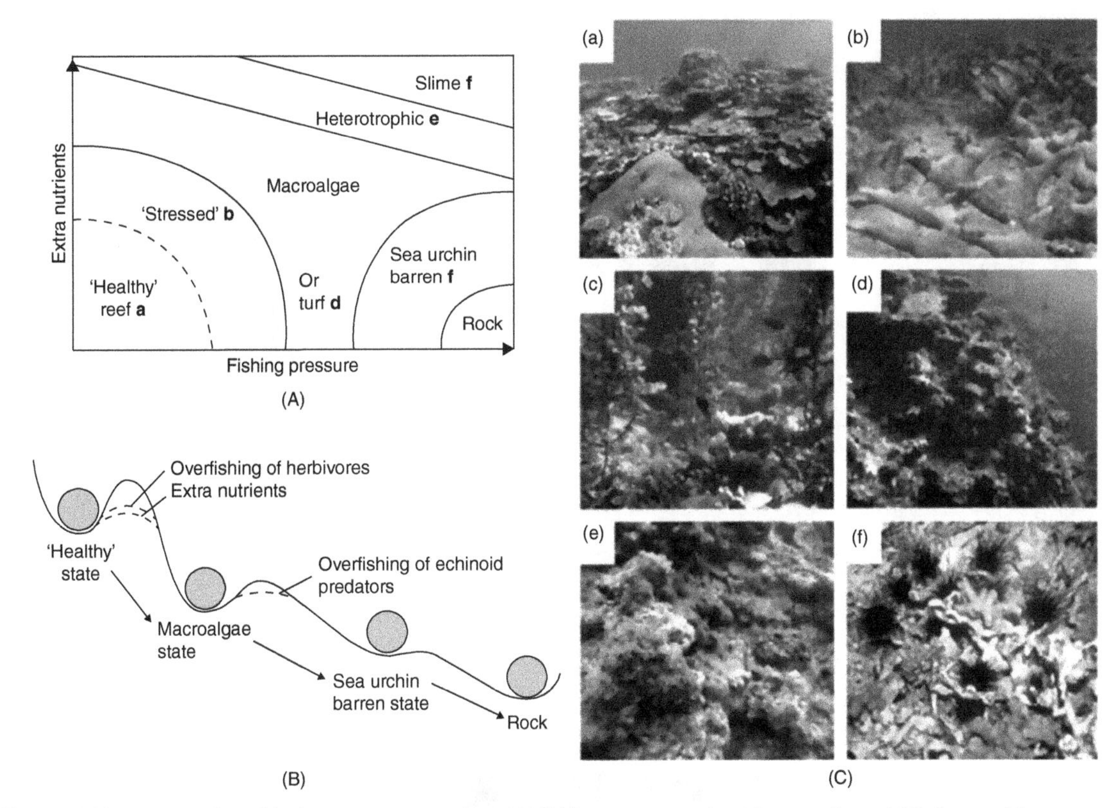

Figure 11.13 Conceptual model of ecosystem states given (A) fishing pressure and nutrient supply, and (B) the transition sequence between them illustrated by (C) a range of reef habitats. (*Source*: Bellwood et al. (2004). Reproduced with permission from Nature Publishing Group.) *(See colour plate section for the colour representation of this figure.)*

prehuman forests were entirely different in composition from any modern equivalent (Wilmshurst et al., 2014).

The same may be true of coral reefs. Their proximity to islands and hence human settlements has led, over thousands of years, to their being stripped of top predators, especially sharks. This has resulted in an increase in the abundance and species richness of planktivorous fish. The stereotypical vision of a coral reef is of shoals of small, brightly coloured fish swarming everywhere. This may be what is presented on wildlife documentaries, but is it really an indication of a healthy reef?

To tell the difference requires finding somewhere that humans have not fully disturbed. Perhaps the last reefs on earth with such credentials are Kingman and Palmyra Atolls in the Northern Line Islands of the Pacific (Sandin et al., 2008), which apart from brief occupancy by a US military base on Palmyra have never seen full colonisation by a human settlement.

Examining the community structure on these atolls reveals something that may seem bizarre to anyone familiar with the conventional appearance of reefs in the Bahamas or the Great Barrier Reef. On Kingman and Palmyra, the biomass of sharks is enormous, making up the majority of the fish community (Fig. 11.14a). This is known as an *inverted biomass pyramid* and is rare in nature, because energetic constraints mean that any food web based on a common resource base must be bottom heavy (Trebilco et al., 2013). In this case the likely explanation is that sharks are wide-ranging top predators, taking resources from much further afield, and their populations are inflated on the reefs where they choose to

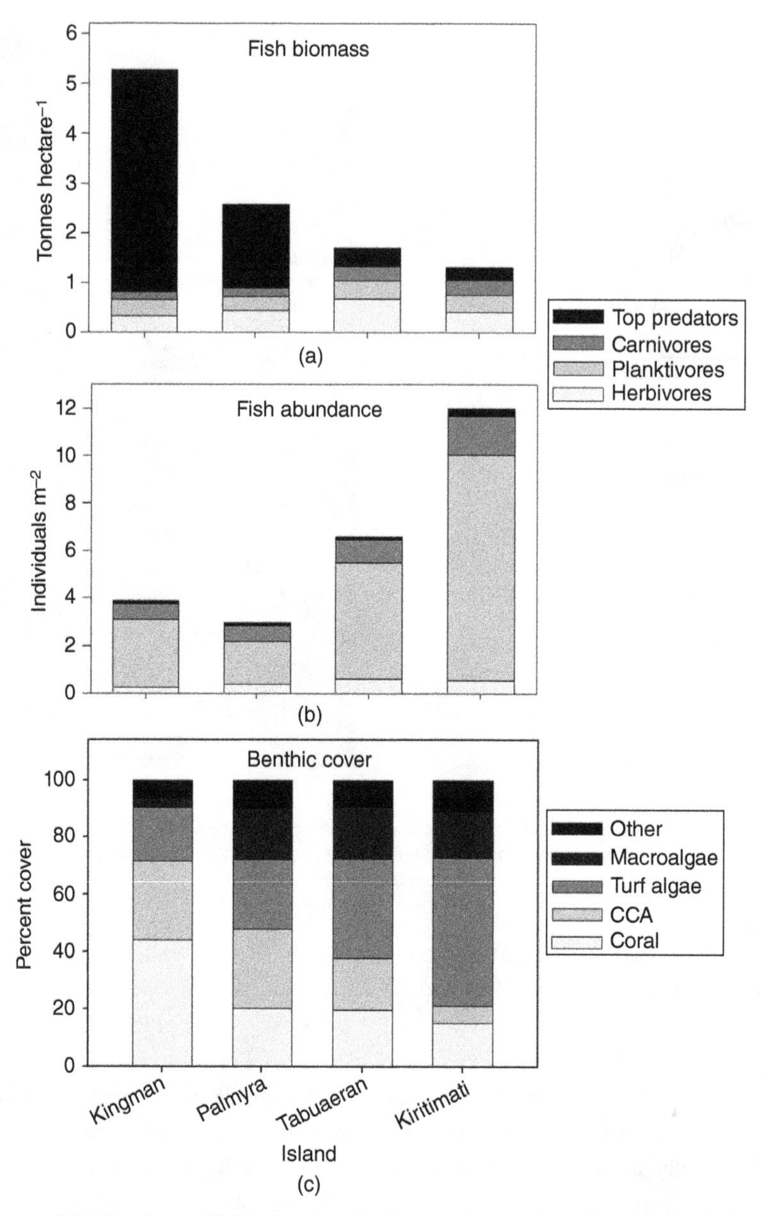

Figure 11.14 (a) Biomass and (b) abundance of fish in four functional groups (top predators, carnivores, planktivores and herbivores), plus benthic cover (c) on reefs in the Northern Line Islands, ordered by increasing levels of human disturbance. CCA: crustose coralline algae. (*Source*: From Sandin et al. (2008). Used under CC-BY-SA 2.5 http://creativecommons.org/licenses/by-sa/2.5/.)

gather. Nonetheless, it is only on those reefs where the largest top predators have been removed that the abundance of small planktivorous fish increases dramatically; disturbed reefs actually have more small fish (Fig. 11.14b). The size distribution of individuals could provide an effective means of measuring the impact of humans and predicting the natural structure of communities (Trebilco et al., 2013).

Note also the amount of the reef covered by coral or crustose coralline algae (CCA; Fig. 11.14c). These indicators of reef health dominate on Kingman, but as the biomass of top predators declines, there is an increasing transition to macroalgae (seaweeds) and turf algae. The recruitment rate of corals is higher, and the prevalence of disease lower, on the less-disturbed atolls (Fig. 11.15).

The apparent greater health of the kingman reef may have as much to do with levels of pollution from human settlements as the structure of the fish communities (Dinsdale et al., 2008), but it at least indicates that absence of sharks might be a useful sign of human impacts. In some cases however the prevalence of coral disease can be directly related to the structure of the community. In the Philippines, Raymundo et al. (2009) found a lower prevalence of disease within marine protected areas (MPAs) when compared to those which were actively fished. The frequency of coral disease was negatively correlated with the taxonomic distinctness of fish communities (Fig. 11.16a; for an explanation of Δ^+, see Section 5.7). Falling Δ^+ was associated with greater abundance of chaetodontids (butterflyfish; Fig. 11.16b), which are unpleasant to eat and are therefore not fished, benefitting from the removal of their competitors. These then feed on corals and spread disease (Fig. 11.16c,d).

Even the absence of sharks may not tell the full story. Large marine turtles are also important grazers of reef systems, whose populations collapsed long before present memory (Goatley et al., 2012).

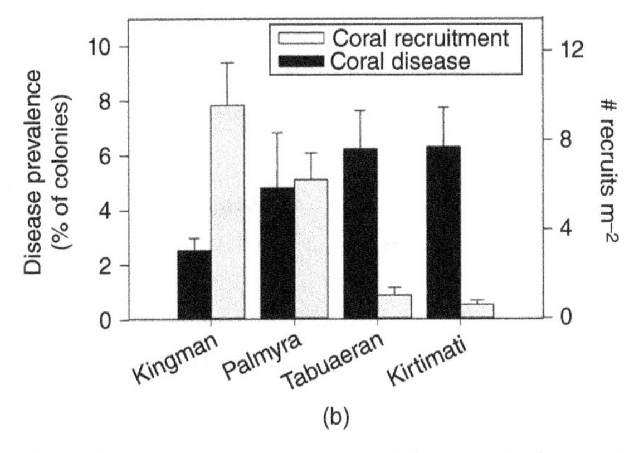

Figure 11.15 Coral recruitment and disease prevalence on reefs in the Northern Line Islands ordered as Fig. 11.14. (*Source*: Sandin et al. (2008). Used under CC-BY-SA 2.5 http://creativecommons.org/licenses/by-sa/2.5/.)

Likewise the modern world may only contain 10% of the great whales that existed before industrial whaling. Their role in moving nutrients from deep to shallow waters, or between regions, has never been replaced (Roman et al., 2014). It is difficult to know how great the impact of these changes has been because we have no basis for comparison.

That removal of top predators and large browsers might have ramifications for all levels of a community should be familiar from the section on trophic cascades in Section 10.9. Trophic cascades are ubiquitous in natural systems. Because hunting tends to remove the largest species, they are among the features of communities most commonly manipulated by humans (Estes et al., 2011). Fished coral reefs around the world tend to have the same abundance of fish as inaccessible reefs, but less than half the total biomass, since it is the large browsing species that are most susceptible to exploitation (Edwards et al., 2014).

How different would reefs or other ecosystems look if their larger species were still present? This is a sobering thought—as a result of shifting baselines, we can never be certain that the systems we see today, or even that we remember, are the same as would occur in the absence of humans.

11.8 Conclusions

Some combinations of species appear capable of persisting through time while most do not. This leads to the concept of stability in communities. Natural systems vary in their resilience or resistance to external disturbances, which can come through gradual environmental change, freak events or the impacts of humans. It is also possible for several alternative stable states to exist in any given location. Transitions between them can often be rapid, unpredictable and catastrophic, with hysteresis preventing an easy return to their prior state. Coral reefs provide a particularly topical example. Finally, remember that observed communities might not be entirely natural, since even apparently healthy systems

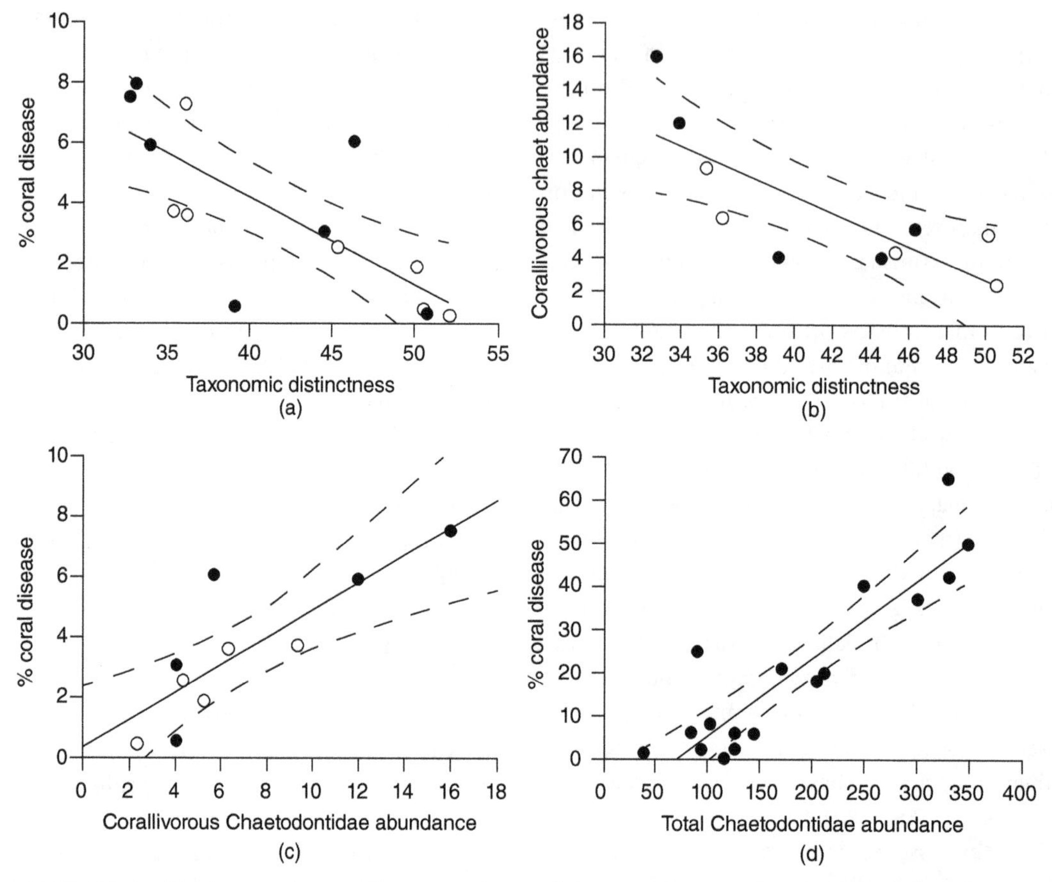

Figure 11.16 Relationships between (a) coral disease prevalence (all sites) and (b) abundance of coral-feeding chaetodontid fish with Δ^+ of fish communities (sites with >50% live coral cover only) and (c,d) between coral disease prevalence and abundance of coral-feeding chaetodontids (butterflyfish). Open circles, MPAs; closed circles, fished sites. (a–c) Philippines, (d) Great Barrier Reef. (*Source*: From Raymundo et al. (2009).)

can show symptoms of pervasive human impacts. Shifting baselines through human history can distort our perception of what communities would look like without our interventions, making it difficult to determine appropriate targets for conservation and restoration. Few people alive today have ever seen a truly 'natural' community.

Following is a brief consideration of Gaia theory, an idea rejected by most ecologists but which can superficially appear to be an obvious extension of the ideas presented in this chapter. It is presented separately to emphasise that this is not a mainstream debate, though one which continues outside the scientific literature.

11.8.1 Recommended reading

Barnosky, A., E. Hadly, J. Bascompte, E. L. Berlow, J. H. Brown, M. Fortelius, W. Getz, J. Harte, A. Hastings, P. Marquet, N. D. Martinez, A. Mooers, P. Roopnarine, G. Vermeij, J. Williams, R. Gillespie, J. Kitzes, C. Marshall, N. Matzke, D. Mindell, E. Revilla, and A. Smith, 2012. Approaching a state shift in Earth's biosphere. *Nature* 486:52–58.

Ling, S. D., C. R. Johnson, S. D. Frusher, and K. R. Ridgway, 2009. Overfishing reduces resilience of kelp beds to climate-driven catastrophic phase shift. *Proceedings of the National Academy of Sciences of the United States of America* 106:22341–22345.

Scheffer, M. and S. R. Carpenter, 2003. Catastrophic regime shifts in ecosystems: linking theory to observation. *Trends in Ecology & Evolution* 18:648–656.

11.8.2 Questions for the future

- What features of communities confer resistance or resilience, and are these properties opposed to one another?
- Can transitions between stable states be predicted or averted?
- Most empirical studies of alternative stable states to date have come from aquatic systems. Do the same phenomena occur on land?

11.9 Coda: the seduction of Gaia

Plenty of evidence exists that local communities exist in stable, self-perpetuating states. Can these processes be extrapolated to the planetary scale? This was the radical suggestion of the Gaia hypothesis, championed by Lovelock (1979).

At first sight the idea appears reasonable. Earth has had a remarkable history, as seen in Chapter 3. The action of single-celled organisms over billions of years increased the oxygen content of the atmosphere, creating one of the necessary conditions for complex multicellular life. Since then, despite repeated perturbation from asteroid impacts, shifting continents and climatic changes driven by trends in Earth's orbit, life has always recovered and been maintained. Is nature self-perpetuating? The idea that the planet is itself a self-regulating entity with its own system of checks and balances, much like any individual organism, is compelling as it connects to an ancient cultural belief in Mother Nature. The name of the hypothesis derives from the Greek earth goddess Gaia.

The logic behind it was explained via the Daisyworld model. Imagine a grey planet inhabited by two species of daisy, one black and one white. They differ in their albedo—the degree to which they reflect light. When incoming light is low, black daisies are at an advantage because they absorb more heat. This will then warm up the planet, favouring the white daisies. If incoming light is high, white daisies will increase in abundance, reflecting more light and thereby cooling the planet down. This interaction will have a stabilising effect, such that even in the face of substantial variation in solar radiation, the planet will remain within the range of temperatures suitable for life (Fig. 11.17).

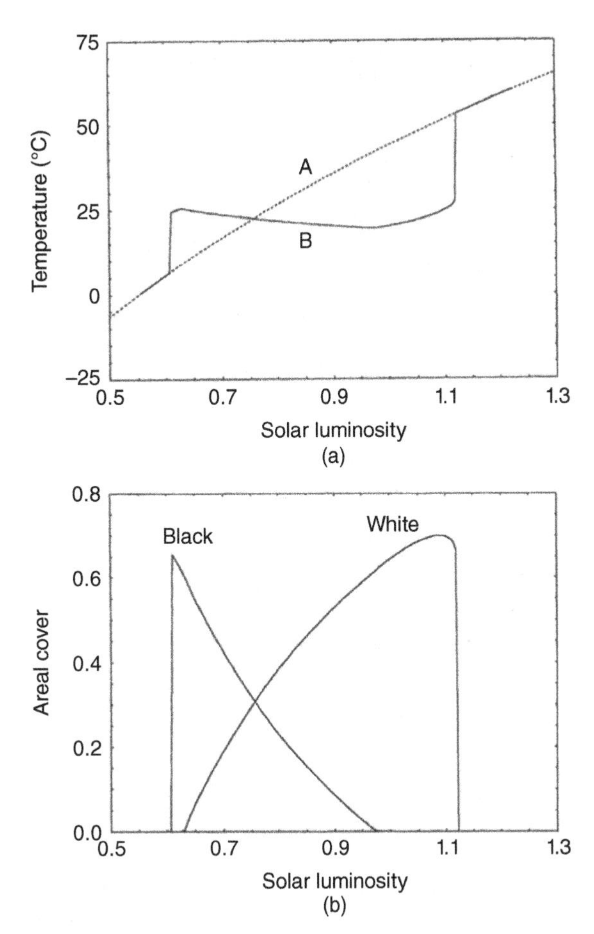

Figure 11.17 The Daisyworld thought experiment. Daisies are either black or white on a grey planet. (a) Planetary temperature increases with solar luminosity; dashed line (A) shows the temperature in the absence of daisies and the solid line (B) shows the temperature in the presence of daisies. (b) Proportion of planet covered by black or white daisies along the range of habitable conditions. (*Source*: Lenton (1998). Reproduced with permission from Nature Publishing Group.)

We can pick this reasoning apart quickly because it makes several assumptions that are hard to defend. The first is that species traits will have exactly the same effect at both local and planetary scales. This is implausible—the processes governing movement of heat in a flower petal and the atmosphere are not necessarily the same. It also implies that no evolution takes place and species remain identical over time. Again this is highly unlikely—why not be a grey daisy? There are more sophisticated Daisyworld-style models that can incorporate some of these processes, but they have left Gaia some way behind.

Has the Earth been regulated by life? The core assertion, that living organisms can influence the environment, is not controversial. The biotic and abiotic characteristics of Earth have continuously changed in response to one another. It's also obvious that life has been maintained for billions of years. It could be that we've simply been lucky, and stabilising feedbacks have occurred by chance. Alternatively, as Gaia theory proposes, stabilising feedbacks might be an inherent property of life. Given that starting conditions for life exist, the biosphere might be maintained by life-environment feedbacks. Possible examples include both atmospheric changes and glacial cycles. This is, however, an untestable hypothesis, as we can't do replicated experiments on multiple planets with and without life.

Many ecologists have been extremely critical of the Gaia hypothesis, and therefore you should treat it with great caution. The major holes in the theory include that it requires cost-free co-operation among species. Why would a species act for the benefit of the biosphere rather than in its own interests? It requires that there be some form of biosphere-level selection, which is impossible. In order for a form of Darwinian selection to take place, biospheres would have to 'reproduce', perhaps by colonising new worlds. Since we have yet to escape our planet and have no evidence that extraterrestrial life exists, the idea of planetary selection is fanciful.

Finally, and most importantly, it assumes that stable subsystems imply global stability. This is a scale error. The mechanisms that balance a household budget bear no relation to those of a national economy, even though both imply the movement of money. In a similar fashion, the internal dynamics of ecosystems might be locally stabilising, but this does not imply that the same processes operate or would even be effective at a planetary level. The danger of the Gaia hypothesis, and the reason it must be resisted, is that it could be used to imply that human-induced climate change is not a serious problem. After all, if the planet regulates itself, what can go wrong? Sometimes apparently innocuous ideas can have damaging implications, and for this reason it's important to know why the Gaia hypothesis has been rejected. For a good balanced review of the evidence for and against Gaia, see the review by Free and Barton (2007).

Just because we have rejected Gaia theory, however, does not mean we can dismiss entirely the idea that Earth exists in some form of stable state, and that human influence might tip it into another (Barnosky et al., 2012; Hughes et al., 2013), but see Brook et al. (2013). Palaeoclimatic evidence suggests that the whole planet has moved between stable states multiple times throughout its history with dramatic consequences for life. Such a switch, if induced by human activity, could be destructive, unpredictable and effectively irreversible. The process may have already begun (Lenton and Williams, 2013).

References

Barnosky, A. D., E. A. Hadly, J. Bascompte, E. L. Berlow, J. H. Brown, M. Fortelius, W. M. Getz, J. Harte, A. Hastings, P. A. Marquet, N. D. Martinez, A. Mooers, P. Roopnarine, G. Vermeij, J. W. Williams, R. Gillespie, J. Kitzes, C. Marshall, N. Matzke, D. P. Mindell, E. Revilla, and A. B. Smith, 2012. Approaching a state shift in Earth's biosphere. *Nature* 486:52–58.

Bellwood, D. R., T. P. Hughes, C. Folke, and M. Nyström, 2004. Confronting the coral reef crisis. *Nature* 429: 827–833.

Brook, B. W., E. C. Ellis, M. P. Perring, A. W. Mackay, and L. Blomqvist, 2013. Does the terrestrial biosphere have planetary tipping points? *Trends in Ecology & Evolution* 28:396–401.

Carpenter, S. R., J. J. Cole, M. L. Pace, R. Batt, W. A. Brock, T. Cline, J. Coloso, J. R. Hodgson, J. F. Kitchell, D. A. Seekell, L. Smith, and B. Weidel, 2011. Early warnings of regime shifts: a whole-ecosystem experiment. *Science* 332:1079–1082.

Celiker, H. and J. Gore, 2014. Clustering in community structure across replicate ecosystems following a long-term bacterial evolution experiment. *Nature Communications* 5:4643.

Clements, F. E., 1916. *Plant Succession: Analysis of the Development of Vegetation.* Carnegie Institute of Washington.

Dakos, V., E. H. van Nes, and M. Scheffer, 2013. Flickering as an early warning signal. *Theoretical Ecology* 6:309–317.

Dantas, V. de. L., M. A. Batalha, and J. G. Pausas, 2013. Fire drives functional thresholds on the savanna-forest transition. *Ecology* 94:2454–2463.

Dinsdale, E. A., O. Pantos, S. Smriga, R. A. Edwards, F. Angly, L. Wegley, M. Hatay, D. Hall, E. Brown, M. Haynes, L. Krause, E. Sala, S. A. Sandin, R. V. Thurber, B. L. Willis, F. Azam, N. Knowlton, and F. Rohwer, 2008. Microbial ecology of four coral atolls in the northern Line Islands. *PLoS ONE* 3:e1584.

Edwards, C. B., A. M. Friedlander, A. G. Green, M. J. Hardt, E. Sala, H. P. Sweatman, I. D. Williams, B. Zgliczynski, S. A. Sandin, and J. E. Smith, 2014. Global assessment of the status of coral reef herbivorous fishes: evidence for fishing effects. *Proceedings of the Royal Society Series B* 281:20131835.

Estes, J. A., J. Terborgh, J. S. Brashares, M. E. Power, J. Berger, W. J. Bond, S. R. Carpenter, T. E. Essington, R. D. Holt, J. B. C. Jackson, R. J. Marquis, L. Oksanen, T. Oksanen, R. T. Paine, E. K. Pikitch, W. J. Ripple, S. A. Sandin, M. Scheffer, T. W. Schoener, J. B. Shurin, A. R. E. Sinclair, M. E. Soulé, R. Virtanen, and D. A. Wardle, 2011. Trophic downgrading of planet Earth. *Science* 333:301–306.

Foley, J. A., M. T. Coe, M. Scheffer, and G. Wang, 2003. Regime shifts in the Sahara and Sahel: interactions between ecological and climatic systems in northern Africa. *Ecosystems* 6:524–532.

Fouchald, P., 2010. Predator-prey reversal: a possible mechanism for ecosystem hysteresis in the North Sea? *Ecology* 91:2191–2197.

Free, A. and N. H. Barton, 2007. Do evolution and ecology need the Gaia hypothesis? *Trends in Ecology & Evolution* 22:611–619.

Goatley, C. H. R., A. S. Hoey, and D. R. Bellwood, 2012. The role of turtles as coral reef macroherbivores. *PLoS ONE* 7:e399979.

Hare, S. R. and N. J. Mantua, 2000. Empirical evidence for North Pacific regime shifts in 1977 and 1989. *Progress in Oceanography* 47:103–145.

Hughes, T. P., S. Carpenter, J. Rockström, M. Scheffer, and B. Walker, 2013. Multiscale regime shifts and planetary boundaries. *Trends in Ecology & Evolution* 28:389–395.

Kéfi, S., V. Dakos, M. Scheffer, E. H. van Nes, and M. Rietkerk, 2013. Early warning signals also precede non-catastrophic transitions. *Oikos* 122:641–648.

Lenton, T. M., 1998. Gaia and natural selection. *Nature* 394:439–447.

Lenton, T. M. and H. T. P. Williams, 2013. On the origin of planetary-scale tipping points. *Trends in Ecology & Evolution* 28:380–382.

Ling, S. D., C. R. Johnson, S. D. Frusher, and K. R. Ridgway, 2009. Overfishing reduces resilience of kelp beds to climate-driven catastrophic phase shift. *Proceedings of the National Academy of Sciences of the United States of America* 106:22341–22345.

Lovelock, J., 1979. *Gaia: A New Look at Life on Earth.* Oxford University Press.

MacArthur, R. H. and E. O. Wilson, 1967. *The Theory of Island Biogeography.* Princeton University Press.

Mumby, P. J., A. Hastings, and H. J. Edwards, 2007. Thresholds and the resilience of Caribbean coral reefs. *Nature* 450:98–101.

Raymundo, L. J., A. R. Halford, A. P. Maypa, and A. M. Kerr, 2009. Functionally diverse reef-fish communities ameliorate coral disease. *Proceedings of the National Academy of Sciences of the United States of America* 106:17067–17070.

Roman, J., J. A. Estes, L. Morissette, C. Smith, D. Costa, J. McCarthy, J. B. Nation, S. Nicol, A. Pershing, and V. Smetacek, 2014. Whales as marine ecosystem engineers. *Frontiers in Ecology and the Environment* 12:377–385.

Sandin, S. A., J. E. Smith, E. E. DeMartini, E. A. Dinsdale, S. D. Donner, A. M. Friedlander, T. Konotchick, M. Malay, J. E. Maragos, D. Obura, O. Pantos, G. Paulay, M. Richie, F. Rohwer, R. E. Schroeder, S. Walsh, J. B. C. Jackson, N. Knowlton, and E. Sala, 2008. Baselines and degradation of coral reefs in the Northern Line Islands. *PLoS ONE* 3:e1548.

Scheffer, M. and S. R. Carpenter, 2003. Catastrophic regime shifts in ecosystems: linking theory to observation. *Trends in Ecology & Evolution* 18:648–656.

Scheffer, M., S. Carpenter, J. A. Foley, C. Folke, and B. Walker, 2001. Catastrophic shifts in ecosystems. *Nature* 413:591–596.

Scheffer, M., S. Szabó, A. Gragnani, E. H. van Nes, S. Rinaldi, N. Kautsky, J. Norberg, R. M. M. Roijackers, and R. J. M. Franken, 2003. Floating plant dominance as a stable state. *Proceedings of the National Academy of Sciences of the United States of America* 100:4040–4045.

Schröder, A., L. Persson, and A. M. De Roos, 2005. Direct experimental evidence for alternative stable states: a review. *Oikos* 110:3–19.

Sheil, D. and D. Murdiyarso, 2009. How forests attract rain: an examination of a new hypothesis. *Bioscience* 59:341–347.

Traveset, A., R. Heleno, S. Chamorro, P. Vargas, C. K. McMullen, R. Castro-Urgal, M. Nogales, H. W. Herrera, and J. M. Olesen, 2013. Invaders of pollination networks in the Galápagos Islands: emergence of novel communities. *Proceedings of the Royal Society Series B* 280:20123040.

Trebilco, R., J. K. Baum, A. K. Salomon, and N. K. Dulvy, 2013. Ecosystem ecology: size-based constraints on the pyramid of life. *Trends in Ecology & Evolution* 28:423–431.

Veraart, A. J., E. J. Faassen, V. Dakos, E. H. van Nes, M. Lürling, and M. Scheffer, 2012. Recovery rates reflect distance to a tipping point in a living system. *Nature* 481:357–359.

Wang, R., J. A. Dearing, P. G. Langdon, E. Zhang, X. Yang, V. Dakos, and M. Scheffer, 2012. Flickering gives early warning signals of a critical transition to a eutrophic lake state. *Nature* 492:419–422.

Wilmshurst, J. M., N. T. Moar, J. R. Wood, P. J. Bellingham, A. M. Findlater, J. J. Robinson, and C. Stone, 2014. Use of pollen and ancient DNA as conservation baselines for offshore islands in New Zealand. *Conservation Biology,* 28:202–212.

Yasuhara, M., T. M. Cronin, P. B. deMenocal, H. Okahashi, and B. K. Linsley, 2008. Abrubt climate change and collapse of deep-sea ecosystems. *Proceedings of the National Academy of Sciences of the United States of America* 105:1556–1560.

CHAPTER 12
Changes through time

12.1 The big question

Nature in the popular imagination is constant, unchanging and immutable. The rapid pace of human life is often contrasted with the peace and stability of natural systems. Nothing could be further from the truth. In fact, perpetual change is a law of nature, and it is important to recognise and understand its patterns through time. Can common features be identified in the way that communities change? If so, is it possible to create predictive models to anticipate how natural systems will alter in the future?

Directional change, known as succession, is common in many communities but varies in its character depending on the system under study. Natural disturbances occur frequently, as well as those induced by humans, and communities are able to recover from them. We should seek to understand all these processes, and the best way to demonstrate this is through forming models of the various mechanisms and testing them against the real world.

12.2 Succession

Succession refers to change over time in the composition of a community which is nonseasonal, directional and continuous. Its narrative is formed by the colonisation and extinction of species at a given site. There are two basic forms. Primary succession occurs on sites where no previous community existed. The bare rock left behind by a retreating glacier and the emergence of a volcanic island from the ocean are both typical examples. These are contrasted with secondary successions that occur within existing communities and describe their transition to another state. An abandoned agricultural field might revert to woodland, or a planktonic community could shift in response to a novel input of nutrients. The two processes behave quite differently from one another. For the remainder of this chapter we focus on primary successions, which are easier to understand since they commence with a blank slate. The direction of a secondary succession is much more contingent on the species that currently occupy a site and how they respond to changing conditions; as a result, generalisations are elusive and predicting what will occur is difficult. The chapter concludes with an example of a secondary succession.

The historical debate between Clementsian and Gleasonian community ecologists outlined in Section 10.2 is reflected in two very different views of the end point of succession. Many argued, after Clements, that every succession would eventually reach a climax community, determined by the environment of the site and the inevitable outcome of interactions among the species present. Others, following Gleason, maintained a more fluid perspective, with succession not necessarily following a prescribed set of stages nor reaching a predictable end, with species

Natural Systems: The organisation of life, First Edition. Markus P. Eichhorn.

both responding to and altering the environment around them. If the environment changed faster than species were able to keep up, a climax would never occur. In the memorable words of their predecessor, Henry Cowles, succession was 'a variable chasing a variable' (Cowles, 1901).

Both viewpoints are maintained in the present day, with Clements' perspective still commonly taken as a starting assumption, perhaps due to its more intuitive character and specific predictions. Much conservation science is predicated on the belief that there is a 'correct' community for a given location, that we can work out what it is and that we should take management action to help it on its way. It is ironic that while the underlying hypothesis of the community as superorganism is no longer accepted, its prediction is so often assumed. A world in which multiple outcomes are possible with constant change feels less amenable to our control.

The two perspectives were brought together by Connell and Slatyer (1977), who added the important insight that the end point reached would depend upon the interactions that occurred among the species that first colonised a site (Fig. 12.1). If some species, known as **pioneers**, were specially adapted to occupy vacant habitats, they might alter the environment in such a way that they enabled more competitive species to establish and inevitably take over. This is the *facilitation* model and leads to a series of discrete community stages. One set of species will take over from another, exactly as Clements envisaged. Alternatively, any species might have an equal chance of getting in at the start, though having a negligible effect on the environment (the *tolerance* model). Species might vary in maturation rates and competitive ability, such that the balance of the community shifts through time, eventually ending in a stable state. Finally, it could be that the first colonists simply hold a site and alter the environment in a manner that favours themselves, preventing other species from entering (the *inhibition* model). So long as these species are not disturbed (perhaps by a predator, or environmental stress), they can hold a site in perpetuity. Only when they die or are held back does the opportunity arise for new species to enter and change the system.

Clements believed that there were identifiable stages along the pathway of a succession, which he called 'seres'. This is again a widely held belief, implicit in the boxes of Fig. 12.1, though one which has seldom been tested using quantitative methods (see Zaplata et al. (2013) for a rare example). Many studies have assumed that looking at communities of different ages (chronosequences) can indicate the pathways of succession, but even the most famous textbook examples have been challenged by long-term monitoring of single sites (Johnson and Miyanishi, 2008), so be careful before accepting stories of community change at face value. Instead of looking for patterns of species associations, we should instead look for evidence of underlying processes.

12.3 Succession and niche theory

The successional niche of a species can be defined according to its place along an axis from high colonisation rates through to strong competitive ability. Colonists are those species with high fecundity (producing many offspring) and propagules which disperse over large areas. Should an individual be fortunate enough to land in an area with abundant resources, colonists are able to grow and reproduce rapidly. They tend to be poor competitors though and fail to survive when resource levels fall (in the terminology of MacArthur and Wilson (1967) they are r-selected). Most plants that are considered weeds fall into this category.

During succession, communities typically shift from species with traits adapted for rapid resource acquisition and usage towards those favouring resource conservation in the face of competition. Competitor species take longer to establish, but once they do, they prove more effective at competing for scarce resources (they have a lower R^*, or are K-selected *sensu* MacArthur and Wilson (1967)). Typically they produce fewer offspring or seeds

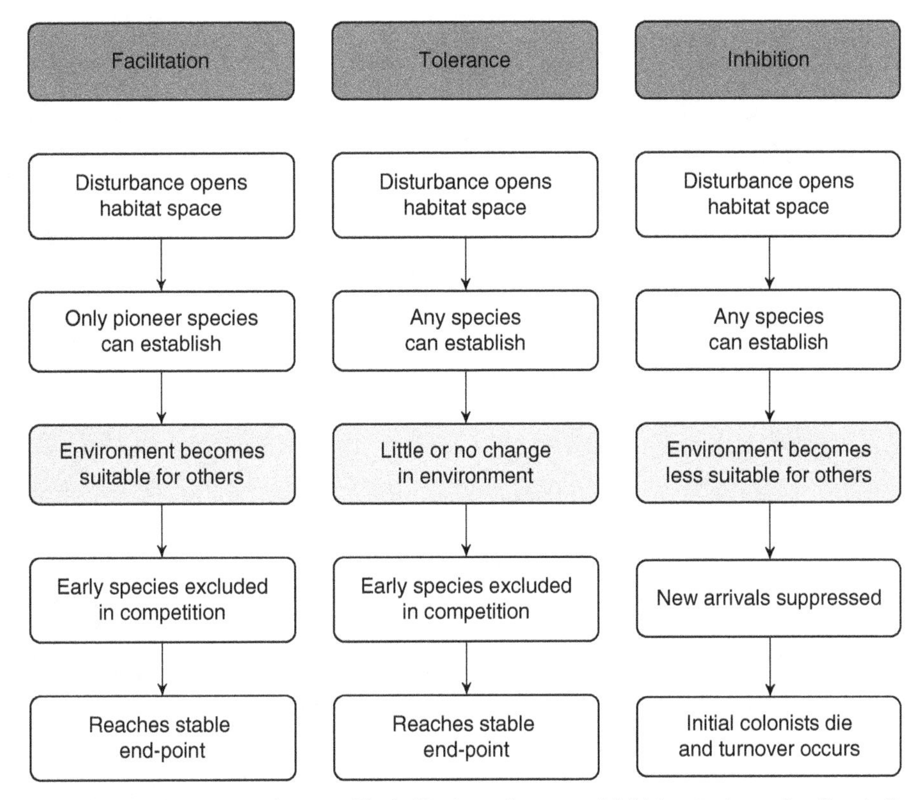

Figure 12.1 Three pathways of succession dominated by facilitation, tolerance or inhibition. Dark grey headings indicate the succession model, while light grey nodes show the unique ecosystem-level implications of each pathway. (*Source*: Adapted from Connell and Slatyer (1977).)

which do not disperse as broadly. Trees are the standard example among plants. It is difficult to be both a good coloniser of vacant habitats and a strong competitor in established communities; this axis of variation among species is therefore known as the **competition–colonisation trade-off**. As with all scales in ecology, it is important to realise that it is continuous, and species should not be placed into two extreme categories.

The net result is that communities tend to transition through time from being dominated by a few fast-growing, weedy species to containing a greater diversity, both in terms of species richness and evenness (e.g. Fig 12.2).

We can now compose a theory of the successional niche that fits with the model of species interactions from Chapter 6. The competition–colonisation trade-off is depicted in Fig. 12.3. In this example

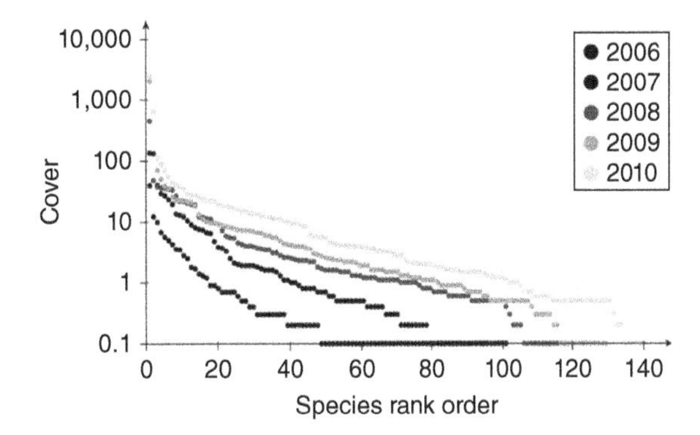

Figure 12.2 Rank/abundance plots of plant species in a vegetation succession on an abandoned mine in eastern Germany from 2006 to 2010; percentage cover estimates summed across 41 plots for each species. (*Source*: Zaplata et al. (2013). Reproduced with permission from University of Chicago Press.)

the first species to arrive when there are abundant resources are the colonists (dotted lines), usually weedy species that have high R^* values. They are

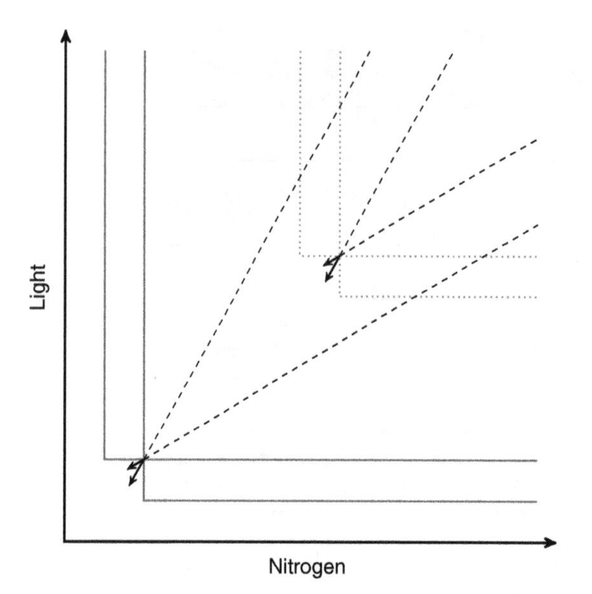

Figure 12.3 The successional niche. Colonists/pioneers (dotted lines) have an early advantage but are supplanted by competitors (solid lines). Zero net growth isoclines (ZNGIs) with impact vectors (arrows) projected back into resource space to indicate the outcome of competition (dashed lines). For details on interpreting these figures, refer to Chapter 6. (*Source*: Adapted from Chase and Leibold (2003).)

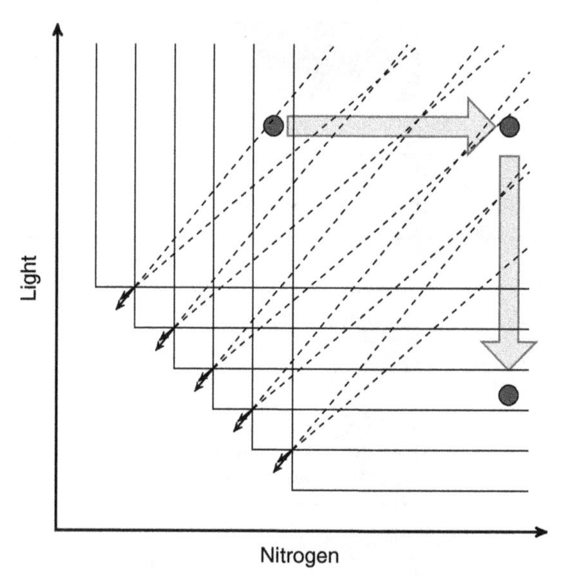

Figure 12.4 Zero net growth isoclines (ZNGI solid lines) with impact vectors (arrows) projected back into resource space to indicate the outcome of competition (dashed lines). Grey circles indicate resource supply points, with arrows showing changes in resource availability through succession. Facilitation causes different species to be favoured as the supply rate of nitrogen increases (e.g. through development of soil microbial communities). Competition for light then leads to a further shift in species composition due to alteration in the supply of light to recruits (e.g. through the casting of shade by forest canopies). (*Source*: Adapted from Chase and Leibold (2003).)

able to form a community, but this will not be able to persist once competitor species (solid lines) arrive, since these have lower R^* and can undercut the colonists. As a result, the levels of resources in the system are dragged down below the tolerance of the colonists, which will go locally extinct (see Chapter 8 in Chase and Leibold, 2003). Both strategies can coexist in the landscape provided that the colonists are able to disperse their offspring to another vacant site before being eliminated. The main difference between primary and secondary successions is that in the former, the competitor species either take some time to arrive or fail to establish, whereas in the latter case they are present throughout but take some time to exert their dominance.

Even once a stable community has formed, further alterations are likely if the supply of resources shifts (i.e. a secondary succession). This process of facilitation is shown in Fig. 12.4. Perhaps new plant colonists increase nitrogen fixation and accumulation in the soil, favouring a new set of species. These then grow taller and overtop them, reducing the availability of light (we saw a similar process in Section 8.3). A subsequent process of competitive displacement might lead to trees with ever taller and denser canopies supplanting each other, preventing other species from being able to survive and grow in the forest understorey. By now the power of the niche theory from Chapter 6 should be apparent; it helps to visualise what is going on in natural systems and provides hypotheses that can be tested in field systems.

12.4 Examples of succession

Boreal forests are plagued by fires which sweep through the landscape on a frequent basis. Sometimes they occur every few decades, other times centuries may pass, but across the world large areas are destroyed every year. They are restored by colonisation of trees and other plants that regenerate the forest structure. Which species take over, however, depends on local site conditions. A well-known case study comes from the area around Fairbanks in Alaska (Fig. 12.5). On cold north-facing slopes, with permafrost only a short distance beneath the surface, the end point of succession is a stand of mature black spruce trees (*Picea mariana*). On south-facing slopes, which receive greater warmth from the sun, a very different forest dominated by white spruce (*Picea glauca*) is the final community. In both cases most species are already present at the very start of succession, either as seeds or surviving tree stumps, and what takes place is a gradual takeover by conifers following the tolerance model.

In truth, even for such a simple and well-studied system, the pathway followed by any individual site is much more complex than Fig. 12.5 suggests, as Chen and Taylor (2012) found when examining long-term datasets on forests in boreal Canada. At any given location, time since the last fire was either insufficient to explain the dynamics of species turnover or entirely irrelevant. Instead the characteristics of each individual species, interacting with local site conditions, determined the community present. On infertile dry sites, *Pinus banksiana* dominated for long periods, whereas on fertile soils it grew too thickly and shaded out its own seedlings, allowing shade-tolerant species such as *Thuja occidentalis* to take over. Some species were present in both early and late successions, such as *Betula* and *Populus* species, which performed equally well at colonising large empty spaces as the small gaps created by the death of large trees in closed forests. Each species has particular requirements dictating where it is able to establish, known as its **regeneration niche** (Grubb, 1977). The pathways in Fig. 12.5 are too simplistic to capture the full scope of dynamics even in these apparently simple systems.

Often the pathway followed by any single community and the eventual outcome are not entirely predictable. A particularly revealing study carried out by Márquez and Kolasa (2013) examined the communities of marine invertebrates that formed in rock pools in Jamaica. Their premise was a simple one. If a set of communities have identical environmental conditions, and the same colonising species with the same relative abundance, then communities should all converge on the same structure and composition. But what happens if you vary the environment, or the input of species?

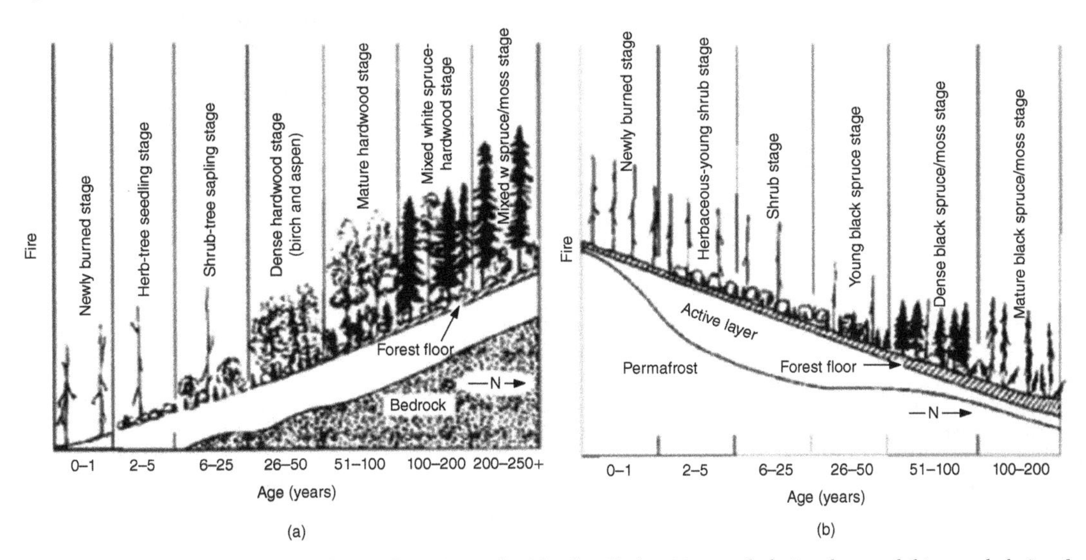

Figure 12.5 Succession following forest fires in the vicinity of Fairbanks, Alaska; (a) a south-facing slope and (b) a north-facing slope. (*Source*: Van Cleve and Viereck (1981). Reproduced with permission from Springer Business and Print Media.)

This could be tested with a simple design. Márquez and Kolasa selected a series of rock pools which received water from both rain and occasional wave splash, giving a range of conditions from salty to freshwater. Over 70 species of invertebrate were found in them. They collected entire communities—every individual organism—from a range of pools and mixed them together to give a homogeneous set of colonists. Natural pools were cleaned out and their water replaced with this mixture to reset them to the beginning of succession.

The next step was even more ingenious. Plastic beakers were prepared and filled with a batch of the mixed community. Some of these had mesh-covered holes in the side, allowing exchange of water—and therefore environmental influence—but no movement of organisms. These were placed into the rock pools. Another set with no holes were also placed in the rock pools, and a final set prepared and left on a bench in a uniform environment. The relative similarity of communities is shown in Fig. 12.6. The original communities (PP) varied greatly but once mixed together in a null community were almost entirely uniform (NC). Given identical starting conditions and the same environment (D), they ended up at the same point. Even when immersed in pools but still sealed there was little variation among them (C). Differences emerged, however, as soon as beakers were opened to local environmental influences (B) or to colonists from the outside world (A). Where a community ends up depends on both the local environment and the input of colonists, in almost equal measure. A deterministic view was not sufficient to explain variation in the end points of these communities.

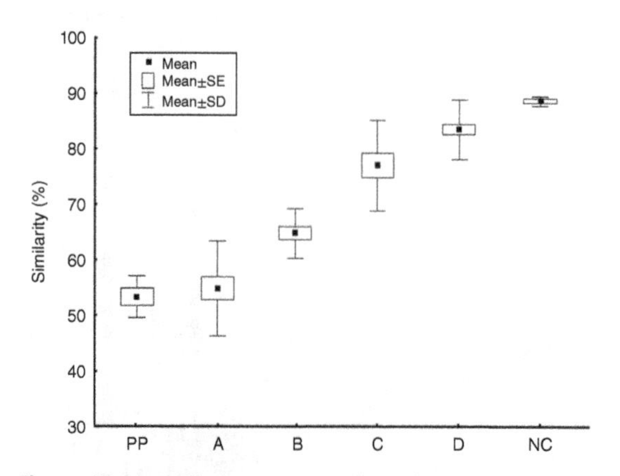

Figure 12.6 Similarity among rock pool invertebrate community treatments. PP: pools prior to treatment; A: natural pools inoculated with the same starting community; B: immersed beaker communities with holes to allow water flow but no dispersal; C: immersed and sealed beaker communities; D: beaker communities in a homogeneous environment; NC: null community used as inoculum. (*Source*: From Márquez and Kolasa (2013). Used under CC-BY-SA 2.5 http://creativecommons.org/licenses/by-sa/2.5/.)

12.5 Disturbance

Natural systems are often prevented from reaching a stable state by disturbance, which describes the suite of processes which disrupt communities through the death of individuals. Disturbance is often entirely natural and varies from extreme events such as Hurricane Katrina in 2005, which is thought to have felled some 320 million trees across the United States, down to minor disruptions that might only dislodge a few individual organisms. On land disturbance events can have abiotic (e.g. wind, fire, flood, landslides) and biotic causes (e.g. pest outbreaks, diseases, large animal damage such as trampling by elephants). Disturbance from extreme events is obvious but smaller-scale impacts are often continuous and part of the natural dynamics of communities.

Humans are responsible for a large proportion of disturbance in the modern world, induced by activities including land clearance for agriculture, logging, wildfires or release of invasive species. The resilience of communities often allows them to cope with a certain level of disturbance, but human impacts are typically characterised by a greater frequency and intensity than natural ones. Finding the limits of the tolerance of natural systems to our activities is a crucial element in determining their **sustainability**.

Two types of community can be characterised on the basis of their response to disturbance (Yodzis, 1986). Some are *dominance controlled*. Here a strict competitive hierarchy exists among species and potentially also trade-offs between competition and colonisation. A predictable succession takes place in exactly the manner described by the facilitation model (see Fig. 12.1). In other cases, communities can be *founder controlled*, as in the inhibition model. Multiple equivalent species exist which are all capable of colonising vacant patches. Once any single species (perhaps by chance) has established itself, the benefits of incumbency mean it can resist displacement from other species. This is sometimes referred to as lottery recruitment.

A notable feature of disturbance is that while it causes major changes in the abundance of particular species, both positive and negative, its effects on aggregate community properties such as overall species richness and diversity are less easy to discern. It is sometimes assumed that disturbance will always reduce diversity, but the compensatory dynamics of communities tend to even out the large shifts shown by individual species (Supp and Ernest, 2014). Extinctions are balanced by recruitment, and some species increase in abundance while others decline.

Table 12.1 Transition probabilities for tree species in New Jersey forests based on frequency of saplings beneath adults.

	Species after 50 years			
Current species	**Grey birch**	**Blackgum**	**Red maple**	**Beech**
Grey birch	0.05	0.36	0.50	0.09
Blackgum	0.01	0.57	0.25	0.17
Red maple	0	0.14	0.55	0.31
Beech	0	0.01	0.03	0.96

Source: Horn (1981). Reproduced with permission.

Table 12.2 Predicted percentage composition of New Jersey forests through a successional sequence starting from 100% grey birch following transition probabilities from Table 12.1.

	Age of forest (years)					
Species	**0**	**50**	**100**	**150**	**200**	**∞**	**Natural**
Grey birch	100	5	1	0	0	0	0
Blackgum	0	36	29	23	18	5	3
Red maple	0	50	39	30	24	9	4
Beech	0	9	31	47	58	86	93

Source: Horn (1981). Reproduced with permission.

12.6 Modelling succession

The real test of our understanding of natural systems is whether we can predict what they will do next. The classic work on this was performed by Horn (1975, 1981), who created a model to understand the replacement of trees in New Jersey forests. His approach was to look beneath adult trees at the abundance of saplings in the understorey. The simple rule was that the proportion of each species among saplings dictated the probability that it would eventually replace the adult tree above.

These values can be calculated for all species combinations and are known as transition probabilities (Table 12.1). Assuming that these probabilities remain constant, that each tree lives for 50 years and that the total number of trees in the canopy is fixed, it's straightforward to predict the eventual composition of the forest. The table indicates that there is only a 5% probability that a spot occupied by a grey birch will remain as the same species, whereas 96% of the time a beech will be replaced by another beech. The probability is not absolute, however, leaving room for other species to persist.

The predicted changes in a forest which starts entirely dominated by grey birch are shown in Table 12.2. Note that each of the columns sums to 100%. After 50 years, the percentages of each species are the same as the transition probabilities from grey birch. After that each species behaves differently, with grey birch disappearing by 150 years and beech gradually coming to dominate by the fourth generation of trees. The model can be run until it stabilises at infinity (∞), at which point the composition looks remarkably close to the actual forests that Horn documented.

This is known as a *Markov chain model*, in which the state at each time step is a predictable consequence of the conditions preceding it, following fixed rules. Its outcome is inevitable, regardless of the starting composition. They can be used to predict not only the final state but also the steps occurring along the way and the time until the system stabilises. They are also dynamic; even when the model reaches a steady state, it still reflects a continuous process of death and replacement within the community, akin to the behaviour of natural systems. This is a very basic case study, but these models can be expanded by incorporating dispersal probabilities, stochastic variation, environmental change, disturbance and any manner of other processes to match the core properties of real systems.

Exactly the same approach can be used for animal communities. Hill et al. (2004) conducted an investigation of the Ammen Rock Pinnacle in the Gulf of Maine, a rocky subtidal habitat. This sheer face of vertical rock walls descends to a depth of 200 m and is coated with a community of 14 invertebrate species. Two of these, a sponge and a bryozoan, are the dominants in terms of surface area covered. As with forests, there is strong competition for space—exactly the conditions required for a Markov chain model to be a good approximation.

Sections of the rock face were photographed regularly to assess the rates and direction of species turnover. This information was used to construct a transition matrix which could predict the composition of an equilibrium community. This proved to be very close to the observed abundances of each species (Fig. 12.7). The amount of bare rock was also included as a category.

As an interesting aside, the rocky subtidal community contained 8% bare rock, which took an average of 1.4 years to be recolonised. This can be compared to the marine intertidal zone, where only 2% bare rock exists and is colonised within around a year, or coral reefs, where >60% bare rock exists in stable communities, and can persist for 3–7 years. These figures represent the balance between rates of disturbance (clearing of rock, perhaps through wave action) and colonisation of empty sites by new animals.

The Markov chain approach is valuable, but it can come unstuck, and this is especially true for founder-controlled communities. Sale (1977, 1979)

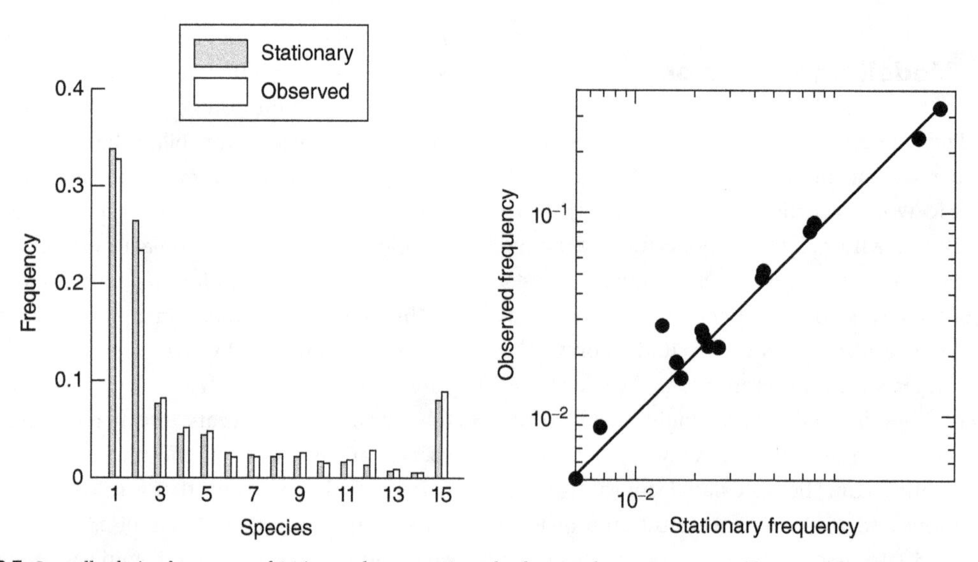

Figure 12.7 Overall relative frequency of 14 invertebrate species (plus bare rock as 15) compared between the outcome of a Markov chain simulation model (stationary frequency) and the observed community on the Ammen Rock Pinnacle. (*Source*: Hill et al. (2004). Reproduced with permission from University of Chicago Press.)

Figure 12.8 Cross section of a wave in an *Abies balsamea* forest. (*Source*: Sprugel (1976). Reproduced with permission from John Wiley & Sons Ltd.)

studied territorial fish on the Great Barrier Reef. The first fish to take a territory occupies it until its death. There is an incredible species richness of these fish, from 900 species in the south up to 1,500 in the northern region. Within any given 3 m quadrat it is possible to find over 50 fish species. These share overlapping diets with no obvious signs of resource partitioning (this example was one of the inspirations behind the development of neutral theory). Most of all, any attempt to produce a Markov chain model led to circular or nonsensical patterns. Territory replacement appears to be no different from random and fish recruitment is effectively a lottery. To summarise, Markov chain models work well in dominance-controlled successions but are not appropriate when communities are founder controlled.

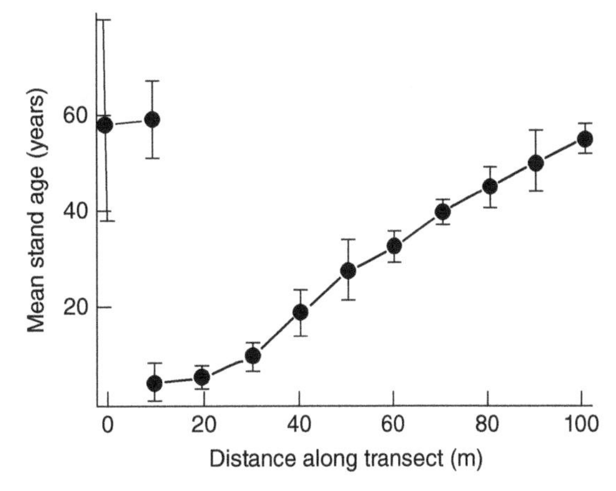

Figure 12.9 Tree ages across a wave in an *Abies balsamea* forest. (*Source*: Sprugel (1976). Reproduced with permission from John Wiley & Sons Ltd.)

12.7 Regeneration

Constant change occurs even within communities that have not been obviously disturbed. At the end state of a Markov chain model, individuals continue dying and being replaced all the time, though the overall abundance of each species remains the same. In natural systems the processes that return a community to its previous state following disturbance, or their internal dynamics, are collectively referred to as **regeneration**. They form special cases of secondary succession and take place continuously.

A classic study by Sprugel (1976) investigated a peculiar pattern observed in *Abies balsamea* forests on mountains in the Adirondack range in North America. Across the forest there were linear waves of tree size, which were also seen to move with time (Fig. 12.8).

It's obvious that the trees couldn't be moving, so what was going on? A clue came from taking tree cores along a transect through any single wave then looking at the ages of the trees as determined by the number of rings in the trunks (Fig. 12.9). This revealed the underlying regeneration process. Waves tended to move in the same direction as the wind. Trees that were subject to the full force of winter winds were often killed by ice accumulation on their needles and branches, while those behind them were protected. Once trees died it took around 60 years for the forest to recover and the trees to grow back before another wave passed through.

This maintained an ongoing process of regeneration, a never-ending secondary succession in which the overall forest structure remained the same on the landscape scale.

12.8 Plants and animals

Most of the classic case studies of succession focus on vegetation. This is no accident. On land plants are the main drivers of succession and community change. They're also easier to study—since they neither move nor run away, monitoring plant communities on a site through time poses less of a challenge than for animals.

Plants make up the overwhelming majority of terrestrial biomass and act as the primary producers that support the rest of the food chain. In older communities such as forests, they also lock away nutrients in **necromass**, dead tissue where it is of no use to the plant itself, but also unavailable to other species. The trunks of most trees are dead apart from a ring of living tissue just below the bark. The necromass is greater than is actually required to support the tree—hence the ability of hollow trees to remain standing. This hoarding of resources is part of why trees are the dominant competitors in plant communities. Plants also form the physical structure upon which other organisms such as animals depend.

Animals usually have little impact on succession because their bodies decompose faster than plants. This matters because dead plant material can play a central role in shaping habitats, while animals simply die and vanish practically overnight. Peat bogs are essentially piles of undecomposed plants, and even in death a tree continues to influence its surroundings. Exceptions include some animals that create the structural elements of habitats, such as corals. Animals are often presented as passive followers of plant communities (at least on land). This generalisation is not always true; those which have important interactions with plants can provide dynamic feedbacks that promote or suppress particular plant species.

In order to show one way in which animals can modulate and control succession, this chapter finishes with a case study from East Africa.

12.9 Case study: Mpala, Kenya

In Laikipia Province, Northern Kenya, Todd Palmer and colleagues have been studying the communities around the research centre at Mpala for many years. The main form of vegetation is savannah, which comprises trees surrounded by grasses. The dominant trees are still referred to as acacias although strictly they are in the genus *Vachellia*. Acacias are particularly notable for their mutualistic interactions with ants through the presence of extrafloral nectaries, organs which secrete sugar-rich fluid to attract ants, while some tree species also provide nesting sites called domatia in specialised hollow twigs (Fig. 12.10a). Ants benefit the trees by defending them against browsing herbivores. There are a number of ant species involved; four are particularly common, though only a single ant species is ever found on any given tree.

The question Palmer was interested in was what would happen if large herbivores were removed. This was not a straightforward exercise; it involved setting up tall electric fences, high enough to exclude giraffes and powerful enough to deter elephants, and leaving them in place for over a decade (Fig. 12.10b). The difficulties in building the fences, sending out long cables and providing a reliable source of electricity in the African bush already make this a remarkable study.

Once large herbivores were excluded, the proportion of trees occupied by each ant species shifted (Fig. 12.11a). The strongest competitor ant species, *Crematogaster sjostedti* (Cs), increased in abundance at the expense of *Crematogaster mimosae* (Cm), which was not only the most common ant species in open savannah habitats but also the best mutualist for the trees. The other two species occurred less frequently. The consequences for the trees could be seen in terms of their mortality rates. Around a third of trees

Figure 12.10 (a) Domatia on *Vachellia drepanolobium*; (b) example of a large animal exclosure used by Palmer et al. (2008). (*Source*: Photographs reproduced with permission from Todd Palmer.)

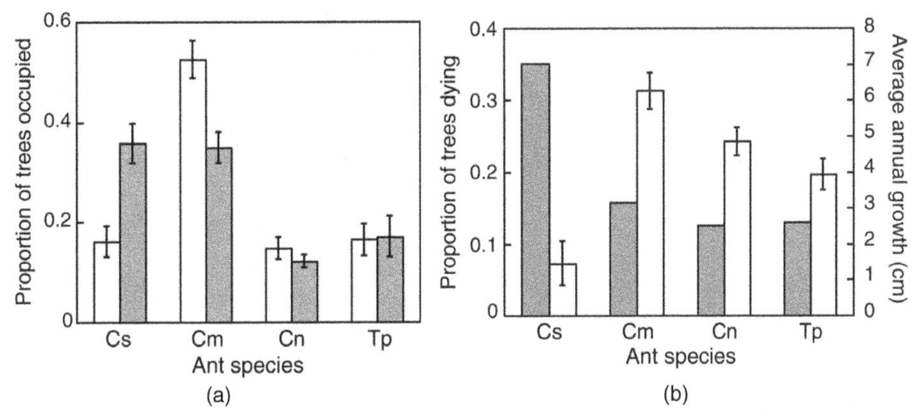

Figure 12.11 (a) Proportion of host trees occupied by the four acacia-ant species in the presence of large herbivores (white bars) and in plots from which large herbivores had been excluded (grey bars) for 10 years; (b) cumulative mortality (grey bars) and annual growth (white bars ± SE) for trees occupied by the four acacia-ant species over an 8-year period. Cs *Crematogaster sjostedti*; Cm *C. mimosae*; Cn *C. nigriceps*; Tp *Tetraponera penzigi*. (*Source*: Palmer et al. (2008). Reproduced with permission from American Association for the Advancement of Science.)

occupied by *C. sjostedti* died during the experimental period, and their growth rate was less than 20% that of trees which had *C. mimosae* ants (Fig. 12.11b).

It transpired that due to reduced grazing by large herbivores in exclosures, the trees had turned off their EFNs and reduced production of domatia. This helped *C. sjostedti* to outcompete *C. mimosae* because it made much less use of nectar resources. The fact that *C. sjostedti* was less willing to deter mammalian herbivores was irrelevant since they were excluded anyway. The real problem was longhorn beetles (Cerambycidae), which laid their eggs in the stems, and whose larvae bored holes through them (Fig. 12.12). This enabled fungal infections to enter the trees and increased their mortality. Trees occupied by *C. sjostedti* were more vulnerable.

Ironically, reducing large animal herbivory lowered tree fitness in the long term by favouring different ant species. Were the large herbivores to disappear entirely, the community would undergo

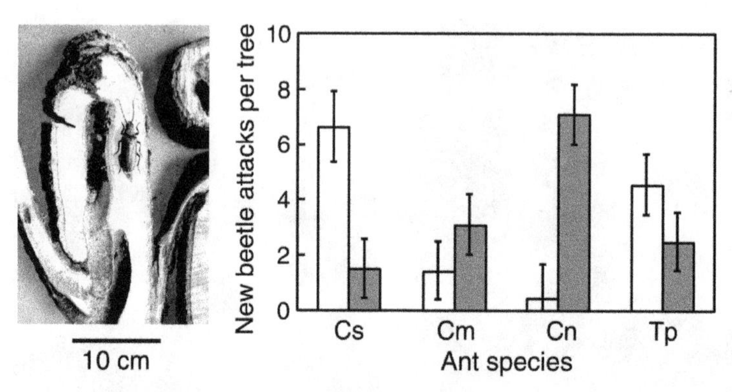

Figure 12.12 Cavities excavated by boring beetle larvae within a longitudinal cross section of a main stem of *V. drepanolobium*. An adult longhorn beetle (Cerambycidae) from the same stem is also shown. Graph shows beetle attack on control (white bars, mean ± SE) and ant-removal trees (grey bars) occupied by the four acacia-ant species labelled as in Fig. 12.11. (*Source*: Palmer et al. (2008). Reproduced with permission from American Association for the Advancement of Science.)

a secondary succession, potentially shifting towards one with fewer trees. This outcome could not be predicted based upon interactions among plants alone. Similar changes take place in systems with strong top-down control of herbivores; see the study by Schmitz (2008) in Section 8.6 for a comparable example.

12.10 Conclusions

Nothing stands still in nature, and communities are constantly changing. Directional change is known as succession, which can be either primary or secondary, depending on whether it begins with an empty habitat or takes place within an existing community. Regeneration refers to the suite of mechanisms by which communities remain stable in composition through time, thereby constituting their resilience. The main processes that control successional dynamics are colonisation, facilitation and competition, which can all be visualised using the niche framework from Chapter 6. Where succession is founder controlled, the outcome is stochastic and unpredictable, contingent on chance colonisation events. When a competitive hierarchy exists, a predictable succession takes place that can be effectively predicted using Markov chain models. Disturbance is a natural process, from which communities often recover their original form through regeneration, often in a never-ending cycle. Although succession on land is usually led by plants, animals can modulate pathways and control the final outcome.

12.10.1 Recommended reading

Connell, J. H. and R. O. Slatyer, 1977. Mechanisms of succession in natural communities and their role in community stability and organization. *American Naturalist* 111:1119–1144.

Johnson, E. A. and K. M. Miyanishi, 2008. Testing the assumptions of chronosequences in succession. *Ecology Letters* 11:419–431.

Rees, M., R. Condit, M. Crawley, S. Pacala, and D. Tilman, 2001. Long-term studies of vegetation dynamics. *Science* 293:650–655.

12.10.2 Questions for the future

- Is it possible to identify which communities are able to reconstitute themselves after disturbance, and which will respond unpredictably?
- How much observed change in present communities is caused by anthropogenic climate change or other human impacts?
- What effects do different trophic levels have on the pathways followed by succession?

References

Chase, J. M. and M. A. Leibold, 2003. *Ecological Niches: Linking Classical and Contemporary Approaches*. Chicago University Press.

Chen, H. Y. H. and A. R. Taylor, 2012. A test of ecological succession hypotheses using 55-year time-series data for 361 boreal forest stands. *Global Ecology and Biogeography* 21:441–454.

Connell, J. H. and R. O. Slatyer, 1977. Mechanisms of succession in natural communities and their role in community stability and organization. *American Naturalist* 111:1119–1144.

Cowles, H. C., 1901. The physiographic ecology of Chicago and vicinity. *Botanical Gazette* 31:73–108.

Grubb, P. J., 1977. The maintenance of species-richness in plant communities: the importance of the regeneration niche. *Biological Reviews* 52:107–145.

Hill, M. F., J. D. Witman, and H. Caswell, 2004. Markov chain analysis of succession in a rocky subtidal community. *American Naturalist* 164:E46–E61.

Horn, H. S., 1975. Markovian processes of forest succession. In M. L. Cody and J. M. Diamond, editors, *Ecology and Evolution of Communities*, pages 196–213. Belknap Press.

Horn, H. S., 1981. Succession. In R. M. May, editor, *Theoretical Ecology: Principles and Applications*, pages 253–271. Blackwell Publishing.

Johnson, E. A. and K. Miyanishi, 2008. Testing the assumptions of chronosequences in succession. *Ecology Letters* 11:419–431.

MacArthur, R. H. and E. O. Wilson, 1967. *The Theory of Island Biogeography*. Princeton University Press.

Márquez, J. C. and J. Kolasa, 2013. Local and regional processes in community assembly. *PLoS ONE* 8:e54580.

Palmer, T. M., M. L. Stanton, T. P. Young, J. R. Goheen, R. M. Pringle, and R. Karban, 2008. Breakdown of an ant-plant mutualism follows the loss of large herbivores from an African savanna. *Science* 319:192–195.

Sale, P. F., 1977. Maintenance of high diversity in coral reef fish communities. *American Naturalist* 111:337–359.

Sale, P. F., 1979. Recruitment, loss and coexistence in a guild of territorial coral reef fishes. *Oeologia* 42: 159–177.

Schmitz, O. J., 2008. Effects of predator hunting mode on grassland ecosystem function. *Science* 319:952–954.

Sprugel, D. G., 1976. Dynamic structure of wave-regenerated *Abies balsamea* forests in the north-eastern United States. *Journal of Ecology* 64:889–911.

Supp, S. R. and S. K. M. Ernest, 2014. Species-level and community-level responses to disturbance: a cross-community analysis. *Ecology* 95:1717–1723.

Van Cleve, K. and L. A. Viereck, 1981. Forest succession in relation to nutrient cycling in the boreal forest of Alaska. In D. C. West, H. H. Shugart, and D. B. Botkin, editors, *Forest Succession: Concepts and Application*, pages 185–221. Springer-Verlag.

Yodzis, P., 1986. Competition, mortality and commuity structure. In J. M. Diamond and T. J. Case, editors, *Community Ecology*, pages 480–491. Harper and Row.

Zaplata, M. K., S. Winter, A. Fischer, J. Kollmann, and W. Ulrich, 2013. Species-driven phases and increasing structure in early-successional plant communities. *American Naturalist* 181:E17–E27.

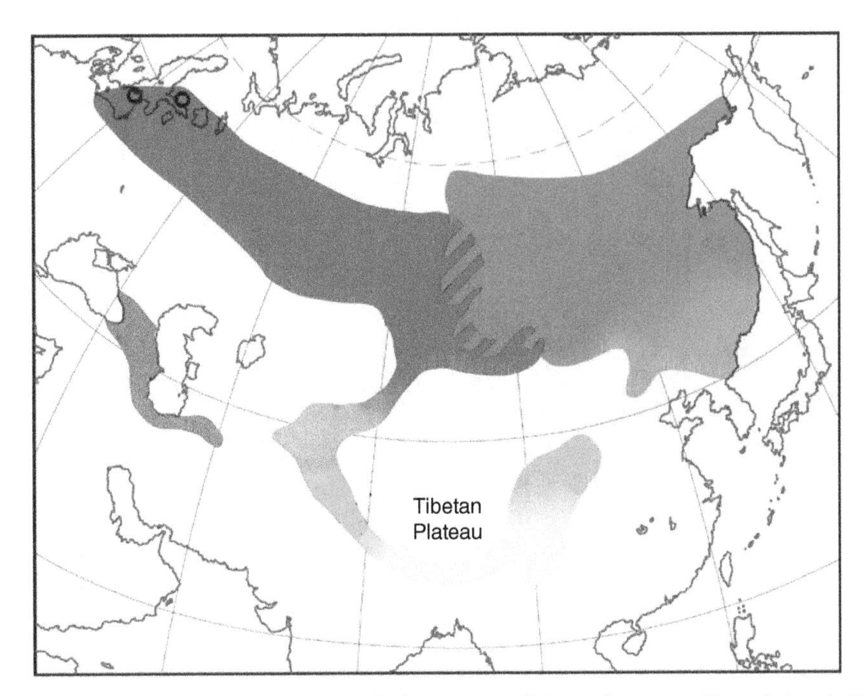

Figure 2.2 Breeding range of greenish warblers in Asia. Shades represent distinct subspecies, with gradations indicating change in morphology. The hatched area in central Siberia is the overlap zone between two distinct subspecies. Sampling sites are indicated by symbols corresponding to mitochondrial clades. (*Source*: Irwin et al. (2005). Reproduced with permission from American Association for the Advancement of Science.)

Natural Systems: The organisation of life, First Edition. Markus P. Eichhorn.

Figure 3.2 An artistic reconstruction of a Carboniferous rain forest dominated by lycopsids. (*Source*: DiMichele et al. (2007). Reproduced with permission from the Geological Society of America.)

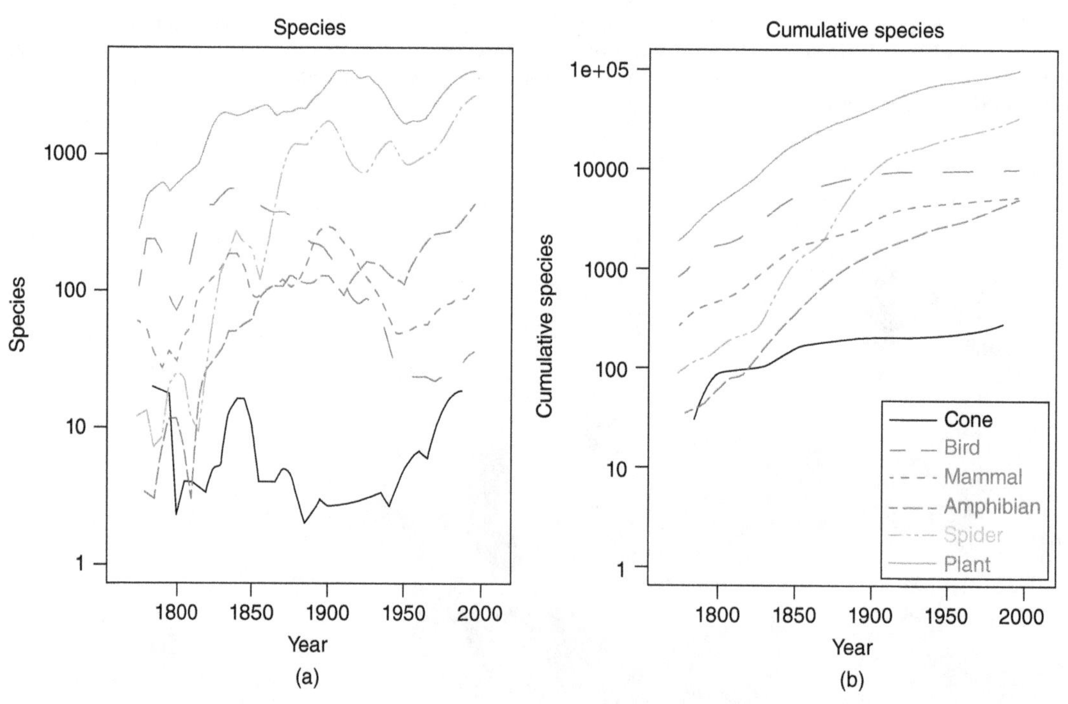

Figure 4.2 Number of species discovered (a) per 5-year interval and (b) cumulatively since 1800. (*Source*: Joppa et al. (2011). Reproduced with permission from Elsevier.)

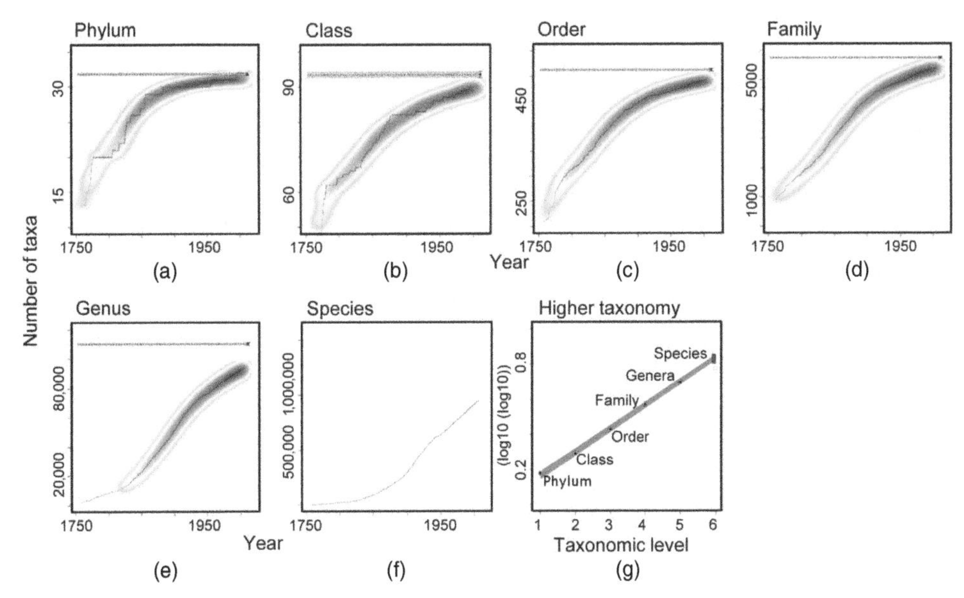

Figure 4.4 Temporal accumulation of animal taxa (solid lines) and model fits to obtain asymptotes for richness at higher taxonomic levels (a–e). Shading represents the frequency of estimates for each year across multiple models; horizontal dashed lines indicate the estimated asymptotic number of taxa, and the horizontal grey area its standard error. (f) shows the number of species described. In (g) asymptotes are extrapolated across taxonomic levels to estimate the total number of species. Black circles represent the consensus asymptotes, gray circles the catalogued number of taxa, and the box at the species level indicates the 95% confidence interval around the predicted number of species. (*Source*: Mora et al. (2011). Used under CC-BY-SA 2.5 http://creativecommons.org/licenses/by-sa/2.5/.)

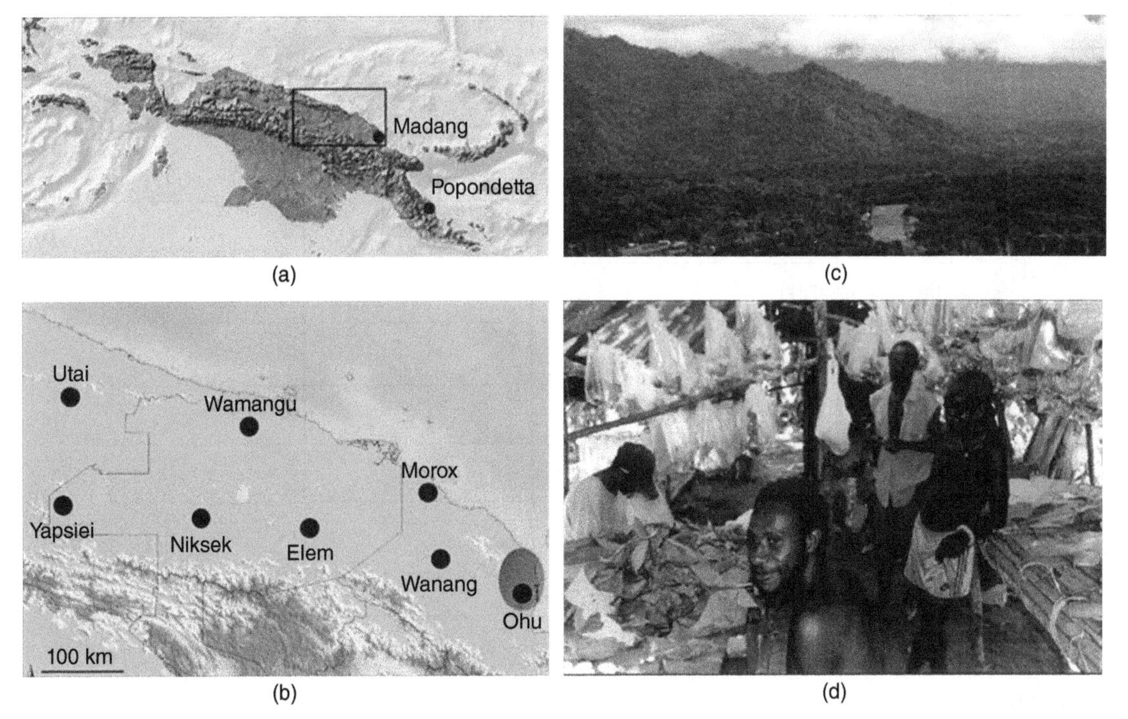

Figure 5.9 (a) Region of Papua New Guinea studied; (b) sampling locations; (c) aircraft landing strip and (d) field laboratory. (*Source*: Novotny et al. (2007). Reproduced with permission from Nature Publishing Group.)

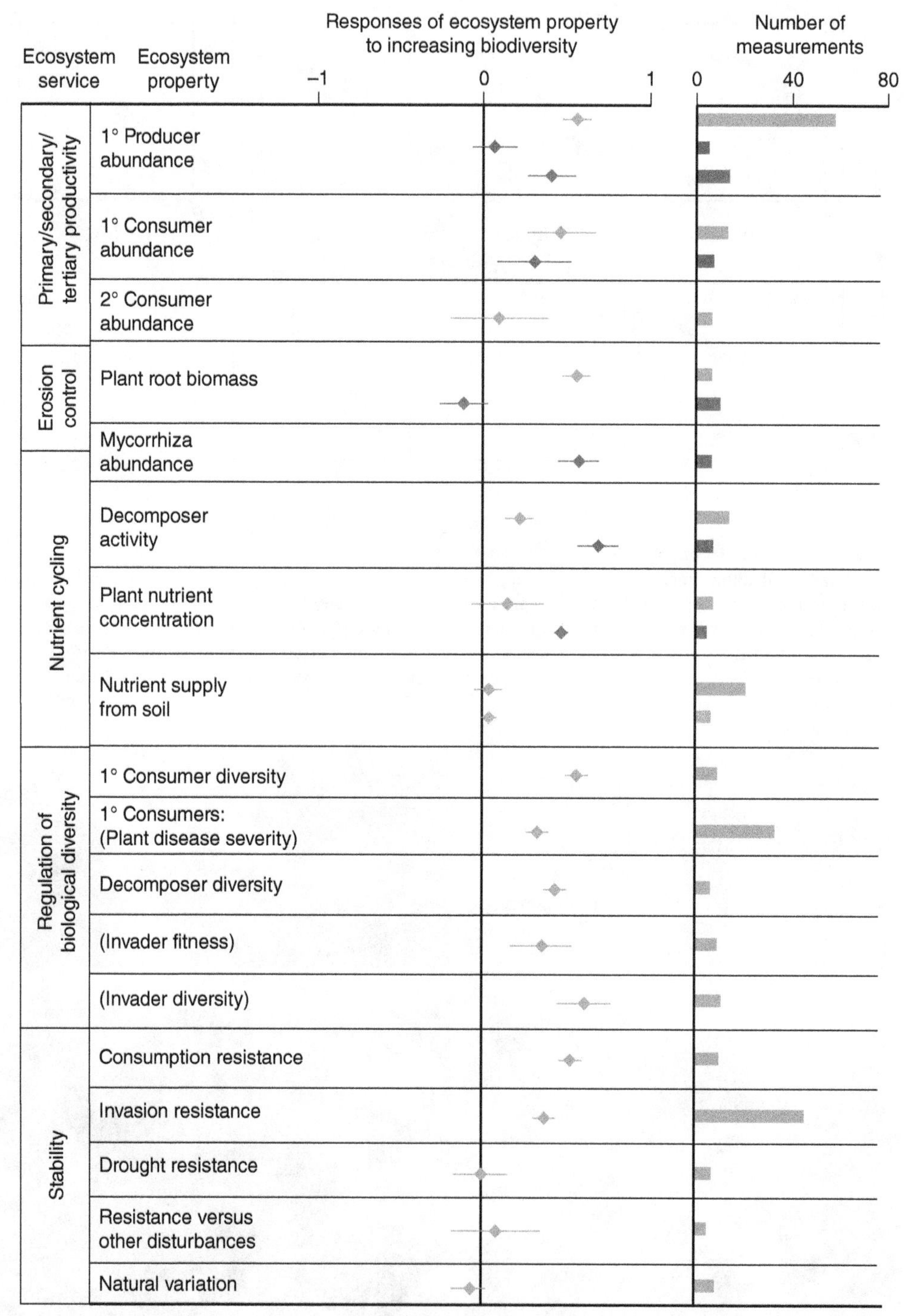

Figure 9.13 Magnitude and direction of effects of diversity on ecosystem processes (mean ± SE) with number of measurements available. Ecosystem properties shown in parentheses were considered of negative value for human well-being, and thus inverse effect sizes are shown. Coloured bars show trophic level manipulated: green, primary producers; blue, primary consumers; pink, mycorrhizae; brown, decomposers; grey, multiple levels. (*Source*: Balvanera et al. (2006). Reproduced with permission from John Wiley & Sons Ltd.)

Figure 10.5 (a) The giant kangaroo rat, *Dipodomys ingens*; (b) a burrow mound; (c) mounds in the Carrizo Plain National Monument, CA; (d) a kangaroo rat exclosure experiment (left side of image). (*Source*: Prugh and Brashares (2012). Reproduced with permission from John Wiley & Sons Ltd.)

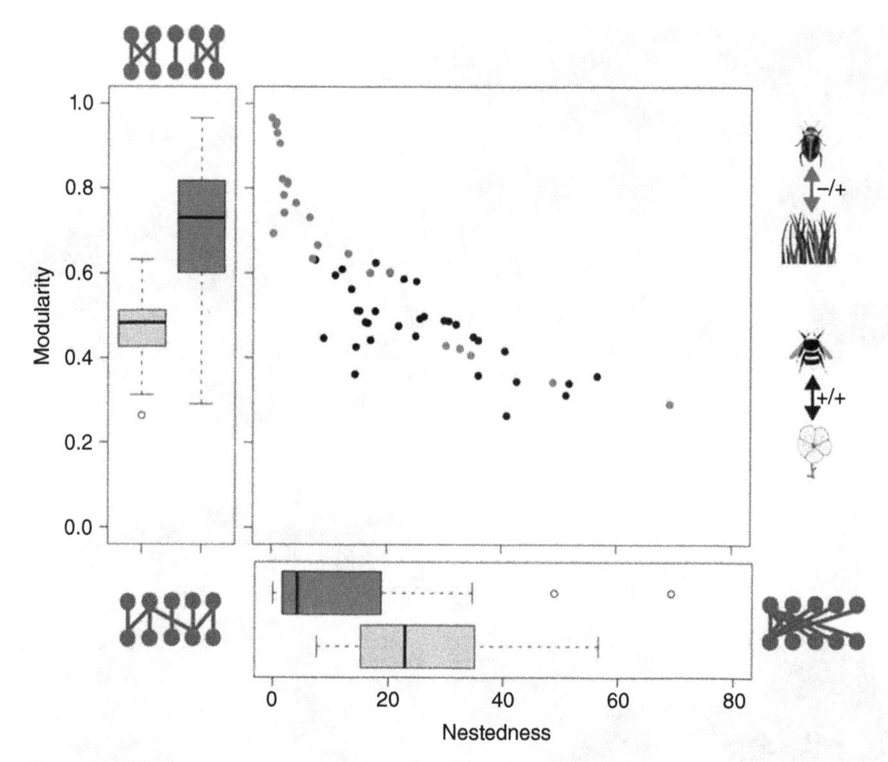

Figure 10.11 Relationship between network nestedness and modularity, with box plots of nestedness (below) and a measure of network modularity (left). Higher values are more nested or modular, respectively. Each dot represents an empirical network, either pollination (black) or herbivory (red), with corresponding box plots showing, median line and interquartile range with bars ±1.5 interquartile range. (*Source*: Thebault and Fontaine (2010). Reproduced with permission from the American Association for the Advancement of Science.)

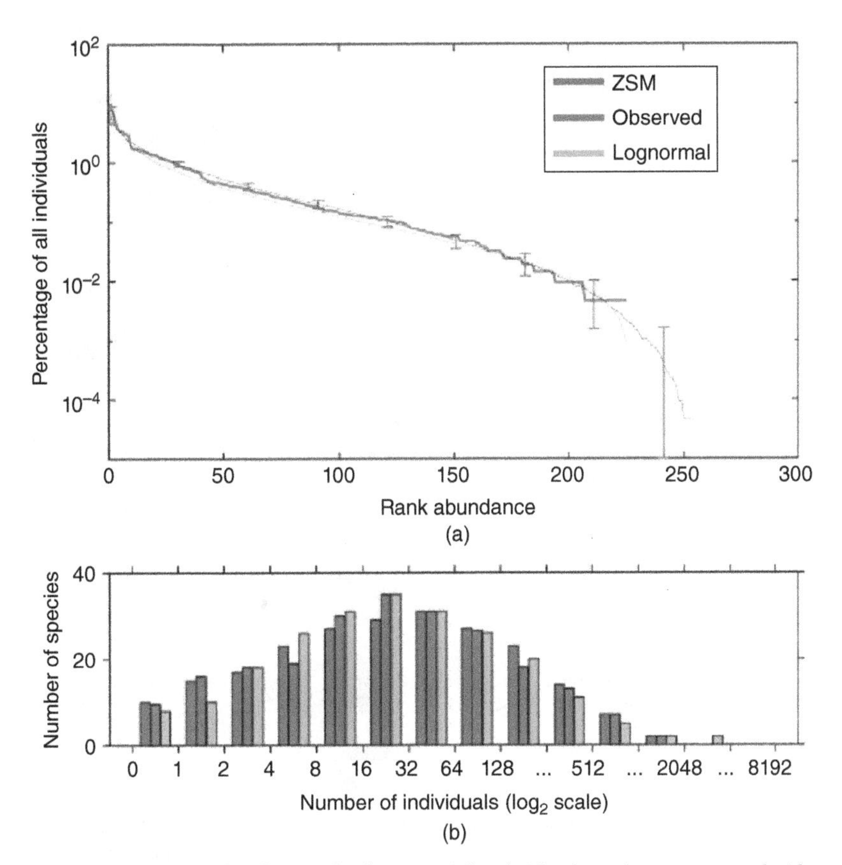

Figure 10.19 Observed species abundance distributions for the Barro Colorado Island tree dataset compared with predictions based on the lognormal and neutral theory (zero sum multinomial (ZSM)). (*Source*: McGill (2003). Reproduced with permission from Nature Publishing Group.)

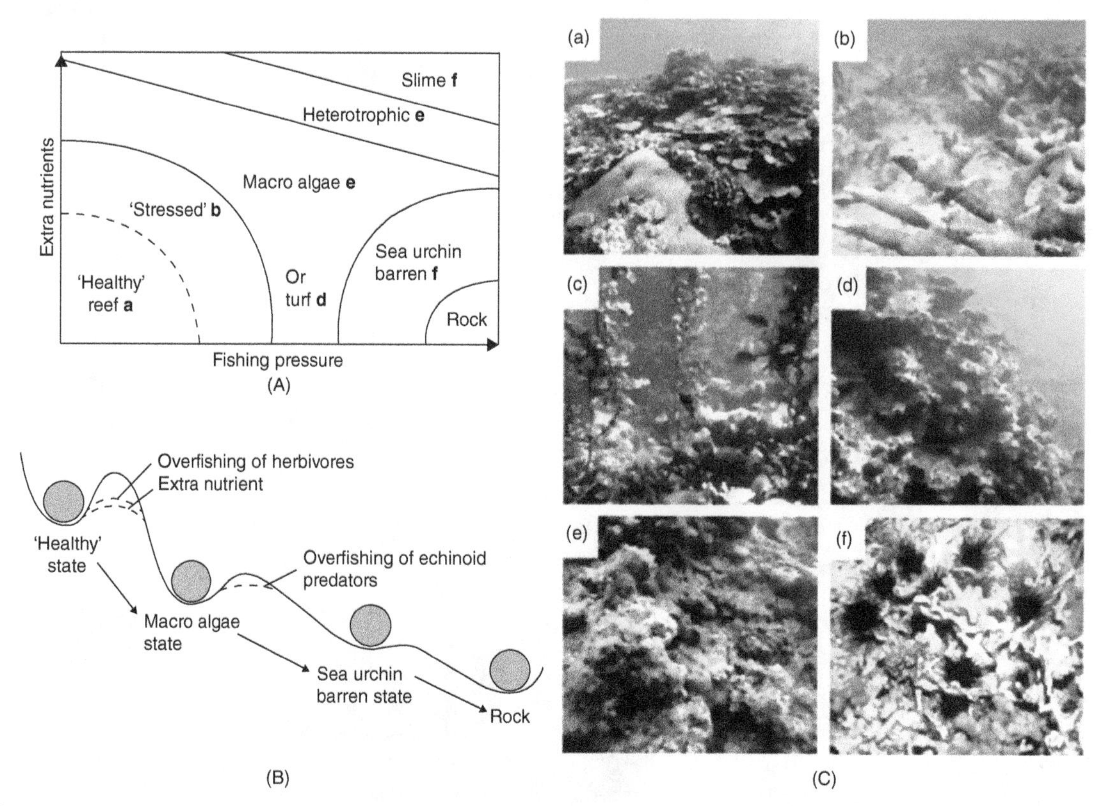

Figure 11.13 Conceptual model of ecosystem states given (A) fishing pressure and nutrient supply, and (B) the transition sequence between them illustrated by (C) a range of reef habitats. (*Source*: Bellwood et al. (2004). Reproduced with permission from Nature Publishing Group.)

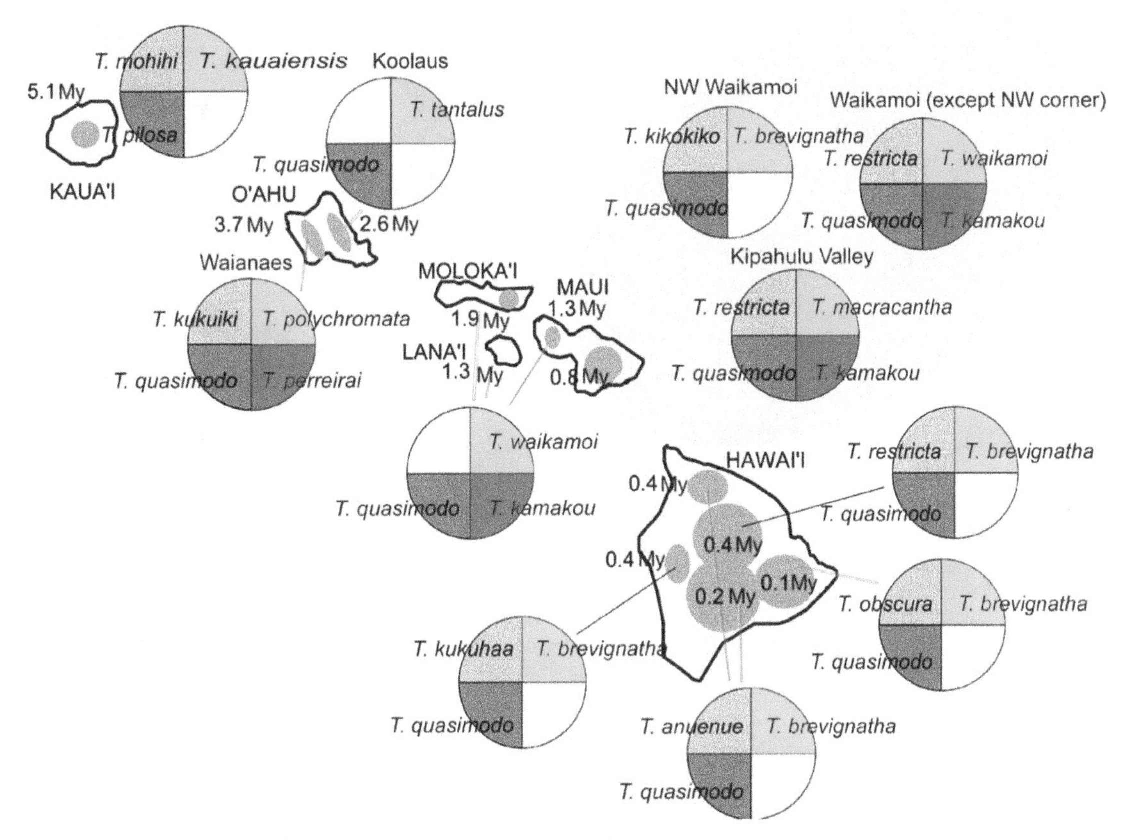

Figure 13.3 Distribution of spider ecomorphs in the genus *Tetragnatha* across the Hawaiian archipelago. Volcanoes are shown as grey circles with their age in millions of years (My). Coloured quadrants indicate whether each of four ecomorphs is present at a site (green, maroon, dark brown or light brown), along with the name of the species present at that site. (*Source*: Gillespie (2004). Reproduced with permission from American Association for the Advancement of Science.)

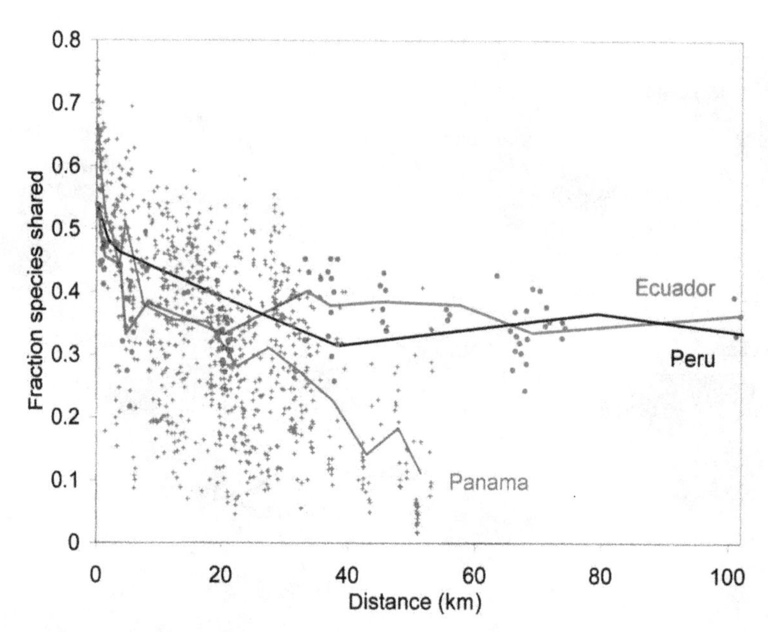

Figure 13.18 Sørensen similarity index between pairs of 1 ha forest plots as a function of distance. Red line and dots, Ecuador; black line, Peru; blue line and crosses, Panama. Points for Peru are omitted for clarity. (*Source*: Condit et al. (2002). Reprodcued with permission from American Association for the Advancement of Science.)

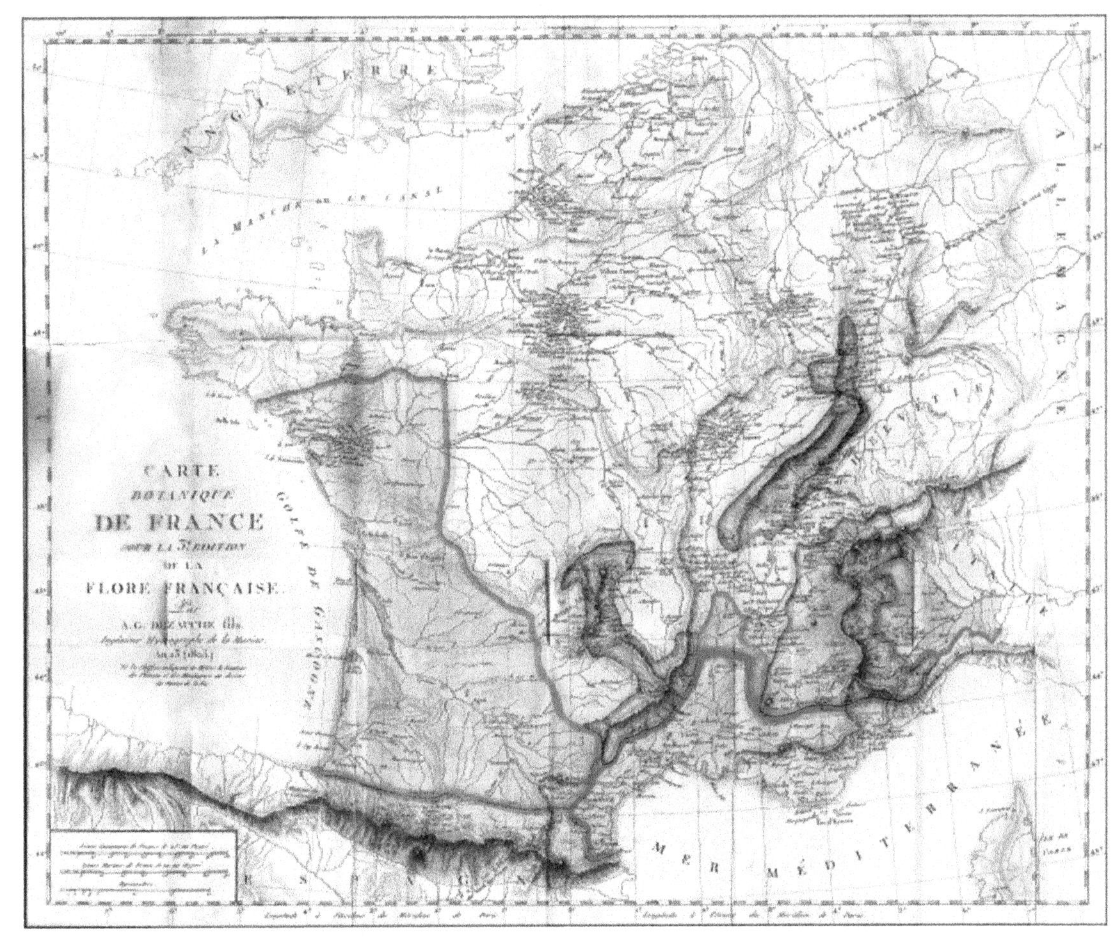

Figure 14.2 Botanical map of France produced in 1805 by Lamarck and de Candolle. Coloured regions denote consistent floras. (*Source*: Ebach and Goujet (2006). Reproduced with permission from John Wiley & Sons Ltd.)

Figure 14.4 Terrestrial zoogeographic realms and regions based on distributions of 21,037 species of amphibians, non-pelagic birds and non-marine mammals. Dashed lines delineate the 20 zoogeographic regions; thick lines group regions into 11 named realms, for which colour differences indicate the amount of phylogenetic turnover; dotted regions have no species records; Antarctica was not included in the analyses. (*Source*: Holt et al. (2013). Reproduced with permission from American Association for the Advancement of Science.)

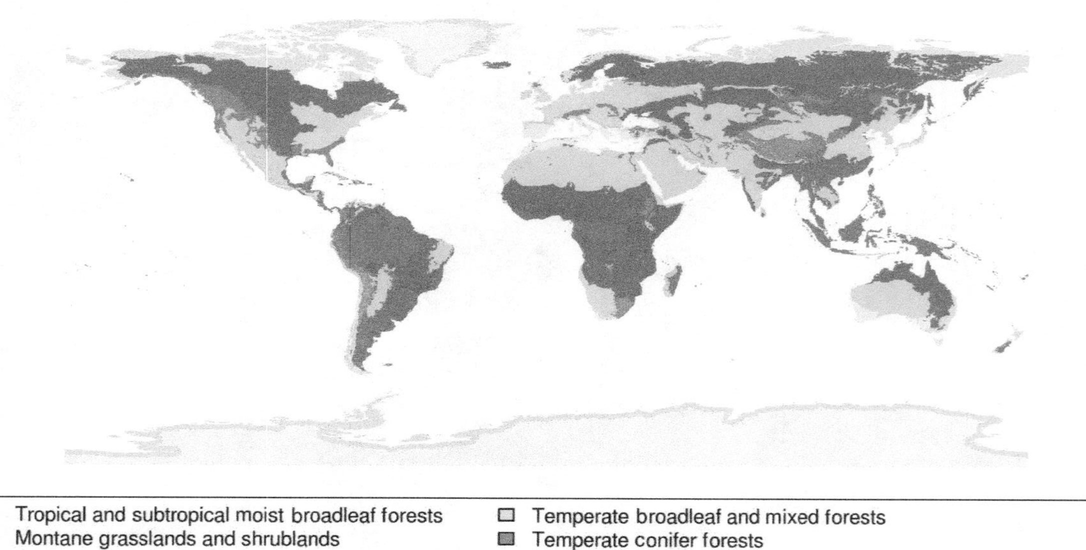

Figure 14.5 Global terrestrial biomes. (*Source*: Data provided by The Nature Conservancy; figure prepared with help from Tim Newbold.)

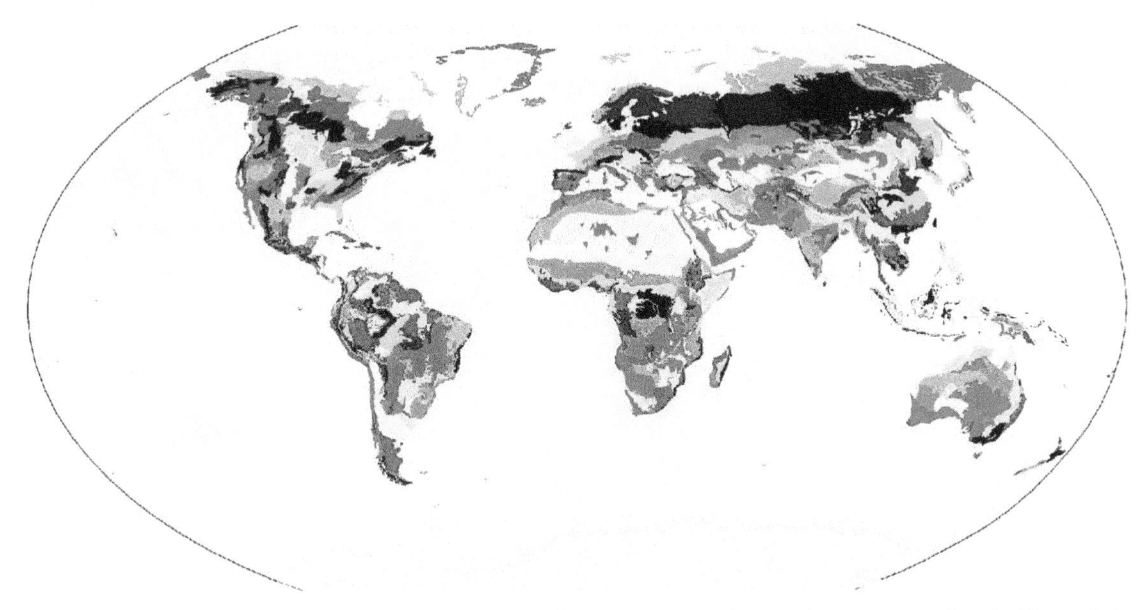

Figure 14.10 Terrestrial ecoregions as identified by Olson et al. (2001). (*Source*: Olson et al. (2001). Reproduced with permission from Oxford Universty Press.)

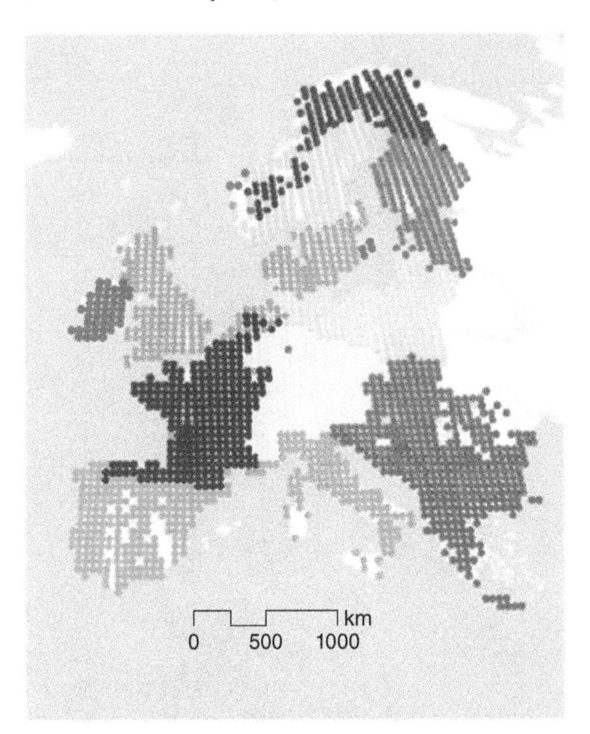

Figure 14.16 Faunal elements for European land mammals (Gagné and Proulx, 2007; Heikinheimo et al., 2007).

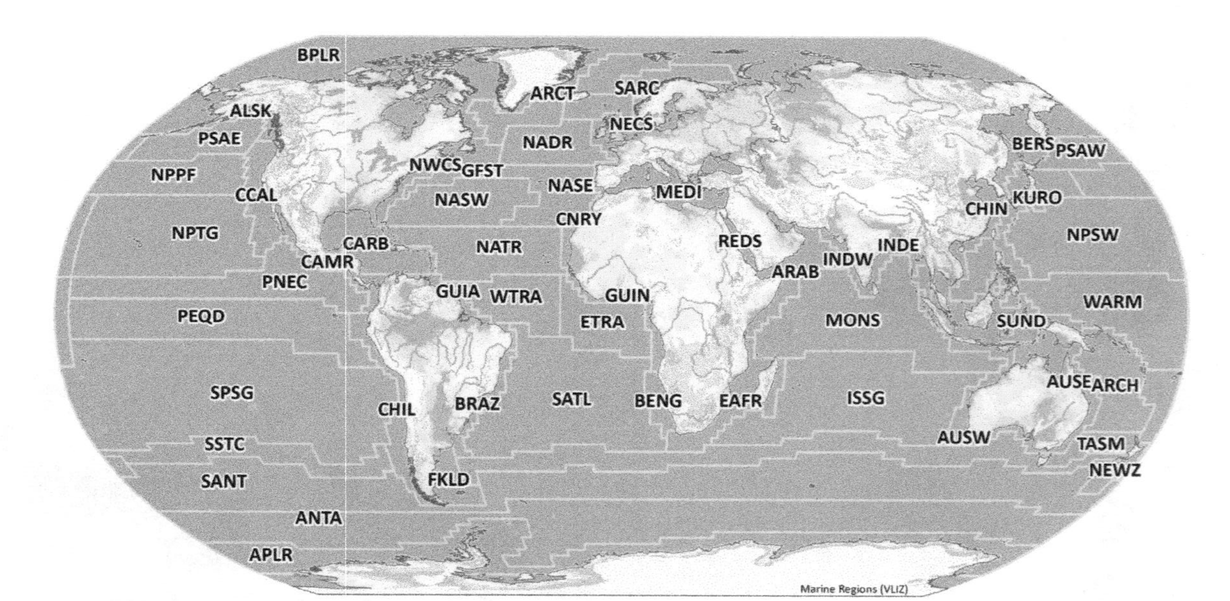

Figure 14.17 Partitioning of the world's oceans into provinces based on phytoplankton distribution. Note that these boundaries are not fixed and alter in response to seasonal and interannual changes in the environment. (*Source*: Longhurst (1998). Reproduced with permission from John Wiley & Sons Ltd.)

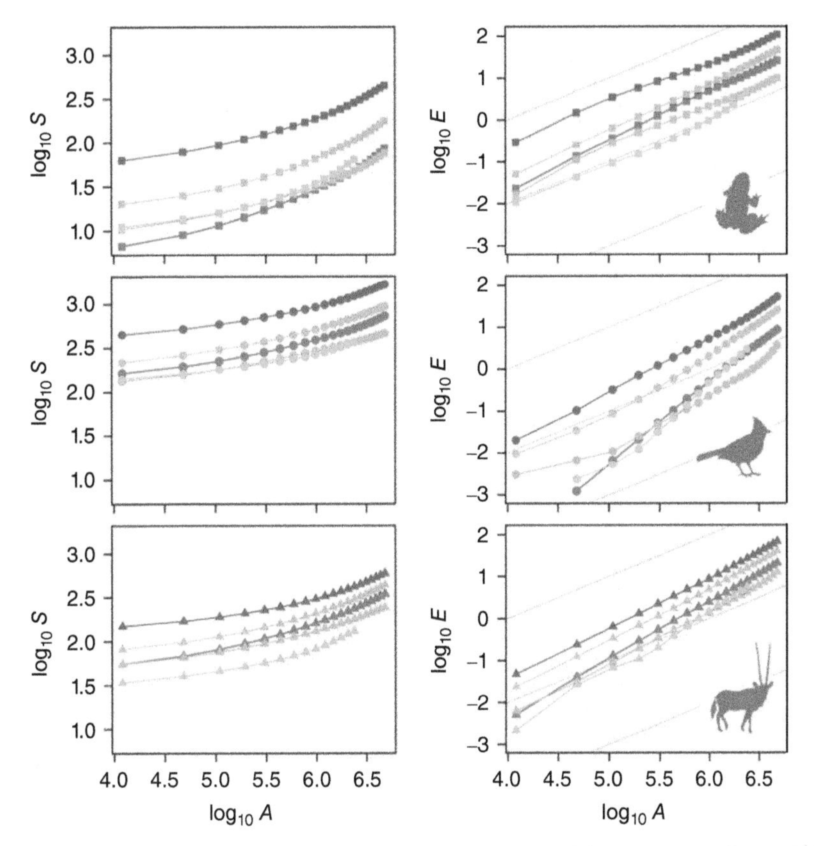

Figure 15.14 Species- and Endemics-Area Relationships (SARs and EARs) for amphibians (squares), birds (circles) and mammals (triangles). Data from five continents (blue: Africa, red: Eurasia, yellow: North America, purple: South America, green: Australia). Grey lines correspond to a power law with a slope of 1; that is, proportionality between area and the number of species. S is the mean number of species, E is the mean number of endemics, A is the area in km^2. (*Source*: Storch et al. (2012). Reproduced with permission from Nature Publishing Group.)

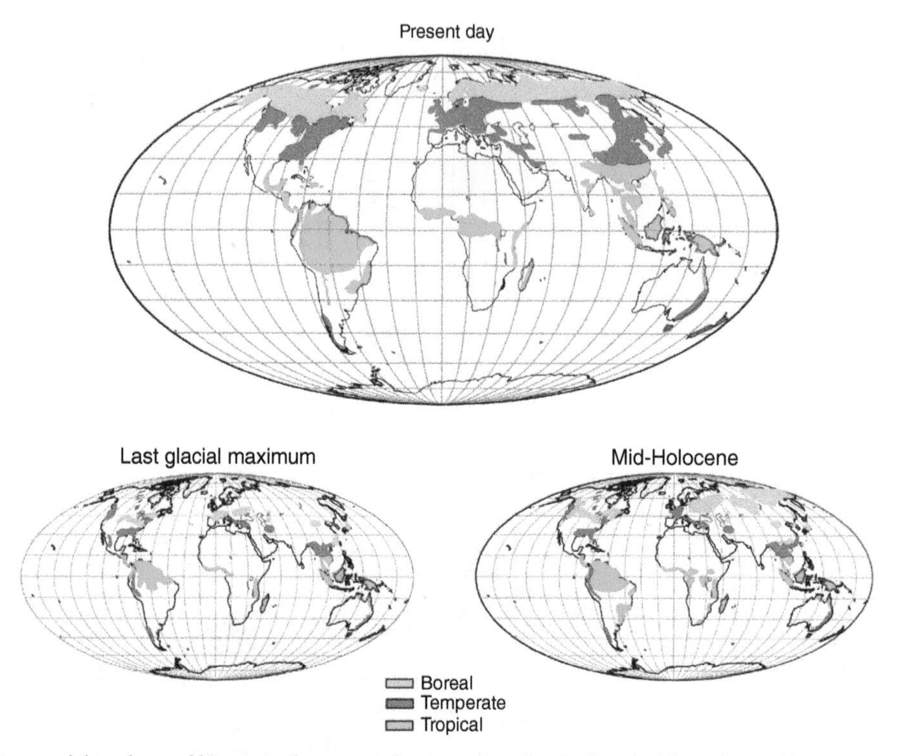

Figure 16.1 Extent of three forested biomes in the present day (upper) and at the last glacial maximum (20,000 ya), mid-Holocene (6,000 ya). (*Source*: Fine and Ree (2006). Reproduced with permission from the University of Chicago Press.)

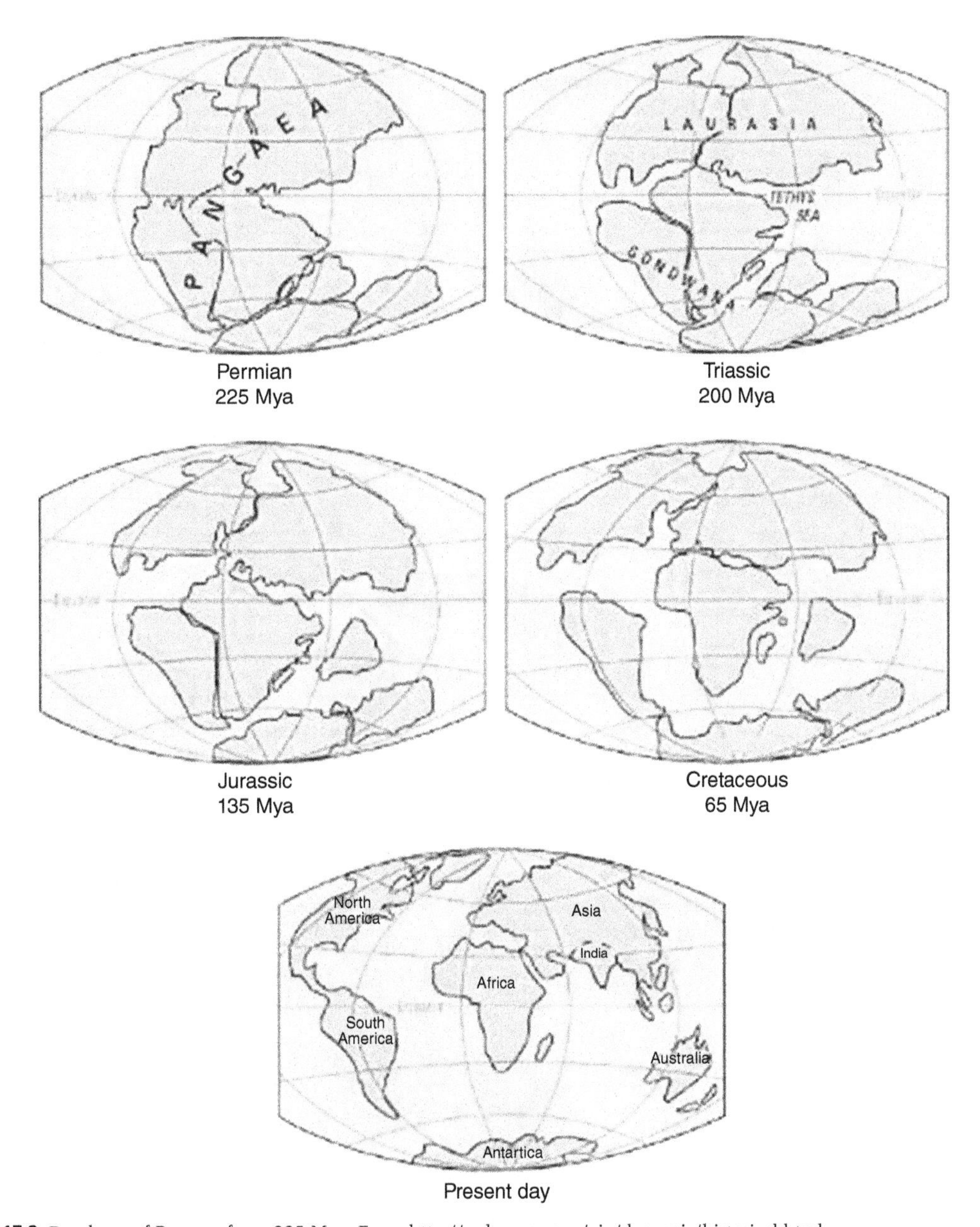

Figure 17.2 Break-up of Pangæa from 225 Mya. From http://pubs.usgs.gov/gip/dynamic/historical.html.

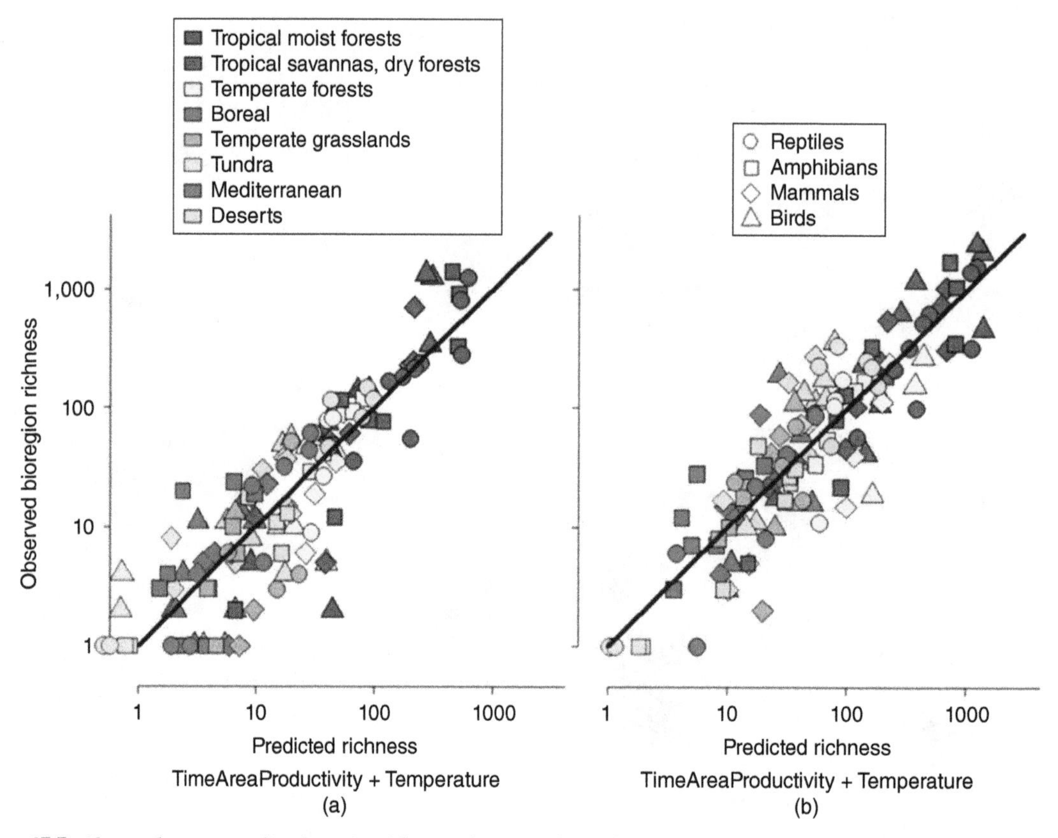

Figure 17.7 Observed versus predicted species richness of terrestrial vertebrates in 32 bioregions and four taxonomic groups (a, endemic species; b, resident species). The model includes two parameters: TimeAreaProductivity and Temperature; lines indicate least squares fit of regressions (r^2 [endemic] = 0.78, r^2 [resident] = 0.78, N = 128). See text for further details. (*Source*: From Jetz and Fine (2012). Used under CC-BY-SA 2.5 http://creativecommons.org/licenses/by-sa/2.5/.)

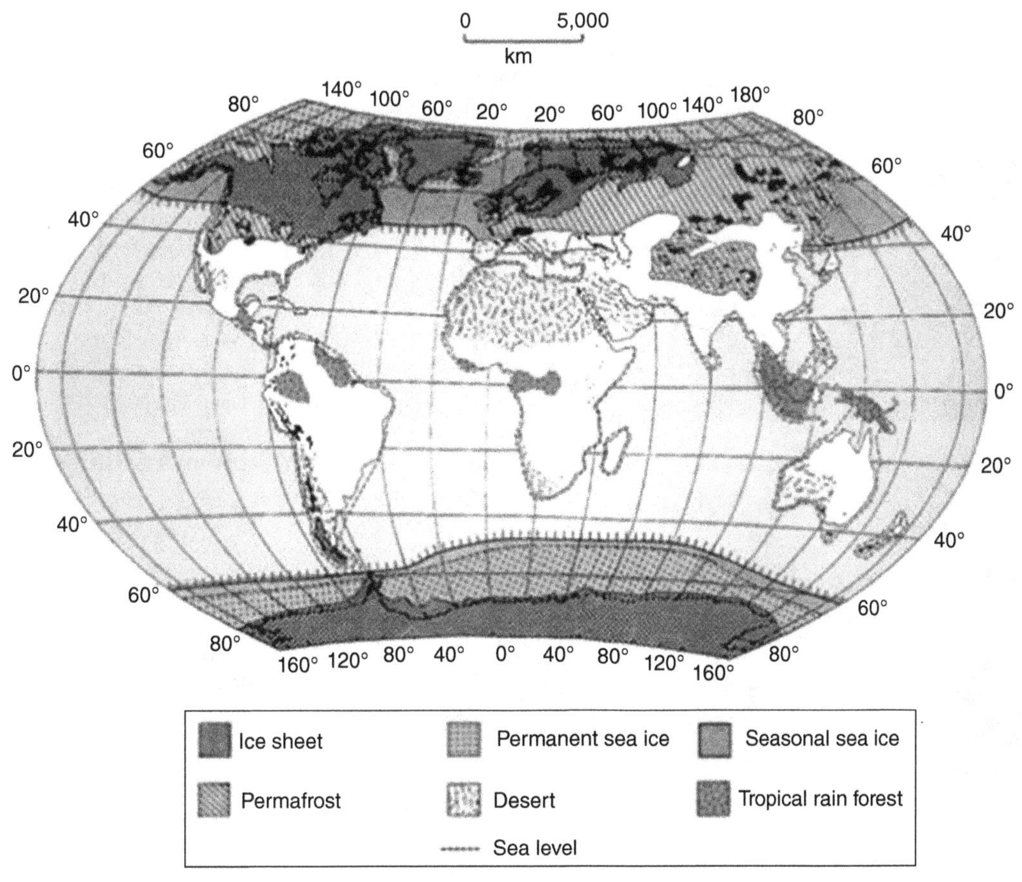

Figure 17.8 Maximum extent of ice and permafrost at the last glacial maximum c. 20,000 years ago. Note changes in the shape of continents owing to lowered sea levels. Reproduced from Hewitt (2000) with permission from the Nature Publishing Group. (*Source*: Modified from Williams et al. (1998) and reproduced with permission from Taylor and Francis in Hewitt (2000).)

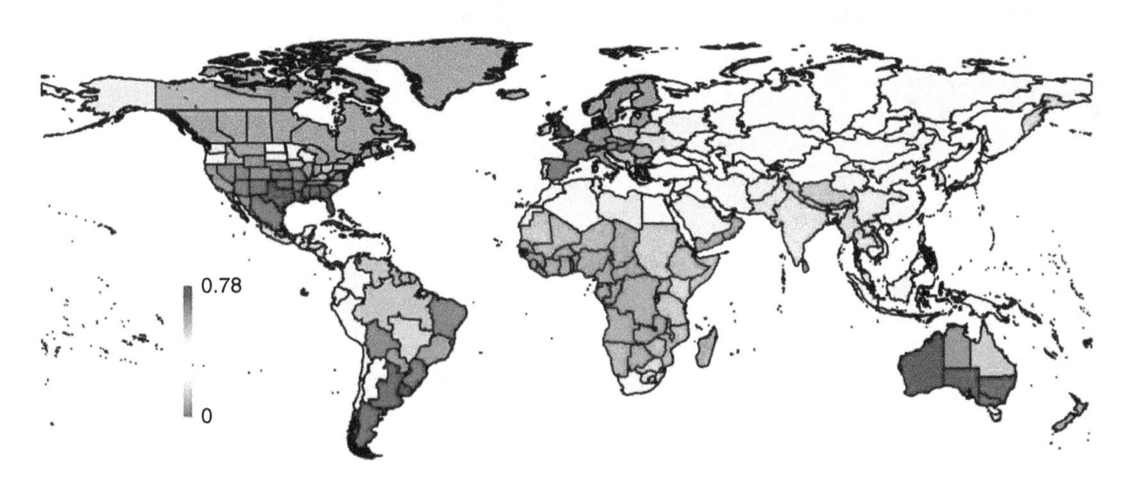

Figure 17.12 Severity of Quaternary large mammal extinctions measured as the proportion of large mammals (≥10 kg) per country which went extinct between 132,000 and 1,000 years ago. Regions in dark grey were excluded from this study. (*Source*: From https://creativecommons.org/licenses/by/4.0/.)

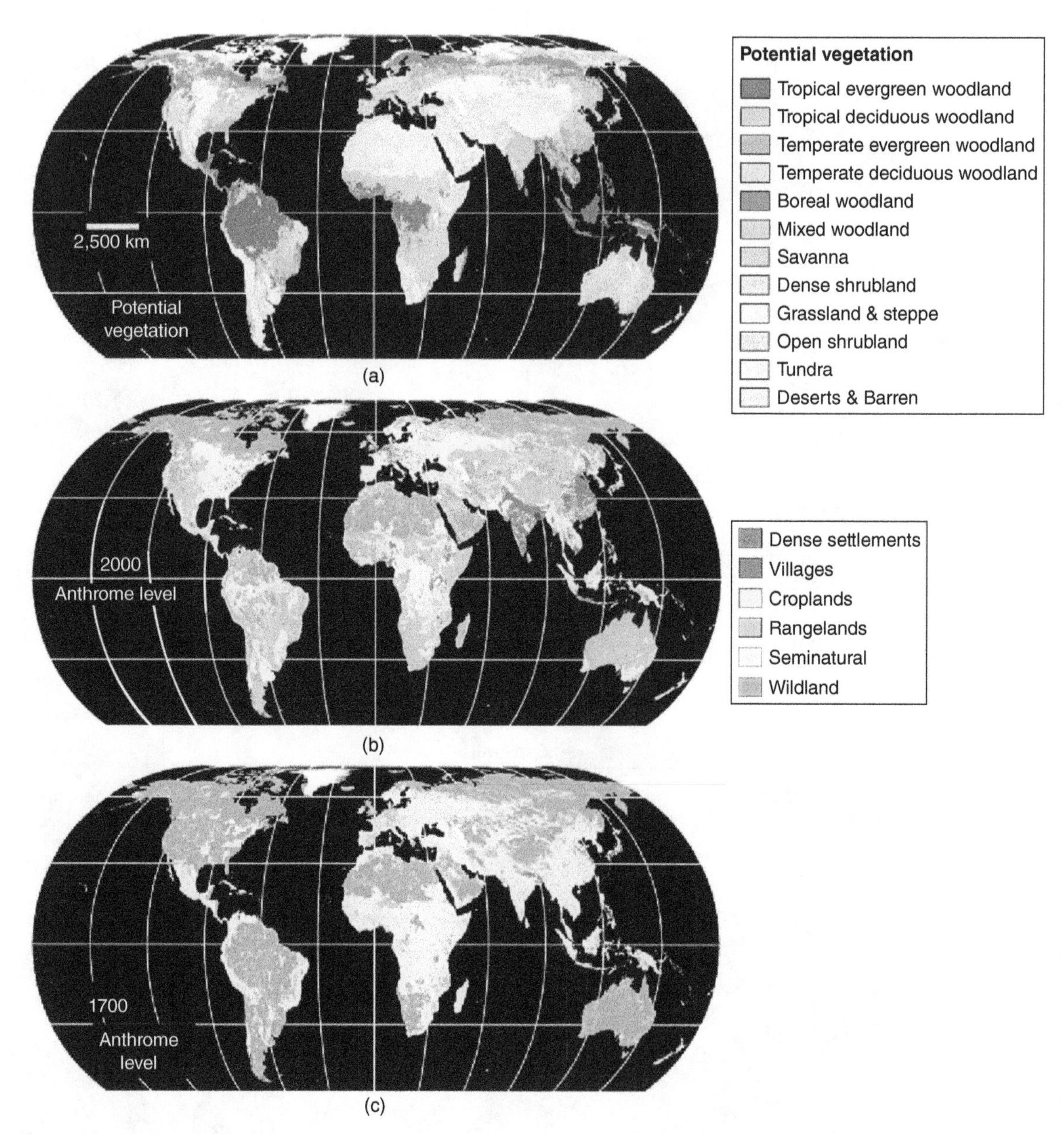

Figure 17.15 (a) Potential natural vegetative biomes with their state in (b) 2000 and (c) 1700. (*Source*: Ellis et al. (2010). Reproduced with permission from John Wiley and Sons Ltd.)

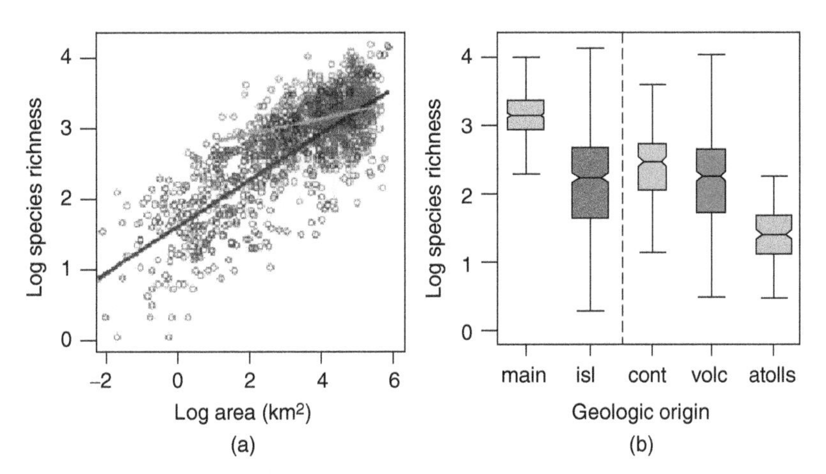

Figure 19.7 (a) Species–area relationship for island (blue) and mainland (brown) plant species richness; (b) variation in plant species richness between mainland areas and islands, dependent on island geological origin (continental, volcanic or atolls). (*Source*: Kreft et al. (2008). Reproduced with permission from John Wiley & Sons Ltd.)

INTERNATIONAL CHRONOSTRATIGRAPHIC CHART

IUGS

www.stratigraphy.org

International Commission on Stratigraphy

v 2015/01

Eonothem / Eon	Erathem / Era	System / Period	Series / Epoch	Stage / Age	numerical age (Ma)
					present
Phanerozoic	Cenozoic	Quaternary	Holocene		0.0117
			Pleistocene	Upper	0.126
				Middle	0.781
				Calabrian	1.80
				Gelasian	2.58
		Neogene	Pliocene	Piacenzian	3.600
				Zanclean	5.333
			Miocene	Messinian	7.246
				Tortonian	11.63
				Serravallian	13.82
				Langhian	15.97
				Burdigalian	20.44
				Aquitanian	23.03
		Paleogene	Oligocene	Chattian	28.1
				Rupelian	33.9
			Eocene	Priabonian	37.8
				Bartonian	41.2
				Lutetian	47.8
				Ypresian	56.0
			Paleocene	Thanetian	59.2
				Selandian	61.6
				Danian	66.0
	Mesozoic	Cretaceous	Upper	Maastrichtian	72.1 ±0.2
				Campanian	83.6 ±0.2
				Santonian	86.3 ±0.5
				Coniacian	89.8 ±0.3
				Turonian	93.9
				Cenomanian	100.5
			Lower	Albian	~ 113.0
				Aptian	~ 125.0
				Barremian	~ 129.4
				Hauterivian	~ 132.9
				Valanginian	~ 139.8
				Berriasian	~ 145.0

Eonothem / Eon	Erathem / Era	System / Period	Series / Epoch	Stage / Age	numerical age (Ma)
					~ 145.0
Phanerozoic	Mesozoic	Jurassic	Upper	Tithonian	152.1 ±0.9
				Kimmeridgian	157.3 ±1.0
				Oxfordian	163.5 ±1.0
			Middle	Callovian	166.1 ±1.2
				Bathonian	168.3 ±1.3
				Bajocian	170.3 ±1.4
				Aalenian	174.1 ±1.0
			Lower	Toarcian	182.7 ±0.7
				Pliensbachian	190.8 ±1.0
				Sinemurian	199.3 ±0.3
				Hettangian	201.3 ±0.2
		Triassic	Upper	Rhaetian	~ 208.5
				Norian	~ 227
				Carnian	~ 237
			Middle	Ladinian	~ 242
				Anisian	247.2
			Lower	Olenekian	251.2
				Induan	252.17 ±0.06
	Paleozoic	Permian	Lopingian	Changhsingian	254.14 ±0.07
				Wuchiapingian	259.8 ±0.4
			Guadalupian	Capitanian	265.1 ±0.4
				Wordian	268.8 ±0.5
				Roadian	272.3 ±0.5
			Cisuralian	Kungurian	283.5 ±0.6
				Artinskian	290.1 ±0.26
				Sakmarian	295.0 ±0.18
				Asselian	298.9 ±0.15
		Carboniferous – Pennsylvanian	Upper	Gzhelian	303.7 ±0.1
				Kasimovian	307.0 ±0.1
			Middle	Moscovian	315.2 ±0.2
			Lower	Bashkirian	323.2 ±0.4
		Carboniferous – Mississippian	Upper	Serpukhovian	330.9 ±0.2
			Middle	Visean	346.7 ±0.4
			Lower	Tournaisian	358.9 ±0.4

Eonothem / Eon	Erathem / Era	System / Period	Series / Epoch	Stage / Age	numerical age (Ma)
					358.9 ± 0.4
Phanerozoic	Paleozoic	Devonian	Upper	Famennian	372.2 ±1.6
				Frasnian	382.7 ±1.6
			Middle	Givetian	387.7 ±0.8
				Eifelian	393.3 ±1.2
			Lower	Emsian	407.6 ±2.6
				Pragian	410.8 ±2.8
				Lochkovian	419.2 ±3.2
		Silurian	Pridoli		423.0 ±2.3
			Ludlow	Ludfordian	425.6 ±0.9
				Gorstian	427.4 ±0.5
			Wenlock	Homerian	430.5 ±0.7
				Sheinwoodian	433.4 ±0.8
			Llandovery	Telychian	438.5 ±1.1
				Aeronian	440.8 ±1.2
				Rhuddanian	443.8 ±1.5
		Ordovician	Upper	Hirnantian	445.2 ±1.4
				Katian	453.0 ±0.7
				Sandbian	458.4 ±0.9
			Middle	Darriwilian	467.3 ±1.1
				Dapingian	470.0 ±1.4
			Lower	Floian	477.7 ±1.4
				Tremadocian	485.4 ±1.9
		Cambrian	Furongian	*Stage 10*	~ 489.5
				Jiangshanian	~ 494
				Paibian	~ 497
			Series 3	Guzhangian	~ 500.5
				Drumian	~ 504.5
				Stage 5	~ 509
			Series 2	*Stage 4*	~ 514
				Stage 3	~ 521
			Terreneuvian	*Stage 2*	~ 529
				Fortunian	541.0 ±1.0

Eonothem / Eon	Erathem / Era	System / Period	numerical age (Ma)
			~ 541.0 ±1.0
Precambrian – Proterozoic	Neoproterozoic	Ediacaran	~ 635
		Cryogenian	~ 720
		Tonian	1000
	Mesoproterozoic	Stenian	1200
		Ectasian	1400
		Calymmian	1600
	Paleoproterozoic	Statherian	1800
		Orosirian	2050
		Rhyacian	2300
		Siderian	2500
Precambrian – Archean	Neoarchean		2800
	Mesoarchean		3200
	Paleoarchean		3600
	Eoarchean		4000
Precambrian – Hadean			~ 4600

Units of all ranks are in the process of being defined by Global Boundary Stratotype Section and Points (GSSP) for their lower boundaries, including those of the Archean and Proterozoic, long defined by Global Standard Stratigraphic Ages (GSSA). Charts and detailed information on ratified GSSPs are available at the website http://www.stratigraphy.org. The URL to this chart is found below.

Numerical ages are subject to revision and do not define units in the Phanerozoic and the Ediacaran; only GSSPs do. For boundaries in the Phanerozoic without ratified GSSPs or without constrained numerical ages, an approximate numerical age (~) is provided.

Numerical ages for all systems except Lower Pleistocene, Permian, Triassic, Cretaceous and Precambrian are taken from 'A Geologic Time Scale 2012' by Gradstein et al. (2012); those for the Lower Pleistocene, Permian, Triassic and Cretaceous were provided by the relevant ICS subcommissions.

Coloring follows the Commission for the Geological Map of the World (http://www.ccgm.org)

CCGM CGMW

Chart drafted by K.M. Cohen, S.C. Finney, P.L. Gibbard

To cite: Cohen, K.M., Finney, S.C., Gibbard, P.L. & Fan, J.-X. (2013; updated) The ICS International Chronostratigraphic Chart. Episodes 36: 199-204.

URL: http://www.stratigraphy.org/ICSchart/ChronostratChart2015-01.pdf

CHAPTER 13

Changes through space

13.1 The big question

While walking in the hills, if you were to find a patch of moss growing on a rock surface, took a sample of the water held by the leaves,[1] and examined it under a microscope, you would witness a bewildering number of minute species swimming around. Should you repeat this on the next rock, or one a little way down the trail, you would find a different set of species. There would be some overlap, of course, but also many differences in the identity and relative abundances of the organisms present. How can such an apparently simple community vary so much between locations?

There are many possible explanations. These include local effects such as variation in microenvironments, competition among species and contingencies in how the communities initially formed. There may also be processes acting on the larger scale, regulating which species are able to disperse into each community. This leads to the concept of metacommunities and the realisation that to understand what is happening in any single patch, we need to appreciate how it is linked to other similar patches in the landscape. It's also crucial to ask which species are not present in a given patch. Explaining why some species are absent from a particular area helps us to better understand the species that do occur.

[1] Strictly speaking they are phyllids rather than leaves.

13.2 Community assembly

We should start by stating a set of truisms in ecology. No species ...

... is found everywhere
... occurs anywhere it could
... dominates wherever it occurs.

There are clearly limits on the abilities of species to spread, and these are not solely related to their environmental tolerance. This observation led ecologists to investigate whether communities possess **assembly rule** that dictate which species are able to invade, the order in which they do so (e.g. predators must follow their prey) and whether they persist once they get there. These depend on local effects (features of the habitat) and interactions among species (niche assembly), which may be competitive, mutualistic or more indirect.

There is good evidence that such rules exist. Along the coastline of Chile, there is great variation in the supply rate of marine invertebrates, depending on local oceanic currents. Caro et al. (2010) documented the composition of recruiting species in the intertidal zone along 800 km of coastline and compared it to the assemblages of adults. Despite great variation in the input of juveniles, the final assemblages converged on a very similar and distinct composition (Fig. 13.1). Somehow, in these intertidal communities, internal processes were determining the final state reached. Note how different this is from the example of Márquez and Kolasa, (2013)

Natural Systems: The organisation of life, First Edition. Markus P. Eichhorn.

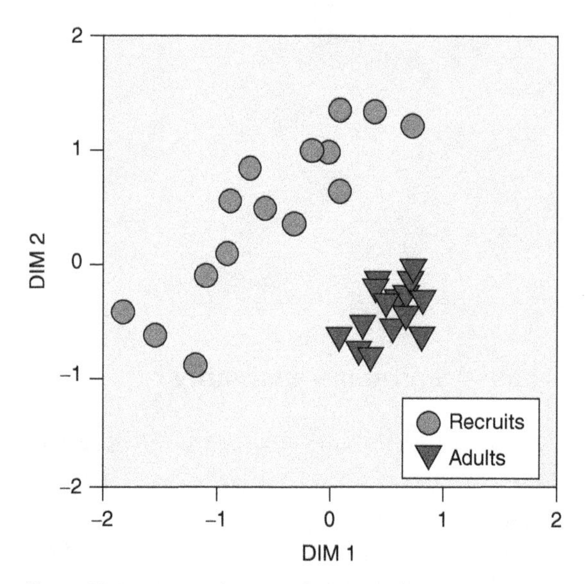

Figure 13.1 NMDS ordination of recruit and adult assemblages of intertidal invertebrates across sites. along the coastline of Chile. (*Source*: From Caro et al. (2010).)

in Section 12.4, where experimental rock pool communities diverged into a range of states despite starting with the same colonists. When bathing birds disperse propagules between rock pools, however, they increase the similarity of their faunas by breaking down dispersal barriers (Simonis and Ellis, 2014).

Comparing across multiple communities, some species are frequently found together, whereas others never combine. The former is easy to explain in terms of a common response to the environment or interactions among the species. Less easy to understand is why some species are never found in the same place. These patterns are often referred to as checkerboards (Diamond, 1975). What might be causing them? There are four possible explanations:

1 Competitive exclusion
2 Historical processes
3 Habitat checkerboards
4 Chance and contingency

We will consider these in turn.

13.2.1 Competitive exclusion

Firstly, could checkerboards arise through competitive exclusion (*sensu* Gause, 1936)? In other words, are the species that fail to co-occur competing for the same niche? Unfortunately it is impossible to demonstrate unequivocally that competition underlies checkerboards because the species are in different patches and therefore cannot be directly competing. Connell (1980) referred to this as the 'ghost of competition past': competition is assumed to be the root cause, but its effects have already played out, and only the end result remains.

Testing for the signal of past competition requires more specific predictions than species distributions alone. There are several possible expectations for sets of co-occurring species if they are the result of past competition. Species might (i) be less related than expected by chance, (ii) show dispersion of body sizes or other traits or (iii) differentiate resources or habitat space among themselves.

Species might be less related than random expectation if we assume that closely related species have similar traits and are therefore more likely to compete strongly with one another. Although this is plausible, remarkably few people have tested the assumption (for an example see Burns and Strauss, 2011). In experimental microcosms, closely related species of bacterial protist are less likely to coexist, and the more related they are, the faster competitive exclusion occurs (Fig. 13.2).

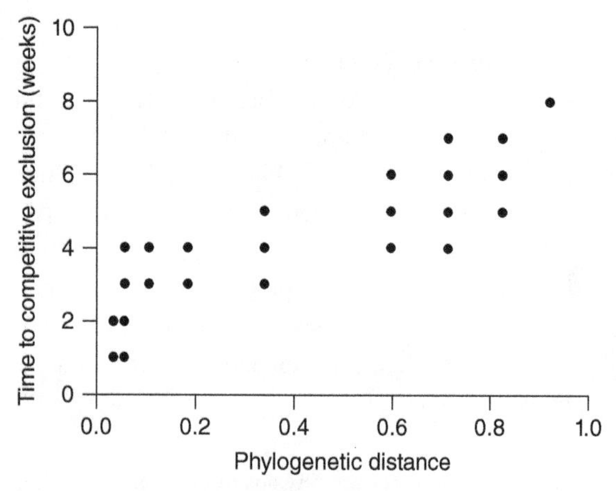

Figure 13.2 Time to competitive exclusion with phylogenetic distance between competing pairs of bacterial protist species in experimental microcosms. (*Source*: Violle et al. (2011). Reproduced with permission from John Wiley & Sons.)

Table 13.1 Summary of studies testing possible features of checkerboard distributions, assuming competition to be the root cause.

	Supporting	Non-significant	Opposing
Species/genus ratio	0	124	56
Body size ratio	7	13	1
Resource overlap	27	13	16

Source: Adapted from Chase and Leibold 2003.

Table 13.1 summarises the number of studies that have found support for the proposition that coexisting species should generally be unrelated. The ratio of species to genera provides a quick guide as to whether there are few sister species. The outcome is very clear—in most cases there is no pattern, and in about a third the opposite happens, making species *more* related than random! Considering most studies expected to find similar species segregating, this is a startling observation, and one we will return to later. Conflicting patterns in relatedness have now been found across many scales and systems and are hard to reconcile with current theory (HilleRisLambers et al. 2012). This remains an important avenue for future research. Perhaps at the landscape scale, the environment filters more similar species, which are those most likely to be able to tolerate the prevailing climate and resource levels. At the scale of a single patch, however, strong competition between close relatives might prevent coexistence (e.g. Carboni et al. 2013). Such an effect is mostly found in experimental studies though, where communities are artificially assembled; given a more natural mix of species, the evidence for resistance to related invaders disappears (Price and Pärtel, 2013).

Body size ratios convey something about resource usage and are often used as a surrogate for niches (e.g. Hutchinson, 1959). We would expect species of similar size to be unable to coexist; the same should be true of other traits. Table 13.1 shows that about a third of cases support this pattern, but two thirds find none (one goes in the opposite direction).

Finally, overlap in resource usage would be predicted to be smaller if coexisting species occupied distinct niches. This was found in only half of cases, and a substantial proportion found the opposite (Table 13.1). Sanders et al. (2007) examined assemblages of ants in California and Oregon, where a wide variety of species occurred. These ant species differed in size and their preferences for habitats and types of food. The study surveyed two habitats, forests and fens, and also considered whether the patch had been recently disturbed by fire. The expectation was that there would be evidence for niche structuring (i.e. size segregation) in the more stable, unburnt forests, whereas this might break down following disturbance, at least until the community settled down again. The results showed that ant assemblages were entirely random with respect to body size in both years of their study. It may be peculiar in a textbook to use an example that found only non-significant results, but it makes an important point. Here and elsewhere we need to invoke processes other than competition to explain variation among communities.

Finding patterns in Table 13.1 so opposed to what appear to be a common-sense set of predictions forces us to reconsider our understanding of the dynamics of species interactions. Ricklefs (2012) examined assemblages of birds, looking for evidence that the morphology of co-occurring species would spread out in niche space. In fact he saw the opposite—each region contained the same overall range of bird morphologies, but within local assemblages they were more similar than expected by chance, rather than different. This extends right down to patterns of association in mixed flocks of birds—species found together tend to be more similar in size and foraging method than expected by chance (Sridhar et al. 2012). Birds of a feather really do flock together.

In response to this paradox, it has been suggested that there are essentially two ways to coexist—to be sufficiently different or to be sufficiently similar. This proposition has been backed up by mathematical models (Scheffer and van Nes, 2006; Vergnon et al. 2012). The smaller the difference between any two competing species, the longer it takes for competitive exclusion to play out, which means that it can be deferred for many generations or coexistence can maintained by another mechanism (e.g. top-down control).

To understand how this works, imagine yourself moving to a new town and joining the local rugby club (or any team sport). The best way to ensure that you get a game is to play in a position that the club currently has vacant. This is seldom possible though; there are only 15 places in the team, each has a defined role, and there is no scope to invent a new kind of player just for you. Instead, you are more likely to succeed by being better than an existing player in a given position (competitive exclusion) or by taking whatever opportunities come up when someone playing in your position is injured or unavailable (neutral coexistence). In the latter case you benefit from taking a defined role and having the same set of traits as the other player. If you do not mind waiting for chances as a squad player, then you can still get plenty of games (this is the storage effect; see Section 8.4.2).

Many species abundance distributions are multimodal (Dornelas and Connolly, 2008), as are size distributions suggesting a limited number of potential roles for species. Perhaps niche structuring mechanisms apply at larger scales and neutral coexistence at smaller scales. There is as yet no direct evidence to back this up, but it points towards an enduring puzzle. Which mechanisms for coexistence are most important, at what scales, and for which groups of organisms?

13.2.2 Historical processes

Patterns caused by direct competition among individuals must be distinguished from those occurring over evolutionary timescales. Where competing species live alongside one another for multiple generations, there are often correlated changes that evolve in each species which reduce the effective competition between them by allowing each to specialise on a particular set of resources or conditions. This process is known as *community-wide character displacement*. There are many documented examples where the traits of species evolve in distinct directions, from the beaks of Galápagos finches as described by Darwin to *Anolis* lizards on Caribbean islands. These well-known case studies involve different, longer-term processes to those found in community assembly. It is important not to conflate the two.

This is not to say that patterns of evolution might not influence which species are found together. Allopatric speciation is when divergence between species occurs in different locations. In this case the lack of coexistence might simply reflect the places where they happened to evolve. Failure to co-occur can purely be an accident of geography.

The interplay between speciation, dispersal and possible competition can sometimes only be revealed through analysis of the evolutionary history of a taxon. Gillespie (2004) studied spiders in the Hawaiian archipelago. There were four **ecomorphs**, varying in colour and morphology, which fed in different parts of forest habitats (e.g. on leaves or in the ground litter), giving each a unique niche. Every site invariably had only one spider of each kind, though the identity of the specific ecomorph differed, and in some cases one or more were missing (Fig. 13.3).

Underlying this pattern were two processes, revealed by the phylogeny of the spiders. The ancestral species (a green ecomorph) had radiated into 20 species, with each alternative ecomorph evolving multiple times, independently on different islands. In some cases spiders had dispersed to an island and then radiated into different ecomorphs, while elsewhere new ecomorphs had evolved and then

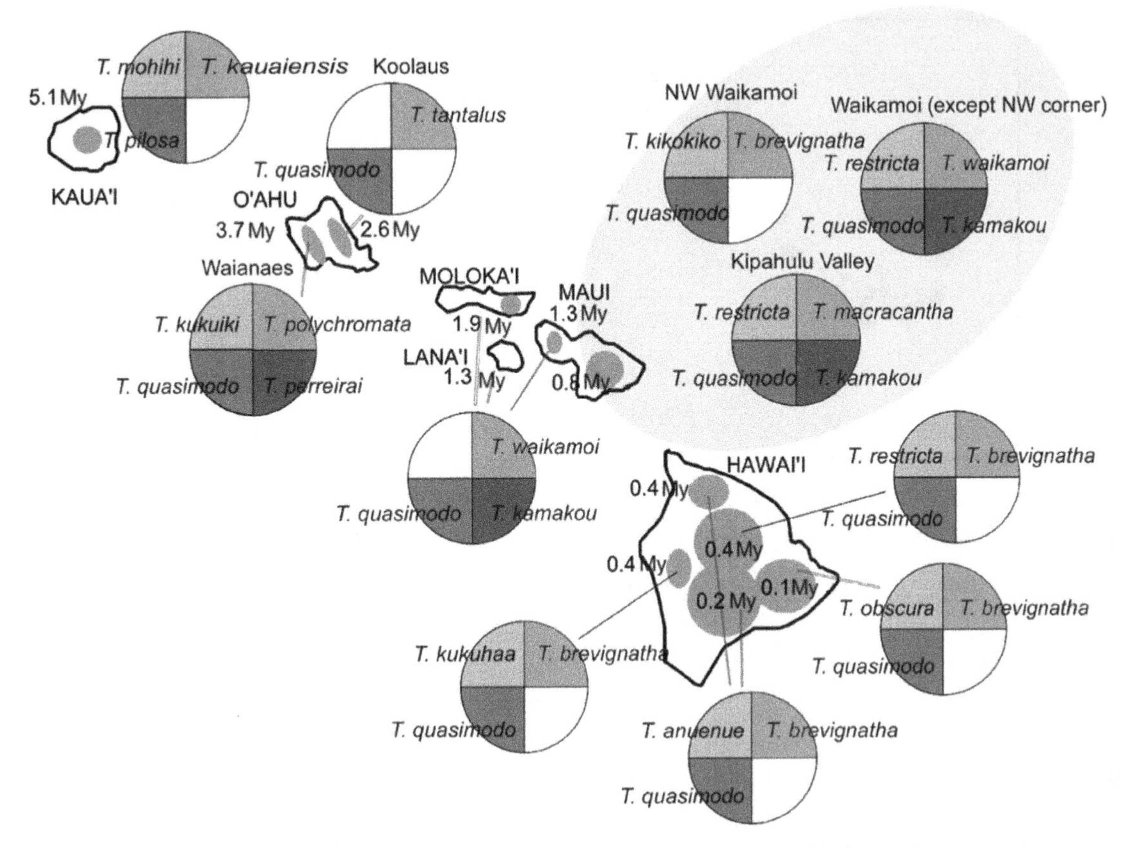

Figure 13.3 Distribution of spider ecomorphs in the genus *Tetragnatha* across the Hawaiian archipelago. Volcanoes are shown as grey circles with their age in millions of years (My). Coloured quadrants indicate whether each of four ecomorphs is present at a site (green, maroon, dark brown or light brown), along with the name of the species present at that site. (*Source*: Gillespie (2004). Reproduced with permission from American Association for the Advancement of Science.) (*See colour plate section for the colour representation of this figure.*)

spread to new islands (and often speciated again). The first species to take any particular role in a given location, whether it arrived though dispersal or speciation, was the one holding the niche, giving rise to checkerboards. Similar patterns are seen in *Anolis* lizards on Caribbean islands (Mahler et al. 2013).

Island systems are more likely to show such patterns, or at least illustrate them most clearly, as there are clearly defined habitat patches with limited dispersal among them. Nevertheless, comparable patterns can be found elsewhere. For example, the communities of figs and their associated insects show remarkable consistency in structure on three continents despite having no common evolutionary origin (Segar et al. 2013). Particular combinations of ecomorphs arise repeatedly through common pathways of evolution. That the species ranges do not overlap is a consequence of their historical origins.

13.2.3 Habitat checkerboards

Another cause of checkerboards might be habitat heterogeneity. If the habitat itself forms a checkerboard, this could explain the distributions of species. On the Urato Islands off the coast of Japan, Yamamoto et al. (2007) investigated what gave rise to different butterfly assemblages. Typically butterflies have host-specific larvae, which means

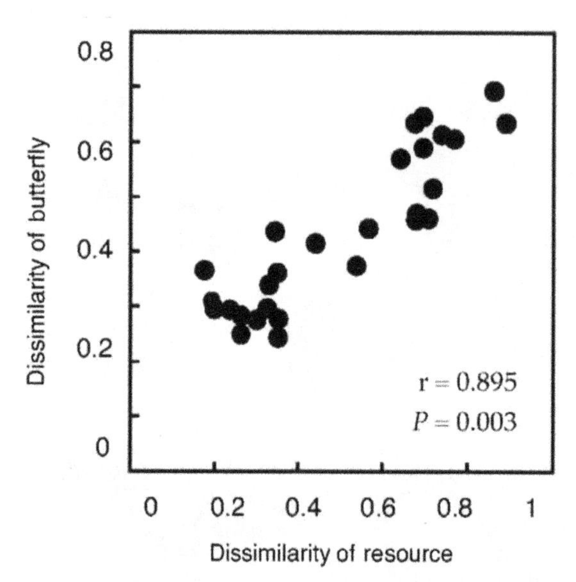

Figure 13.4 Correlation between dissimilarity of plant communities and the butterfly assemblages found alongside them. (*Source*: Yamamoto et al. (2007). Copyright (2007) National Academy of Sciences, USA.)

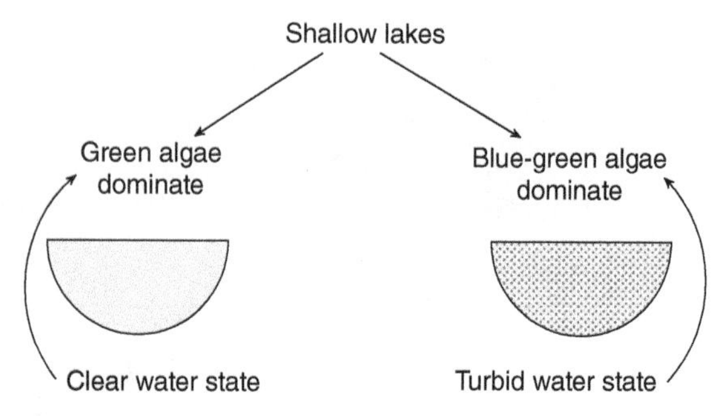

Figure 13.5 Alternative stable states in shallow Dutch lakes (Scheffer et al. 1997).

the abundance of any species should be tied to the availability of its host plant. Variation in vegetation was the main cause of changes in the butterflies present (Fig. 13.4). There was no effect of distance between the islands, which were all within 6 km of one another, and therefore it is likely that butterflies were able to disperse among them relatively easily. Whether a butterfly species was present on a given island mostly depended on the vegetation that was present.

Patterns in the availability of suitable habitats are probably a common cause of checkerboards for specialist species which depend upon a particular food source, host or mutualist, but cannot explain the majority of cases.

13.2.4 Chance and contingency

Finally we turn to the issue of contingency, which refers to differences arising as a result of chance historical events. Recall that communities can exist in multiple stable states (see Chapter 11). In different locations, it may be that the state which occurs is determined by stochastic processes. Scheffer et al. (1997) found that shallow lakes in Holland had either a clear water state, with plants growing up from the bottom and healthy fish communities, or a turbid fish-free state, in which blue-green algae filled the water and prevented plants from growing. Whether green or blue-green algae came to dominate depended upon minor variations in initial conditions and composition. The presence of species in one set of communities rather than the other was the result of the pathway that a particular patch happened by chance to follow (Fig. 13.5).

Complicated patterns emerge when local communities switch between different states, known as *patch dynamics*. Often this is a process of continuous turnover, the result of periodic disturbances followed by recovery. Across the landscape it creates a mishmash of patches in different states proceeding through asynchronous cycles. This increases habitat diversity on the landscape scale, provided that species are able to disperse between patches, and a large number of species coexisting at the γ level.

The fynbos of South Africa contains an incredible diversity of plants. Moving from one shallow valley to the next reveals near-complete turnover of plant species. How can so many coexist in a single landscape? A clue was discovered by Thuiller et al. (2007). They revisited sites 30 years after the vegetation had been described in 1966. Remarkably, when they added all their new samples together, the relative abundances of each species had stayed almost exactly the same: common species were still common, while rare species were still rare

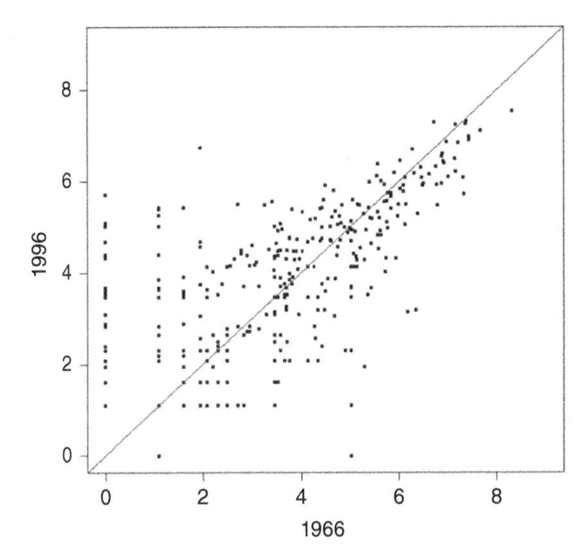

Figure 13.6 Relationship between complete metacommunity species abundances (log total number of individuals for each species across all sites) in 1966 and 1996. (*Source*: From Thuiller et al. (2007). Used under CC-BY-SA 2.5 http://creativecommons.org/licenses/by-sa/2.5/.)

(Fig. 13.6). This only applied at the landscape scale though. When looking at any individual site, 74% of sites had a turnover of over half the species that were present in 1966, and none had turnover of less than 30% of the species. This was despite local α species richness remaining almost identical.

The key to understanding this is that the fynbos is a fire-prone system. Fires break out largely at random and at any time of year. Over the intervening 30 years, sites had burnt at different times and were in various stages of recovery. The weather directly after the fire—whether it rained or there was a dry spell—also strongly influenced species composition by selecting for particular recruits. Throughout the whole landscape many species were able to coexist by dispersing from one valley to another, so long as there was one patch somewhere that they could colonise.

Wrapping up this section, there is some evidence for the importance of competitive exclusion, historical processes, habitat heterogeneity and contingency in creating checkerboard patterns of species distributions. None provides an exclusive explanation. In truth all four processes can occur, often simultaneously, and jointly act to determine how any single community is assembled. A final consideration of patch dynamics, however, points towards processes occurring on the larger scale of the whole landscape or region.

13.3 Metacommunities

Communities do not exist in isolation. Each forms part of a network of patches linked by dispersal. Without these connections, a patch would remain empty and uncolonised, so the source of the entrants to the community needs to be recognised. This is the regional species pool, of which any local community is only a small sample. How can we connect the regional pool of all possible species to the observed local composition? An outline is shown in Fig. 13.7 and forms the basis for our thinking on the topic.

Imagine a regional pool of all available species. The content of this pool will change over time as new species evolve (or invade) while other species go extinct. Of this pool of species, a certain proportion is able to live in any specific community. There is therefore a *habitat species pool* formed by environmental selection; only the species that can tolerate the conditions and obtain sufficient resources will be able to enter. This goes some way towards explaining

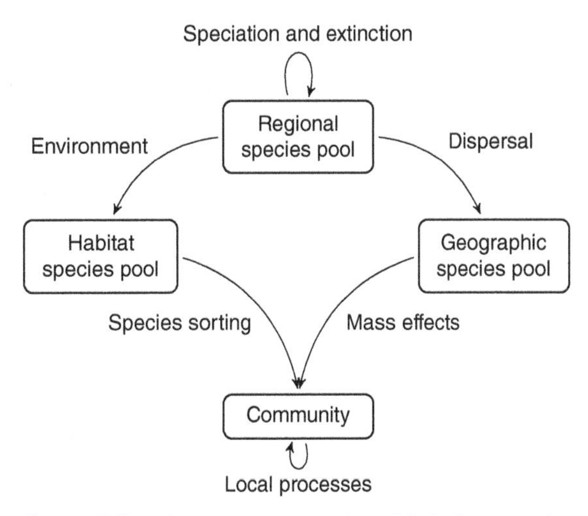

Figure 13.7 Schematic representation of links between the regional species pool and local communities. See text for definition of terms.

why localities with environments closer to the regional average contain a greater number of species (Belmaker and Jetz, 2012).

Soil bacteria provide an instructive example. It is often assumed that because bacteria are small and widely dispersed, their community structure should be more strongly determined by local conditions, a paradigm summed up by the famous dictum of Baas Becking that 'everything is everywhere, but the environment selects' (Baas Becking, 1934). This was tested by Ranjard et al. (2013), who combined data on soil characteristics, local climate and land use with data on soil bacterial communities across the whole of France. This impressive accomplishment largely supported Baas Becking's assertion as turnover in environmental characteristics led to similar change in bacterial communities (Fig. 13.8).

In south-western Australia, a series of dunes has formed over 2 My (Laliberté et al. 2014). The vegetation on each contains a distinct composition of plant species. All dunes are within 10 km of one another, which means that plants should easily be able to disperse among them. Nor can any differences be attributed to succession given the long timescales and frequent fires. Instead, patterns of diversity correlate with pH of the soil, which filters members from the regional species pool. Local effects on diversity, such as those discussed in Chapter 7, are much less important.

Alternatively, there may be species which could live in a particular location but are unable to reach it. There is therefore also a *geographic species pool* where the filter is dispersal ability (Fig. 13.7). Only species that are members of both the habitat and geographic species pool are capable of entering the community. Following this a process of *species sorting* takes place, governed by assembly rules, combined with mass effects of dispersal, depending upon the relative abundance of colonists, to decide what the final composition of the community will be. There is a degree of chance involved as to whether a given species will reach any particular patch and when it will do so.

Species sorting can be illustrated using the standard view of niches from Chapter 6. First consider six species that trade off in their competitive abilities for two resources (Fig. 13.9). Variation in the supply point will alter which particular combination of

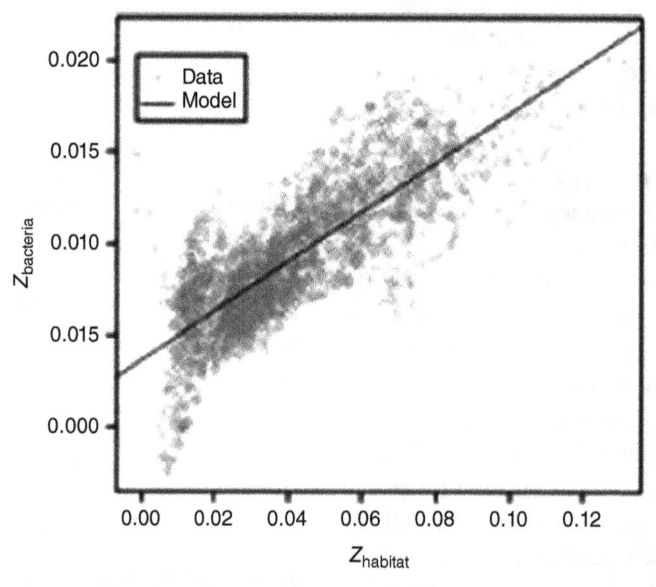

Figure 13.8 Turnover of soil bacterial communities ($Z_{bacteria}$) with that of habitat characteristics ($Z_{habitat}$) across mainland France, with a fitted linear regression line. (*Source*: Ranjard et al. (2013). Reproduced with permission from Nature Publishing Group.)

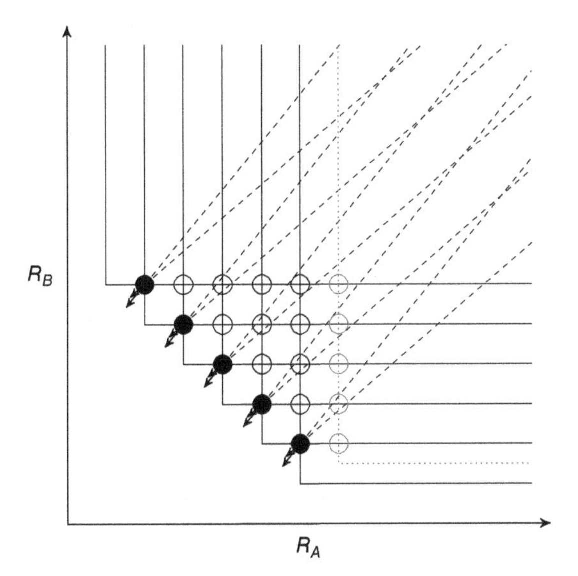

Figure 13.9 Species sorting among six species with differing zero net growth isoclines (ZNGIs solid lines) in a local community. The species trade off in their minimum requirements for two essential resources (R_A, R_B). Impact vectors (arrows) are projected back into resource space to indicate the outcome of competition (dashed lines). Solid circles represent stable pairs of coexisting species. Open circles indicate stable coexistence points that can be invaded by another species. A seventh species (dotted ZNGI) is excluded. See text for further details. (*Source*: Adapted from Chase and Leibold (2003).)

species is favoured at any given site. But what if all six of the species are present in the habitat species pool? This is where species sorting comes into play.

Depending on the supply point of resources, the species that coexist in the final community will be those that, in combination, draw down resources to the lowest level. This is illustrated in Fig. 13.9. Wherever two ZNGIs cross, there is a possible coexistence point. The open circles represent communities that are vulnerable to invasion by one or more species with stronger competitive abilities (lower R^*). Ultimately there are only five points for stable coexistence, shown by the solid circles. This process selects species combinations from the regional pool.

Note a very important prediction of this theory: the species that will coexist locally are actually the *most similar* in terms of their niche requirements. This is the complete opposite of what might be expected from basic competitive displacement theory, which would argue that the most different species should coexist. This also helps explain the pattern seen in Table 13.1—more closely related (and hence more similar) species are *more* likely to be found together.

This outcome assumes sufficient time without disturbance for competitive interactions to play out in full and also the ability of all species to disperse into the community. There may well be dud species which don't cross all the other ZNGIs and therefore possess no stable coexistence points. They are expected to go extinct once a stronger competitor enters but can persist on a regional level if they are good colonisers (the competition–colonisation trade-off; Fig. 12.3). Similarly an aggressive invasive species can make everything else extinct if it is a stronger resource competitor than the native species.

The niche framework can also represent other forms of species sorting, for example, a set of species varying in their resource requirements and tolerance of predation (Fig. 13.10). Once again, the species

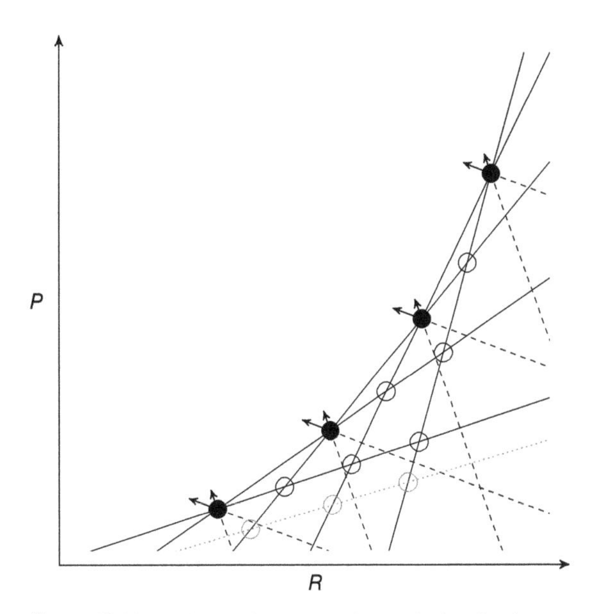

Figure 13.10 Species sorting among six species in a local community based upon a trade-off between resource requirements R and tolerance of predation P. Markings as in Fig. 13.9. Five species have ZNGIs which cross (solid lines) giving stable coexistence points (solid circles). One further species (dotted line) has only unstable coexistence points (open circles) and will be excluded. (*Source*: Adapted from Chase and Leibold (2003).)

capable of persisting in stable coexistence are those with the most similar characteristics, rather than the most different, as a naïve interpretation might have anticipated. Species sorting here might take place along gradients of both resources and predator density.

This seems reasonable in theory, but does it work in practice? A nice case study comes from the work of Paul Fine in the Peruvian Amazon (Fine et al. 2004). There were two main types of soil, either highly fertile clay or infertile white sand. The forests that grew on each had completely different compositions of trees. One might assume, as others had done, that this was the result of trees being specialised to grow on either soil type. But was this the whole story? The only way to find out was to conduct an experiment.

Fine set up a reciprocal transplant trial, which involved taking seedlings of trees that were confined to white sand soils and growing them on clay, and *vice versa*. Seedling performance on home soils could then be compared with the away habitat. Half of the seedlings were enclosed in cages that protected them from insect herbivores. It was expected that this might be important because investment in plant defences diverts resources from growth, which could be a crucial factor on white sand soils where resources were more limited.

The results are shown in Fig. 13.11. On clay soils, the species adapted to growing on clay (grey bars) grew much faster—they were better resource competitors. Protecting them from herbivores only increased their advantage. These were the dominant species on clay, which explained why the species from white sand habitats were excluded.

Something very different occurred on white sand soils though. When protected from herbivores, the species from clay soils grew faster than the local specialists and appeared capable of outcompeting them. Once exposed to predation, however, the roles were reversed. Without the same supply of resources, the

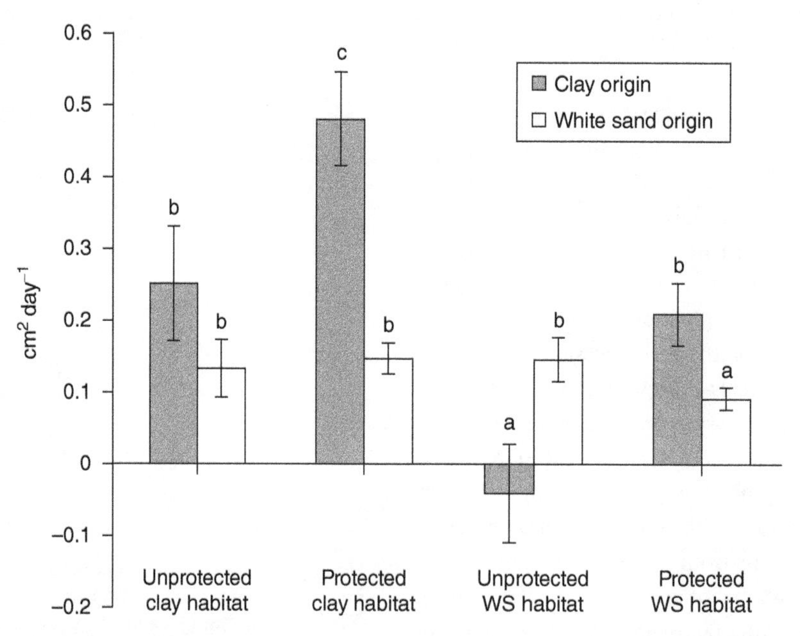

Figure 13.11 Leaf area growth rate of tree seedlings (mean ± SE) from species restricted to clay soils (grey bars) or white sand soils (white bars) in a reciprocal transplant experiment, with plants either unprotected or protected from herbivores. Bars with different letters are significantly different from one another. (*Source*: Fine et al. (2008). Reproduced with permission from American Association for the Advancement of Science.)

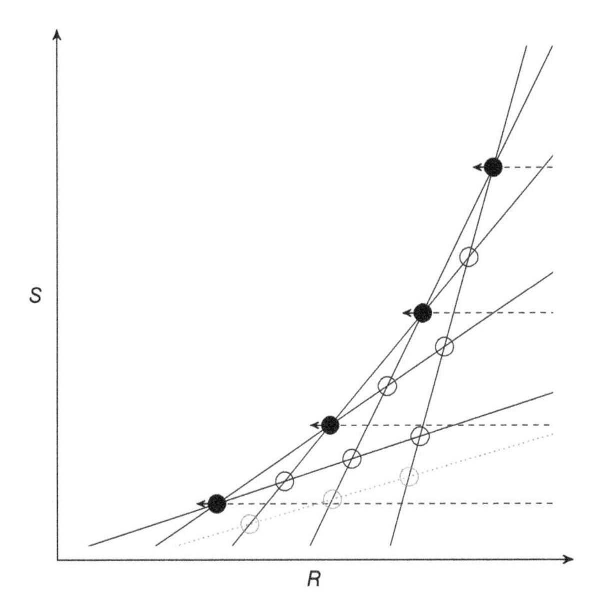

Figure 13.12 Species sorting among five species in a local community based upon a trade-off between resource requirements *R* and tolerance of stress *S*. Markings as in Fig. 13.9. Since the species have no effect on levels of stress (the impact vectors are horizontal), sequential replacement takes place along a stress gradient (vertical axis). Four species have ZNGIs which cross (solid lines) giving stable coexistence points (solid circles). One further species (dotted line) has only unstable coexistence points (open circles) and will be excluded. (*Source*: Adapted from Chase and Leibold (2003).)

species from clay soils couldn't cope with high levels of insect herbivory and failed to grow at all.

This is an especially apposite example as it captures a number of the key ideas so far in the book: the replacement of species along gradients of resources and predation, the importance of top-down control in maintaining less competitive species and the role of habitat heterogeneity in boosting landscape-level diversity.

For the sake of completion, Fig. 13.12 illustrates gradient replacement where species trade off in their ability to acquire resources or tolerate stress. It is common to see serial replacement of species, leading to bands in which different species dominate. A common example are zones of vegetation on mountains, where an ability to tolerate harsher conditions at high altitudes is set against greater resources lower down, leading to defined bands of vegetation at each level. See Fig. 6.15 for another example.

The trade-off between tolerance of stress and resource acquisition underlies changes in communities along such gradients. Violle et al. (2010) took communities of freshwater protists and disturbed them using ultrasonic waves, which break membranes and kill cells. These were applied at levels from mild disruption through to severe treatments that wiped out most species. As shown in Fig. 13.13, the strongest competitors were the least capable of tolerating disturbance, and *vice versa*. Which species survived depended on the balance between resource supply and disturbance intensity.

Whenever the opportunity exists for a trade-off along niche axes, multiple species can specialise on different parts of the gradient. The ultimate extension of the theory is to ask whether species can trade off their ability to tolerate predation (P^*) and stress (S^*). As a demonstration, Staver et al. (2012) examined nine acacia species in South Africa and showed that areas with frequent fires (stress) tended to have lower densities of browsing herbivores (predation, measured using dung counts). Species adapted to either the presence of herbivores or fire differed in

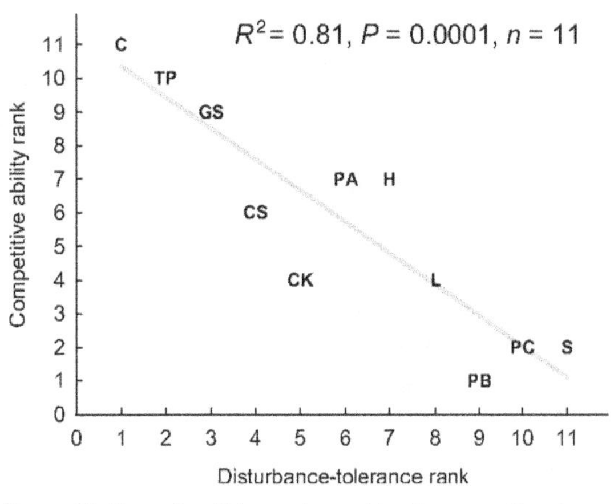

Figure 13.13 Trade-off in species ranking between disturbance tolerance and competitive ability exhibited by protist species based on single-species and two-species experiments, respectively. Capital letters correspond to species names, and a linear regression line and statistics are shown. (*Source*: From Violle et al. (2010).)

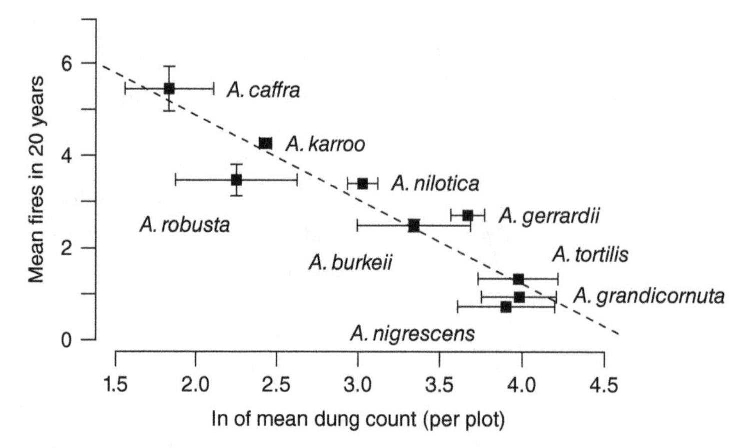

Figure 13.14 Occurrence of nine *Acacia* species along a gradient between fire frequency and herbivore abundance (measured by dung count); species means ± SE. (*Source*: Staver et al. (2012). Reproduced with permission from Johnn Wiley & Sons Ltd.)

their branch architecture, bark thickness and starch storage capacities, meaning that each could specialise at a different point along this gradient (Fig. 13.14). There is no need to invoke resources (R^*) to account for their distributions.

13.4 Dispersal limitation

So much for the species sorting that takes place as they jostle for a position in communities according to the prevailing environmental conditions. But what if a species cannot reach a community in the first place? This is referred to as dispersal limitation. It is worth recalling that *all* species are limited in their dispersal to some extent; no species can get everywhere. Those species in the regional species pool that could potentially be in any given system but are absent have been termed its 'dark diversity' (Pärtel et al. 2011); these unseen species are nevertheless an important element in understanding the composition of any given site. We can learn as much about community assembly from the species that are absent as from those which are present.

Within communities, species tend to be clumped in space, especially those like plants or benthic fauna which are sessile (i.e. they cannot move). This has two major implications for the structure of natural systems. The first is that from one location to another the composition of communities will automatically change, not necessarily because the environment has altered, but simply because of the limited dispersal abilities of the species involved. Communities structured in this way are said to exhibit *dispersal assembly* or to be predominantly influenced by mass effects (i.e. diffusion of individuals).

When members of the same species are close together in space, this increases the amount of intraspecific competition (within species) and decreases interspecific competition. It can be a powerful factor in maintaining diversity as each species limits itself more than others, thereby preventing competitive exclusion from taking place or at least slowing it down considerably.

The implications of this can be demonstrated by playing a simple game of *ro–sham–bo* (or rock–paper–scissors), only using bacteria. Kerr et al. (2002) took three strains of the bacterium *Escherichia coli* and grew them on standard agar. One strain produced a toxin (colicin) to which a second strain was sensitive, while a third strain was resistant. By not investing in the traits required to tolerate colicin, the sensitive strain had a growth advantage over the resistant strain. This set up a perfect system to examine how dispersal limitation could permit coexistence.

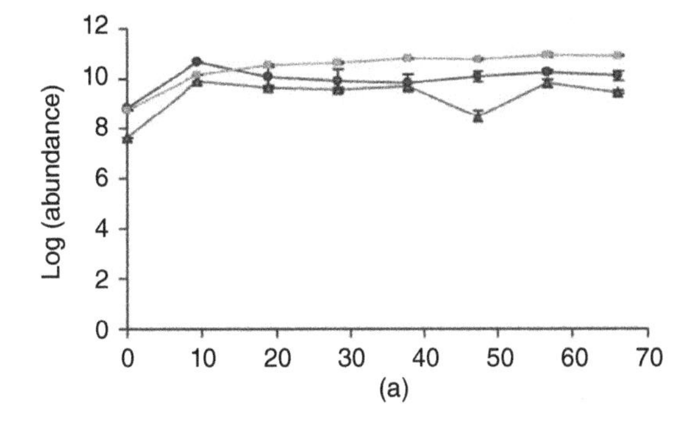

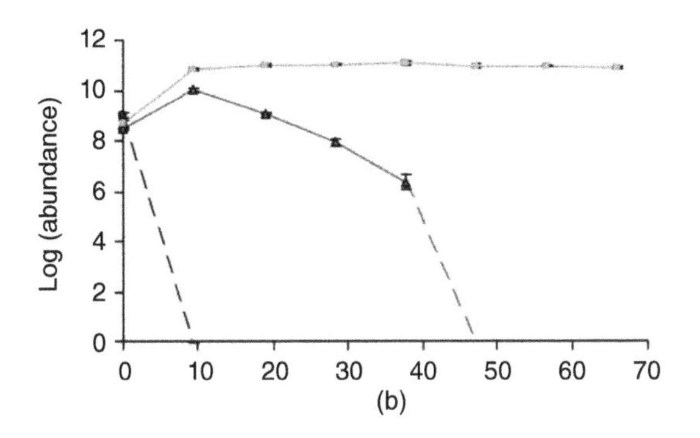

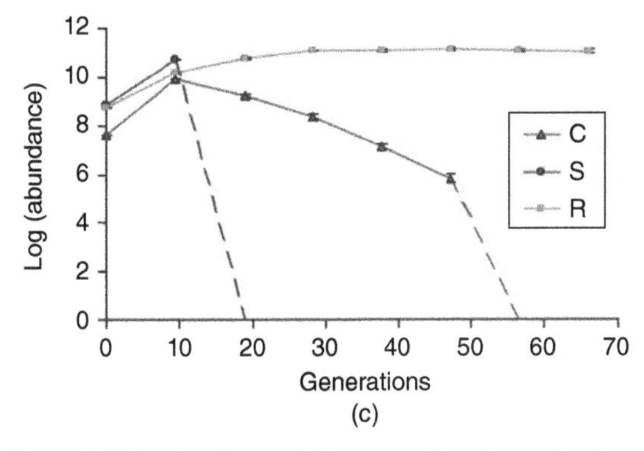

Figure 13.15 Abundance of three *E. coli* strains under three treatments: (a) undisturbed agar plates, (b) mixed aqueous suspension in a flask and (c) mixed colonies on an agar plate. Strains were colicin producing (C), sensitive to colicin (S) or resistant (R). (*Source*: Kerr et al. (2002). Reproduced with permission from Nature Publishing Group.)

Three different experiments were run (Fig. 13.15). In the first, agar plates were seeded with colonies located randomly on a grid and then left to their own devices. Over time the three strains remained at approximately the same level of abundance with no competitive exclusion. Two further experiments confirmed that dispersal limitation was responsible for this. When the three strains were grown in a flask of aqueous agar, the resistant strain survived at the expense of the other two. The same result was obtained using an agar plate that was regularly mixed to prevent spatial segregation. Only when there was spatial structure in the distribution of each type were they able to coexist.

Dispersal limitation provides novel opportunities for coexistence at the regional scale, even without the presence of a competition–colonisation trade-off. Consider, for example, two fish species which are identical in every way, except that one is able to disperse over greater distances. In a metacommunity with unevenly spaced patches, the species that disperses only short distances will dominate in regions where patches are close together, while the species that can swim further will fill the areas with more widely spread patches. This is one way in which high fish diversity on coral reef systems might be maintained (Bode et al. 2011).

The same principle works even on microscopic scales. Yawata et al. (2014) studied two populations of a marine plankton. One was adapted to colonise resource particles by attaching to them and forming a biofilm. It was therefore able to monopolise those resources it reached. The other population was capable of detecting new resource particles and swimming towards them, ensuring that it reached them first. These two strategies are able to coexist alongside one another, providing another possible contributor to the paradox of the plankton (*sensu* Hutchinson, 1961, see Chapter 8).

The idea can be extended to study entire ecosystems, for example, the suite of invertebrate species which inhabit carpets of moss on rocks which were mentioned at the beginning of the chapter. Staddon et al. (2010) collected whole patches of moss and divided them in the laboratory into four treatments. There was either a continuous single bed of moss, or it was cut into four islands. The islands were either connected by corridors (narrow strips allowing the

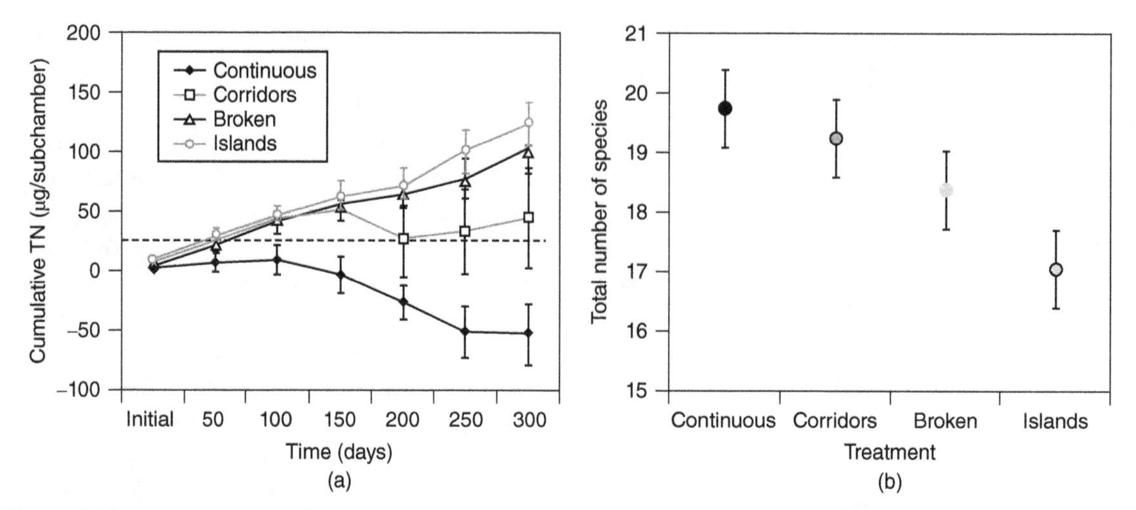

Figure 13.16 (a) Cumulative total nitrogen (TN) and (b) species richness per fragment in experimental moss micro-ecosystems differing in level of connectivity; means ± SE. (*Source*: Staddon et al. (2010). Reproduced with permission from John Wiley & Sons Ltd.)

dispersal of invertebrates along them), corridors with small breaks in them or entirely separate.

Fragmentation of the ecosystem into patches caused variation in ecosystem processes, with less well-connected patches retaining more nitrogen (Fig. 13.16a) and also fixing less carbon. Alongside this, splitting the habitat into islands reduced their species richness, but providing corridors between patches maintained species, because even if a species went extinct in one patch, it could be recolonised from another (Fig. 13.16b). These effects could be traced to the movement of predators, which mediated trophic cascades in each patch. When predators were able to move through the metacommunity, their populations were buffered against local extinction in any single patch, and they mobilised more nitrogen by feeding on grazers. Fewer predators meant more nitrogen being locked up in microbes and grazing invertebrates.

The existence of metacommunities with dispersal of species among them therefore has implications for many aspects of natural systems, from species richness through to the organisation of food webs and the rates of ecosystem processes.

Taking these ideas out of the laboratory, clustered patterns and dispersal limitation are frequently observed in nature. Tropical forest trees are a prime example as they have large seeds that disperse only short distances. This is an unavoidable constraint because little light penetrates into the understorey of forests, which are therefore resource poor. In order for the seedlings to survive, they need to be provisioned with enough carbohydrates to establish themselves, necessitating heavy seeds. As a result, tropical tree species are highly clustered within the forest (Fig. 13.17), with most individuals found within 50 m of another member of the same species.

The outcome is that, over increasing geographical distances, tree communities change even within apparently continuous forest. Figure 13.18 indicates similarity of tree communities between 1 ha forest plots in three regions. Plots close together are similar, but this drops off sharply, so that after 20 km little more than a third of species are shared. Turnover in Panama is much greater (lower similarity among plots), probably because there are strong rainfall and altitudinal gradients across the country, and the pattern is therefore confounded by environmental changes that go along with geographic distance.

This is an example of the more general principle of **distance decay**, whereby communities that are further apart are expected to be increasingly different

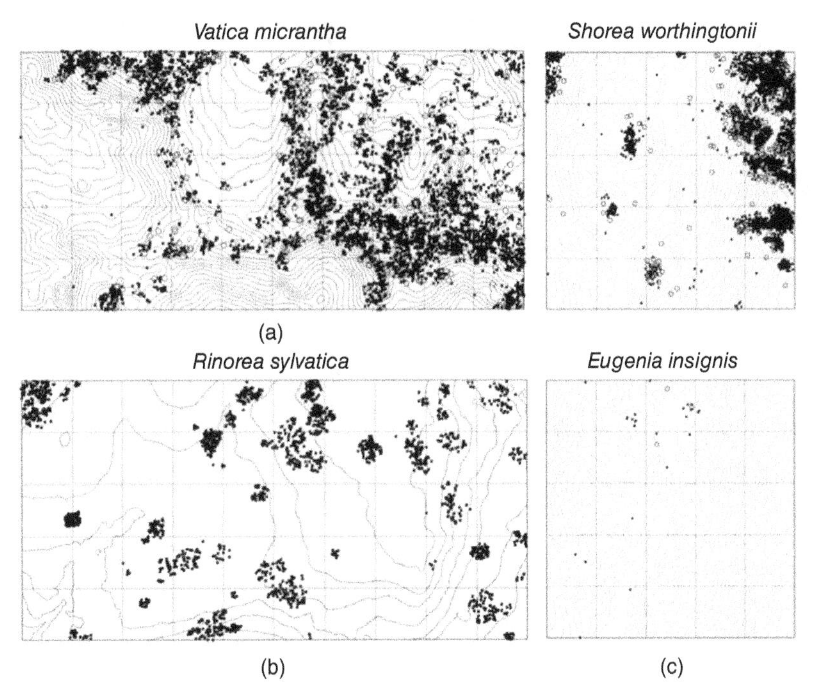

Figure 13.17 Distribution maps for some species from three tropical rain forests: (a) Lambir Hills in Malaysian Borneo, (b) Barro Colorado Island in Panama and (c) Sinharaja in Sri Lanka. Solid circles, trees <10 cm diameter at breast height; open circles, trees ≥10 cm diameter; grid squares each 1 ha; contours show 5 m elevation lines. (*Source*: Condit et al. (2002). Reproduced with permission from American Association for the Advancement of Science.)

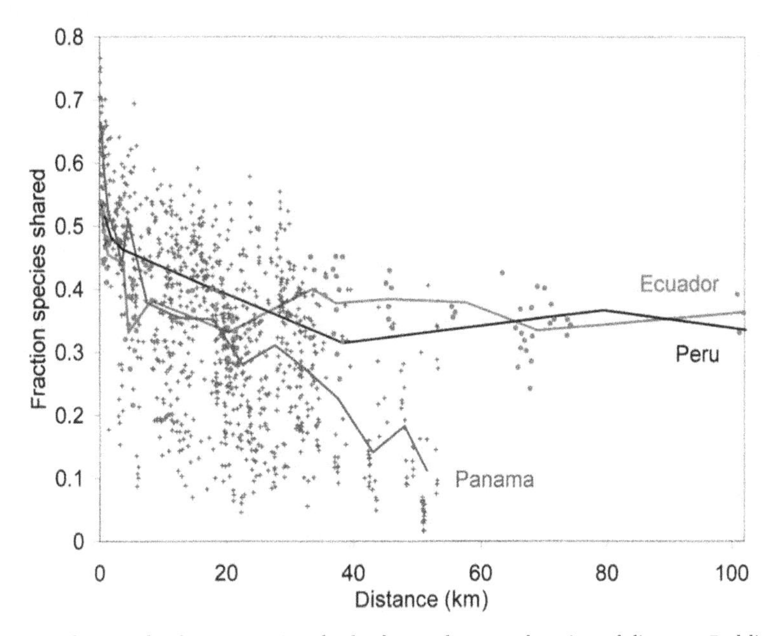

Figure 13.18 Sørensen similarity index between pairs of 1 ha forest plots as a function of distance. Red line and dots, Ecuador; black line, Peru; blue line and crosses, Panama. Points for Peru are omitted for clarity. (*Source*: Condit et al. (2002). Reprodcued with permission from American Association for the Advancement of Science.) *(See colour plate section for the colour representation of this figure.)*

as a result of the distance between them (Nekola and White, 1999). It can be the result of changes in the environment, restricted dispersal or both.

13.5 Combining environment and dispersal

Can the effects of environmental change and dispersal limitation be separated? Tuomisto et al. (2003) studied two groups of plants, pteridophytes (ferns) and melastomes (a family of understorey shrubs) in the Amazon rain forest. The assemblages of these plants varied over large areas, and her underlying question was which root cause mattered more, habitat heterogeneity or dispersal limitation. By collecting large amounts of environmental data, she was able to partition the variation in forest floras between these two trends and found that in combination they accounted for 70–75% of the differences between sites in their species composition. The rest was unknown and may depend on unmeasured processes or some essentially random (or neutral?) effects.

The balance of which forces matter most varies among taxonomic groups and regions. Beaudrot and Marshall (2011) examined primate assemblages across four tropical regions (Fig. 13.19). They found evidence of declines in similarity with geographical distance in all four, but environmental variation was only significant in three and was the most important predictor in Madagascar alone. Elsewhere the composition of primate assemblages was determined more by distance, which is most likely caused by dispersal limitation (Table 13.2).

A large number of studies have now tried to separate the two hypotheses for how communities form, whether they are dominated by species sorting or mass effects of dispersal. The outcome depends on the group of organisms involved; strong dispersers are likely to respond more to local environments, whereas weak dispersers will be spatially constrained. For example, in Ugandan forests, trees with fruits that are consumed by primates show a greater degree of dispersal limitation than those eaten by birds, reflecting the reduced likelihood of being carried long distances (Beaudrot et al. 2013). In contrast, among Andean butterflies, the map distance between any two communities is of no importance at all in explaining their composition (Chazot et al. 2014). Instead, the further apart two communities are in altitude, the greater the phylogenetic dissimilarity between them. This suggests a powerful influence of environmental differences on butterflies (perhaps through effects on their host plants) but that these well-dispersed species are less limited by geographical distance.

Cottenie (2005) assembled 158 data sets from systems across the world and worked out how much of the variation between samples in each system could be attributed to environmental or geographic components. He found that on average 48% of variation among related communities could be accounted for by environment and geography, of which 22% was purely environment, 16% purely geography, and the remaining 10% due to covariance in both. (Note that the validity of these analyses has been questioned (Gilbert and Bennett, 2010); they are nevertheless included here as the best available indicators.) A more recent analysis has put the average contribution of the environment at 26% while noting that the range of values spans 0–88%, so it is important not to treat the average as an expectation (Soininen, 2014). It is also likely to be an underestimate, because is it never possible to include all environmental variables.

Both environment and distance therefore work together and in combination account for around half of the variation among communities in space. This means that local interactions can cause no more than half of the remaining variation (Fig. 13.7), bearing in mind that sampling errors and random chance also play a part in patterns of turnover. The regional metacommunity level is vital and the only way to properly understand the composition of local communities.

These average figures vary considerably between systems but provide a rough guide to how the

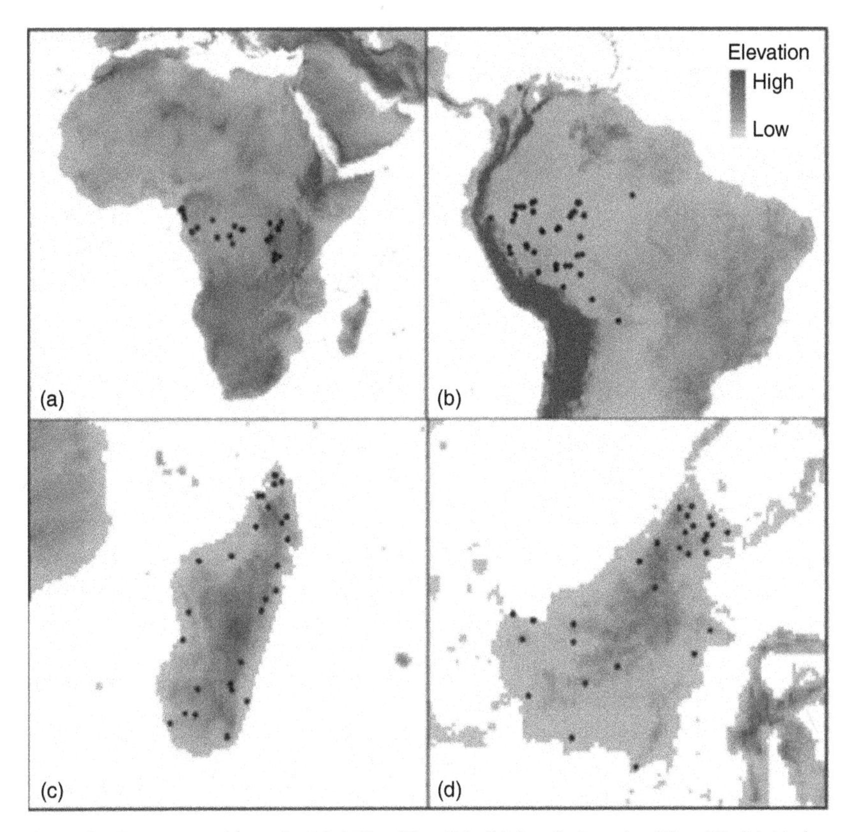

Figure 13.19 Locations of primate assemblages in (a) Africa ($N = 23$), (b) South America ($N = 45$), (c) Madagascar ($N = 28$) and (d) Borneo ($N = 28$) shaded by relative elevation. (*Source*: Beaudrot and Marshall (2011). Reproduced with permission from John Wiley & Sons Ltd.)

Table 13.2 Primate assemblage similarity against geographical distance and environmental change among sites within the four geographical regions in Fig. 13.19.

Distance	Africa		South America		Madagascar		Borneo	
	r	*P*	*r*	*P*	*r*	*P*	*r*	*P*
Distance	**−0.65**	**<0.001**	**−0.36**	**<0.001**	−0.28	<0.001	**−0.22**	**<0.001**
Environment	−0.30	<0.001	−0.02	0.014	**−0.43**	**<0.001**	−0.01	0.306

Correlation values from partial Mantel tests (*r*) and significance values (*P*), with the strongest predictor shown in bold.
Source: Beaudrot et al. (2011). Reproduced with permission from John Wiley & Sons Ltd.

regional species pool is filtered into local communities. Before thinking that the problems is solved however, and as an instructive example, Dexter et al. (2012) examined the decay in similarity of communities of tropical trees in the genus *Inga* on two soil types across 250 km of the Peruvian Amazon. He found exactly the expected pattern—more distant pairs of sites were more dissimilar (though even close ones were never quite identical, perhaps due to sampling effects). The reason had nothing to do with environmental change. Instead, their transect spanned an ancient division between two floras, which had since expanded and met in the middle. Likewise, bird communities change on either side of the Amazon River, despite a lack of environmental differences, due to the presence of a barrier to their dispersal (Pomara et al. 2014). The effect of distance is not always linear or continuous. There are two lessons to learn from this. The first is that we cannot always assume that correlation implies causation—just because communities change in space along with other variables does not mean that the two are linked. The other is that, even within metacommunities, there are still processes acting at larger scales that must be considered, which leads us on to patterns in biogeography.

13.6 Conclusions

Apparently similar habitats can contain remarkably different sets of species. Local composition is determined by assembly rules such as competition, predation and facilitation that dictate which species occupy any particular community. The traditional view was that competition within guilds structured communities, and it is certainly true that many species co-occur less frequently than would be expected by chance. The reasons for this can include competitive exclusion, historical processes (e.g. patterns of speciation), habitat heterogeneity and contingency. Communities are subsets of regional assemblages and cannot be understood in isolation as they are nested within metacommunities, structured by top-down processes that filter species from the regional pool. Patches within metacommunities are often dynamic and cycle asynchronously over time. The species present in any single patch are determined by a combination of niche and dispersal assembly. While each system will be different, on average regional forces account for around half of variation in their composition, after which local interactions through species sorting are influential. In order to fully understand the community level of organisation, we need to take both the top-down and bottom-up perspectives into account.

13.6.1 Recommended reading

Cottenie, K., 2005. Integrating environmental and spatial processes in ecological community dynamics. *Ecology Letters* 8:1175–1182.

HilleRisLambers, J., P. B. Adler, W. S. Harpole, J. M. Levine, and M. M. Mayfield, 2012. Rethinking community assembly through the lens of coexistence theory. *Annual Review of Ecology, Evolution, and Systematics* 43:227–248.

Leibold, M. A., 1998. Similarity and local co-existence of species in regional biotas. *Evolutionary Ecology* 12:95–110.

Leibold, M. A., M. Holyoak, N. Mouquet, P. Amarasekare, J. M. Chase, M. F. Hoopes, R. D. Holt, J. B. Shurin, R. Law, D. Tilman, M. Loreau, and A. Gonzalez, 2004. The metacommunity concept: a framework for multi-scale community ecology. *Ecology Letters* 7:601–613.

13.6.2 Questions for the future

- Why are some assemblages phylogenetically dispersed (species are less related than expected by chance), while others are clustered?
- What determines the balance between niche assembly and dispersal assembly?
- Why does community assembly often lead to species with more similar traits than expected, whereas evolution over longer time periods causes traits of competing species to diverge?

References

Baas Becking, L. G. M., 1934. *Geobiologie of inleiding tot de milieukunde.* W. P. Van Stockum & Zoon.

Beaudrot, L. H. and A. J. Marshall, 2011. Primate communities are structured more by dispersal limitation than by niches. *Journal of Animal Ecology* 80:332–341.

Beaudrot, L., M. Rejmánek, and A. J. Marshall, 2013. Dispersal modes affect tropical forest assembly across trophic levels. *Ecography* 36:984–993.

Belmaker, J. and W. Jetz, 2012. Regional pools and environmental controls of vertebrate richness. *American Naturalist* 179:512–523.

Bode, M., L. Bode, and P. R. Armsworth, 2011. Different dispersal abilities allow reef fish to coexist. *Proceedings of the National Academy of Sciences of the United States of America* 108:16317–16321.

Burns, J. H. and S. Y. Strauss, 2011. More closely related species are more ecologically similar in an experimental test. *Proceedings of the National Academy of Sciences of the United States of America* 108:5302–5307.

Carboni, M., T. Münkemüller, L. Gallien, S. Laverge, A. Acosta, and W. Thuiller, 2013. Darwin's naturalization hypothesis: scale matters in coastal plant communities. *Ecography* 36:560–568.

Caro, A. U., S. A. Navarrete, and J. C. Castilla, 2010. Ecological convergence in a rocky intertidal shore metacommunity despite high spatial variability in recruitment regimes. *Proceedings of the National Academy of Sciences of the United States of America* 107:18528–18532.

Chase, J. M. and M. A. Leibold, 2003. *Ecological Niches: Linking Classical and Contemporary Approaches*. Chicago University Press.

Chazot, N., K. R. Willmott, P. G. Santacruz Endara, A. Toporov, R. I. Hill, C. D. Jiggins, and M. Elias, 2014. Mutualistic mimicry and filtering by altitude shape the structure of Andean butterfly communities. *American Naturalist* 183:26–39.

Condit, R., P. S. Ashton, P. Baker, S. Bunyavejchewin, S. Gunatilleke, N. Gunatilleke, S. P. Hubbell, R. B. Foster, A. Itoh, J. V. LaFrankie, H. S. Lee, E. Losos, N. Manokaran, R. Sukumar, and T. Yamakura, 2000. Spatial patterns in the distribution of tropical tree species. *Science* 288:1414–1418.

Condit, R., N. Pitman, E. G. Leigh Jr., J. Chave, J. Terborgh, R. B. Foster, P. Núñez, S. Aguilar, R. Valencia, G. Villa, H. C. Muller-Landau, E. Losos, and S. P. Hubbell, 2002. Beta-diversity in tropical forest trees. *Science* 295:666–669.

Connell, J. H., 1980. Diversity and the coevolution of competitors, or the ghost of competition past. *Oikos* 35:131–138.

Cottenie, K., 2005. Integrating environmental and spatial processes in ecological community dynamics. *Ecology Letters* 8:1175–1182.

Dexter, K. G., J. Terborgh, and C. W. Cunningham, 2012. Historical effects on beta diversity and community assembly in Amazonian trees. *Proceedings of the National Academy of Sciences of the United States of America* 109:7787–7792.

Diamond, J. M., 1975. Assembly of species communities. In M. L. Cody and J. M. Diamond, editors, *Ecology and Evolution of Communities*, pages 342–444. Belknap.

Dornelas, M. and S. R. Connolly, 2008. Multiple modes in a coral abundance distribution. *Ecology Letters* 11:1008–1016.

Fine, P. V. A., I. Mesones, and P. D. Coley, 2004. Herbivores promote habitat specialization by trees in Amazonian forests. *Science* 305:663–665.

Gause, G. F., 1936. *The Struggle for Existence*. Williams and Wilkins.

Gilbert, B. and J. R. Bennett, 2010. Partitioning variation in ecological communities: do the numbers add up? *Journal of Animal Ecology* 47:1071–1082.

Gillespie, R., 2004. Community assembly through adaptive radiation in Hawaiian spiders. *Science* 303:356–359.

HilleRisLambers, J., P. B. Adler, W. S. Harpole, J. M. Levine, and M. M. Mayfield, 2012. Rethinking community assembly through the lens of coexistence theory. *Annual Review of Ecology, Evolution, and Systematics* 43:227–248.

Hutchinson, G. E., 1959. Homage to Santa Rosalia, or why are there so many kinds of animals? *American Naturalist* 93:145–159.

Hutchinson, G. E., 1961. The paradox of the plankton. *American Naturalist* 95:137–145.

Kerr, B., M. A. Riley, M. W. Feldman, and J. M. Bohannan, 2002. Local dispersal promotes biodiversity in a real-life game of rock-paper-scissors. *Nature* 418:171–174.

Laliberté, E., G. Zemunik, and B. L. Turner, 2014. Environmental filtering explains variation in plant diversity along resource gradients. *Science* 6204:1602–1605.

Mahler, D. L., T. Ingram, L. J. Revell, and J. B. Losos, 2013. Exceptional convergence on the macroevolutionary landscape in island lizard radiations. *Science* 341:292–295.

Márquez, J. C. and J. Kolasa, 2013. Local and regional processes in community assembly. *PLoS ONE* 8:e54580.

Nekola, J. C. and P. S. White, 1999. The distance decay of similarity in biogeography and ecology. *Journal of Biogeography* 26:867–878.

Pärtel, M., R. Szava-Kovats, and M. Zobel, 2011. Dark diversity: shedding light on absent species. *Trends in Ecology & Evolution* 26:124–128.

Pomara, L. Y., K. Ruokolainen, and K. R. Young, 2014. Avian species composition across the Amazon river: the roles of dispersal limitation and environmental heterogeneity. *Journal of Biogeography* 41:784–796.

Price, J. N. and M. Pärtel, 2013. Can limiting similarity increase invasion resistance? A meta-analysis of experimental studies. *Oikos* 122:649–656.

Ranjard, L., S. Dequiedt, N. Chemidlin Prévost-Bouré, J. Thioulouse, N. P. A. Saby, M. Lelievre, P. A. Maron, F. E. R. Morin, A. Bispo, C. Jolivet, D. Arrouays, and P. Lemanceau, 2013. Turnover of soil bacterial diversity driven by wide-scale environmental heterogeneity. *Nature Communications* 4:1434.

Ricklefs, R. E., 2012. Species richness and morphological diversity of passerine birds. *Proceedings of the National Academy of Sciences of the United States of America* 109:14482–14487.

Sanders, N. J., N. J. Gotelli, S. E. Wittman, J. S. Ratchford, A. M. Ellison, and E. S. Jules, 2007. Assembly rules of ground-foraging ant assemblages are contingent on disturbance, habitat and spatial scale. *Journal of Biogeography* 34:1632–1641.

Scheffer, M. and E. H. van Nes, 2006. Self-organized similarity, the evolutionary emergence of groups of similar species. *Proceedings of the National Academy of Sciences of the United States of America* 103:6230–6235.

Scheffer, M., S. Rinaldi, A. Gragnani, L. R. Mur, and E. H. vanNes, 1997. On the dominance of filamentous cyanobacteria in shallow, turbid lakes. *Ecology* 78:272–282.

Segar, S. T., R. A. S. Pereira, S. G. Compton, and J. M. Cook, 2013. Convergent structure of multitrophic communities over three continents. *Ecology Letters* 16:1436–1445.

Simonis, J. L. and J. C. Ellis, 2014. Bathing birds bias β-diversity: frequent dispersal by gulls homogenizes fauna in a rock-pool metacommunity. *Ecology* 95:1545–1555.

Soininen, J., 2014. A quantitative analysis of species sorting across organisms and ecosystems. *Ecology* 95:3284–3292.

Sridhar, H., U. Srinivasan, R. A. Askins, J. C. Canales-Delgadillo, C.-C. Chen, D. N. Ewert, G. A. Gale, E. Goodale, W. K. Gram, P. J. Hart, K. A. Hobson, R. L. Hutto, S. W. Kotagama, J. L. Knowlton, T. M. Lee, C. A. Munn, S. Nimnuan, B. Z. Nizam, G. Péron, V. V. Robin, A. D. Rodewald, P. G. Rodewald, R. L. Thomson, P. Trivedi, S. L. V. Wilgenburg, and K. Shanker, 2012. Positive relationships between association strength and phenotypic similarity characterize the assembly of mixed-species bird flocks worldwide. *American Naturalist* 180:777–790.

Staddon, P., Z. Lindo, P. D. Crittenden, F. Gilbert, and A. Gonzalez, 2010. Connectivity, non-random extinction and ecosystem function in experimental metacommunities. *Ecology Letters* 13:543–552.

Staver, A. C., W. J. Bond, M. D. Cramer, and J. L. Wakeling, 2012. Top-down determination of niche structure and adaptation among African acacias. *Ecology Letters* 15:673–679.

Thuiller, W., J. A. Slingsby, D. S. J. Privett, and R. M. Cowling, 2007. Stochastic species turnover and stable coexistence in a species-rich, fire-prone plant community. *PLoS ONE* 2:e938.

Tuomisto, H., K. Ruokolainen, and M. Yli-Halla, 2003. Dispersal, environment, and floristic variation of Western Amazonian forests. *Science* 299:241–244.

Vergnon, R., E. H. van Nes, and M. Scheffer, 2012. Emergent neutrality leads to multimodal species abundance distributions. *Nature Communications* 3:663.

Violle, C., D. R. Nemergut, Z. Pu, and L. Jiang, 2011. Phylogenetic limiting similarity and competitive exclusion. *Ecology Letters* 14:782–787.

Violle, C., Z. Pu, and L. Jiang, 2010. Experimental demonstration of the importance of competition under disturbance. *Proceedings of the National Academy of Sciences of the United States of America* 107:12925–12929.

Yamamoto, N., J. Yokoyama, and M. Kawata, 2007. Relative resource abundance explains butterfly biodiversity in island communities. *Proceedings of the National Academy of Sciences of the United States of America* 104:10524–10529.

Yawata, Y., O. X. Cordero, F. Menolascina, J.-H. Hehemann, M. F. Polz, and R. Stocker, 2014. Competition-dispersal tradeoff ecologically differentiates recently speciated marine bacterioplankton populations. *Proceedings of the National Academy of Sciences of the United States of America* 111:5622–5627.

PART IV
Biogeography

CHAPTER 14
Global patterns of life

14.1 The big question

The urge to describe large-scale patterns in nature captivated many ancient writers; it would be reasonable to claim that biogeography as a discipline predates ecology, if not most branches of science. The core question is whether we can derive a series of general rules that explain the distribution of life across the entire globe. To achieve this requires the assimilation of many branches of evidence that do not typically fall within the purview of ecologists.

14.2 Biogeography

Biogeography is a peculiar field as it lies at the interface between ecology, geography, systematics and geology (Fig. 14.1). There are few academic departments of biogeography anywhere in the world, nor even a large number of scientists who would claim it to be their main focus. Instead the expertise required to answer biogeographical questions tends to reside in departments of Biology, Geography, Geology and further afield in museums and national archives. There are still roles for the field collector and explorer as there remain many areas of the world where the flora and fauna remain poorly described. The immense data requirements and necessity of collaboration made the greatest puzzles intractable until recent years. The main research outlet, the Journal of Biogeography, only came into existence in 1974, and the International Biogeography Society was founded in 2001. Biogeography has now become a rapidly growing area of research.

It was recognised even in antiquity that regions of the world differed in their biotic characteristics. Early records from classical authors allude to what was known, as well as the allure of the unknown. It was however only in the age of the great explorers, from the eighteenth century onwards, that scientists began to take a systematic approach to documenting this variation at a global scale. One of the first was Buffon (1761), who noted that the Old and New World contained very different types of large mammals. This tenet is now known as Buffon's Law.

The real father of biogeography, following Buffon's lead, was Alexander von Humboldt. The treatises based on his explorations (beginning with von Humboldt, 1808) were central pillars of scholarship in his day, so much so that Charles Darwin claimed to know parts of Humboldt's South American writings by heart, taking them with him on the Beagle. Humboldt's work made clear that not only the large fauna differed among regions, but also many other animal groups, as well as flowering plants.

That foreign lands were not merely different, but variable, inspired others to begin collecting information on where particular types of life might be found. de Candolle (1820) wrote of plant 'formations' whose distributions were driven by variation in the climate. His model was largely verbal, but along with the oft-maligned Lamarck, he can lay claim to having

Natural Systems: The organisation of life, First Edition. Markus P. Eichhorn.

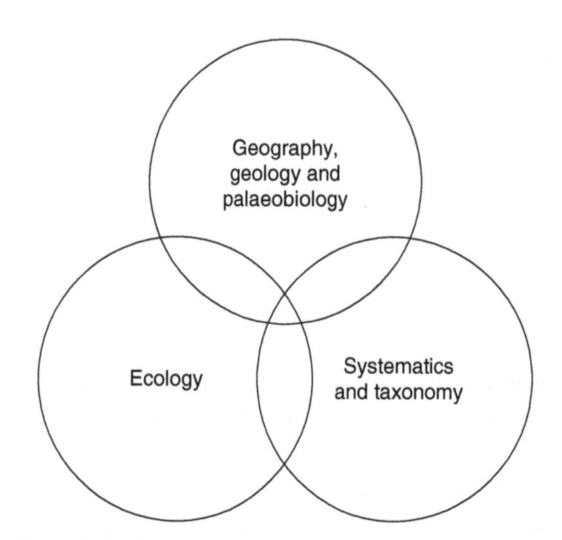

Figure 14.1 Elements of biogeography.

in 1805 produced the first ever biogeographical map, a chart intended as a guide to the patterns of life, not solely as an exercise in cartography (Ebach and Goujet, 2006). This split France into four regions based on their flora (Fig. 14.2).

Biogeographers since have been preoccupied with carrying out this exercise on ever larger scales. Soon the outline of the world was consolidated by botanists into six broad floral kingdoms. Zoologists preferred to recognise six zoological regions following the eminent naturalist Alfred Russel Wallace, co-originator of the theory of evolution by natural selection, who primarily based his divisions on the distributions of mammals (Fig. 14.3). These have been recently revised on the basis of more information (Cox, 2001). The changes have tended to bring the perspectives from animal and plant geographers into closer alignment, though a few important differences remain.

Early maps, based largely on informed opinion, have lately been superseded by quantitative methods similar to those used to identify and delineate communities at the local scale (Kreft and Jetz, 2010). More recently attempts have been made to extend Wallace's original focus on mammals and combine information from as wide a range of animal taxa as possible. Holt et al. (2013) collated information on distribution of over 21,000 species to divide the world into 11 realms which could be further subdivided into 20 regions (Fig. 14.4). Absent from this latter scheme is the *cryosphere*, which refers to those systems formed on ice. These make up around 10% of the land surface and were at least three times larger at the Last Glacial Maximum, so cannot be ignored (Anesio and Laybourn-Parry, 2012).

One of the most contended boundaries is that between the Indo-Pacific and Australian biotas. A discontinuity was documented by Wallace in a letter to his friend Henry Bates in South America, whose name is now attached to the concept of Batesian mimicry. In 1858, he wrote[1]:

> In this Archipelago there are two distinct faunas rigidly circumscribed, which differ as much as those of South America and Africa, ... yet there is nothing on the map or on the face of the islands to mark their limits.

This observation was later published in a classic and very readable account that described the stark contrast between the 'Indian' and 'Australian' regions (Wallace, 1860). What made the boundary most remarkable was that it appeared to fall definitively between the two islands of Bali and Lombok, which are barely 35 km apart, and would have been closer still with lower sea levels during the ice ages.

The division was most baffling in birds, which one might imagine to be capable of crossing such a short stretch of water. Yet although barbets were found throughout the Oriental islands southwards as far as Bali, they stopped there. Similarly cockatoos were present from Australia northwards and westwards but only until Lombok. The divide was mostly respected by the fauna, while lines for the flora were blurred.

There have been a number of attempts to redraw the position of Wallace's Line, as it became known, depending on the clade of organisms under study. In truth there is no absolute dividing line, and instead the central islands of Southeast Asia fit clearly into

[1] From the collected letters of Alfred Russel Wallace (Marchant, 1916).

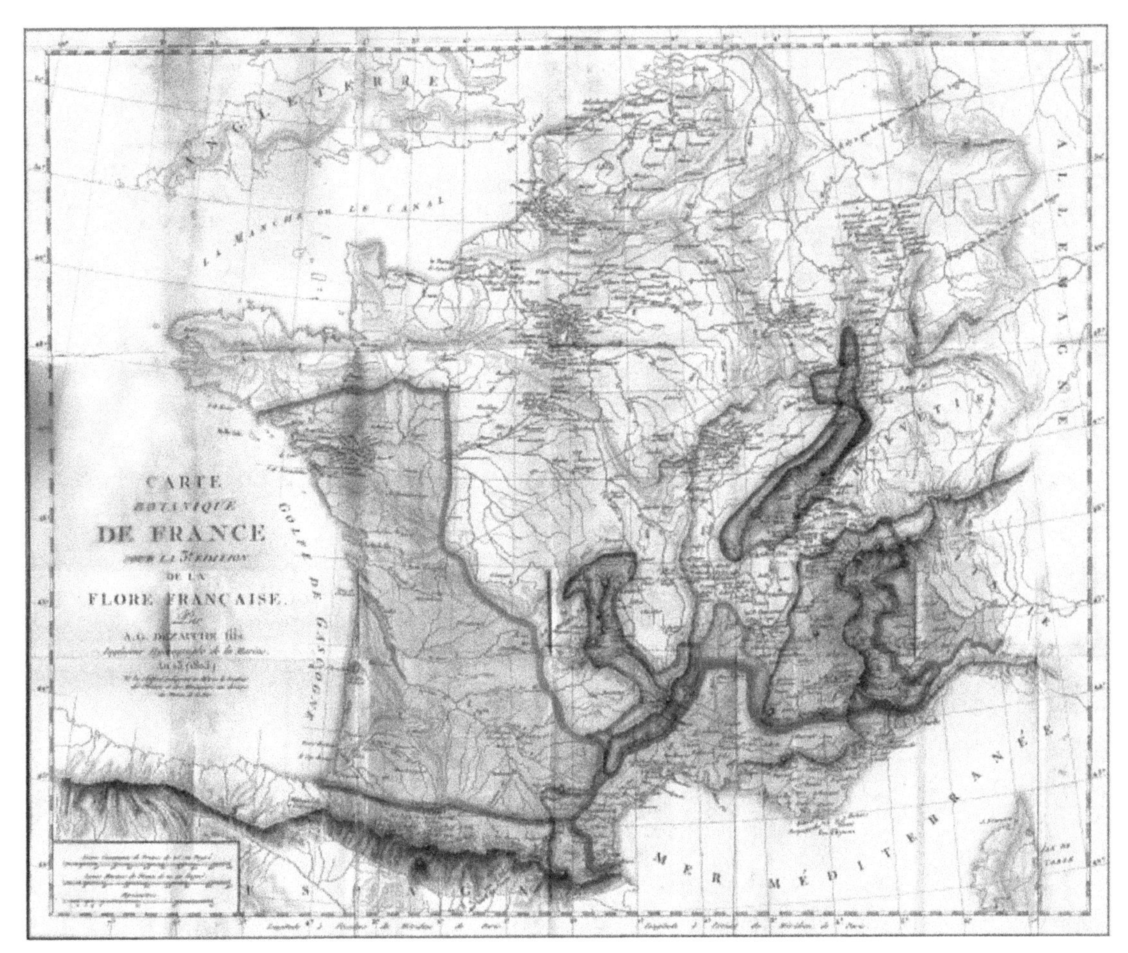

Figure 14.2 Botanical map of France produced in 1805 by Lamarck and de Candolle. Coloured regions denote consistent floras. (*Source*: Ebach and Goujet (2006). Reproduced with permission from John Wiley & Sons Ltd.) (*See colour plate section for the colour representation of this figure.*)

neither region, being composites of both. As a result this area is now generally known as Wallacea (though in fact the pattern had been noted before Wallace by Müller (1846)). The reasons for the divide will be explored in Chapter 17.

This example illustrates one of the problems of the traditional approach to biogeography. The desire to partition the world into distinct zones collides with the resistance of natural systems to absolute classification. To even begin to draw lines on a map depended on expert judgement and accumulated knowledge, hunches or hearsay about the often fluid distributions of taxa and where disjunctions occurred. The tendency for botanists and zoologists to work independently makes it somewhat surprising that the kingdoms or regions they recognised were broadly similar.

14.3 Phytogeography

Botanists had noticed something else, perhaps aided by their closer ties to geology and climatology. Certain forms of vegetation recurred and did so independently of the base rocks, landscape or region (Schimper, 1903). These were termed the biomes, of which a varying number are recognised

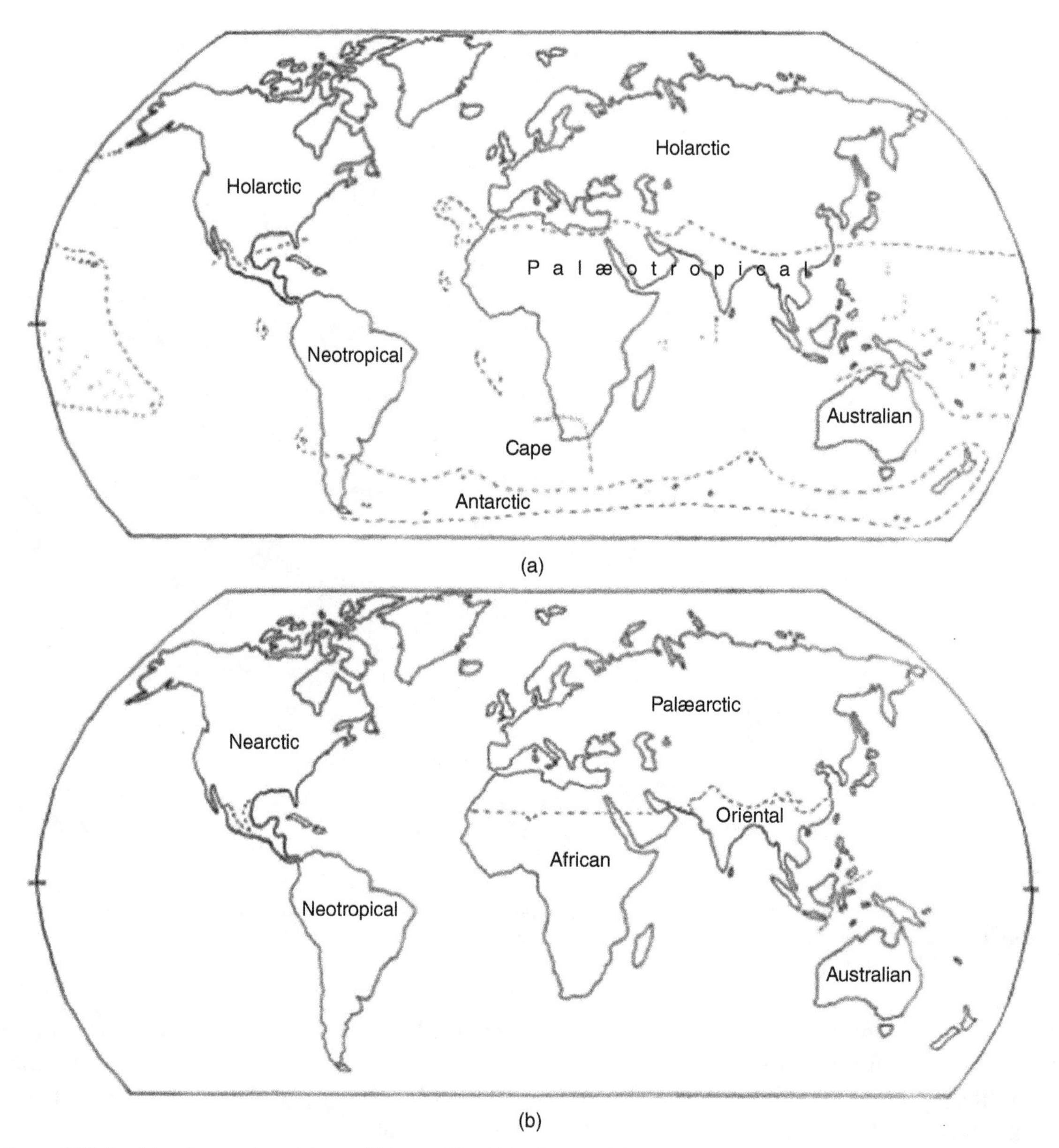

Figure 14.3 Traditional separation of the world into (a) floral kingdoms and (b) zoological regions. (*Source*: Cox (2001). Reproduced with permission from John Wiley & Sons Ltd.)

depending on the author, but around 12–16 is typical (Fig. 14.5). The biomes vary greatly in habitat structure, complexity and composition, from arid deserts to tropical rain forests. The locations of biomes are driven primarily by the climate, in particular annual temperature and rainfall (Fig. 14.6; Wright et al., 2004), as well as groundwater depth (Fan et al., 2013). Mean temperature and amount of precipitation are most important, though seasonality in them can also be a crucial factor.

The environmental determinism of biomes is not entirely one-way. As was mentioned in Section 11.3, evaporation and condensation by forest canopies cause pressure changes which are thought to carry

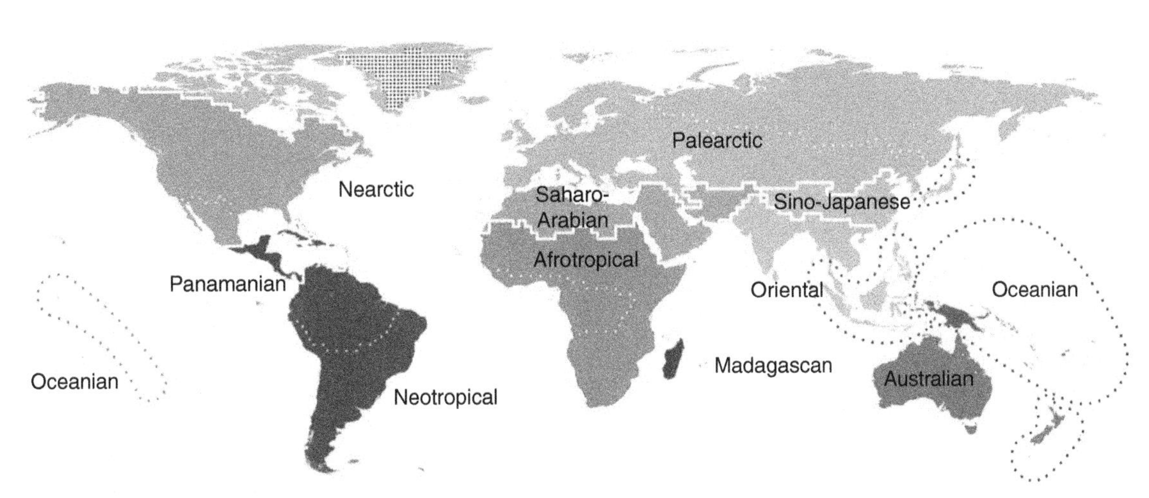

Figure 14.4 Terrestrial zoogeographic realms and regions based on distributions of 21,037 species of amphibians, non-pelagic birds and non-marine mammals. Dashed lines delineate the 20 zoogeographic regions; thick lines group regions into 11 named realms, for which colour differences indicate the amount of phylogenetic turnover; dotted regions have no species records; Antarctica was not included in the analyses. (*Source*: Holt et al. (2013). Reproduced with permission from American Association for the Advancement of Science.) (*See colour plate section for the colour representation of this figure.*)

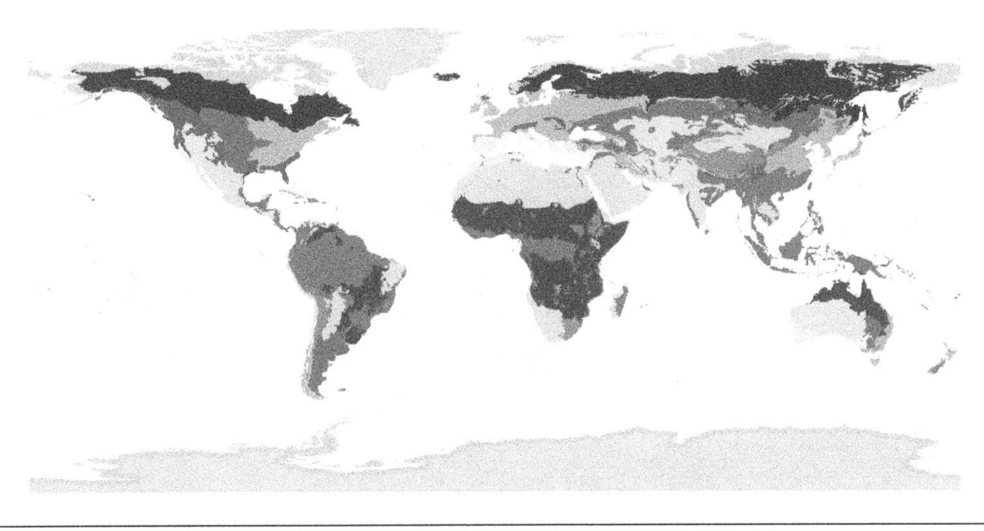

Figure 14.5 Global terrestrial biomes. (*Source*: Data provided by The Nature Conservancy; figure prepared with help from Tim Newbold.) (*See colour plate section for the colour representation of this figure.*)

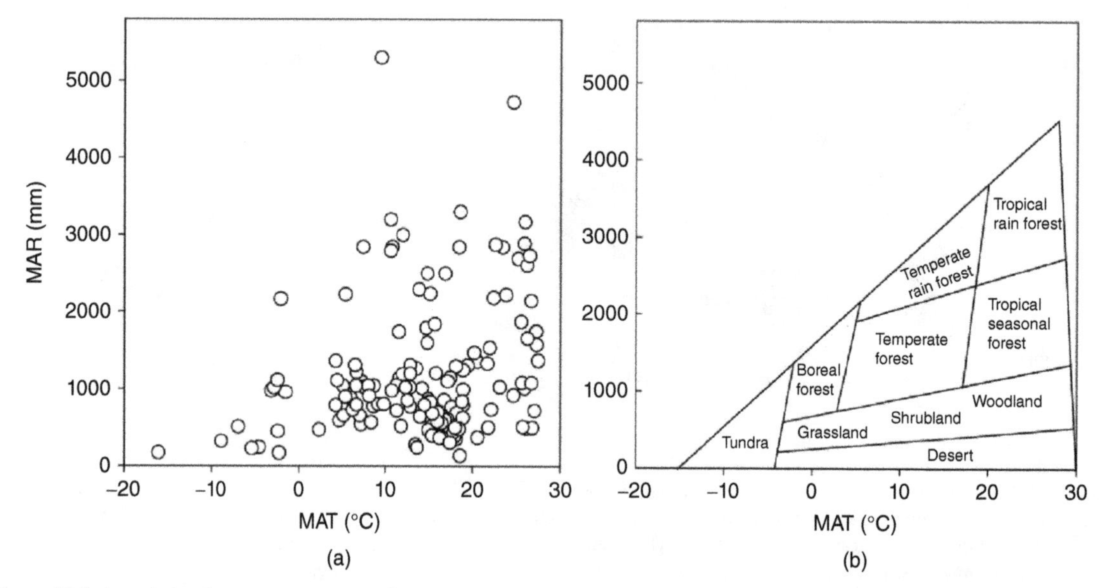

Figure 14.6 Association between mean annual temperature (MAT), mean annual rainfall (MAR) and vegetative biome present; (a) from 175 sites with a global distribution, (b) schematic representing typical biome present. (*Source*: Wright et al. (2004). Reproduced with permission from Nature Publishing Group.)

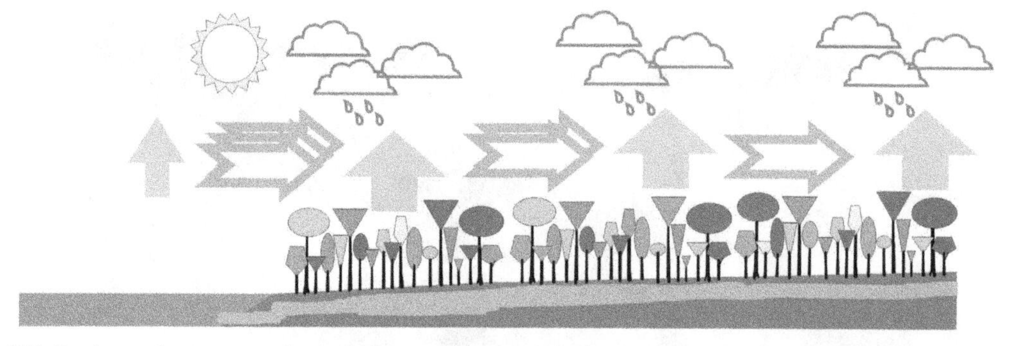

Figure 14.7 Continuous forest cover maintains high rates of transpiration, drawing wet air into the centre of continents from coastal areas and acting as a conveyor for rain clouds. (*Source*: Sheil and Murdiyarso (2009). Reproduced with permission from Oxford University Press.)

moisture from oceans to the interior of continents, changing the regional climate and thereby maintaining vegetation (Fig. 14.7). Over two-thirds of precipitation that falls on land is returned to the atmosphere, mostly through transpiration, and returns to the land again as rain. Trees not only transpire around ten times more water vapour than herbaceous vegetation; they also produce volatile organic compounds that stimulate the formation of water droplets and create further rainfall (Sheil, 2014). In the boreal forests of Eurasia, when leaves are on the trees in the summer, rainfall is carried thousands of miles inland. In the winter, however, the trees lose their leaves, transpiration ceases and the interior of the continent becomes much drier (Fig. 14.8).

Across large areas of the world, patches of forest alternate with savannahs and grasslands, with the amount of tree cover determined by a combination of the amount of rainfall, its seasonality and the frequency of fires that favour more resilient grasses. The three alternative vegetation types switch from

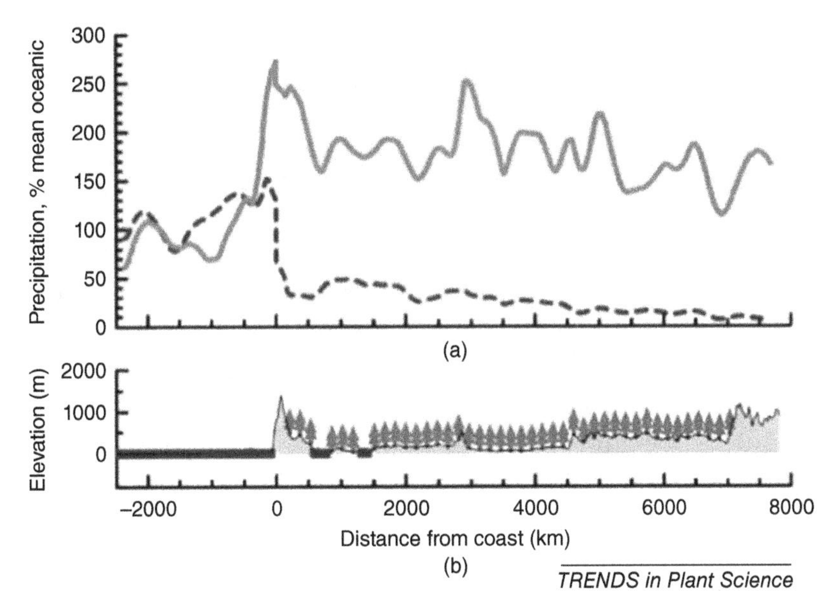

Figure 14.8 (a) Summer (July; continuous line) and winter (January; dashed line) precipitation in a Eurasian boreal forest transect at 61°N plotted as percentage of mean Atlantic oceanic precipitation versus distance from the Atlantic Coast (prevailing winds are inland from this coast). (b) Elevation (black over grey), forest cover (tree symbols) and open water (bold lines). (*Source*: Sheil (2014). Reproduced with permission from John Wiley and Sons Ltd.)

one to another and are often self-sustaining—fires propagate better in open grasslands than in forests, maintaining each in their present states.

Comparing across regions, Hirota et al. (2011) found that forests, savannahs and grasslands occur in distinct clusters within climate space (Fig. 14.9). This is evidence for the existence of alternative stable states at the level of whole biomes, with the options for any single site depending on the actual amount of annual rainfall. Wet sites (> 3,000 mm pa) are almost always forested, while dry sites (< 500 mm pa) are typically treeless, but at intermediate levels there can either be no trees, complete tree cover or a savannah with some trees. This implies that alterations in rainfall distribution with climate change are likely to cause dramatic changes in the distribution of vegetation rather than gentle transitions.

The study also brings up the importance of fire as a structuring force in determining the locations of biomes. Modelling global vegetation suggests that the cover of closed-canopy forests would double from

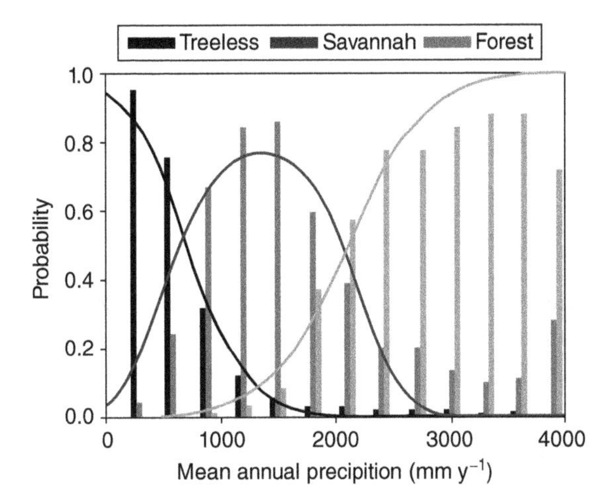

Figure 14.9 The probability of a given site in tropical and subtropical regions of Africa, Australia and South America being covered by forest (≥60% tree cover), savannah (5–60%) or treeless (< 5%) with regard to mean annual precipitation. Bars indicate the relative frequency of the vegetation types in each rainfall class, with curves representing fitted models. (*Source*: Hirota et al. (2011). Reproduced with permission from American Association for the Advancement of Science.)

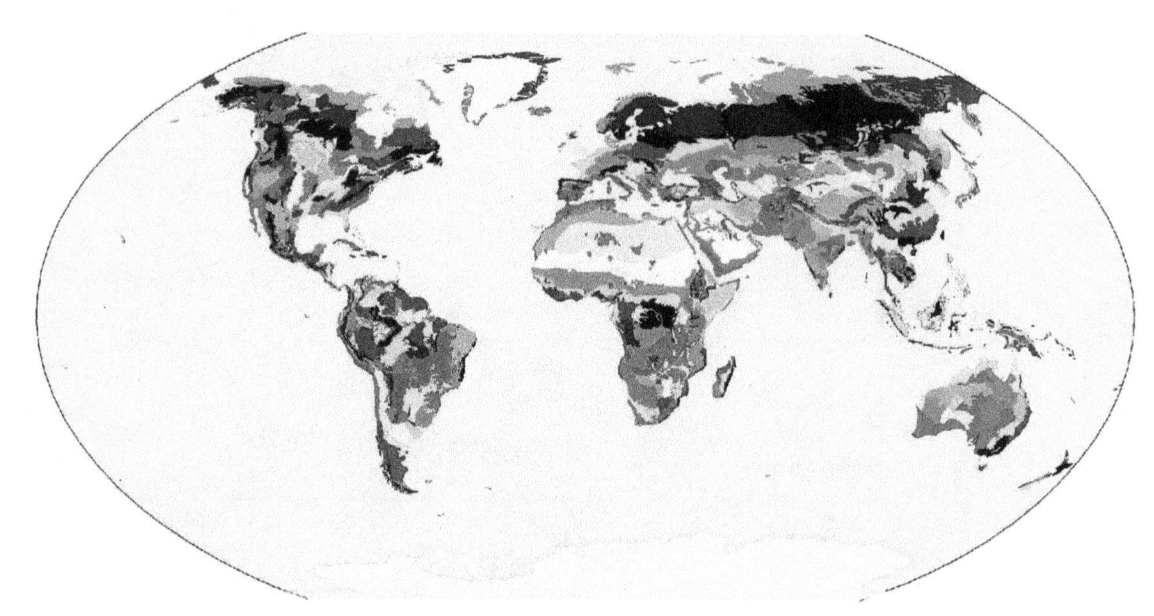

Figure 14.10 Terrestrial ecoregions as identified by Olson et al. (2001). (*Source*: Olson et al. (2001). Reproduced with permission from Oxford University Press.) (*See colour plate section for the colour representation of this figure.*)

27% to 56% of land area in the absence of fire (Bond et al., 2005). Over large parts of the world, the natural vegetation type is uncertain and cannot be predicted from climate alone. The world is not as green as it could be; large parts are brown or black (Bond, 2005). It is even possible to divide the world into pyromes based on fire regimes (Archibald et al., 2013).

14.4 Ecoregions

The kingdoms and biomes sketched on maps represented two different schools of thought as to how best to identify the major boundaries in the natural world. It took a surprisingly long time for a combined approach to consolidate. Many biogeographers now prefer to divide the planet into **ecoregions** (Fig. 14.10). These subunits broadly represent the distributions of distinct (meta)communities, aiming to describe what the world would look like without human-induced land-use change (although in some cases this requires informed speculation).

There are 867 recognised terrestrial ecoregions with a median size of 56,300 km^2. The smallest is the miniscule Clipperton Island, with an area of only 9 km^2, while the largest is the enormous Sahara ecoregion, covering approximately 4.6 million km^2 with little internal variation in either flora or fauna. Since publication of this scheme, ecoregions have become important units in conservation, because they represent more biologically meaningful categories than national boundaries. Their features and characteristic organisms can be explored on the excellent Wildfinder[2] website.

What determines the composition of each ecoregion? As with metacommunities, it is possible to decompose their variation into patterns with geographical distance (caused by dispersal limitation) versus environmental change (driven by niche requirements; Qian and Ricklefs, 2012). Both turn out to be important, although it is hard to provide an accurate weighting because effectively characterising the environment is more difficult than measuring map distance. The further apart two ecoregions are, the more different their biota (Fig. 14.11). The composition of groups with limited powers of dispersal such as amphibians and reptiles changes

[2] http://www.worldwildlife.org/science/wildfinder/.

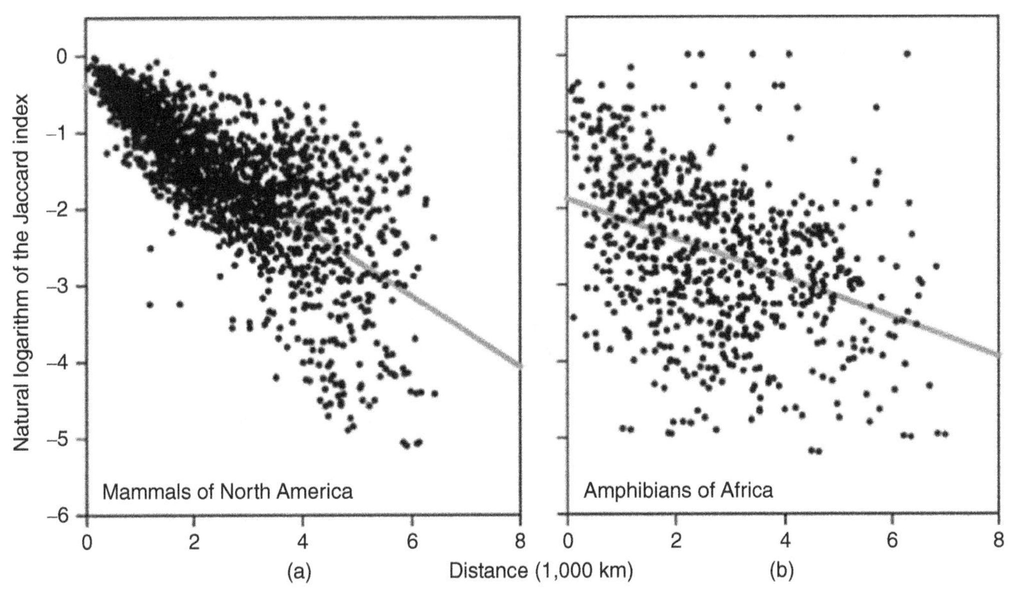

Figure 14.11 Decrease in ln *J*, a measure of similarity, with geographic distance between ecoregions for (a) mammals in North America (including bats) and (b) amphibians in Africa. (*Source*: Qian and Ricklefs (2012). Reproduced with permission from John Wiley & Sons Ltd.)

more with distance between ecoregions than volant taxa such as birds.

14.5 Empirical approaches

Even the ecoregions, while based on the expertise and knowledge of a collection of outstanding scientists, and produced using strict criteria, are still at heart arbitrary boundaries. As with ecological communities, we need to be certain that the categories represent genuine entities in nature. Once again this depends upon multivariate statistics, though at this scale they are usually performed on presence/absence matrices generated from species lists. Abundance data are seldom available or meaningful for geographical regions. Ordination schemes can be used to plot the positions of samples in virtual space and cluster analysis employed to identify the distinct subunits (see Section 10.4).

Armen Takhtajan (1910–2009) was one of the senior figures in plant biogeography in the last century. His encyclopædic knowledge formed the basis for his book 'The Floristic Regions of the World' (Takhtajan, 1986) in which he split the globe into 37 regions based on their plant species, climate and geography. He was notably sceptical of quantitative approaches. His assertions were therefore ripe for testing once the data became available. Conran (1995) did so for the Liliiflorae, the superfamily of plants with bulbs. These originated in the Cenozoic and are conventionally split into 48–56 families. Conran further divided them into 91 sub-taxa and recorded their presence or absence in 28 regions of the world. The outcome of the multivariate analysis is shown in Fig. 14.12.

Reassuringly, the regions recognised by Takhtajan appear to be reasonably distinct and are even clustered within the continents that contain them. Even more remarkable is the similarity between this pattern and the actual layout of the globe. But just because Takhtajan turned out to be broadly correct does not mean we can avoid quantitative approaches; not everyone can afford to spend a lifetime acquiring the ability to make similar judgements.

The same techniques can be used to investigate the distributions of animals. Procheş (2005) selected

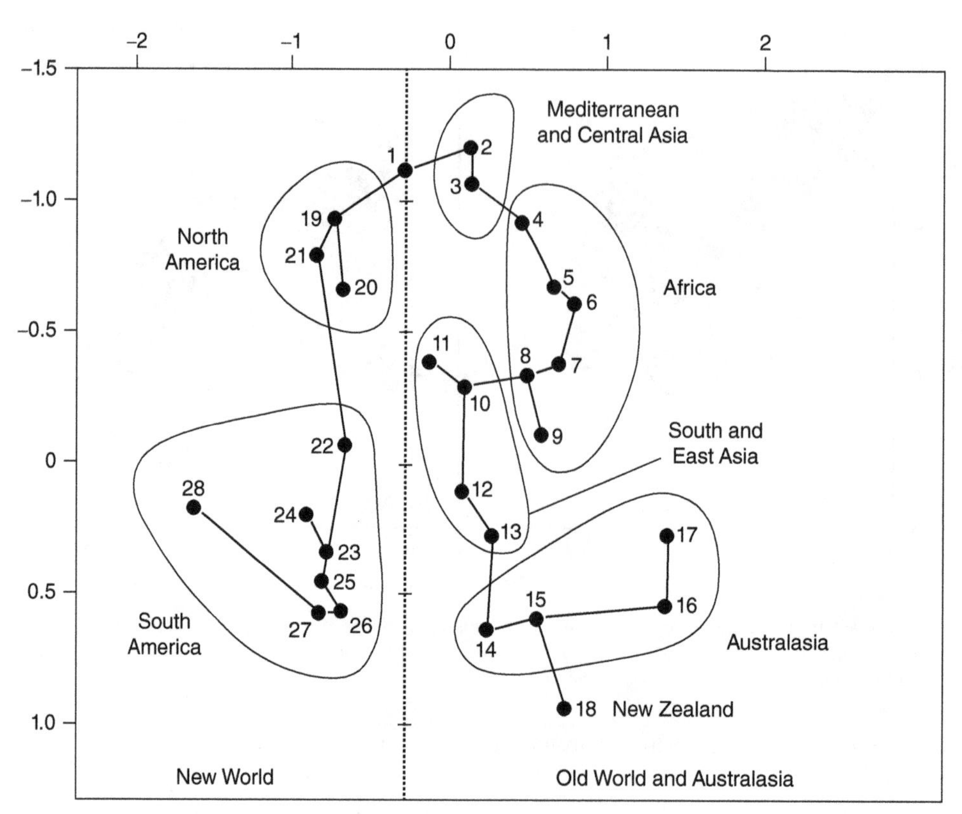

Figure 14.12 Relationships between the Liliiflorae floras following regions in Takhtajan (1986), with axes derived from a multidimensional scaling analysis in Conran (1995). (1) is circumboreal; all others are grouped within continents. (*Source*: Cox (2001). Reproduced with permission from John Wiley & Sons Ltd.)

around 1,000 species of bats, which are globally distributed on all ice-free land masses and where a wealth of good quality information on their presence (or absence) is available. Between 9 and 11 clusters were recognised, which projected nicely onto a map of the world, providing a close match to the standard faunal regions (Fig. 14.13).

Similar techniques can also be employed at regional scales. As with communities, the quantitative approach is more informative than simply drawing lines because it provides greater insight into the degree of difference among subunits and where boundaries might be blurred. For European plants, once again the formative work was done by Takhtajan (1986), who split the continent into 19 vegetative zones with misleadingly firm lines between them (Fig. 14.14). These were judged on the basis of coincident range edges and the presence of endemic species but relied on his opinion rather than a quantitative approach.

In recent years the *Flora Europæa* project has aimed to provide distribution maps for every plant species known to occur within Europe. By 2007 the first 12 volumes had been published, encompassing 2,793 unique taxa, approximately 20% of the total. This was enough for Finnie et al. (2007) to attempt a quantitative test of Takhtajan's floral provinces. They used co-occurrence data for all the species available and came up with 18 consistent floristic elements—almost the same number as Takhtajan. The distributions of four specimen elements are shown in Fig. 14.15. Every species was assigned to one element, and the proportion of those species found in each location determined. Dark colours

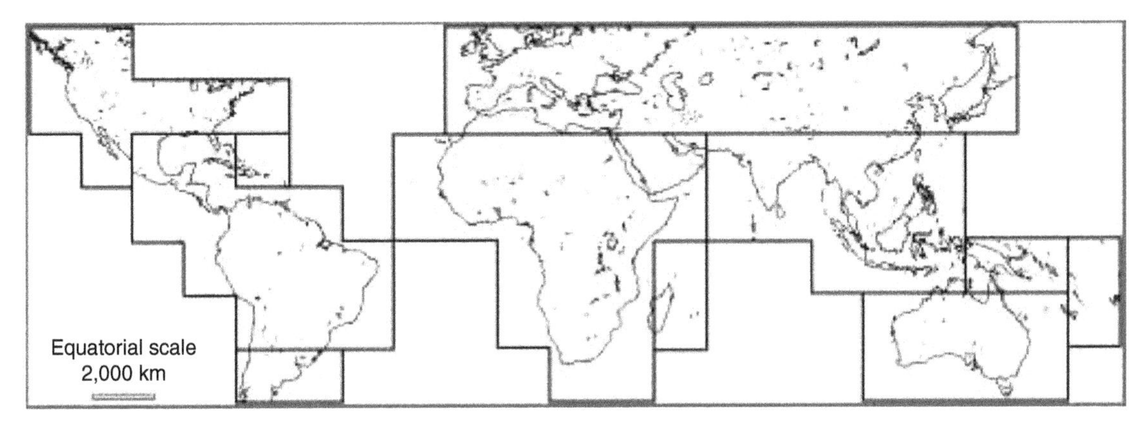

Figure 14.13 Biogeographical regions as delimited by multivariate analyses of the global bat fauna. (*Source*: Proches (2005). Reproduced with permission from John Wiley & Sons Ltd.)

indicate where that particular vegetative element was most characteristic of the area, but there is substantial overlap. Many species are spread over wide areas, often in those with similar environments (e.g. Scandinavia and the Alps). Some elements are more tightly constrained than others. This approach is thereby more informative than drawing lines on the map. Sadly such a data-intensive method is not possible for the entire world; there remain large regions where our botanical knowledge is still poor (Kier et al., 2005).

A comparable regional analysis can be performed for European land mammals (Fig. 14.16). Once again the ability to extend this to other regions is often limited by the availability and quality of data on species distributions.

14.6 The oceans

Up to this point we have been primarily concerned with the terrestrial realm, ignoring two-thirds of Earth's surface. In recognising this bias we are only acknowledging a broader problem, which is that the marine realm is relatively poorly understood. This is due to its great extent, not merely in area but in three dimensions, and the limited intensity of sampling relative to its scale. The seabed poses particular challenges, partly because it is so vast but also due to the technical difficulties of reaching an environment that extends to over 11 km below the ocean surface. The abyssal seafloor (3–6 km depth) makes up 54% of the planet's surface (Gage and Tyler, 1991).

Although the oceans constitute twice as much surface area, their total productivity only equals that of the land (Field et al., 1998). Also apparent is their much lower diversity, with only 15% of global species richness. There are unlikely to be more than a million species in the oceans (Appeltans et al., 2012), as compared to up to 10 million in terrestrial and freshwater habitats (see Chapter 2). Per unit area, species richness is around 14 times higher on land, and if considered by habitat volume, of which the seas make up 99.8% of the global total, species density is around 3,600 times greater on land (Dawson, 2012). In other words, there are remarkably few fish in the sea. About 96% of all 'fish' come from a single group, the actinopterygians, which have similar species richness in freshwater (15,150) and marine (14,470) environments, with comparable rates of diversification in each, despite the enormous discrepancy in volume. All the marine species descend from a single freshwater ancestor that recolonised the oceans after 300 Mya, perhaps following a mass extinction (Vega and Wiens, 2012).

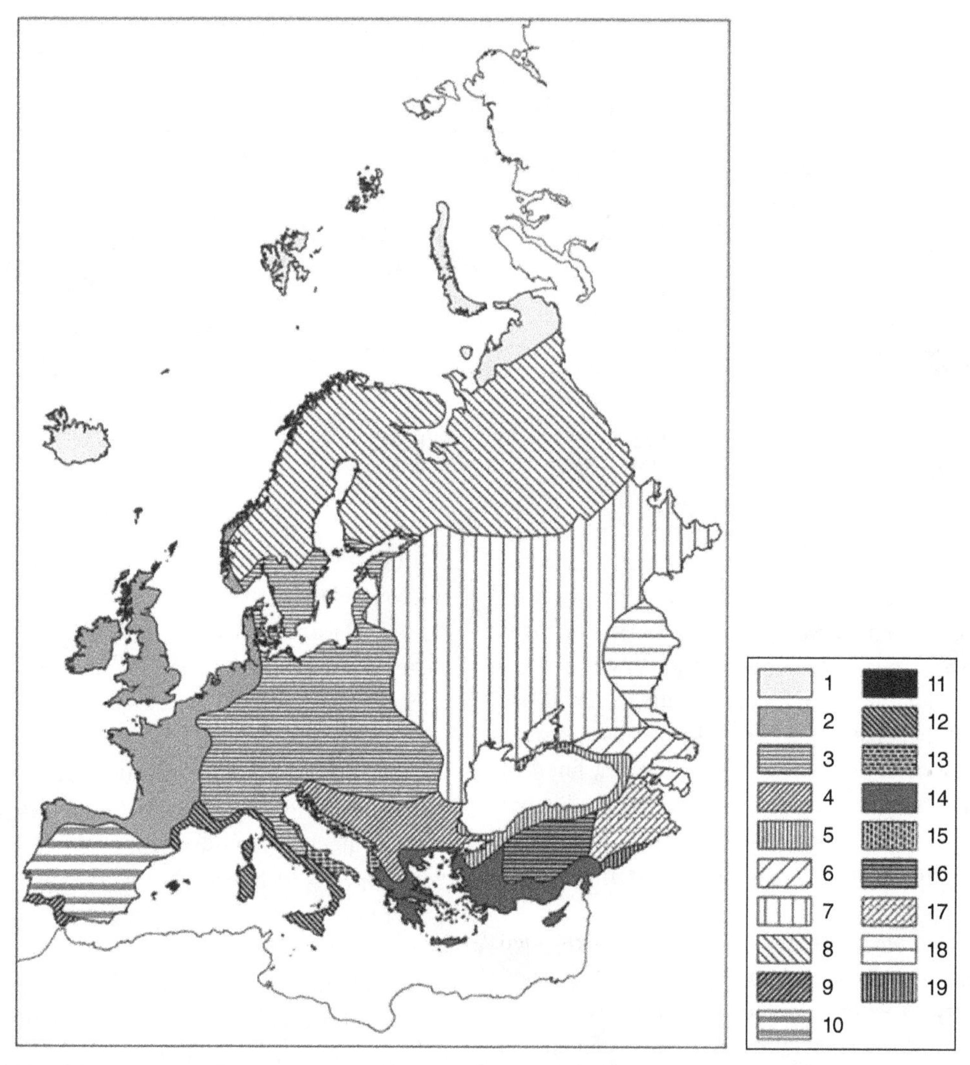

Figure 14.14 Takhtajan's (1986) floral provinces within Europe. 1, Arctic; 2, Atlantic-European; 3, Central European; 4, Illyrian or Balkan; 5, Euxine; 6, Caucasian; 7, Eastern European; 8, Northern European; 9, South-western Mediterranean; 10, Iberian; 11, Balearic; 12, Liguro-Tyrrhenian; 13, Adriatic; 14, East Mediterranean; 15, Crimean-Novorossiysk; 16, Central Anatolian; 17, Armeno-Iranian; 18, Turanian or Irano-Caspian; 19, Mesopotamian. A further Macaronesian Province is not mapped. (*Source*: Finnie et al. (2007). Reproduced with permission from John Wiley & Sons Ltd.)

The reasons for why species richness is greater on land remain hotly disputed. Terrestrial diversity began to overtake that in the oceans around 110 Mya in the mid-Cretaceous (Vermeij and Grosberg, 2010). This means that for 80% of the time during which multicellular life has existed, the oceans were more species rich. We therefore cannot invoke intrinsic differences in the environment to explain the contrast; after all, were someone to survey the planet in the Devonian, they would conclude without any argument that life was more diverse in the oceans.

Given that 98% of marine species are benthic, understanding patterns on the seabed is crucial to developing a biogeographical framework. The sea surface is much more manageable, though only

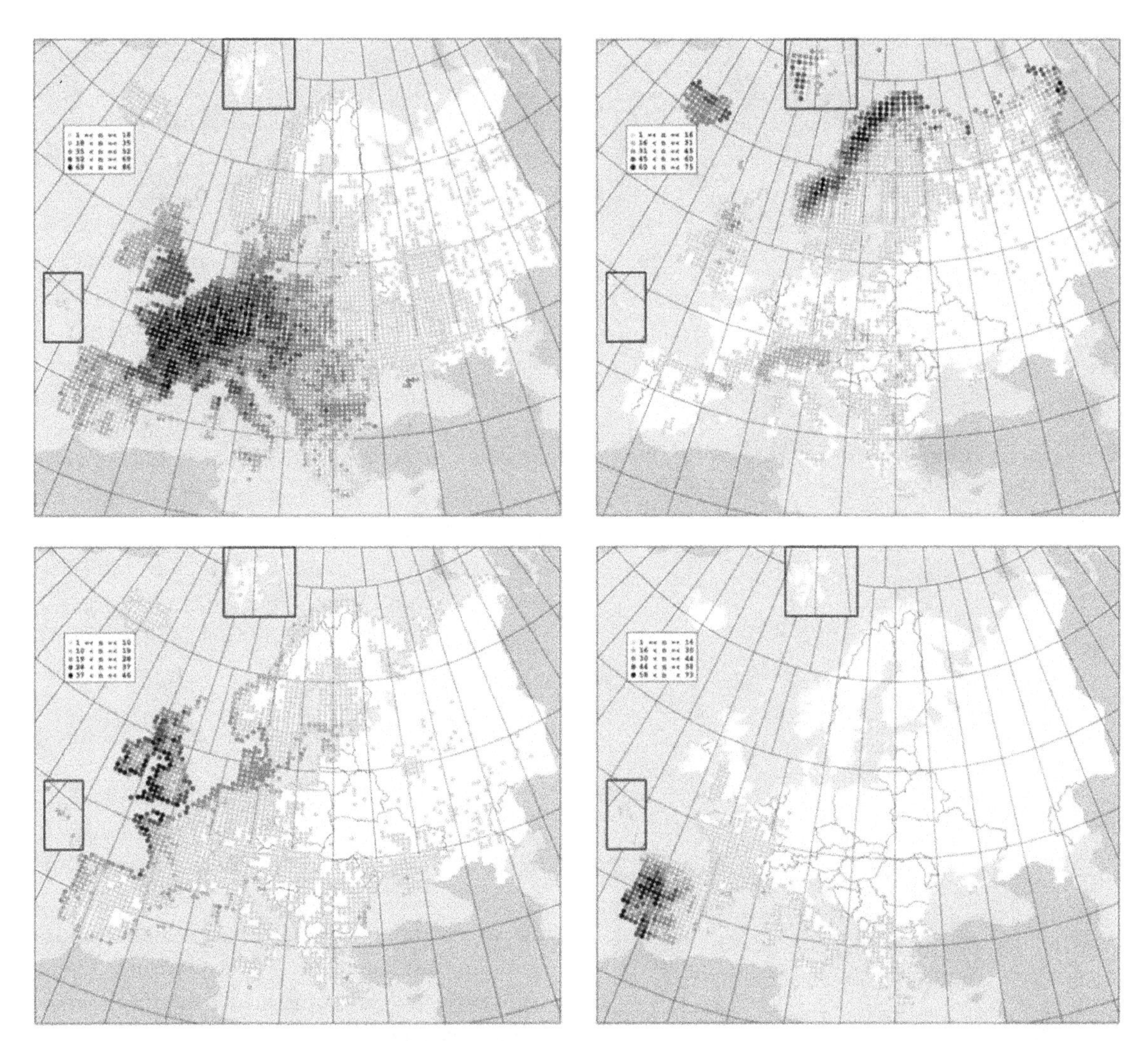

Figure 14.15 Four illustrative European floristic elements. Shading indicates the proportion of species within each element present in any particular locality (circles). Solid grey land masses were not included in the study. (*Source*: Finnie et al. (2007). Reproduced with permission from John Wiley & Sons Ltd.)

because we are able to float on it. The ocean can be split into environmental regions based on surface temperature and currents, and these are reflected by the pelagic (floating) algae found within them. On the basis of these algae, it is possible to roughly divide the surface waters into four biomes and 51 provinces in an analagous fashion to the land (Longhurst, 1998). This divides the oceans into a polar biome, two zones based on predominant wind directions (westerlies and trades) and coastal boundaries (Fig. 14.17).

A similar attempt has been made to identify subdivisions for coastal and shelf areas of the oceans, leading to the designation of 12 realms, 62 provinces and 232 ecoregions (Spalding et al., 2007). A further scheme for pelagic and deep benthic systems would be desirable.

In the last few years there has been an explosion of information from the Census of Marine Life project.[3] This is currently being synthesised and new results

[3] http://www.coml.org/.

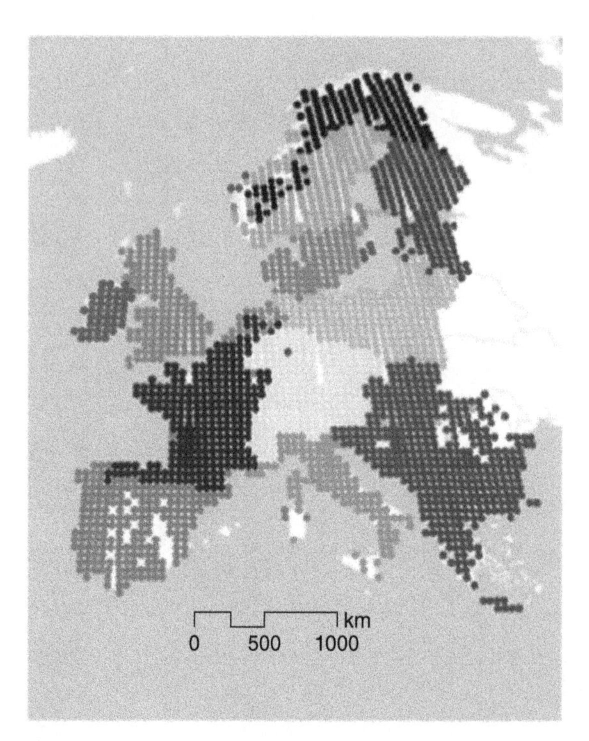

Figure 14.16 Faunal elements for European land mammals (Gagné and Proulx, 2007; Heikinheimo et al., 2007). (*See colour plate section for the colour representation of this figure.*)

are published all the time. Combined with a renewed interest in deep-sea exploration, where every mission uncovers an array of new species, the prospects for finally understanding the oceans have never been so good. One of the remaining problems is the unevenness of sampling. We have been relatively good at documenting the biota of continental shelves and the ocean surface, unsurprisingly, but our knowledge of the deep pelagic parts of the ocean remains extremely poor (Webb et al., 2010).

14.7 Fresh water

A final type of habitat is fresh water. While vital for all terrestrial life, it makes up a miniscule proportion of global surface area. In total there are only 1.5 million km^2 of lakes and rivers, of which a few large bodies comprise the overwhelming majority. Wetlands are much larger, a total of around 2.6 million km^2, but these are usually considered alongside other terrestrial vegetation. A very large fraction of earth's surface is permanently covered by ice and snow (16 million km^2) and as a result contains little life.

The nature of freshwater habitats ensures that they are highly fragmented, with dispersal assembly often playing a major role in the structure of communities. This has also led to high rates of speciation, causing individual bodies of fresh water to be extremely variable in composition. The environmental conditions within them are predominantly influenced by the surrounding land, particularly in terms of nutrient inputs. As a result of all these factors, freshwater systems should be viewed as a special case, and no biogeographical synthesis is likely to be informative (e.g. Logez et al., 2013). They can often be treated as elements of the broader terrestrial landscape.

14.8 Conclusions

A whole field, biogeography, exists to describe the global distribution and patterns of life, though it is still in the process of consolidation. The challenges are the multidisciplinary nature of any research and the large data requirements. Biogeographical questions are becoming increasingly tractable though, with explanations arising at the interface between ecology, geography, geology and systematics. Judgements regarding how best to describe the patterns of life on land can either be based on the distributions of particular taxa or the presence of distinctive habitats. These approaches do not always overlap, and a combined system of ecoregions provides an improvement, including information on plants, animals and their environment. Biogeography has moved on from colouring in maps on the basis of expert judgement and now uses quantitative approaches to back up hunches about where the boundaries lie. Freshwater systems are aberrant and need special consideration. Finally, a wealth of data are newly available for marine systems but have yet to be fully synthesised in a manner that will allow us

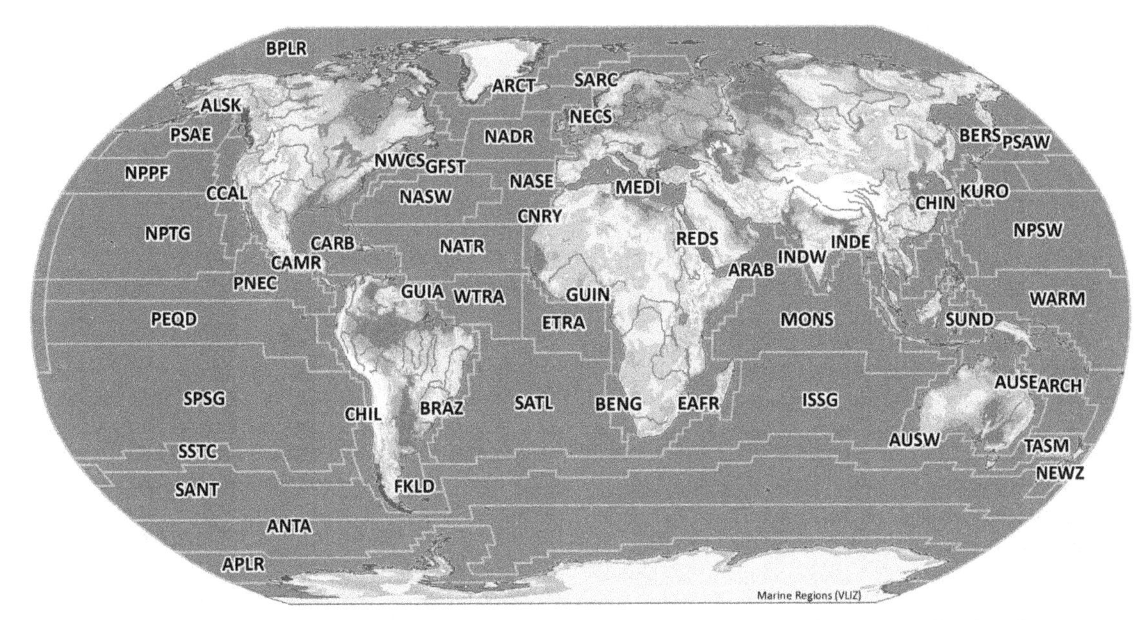

Figure 14.17 Partitioning of the world's oceans into provinces based on phytoplankton distribution. Note that these boundaries are not fixed and alter in response to seasonal and interannual changes in the environment. (*Source*: Longhurst (1998). Reproduced with permission from John Wiley & Sons Ltd.) (*See colour plate section for the colour representation of this figure.*)

to ask whether the same processes are at work both on land and in the sea.

14.8.1 Recommended reading

Holt, B. G., J.-P. Lessard, M. K. Borregaard, S. A. Fritz, M. B. Araújo, D. Dimitrov, P.-H. Fabre, C. H. Graham, G. R. Graves, K. A. Jønsson, D. Nogués-Bravo, Z. Wang, R. J. Whittaker, J. Fjeldså, and C. Rahbek, 2013. An update of Wallace's zoogeographic regions of the world. *Science* 339:74–78.

Kreft, H. and W. Jetz, 2010. A framework for delineating biogeographical regions based on species distributions. *Journal of Biogeography* 37:2029–2053.

Olson, D. M., E. Dinerstein, E. D. Wikramanayake, N. D. Burgess, G. V. N. Powell, E. C. Underwood, J. A. D'Amico, I. Itoua, H. E. Strand, J. C. Morrison, C. J. Loucks, T. F. Allnutt, T. H. Ricketts, Y. Kura, J. F. Lamoreux, W. W. Wettengel, P. Hedao, and K. R. Kassem, 2001. Terrestrial ecoregions of the world: a new map of life on Earth. *Bioscience* 51: 933–938.

Many of the papers cited in this chapter, plus other relevant readings, are included in a special online issue of Journal of Biogeography entitled '100 Years after Alfred Russel Wallace'.

14.8.2 Questions for the future

- What determines whether boundaries between biogeographical regions are sharp or diffuse?
- Why do some taxa respect biogeographical boundaries, while others cross them freely?
- Can existing data be used to predict which species will be present in poorly explored regions?
- Are species the best units for demarcating biogeographical units, or could we use other properties such as the distribution of traits (e.g. van Bodegum et al., 2014)?

References

Anesio, A. M. and J. Laybourn-Parry, 2012. Glaciers and ice sheets as a biome. *Trends in Ecology & Evolution* 27:219–225.

Appeltans, W., S. T. Ahyong, G. Anderson, M. V. Angel, T. Artois, N. Bailly, R. Bamber, A. Barber, I.

Bartsch, A. Berta, M. Błaewicz-Paszkowycz, P. Bock, G. Boxshall, C. B. Boyko, S. N. Brandão, R. A. Bray, N. L. Bruce, S. D. Cairns, T.-Y. Chan, L. Cheng, A. G. Collins, T. Cribb, M. Curini-Galletti, F. Dahdouh-Guebas, P. J. F. Davie, M. N. Dawson, O. De Clerck, W. Decock, S. D. Grave, N. J. de Voogd, D. P. Domning, C. C. Emig, C. Erséus, W. Eschmeyer, K. Fauchald, D. G. Fautin, S. W. Feist, C. H. Fransen, H. Furuya, O. Garcia-Alvarez, S. Gerken, D. Gibson, A. Gittenberger, S. Gofas, L. Gómez-Daglio, D. P. Gordon, M. D. Guiry, F. Hernandez, B. W. Hoeksema, R. R. Hopcroft, D. Jaume, P. Kirk, N. Koedam, S. Koenemann, J. B. Kolb, R. M. Kristensen, A. Kroh, G. Lambert, D. B. Lazarus, R. Lemaitre, M. Longshaw, J. Lowry, E. Macpherson, L. P. Madin, C. Mah, G. Mapstone, P. A. McLaughlin, J. Mees, K. Meland, C. G. Messing, C. E. Mills, T. N. Molodtsova, R. Mooi, B. Neuhaus, P. K. L. Ng, C. Nielsen, J. Norenburg, D. M. Opresko, M. Osawa, G. Paulay, W. Perrin, J. F. Pilger, G. C. B. Poore, P. Pugh, G. B. Read, J. D. Reimer, M. Rius, R. M. Rocha, J. I. Saiz-Salinas, V. Scarabino, B. Schierwater, A. Schmidt-Rhaesa, K. E. Schnabel, M. Schotte, P. Schuchert, E. Schwabe, H. Segers, C. Self-Sullivan, N. Shenkar, V. Siegel, W. Sterrer, S. Stöhr, B. Swalla, M. L. Tasker, E. V. Thuesen, T. Timm, M. A. Todaro, X. Turon, S. Tyler, P. Uetz, J. van der Land, B. Vanhoorne, L. P. van Ofwegen, R. W. M. van Soest, J. Vanaverbeke, G. Walker-Smith, T. C. Walter, A. Warren, G. C. Williams, S. P. Wilson, and M. J. Costello, 2012. The magnitude of global marine species diversity. *Current Biology* 22:2189–2202.

Archibald, S., C. E. R. Lehmann, J. L. Gómez-Dans, and R. A. Bradstock, 2013. Defining pyromes and global syndromes of fire regimes. *Proceedings of the National Academy of Sciences of the United States of America* 110: 6442–6447.

van Bodegum, P. M., J. C. Douma, and L. M. Verheijen, 2014. A fully traits-based approach to modeling global vegetation distribution. *Proceedings of the National Academy of Sciences of the United States of America* 111:13733–13738.

Bond, W. J., 2005. Large parts of the world are brown or black: a different view on the 'Green World' hypothesis. *Journal of Vegetation Science* 16:261–266.

Bond, W. J., F. I. Woodward, and G. F. Midgley, 2005. The global distribution of ecosystems in a world without fire. *New Phytologist* 165:525–538.

Buffon, G. L. L., 1761. *Histoire Naturelle, Generale et Particuliere*, vol. 9. Impremerie Royal.

de Candolle, A. P., 1820. Essai Élémentaire de Géographie Botanique. In *Dictionnaire des Sciences Naturelles*, vol. 18. F. G. Levrault.

Conran, J. G., 1995. Family distributions in the Liliiflorae and their biogeographical implications. *Journal of Biogeography* 22:1023–1034.

Cox, C. B., 2001. The biogeographic regions reconsidered. *Journal of Biogeography* 28:511–523.

Dawson, M. D., 2012. Species richness, habitable volume, and species densities in freshwater, the sea, and on land. *Frontiers of Biogeography* 4:105–116.

Ebach, M. C. and D. F. Goujet, 2006. The first biogeographical map. *Journal of Biogeography* 33:761–769.

Fan, Y., H. Li, and G. Miguez-Macho, 2013. Global patterns of groundwater table depth. *Science* 339:940–943.

Field, C. B., M. J. Behrenfeld, J. T. Randerson, and P. Falkowski, 1998. Primary production of the biosphere: integrating terrestrial and oceanic components. *Science* 281:237–240.

Finnie, T. R., C. D. Preston, M. O. Hill, P. Uotila, and M. J. Crawley, 2007. Floristic elements in European vascular plants: an analysis based on Atlas Florae Europaeae. *Journal of Biogeography* 34:1848–1872.

Flanders Marine Institute (VLIZ), 2009. Longhurst Biogeographical Provinces. Available online at http://www.marineregions.org/. Consulted on 2015-01-07.

Gage, J. D. and P. A. Tyler, 1991. *Deep-Sea Biology: A Natural History of Organisms at the Deep-Sea Floor*. Cambridge University Press.

Gagné, S. A. and R. Proulx, 2007. Accurate delineation of biogeographical regions depends on the use of an appropriate distance measure. *Journal of Biogeography* 36:561–562.

Heikinheimo, H., M. Fortelius, J. Eronen, and H. Mannila, 2007. Biogeography of European land mammals shows environmentally distinct and spatially coherent clusters. *Journal of Biogeography* 34:1053–1064.

Hirota, M., M. Holmgren, E. H. van Nes, and M. Scheffer, 2011. Global resilience of tropical forest and savanna to critical transitions. *Science* 334:232–235.

Holt, B. G., J.-P. Lessard, M. K. Borregaard, S. A. Fritz, M. B. Araújo, D. Dimitrov, P.-H. Fabre, C. H. Graham, G. R. Graves, K. A. Jønsson, D. Nogués-Bravo, Z. Wang, R. J. Whittaker, J. Fjeldså, and C. Rahbek, 2013. An update of Wallace's zoogeographic regions of the world. *Science* 339:74–78.

von Humboldt, A., 1808. *Ansichten de Natur mit Wissenschaflichen Erlauterungen*. J. G. Cotta.

Kier, G., J. Mutke, E. Dinerstein, T. H. Ricketts, W. Küper, H. Kreft, and W. Barthlott, 2005. Global patterns of plant diversity and floristic knowledge. *Journal of Biogeography* 32:1107–1116.

Kreft, H. and W. Jetz, 2010. A framework for delineating biogeographical regions based on species distributions. *Journal of Biogeography* 37:2029–2053.

Logez, M., P. Bady, A. Melcher, and D. Pont, 2013. A continental-scale analysis of fish assemblage functional structure in European rivers. *Ecography* 36:80–91.

Longhurst, A., 1998. *Ecological Geography of the Sea*. Academic Press.

Makarieva, A. M., V. G. Gorschkov, and B.-L. Li, 2013. Revisiting forest impact on atmospheric water vapor transport and precipitation. *Theoretical and Applied Climatology* 111:79–96.

Marchant, J., 1916. *Alfred Russel Wallace: Letters and Reminiscences*. Cassell and Company, Ltd.

Müller, S., 1846. Über den Charakter der Thierwelt auf den Inseln des indischen Archipels, ein Beitrag zur zoologischen Geographie. *Archiv für Naturgeschichte* 12:109–128.

Olson, D. M., E. Dinerstein, E. D. Wikramanayake, N. D. Burgess, G. V. N. Powell, E. C. Underwood, J. A. D'Amico, I. Itoua, H. E. Strand, J. C. Morrison, C. J. Loucks, T. F. Allnutt, T. H. Ricketts, Y. Kura, J. F. Lamoreux, W. W. Wettengel, P. Hedao, and K. R. Kassem, 2001. Terrestrial ecoregions of the world: a new map of life on Earth. *Bioscience* 51:933–938.

Procheş, S., 2005. The world's biogeographical regions: cluster analyses based on bat distributions. *Journal of Biogeography* 32:607–614.

Qian, H. and R. E. Ricklefs, 2012. Disentangling the effects of geographic distance and environmental dissimilarity on global patterns of species turnover. *Global Ecology and Biogeography* 21:341–351.

Schimper, A. F. W., 1903. *Plant Geography on a Physiological Basis*. Clarendon Press.

Sheil, D., 2014. How plants water our planet: advances and imperatives. *Trends in Plant Science* 19:209–211.

Sheil, D. and D. Murdiyarso, 2009. How forests attract rain: an examination of a new hypothesis. *Bioscience* 59:341–347.

Spalding, M. D., H. E. Fox, G. R. Allen, N. Davidson, Z. A. Ferdaña, M. Finlayson, B. S. Halpern, M. A. Jorge, A. Lombana, S. A. Lourie, K. D. Martin, E. McManus, J. Molnar, C. A. Rechhia, and J. Robertson, 2007. Marine ecoregions of the world: a bioregionalization of coastal and shelf areas. *Bioscience* 57:573–585.

Takhtajan, A. L., 1986. *The Floristic Regions of the World*. University of California Press.

Vega, G. C. and J. J. Wiens, 2012. Why are there so few fish in the sea? *Proceedings of the Royal Society Series B* 279:2323–2329.

Vermeij, G. J. and R. K. Grosberg, 2010. The great divergence: when did diversity on land exceed that in the sea? *Integrative and Comparative Biology* 50:675–682.

Wallace, A. R., 1860. On the zoological geography of the Malay Archipelago. *Journal of the Linnaean Society of London* 4:172–184.

Webb, T. J., E. V. Berghe, and R. O'Dor, 2010. Biodiversity's big wet secret: the global distribution of marine biological records reveals chronic under-exploitation of the deep pelagic ocean. *PLoS ONE* 5:e10223.

Wright, I. J., P. B. Reich, M. Westoby, D. D. Ackerley, Z. Baruch, F. Bongers, J. Cavender-Bares, T. Chapin, J. H. C. Cornelissen, M. Diemer, J. Flexas, E. Garnier, P. K. Groom, J. Gulias, K. Hikosaka, B. B. Lamont, T. Lee, W. Lee, C. Lusk, J. J. Midgley, M.-L. Navas, U. Niinemets, J. Oleksyn, N. Osada, H. Poorter, P. Poot, L. Prior, V. I. Pyankov, C. Roumet, S. C. Thomas, M. G. Tjoelker, E. J. Veneklaas, and R. Villar, 2004. The worldwide leaf economics spectrum. *Nature* 428:821–827.

CHAPTER 15

Regional species richness

15.1 The big question

When considering the drivers of local species richness in Chapter 7, it was noted that the size of the regional species pool accounts on average for around 75% of the observed variation. The importance of regional effects continues down to remarkably small scales; on coral reefs the relationship holds not only within sites but also on the level of individual transects (Fig. 7.5). The regional species pool acts to provide the colonists for any individual community, as demonstrated in Chapter 13. Recognising this link only shifts the problem of understanding patterns in species richness up a scale. What sets the size of the regional pool?

First we should acknowledge the scale of the issue. Consider two countries, Costa Rica and Canada, which seem different in every conceivable way (Medellín and Soberón, 1999). Costa Rica is tiny at 51,100 km^2, whereas Canada covers a vast 9,970,610 km^2. In Costa Rica there are at least 218 species of reptiles, 796 birds and 203 mammals. Yet Canada, in an area almost 200 times greater, has only 32 reptile species, 434 birds and 94 mammals. If area is such a powerful force in determining species richness at small scales, then how can Canada contain at best only half the species richness of Costa Rica? As with communities, we can learn as much from asking why so many species are not there as from studying the ones which are (Pärtel, 2014).

A further surprising outcome of large-scale analyses is that there is a reasonable degree of congruence in the richness patterns of taxa at the regional scale (Qian and Ricklefs, 2008). This differs markedly from the lack of covariance at the local scales that were considered in Section 7.5. The trends are not absolute, and vary depending on which taxa are compared, but can be found in many regions (e.g. Fig. 15.1).

Six sets of factors have been linked to regional patterns in species richness (Field et al., 2009). These are based on correlations with:

1. Climate and net primary productivity (NPP)
2. Landscape heterogeneity
3. Edaphics (i.e. soil conditions) and nutrients
4. Area
5. Biotic interactions
6. Dispersal and history

Every one of these has been examined multiple times with some support. Which is most important? Below we will assess the rationale and evidence behind each, but first a more general point needs to be made, which is that the scale at which any of these processes is studied is vital to the interpretation (Whittaker et al., 2001). The **grain size** of a study refers to the minimum area of a unit. While patterns at small scales, perhaps over a few square metres, are dominated by interactions among individual organisms, the higher organisation of nature depends upon processes whose influence might extend over many thousands of hectares or even an entire ecoregion. When assessing the evidence for and against any particular driver of observed patterns, always be sure that you are comparing like-with-like.

Natural Systems: The organisation of life, First Edition. Markus P. Eichhorn.

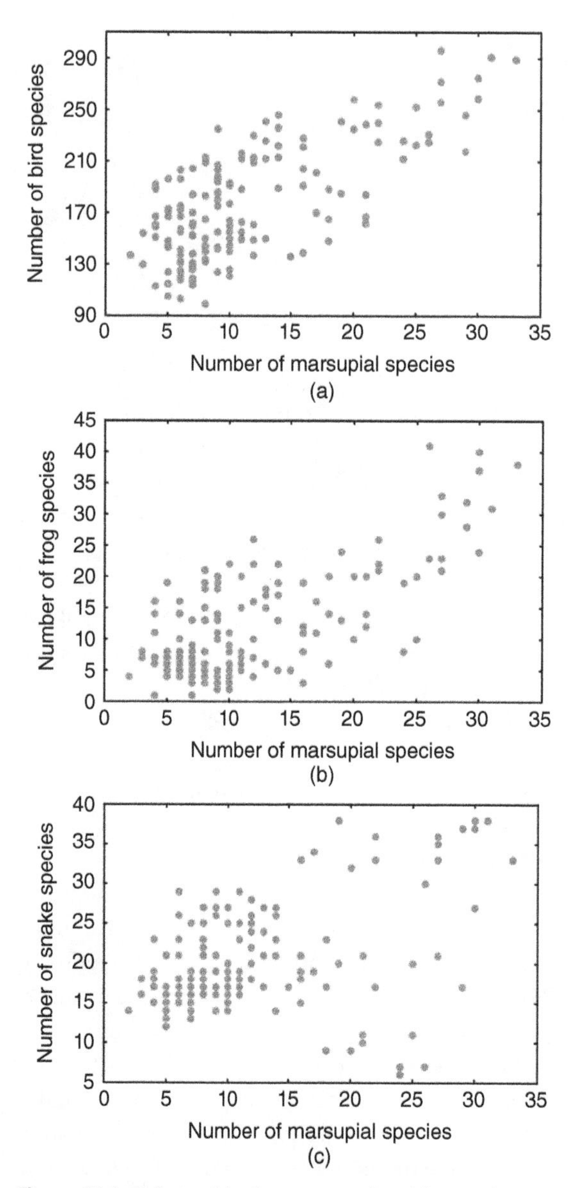

Figure 15.1 Relationships between species richness of vertebrate groups in 240 km grid squares across Australia. Marsupial species richness is plotted against species richness of (a) birds, (b) frogs and (c) snakes. (*Source*: Gaston (2000). Reproduced with permission from Nature Publishing Group.)

15.2 Climate and productivity

The idea that input of energy might influence patterns of species richness was discussed in Section 8.3. While there was no strong evidence that it did so at the level of local communities, or at least not in a consistent direction, the issue needs addressing again when considering larger scales. Greater energy availability might increase population sizes and reduce the likelihood of extinctions. Perhaps reduced energy availability constrains species richness by limiting the length of food chains or provides fewer opportunities for specialist species to survive. It may also set physiological limits on where organisms without adaptations to cold conditions are able to survive. For a full review of potential mechanisms, see Evans et al. (2005); at present there is not enough evidence to favour any one hypothesis over the others, and it may be that several processes act simultaneously.

There is certainly a positive link between plant species richness and average annual temperature. Temperature, in combination with precipitation levels, determines NPP. Looking at the trees of Europe and North America, a strong correlation exists between their species richness and NPP (Fig. 15.2). Even when grain sizes are small, once

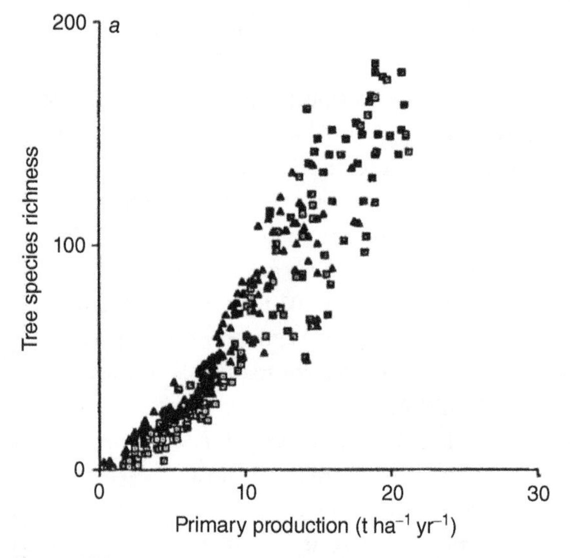

Figure 15.2 Species richness of trees in mesic areas of Europe (triangles) and North America (squares) related to primary productivity. Points represent quadrats of size 2.5° × 2.5° below 45°N, 3.5° × 2.5° from 45 to 60°N and 5° × 2.5° above 60°N; this gives a mean quadrat area of 72,000 km^2. (*Source*: Adams and Woodward (1989). Reproduced with permission from Nature Publishing Group.)

combined across larger geographical extents, clear patterns emerge. A carefully designed test by Wang et al. (2009) examined the link between tree species richness and temperature in 318 montane forest plots from across eastern China, each of which was only 600 m^2 in size, but which were separated by large distances. The relationship between species richness and temperature was strongly positive ($R^2 = 0.65$), suggesting something that requires an explanation. The same remains true through time in the fossil record; when temperatures in a given place rise or fall, plant species richness tracks it remarkably quickly (Vázquez-Rivera and Currie, 2014).

It is not solely a matter of temperature though. A consistent pattern emerges across the world that plant richness depends on both energy (measured as temperature or potential evapotranspiration (PET)) and the availability of water (Francis and Currie, 2003). Generally richness increases with heat but only when sufficient water is available. Kreft and Jetz (2008) examined global patterns of plant species richness and their correlations with a wide range of climatic variables (Fig. 15.3). There is a major influence of PET on plant species richness but only up to around 500 mm per year. Above this level the relationship breaks down, and instead the number of wet days per year becomes important. In other words, energy input is the dominant factor up to a certain point, after which water limitation takes over. A combined model incorporating both these relationships, plus the effect of topographic heterogeneity, is able to account for 70% of patterns in global plant species richness. Even so there is more to add. Outliers include the Cape of South Africa, which contains twice as many plant species as predicted by the model.

Globally there is variation in which particular environmental forces correlate most strongly with species richness (Hawkins et al., 2003). In the tropics and other hot regions, the availaility of water dominates relationships, while at higher latitudes either the interaction between energy and water (for plants) or energy alone (for animals) has the greatest influence.

Global bird species richness correlates strongly with actual evapotranspiration (AET) (Storch et al., 2006). This is true even on the scale of changing seasons. Among American birds, the species richness of any site tracks changes in the temperature across the year, rather than the number of species that could potentially cope with those temperatures (Boucher-Lalonde et al., 2014). Although it is possible that animal species richness might simply follow

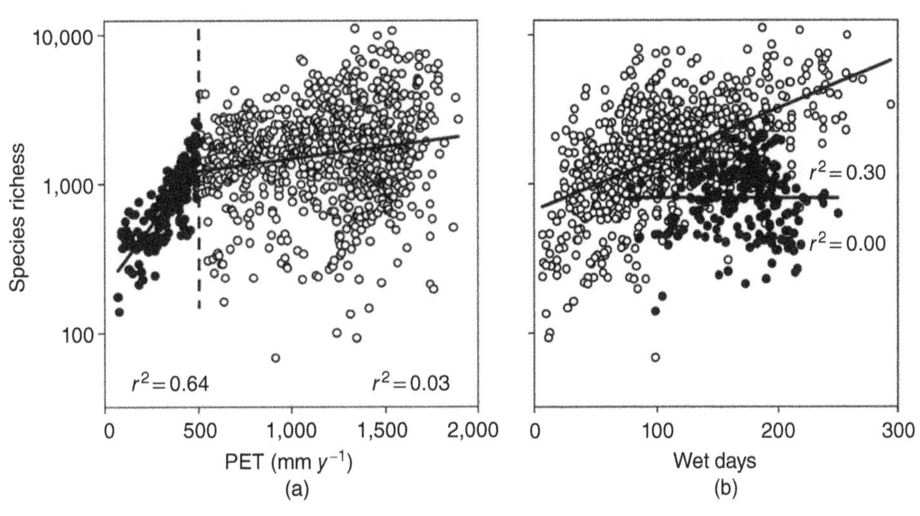

Figure 15.3 Relationship between environmental predictors and plant species richness in low (<500 mm y^{-1}, black dots) and high (>500 mm y^{-1} white) rainfall regions standardised to 10,000 km^2; (a) for potential evapotranspiration (PET), (b) for number of wet days per year. (*Source*: Kreft and Jetz (2007). Copyright (2007) National Academy of Sciences, USA.)

that of plants—a diversity of producers maintaining a diversity of consumers—this simple linkage does not seem to operate on biogeographical scales as clearly as within communities (see Section 8.4). Instead, for vertebrates at least, the environment has a dominant and direct effect (Jetz et al., 2009).

In the oceans, sea surface temperature is the main predictor of species richness rather than productivity, although endotherms are an exception to this and show the reverse trend (Tittensor et al., 2010). Patterns in terrestrial animal species richness follow similar rules. In general, ectotherm richness is most closely linked with temperature, whereas high endotherm richness is associated with greater primary productivity (Buckley et al., 2012).

15.3 Other processes

Given the overwhelming strength of the correlation between energy input (or NPP) and regional species richness, any other trends can only be of secondary importance. Kreft and Jetz (2008) in the study shown in fig. 15.3 also found a general trend for areas with higher soil fertility or nutrient input to contain more plant species. These were minor and inconsistent effects though.

There are certainly positive relationships between species richness and landscape heterogeneity, a widespread and consistent pattern throughout the world and across taxa (Kreft and Jetz, 2008; Stein et al., 2014). The underlying processes might include an increased variety of habitats and resources, the existence of refuges from harsh environmental conditions and a greater probability of speciation in diverse and patchy environments (Stein et al., 2014). For example, bird species richness is greatest in montane regions near the equator, where habitats range from tropical rain forest in the lowlands to permanently snow-capped summits (Davies et al., 2007).

There is a continued relationship between species richness and area at the regional scale, though this is overwhelmed by other forces as equal areas can differ greatly in richness between regions. Biotic interactions however are largely irrelevant at the regional scale. This might seem surprising following the chapters on community ecology, but at the level of entire regions, coexistence of multiple species with similar niches is easy because they may never meet one another. Checkerboard patterns within regions (Section 13.2) attest to the segregation of species that might otherwise outcompete one another if placed together.

As for the postulate that dispersal might play a role, there is certainly an important effect of isolation in limiting the richness of island biotas, which will receive special attention in Chapter 19. Otherwise, however, dispersal appears to be less important in generating regional richness patterns (though see Chapter 18 for some exceptions).

In terms of historical processes, there are a few special cases where regional patterns of species richness can be unambiguously attributed to past events. One example relates to river basins, where the connections between them have changed as sea levels rose following the last ice age (see Section 17.7). At times of lower sea level, many rivers which now flow separately into the sea had a confluence on land and therefore were able to share species. This is particularly important for freshwater systems in which many species struggle to cross a saltwater barrier between river drainages. Dias et al. (2014) found that drainage basins which were connected at the last glacial maximum (LGM) had greater species richness than those which have always been isolated.

Why do some processes that are crucial at local scales cease to be important at regional scales, and *vice versa*, and where should the line between them be drawn? This link between conventional ecology and biogeography is a fertile area of study and we remain uncertain (Jenkins and Ricklefs, 2011). McGill (2010) made an effort to sketch out the distinction between the two, but this is only an outline, and many ecologists will find points to contend with in this scheme (Fig. 15.4).

Regardless of the details of Fig. 15.4, it highlights several points. Climate is important at both small scales (microclimate) and in biogeography. Species

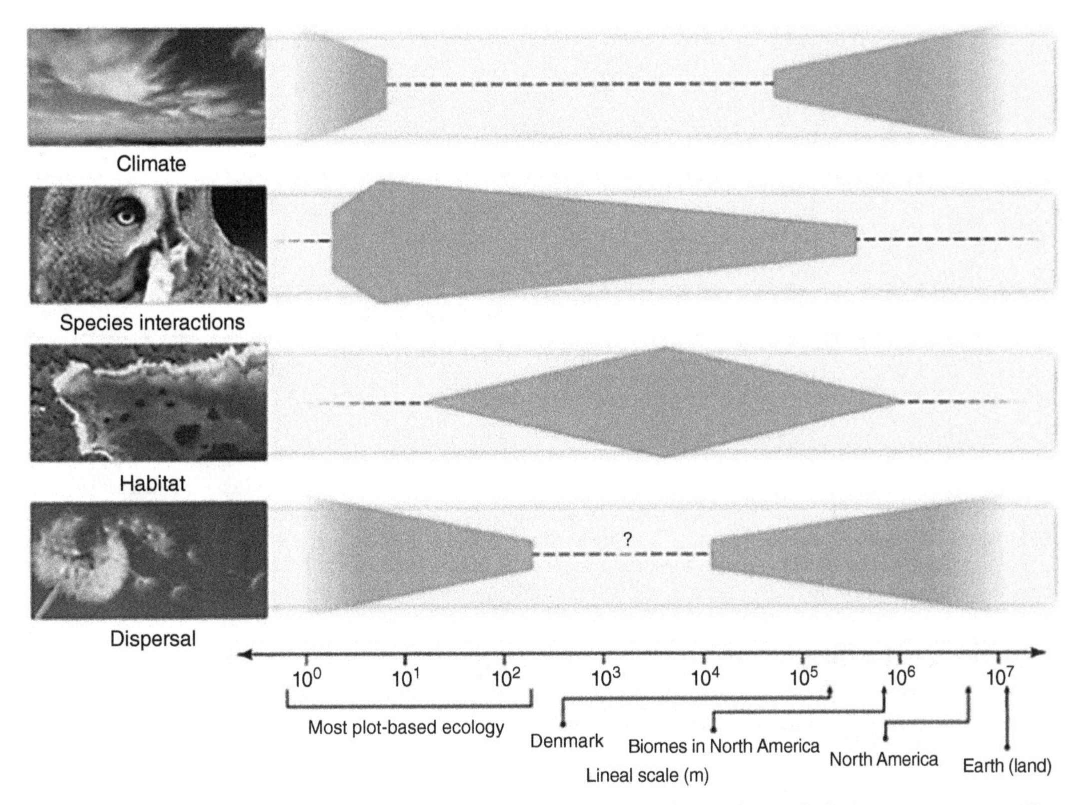

Figure 15.4 Four processes thought to influence the distribution of organisms and the scales at which they act. (*Source*: McGill (2010). Reproduced with permission from American Association for the Advancement of Science.)

interactions matter greatly at local scales then peter out in larger areas. The influence of habitat comes into play at the intermediate, metacommunity stage, when species are sorted from the regional species pool (Section 13.5). The most contentious schematic might be for dispersal, where small-scale individual movements at one end are conflated with historical trends at the other, and neglecting mass effects in the middle (though as the figure accepts, there are limited data on this). Much remains to be learnt about linkages across scales.

Out of our six starting hypotheses for regional species richness patterns, a provisional decision can now be made about which are generally supported and which can be discarded in most cases. Field et al. (2009) took an innovative approach to this problem. They chose almost 400 published studies in which at least two of the hypotheses had been compared against one another, then recorded which one 'won' or rather received the most support from the data. They then compared the number of 'wins' to what would be expected by chance given the numbers of comparisons (Fig. 15.5). Climate or productivity were the dominant correlates in 136 of 280 tests conducted, but when examining the largest grain sizes and extents, this rose to around 75%, making it the overwhelming relationship for both plants and animals. Nutrients, area and biotic interactions were sometimes favoured, but no more often than chance alone would suggest (though in fact biotic interactions were most important at intermediate grain sizes, from 10 to 500 km^2, demonstrating that study scale provides essential context). As for the effects of landscape heterogeneity or dispersal, these were actually less often the dominant relationship than expected, though still important in some cases.

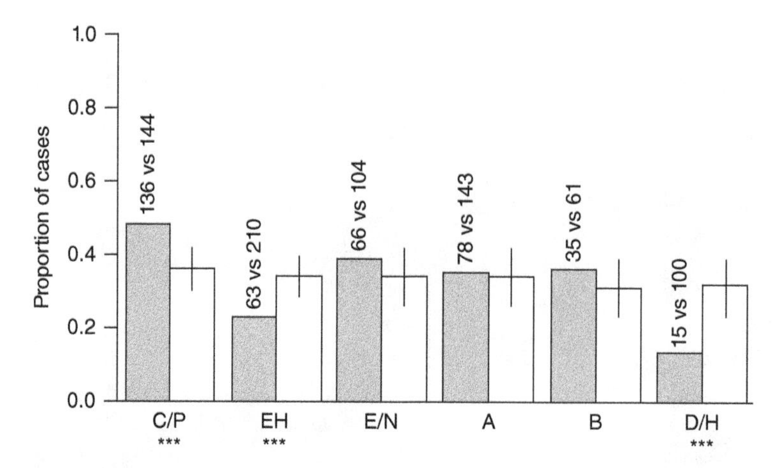

Figure 15.5 Cases in which the tested hypothesis for regional species richness variation were supported over alternatives (grey bars; numbers indicate the number of cases in which it was supported or unsupported for all tests) versus null expectation (white bars, lines show 95% boundaries from 1000 randomisations). C/P, climate or productivity; EH, environmental heterogeneity; E/N, edaphics (i.e. soil type) or nutrients; A, area; B, biotic interactions; D/H, dispersal or history. ***$P < 0.001$ for comparisons between bars within a hypothesis test. (*Source*: Field et al. (2009). Reproduced with permission from John Wiley & Sons Ltd.)

15.4 Scale and productivity

This proposed link between productivity and species richness may seem paradoxical when looking back at the failure to find any such relationship at local levels (Section 8.3). Classically it was believed that there was a hump-shaped relationship between productivity and species richness within communities (Grime, 1973), a paradigm which is now disputed (Adler et al., 2011), yet here the evidence points towards a linear positive relationship at regional scales. How can this be the case? The first point to note is that the regional relationship averages across responses within areas, which means that patterns at the two scales might operate independently (Fig. 15.6).

This exact pattern was found in a study by Chase and Leibold (2002, Fig. 15.7). They assessed species richness in ponds of both producers (vascular plants and algae) and benthic fauna. They then examined the patterns that were obtained when comparing between ponds in the same watershed (local scale) and between watersheds (regional scale). For both plants and animals the same pattern emerges; the relationship between productivity and species richness differs depending on the scale at which it is studied (Pianka, 1966; Waide et al., 1999).

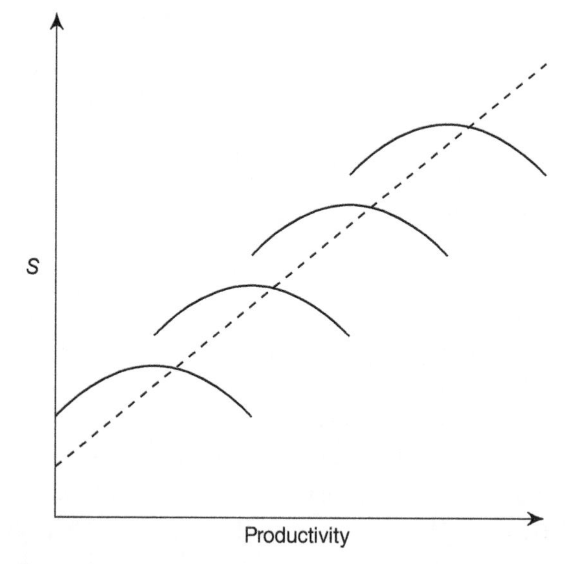

Figure 15.6 Hypothesised difference in the relationship between productivity and species richness *S* at local scales (solid lines) and regional scales (dotted line).

The reason for this emerges from examining β diversity (Fig. 15.8), which shows a positive relationship with regional productivity. Note that the three scales of diversity (α, β and γ); see Section 5.3 are not necessarily positively correlated. An increase in γ diversity (i.e. at the regional scale) can arise

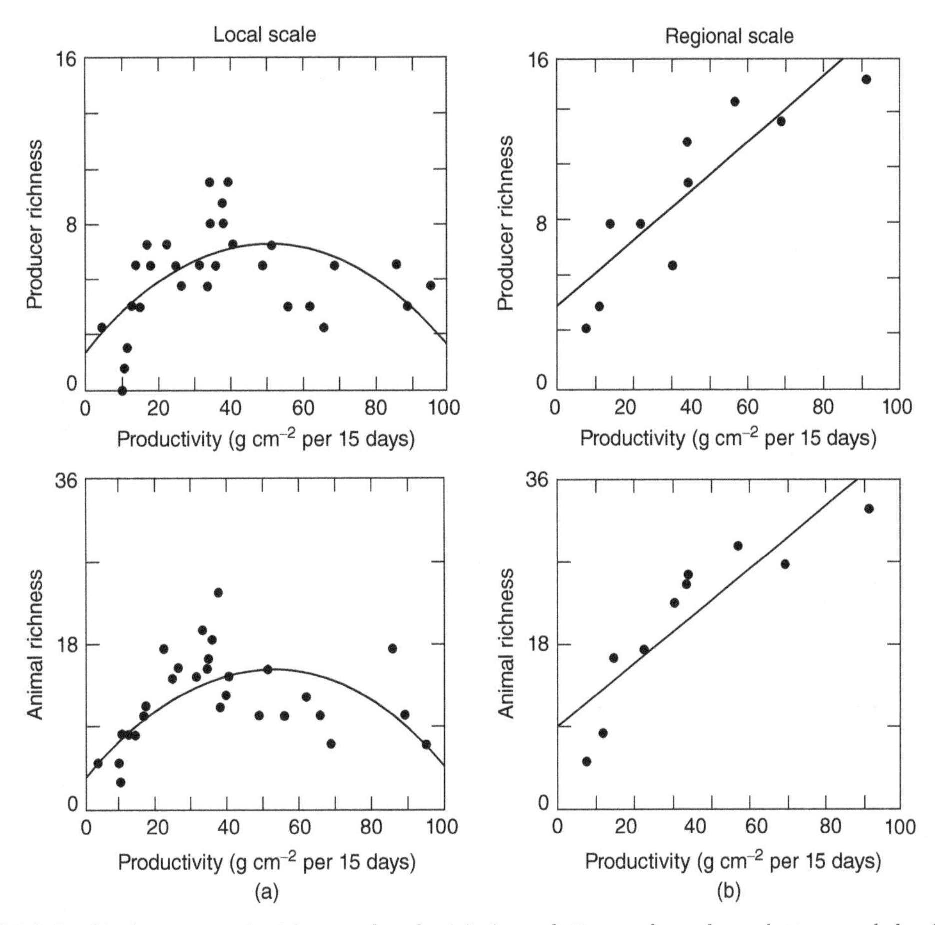

Figure 15.7 Relationships between species richness and productivity in ponds. Top panels, producers; bottom panels, benthic animals. (a) Local scale, within ponds; (b) regional scale, between watersheds. (*Source*: Chase and Leibold (2002). Reproduced with permission from Nature Publishing Group.)

due to an increase in either α or β diversity. In fact, as shown in Table 15.1, an increase in γ diversity can even occur with a simultaneous fall in α. In the first row there is high α diversity but low β; in the second the opposite. This reinforces a lesson from Chapter 5—high local species richness does not automatically imply high regional richness if the same species occur everywhere.

Is variation in regional species richness around the world composed of trends in α diversity, β diversity or a combination of both? This was investigated in marine epifauna by Witman et al. (2004), who recorded the animals attached to vertical rock walls in a range of locations using photographic transects. There were 12 survey locations from 63°N to 62°S. An order of magnitude difference in estimated species richness was observed between the lowest (Gulf of Maine, 28.2 species) and highest (Palau, 303.7 species) diversity sites. In general, greater α species richness was found in tropical samples than temperate.

There was also a strong correlation between local and regional species richness, accounting for 73–76% of the variation among samples (Fig. 15.9a). This recapitulates the pattern seen in Section 7.3. More important for now is to examine the trends in β diversity, here represented as the fraction of the regional species pool found in any one sample

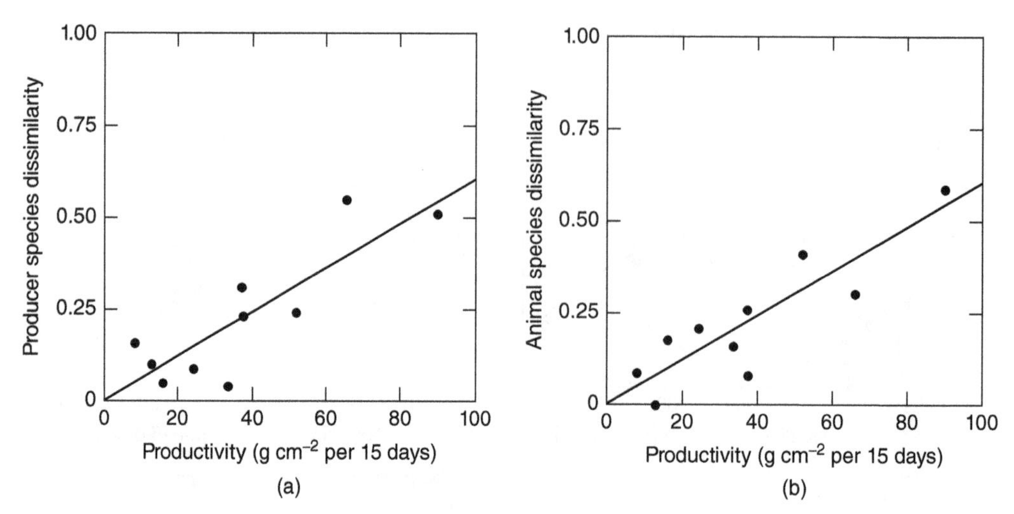

Figure 15.8 Dissimilarity in species composition (β diversity) between the ponds within each watershed for (a) producers and (b) benthic animals measured as 1—Jaccard's index of similarity. (*Source*: Chase and Leibold (2002). Reproduced with permission from Nature Publishing Group.)

Table 15.1 Local (α) and regional (γ) richness for eight putative species (A–H) depending on β diversity.

Patch 1	Patch 2	Regional *S*
A B C D E	A B C D E	5
A B C D	E F G H	8

(Fig. 15.9b). This is lowest in the tropics, where only 50% of available species are found in any given place, whereas closer to the poles it rises to 80%. Not only are the tropics richer in species, but the turnover among locations is greater too. Both α and β diversity are higher.

Intriguingly, it seems that community assembly might work differently in regions with higher productivity. Returning to the experimental pools set up by Chase and Leibold, ponds with higher productivity became more varied (i.e. they had higher β diversity). This occurred due to a shift from deterministic community assembly with limited resources (dominance control) to increased stochasticity resulting in multiple stable states—founder control (Fig. 15.10; Chase, 2010). This was despite the α species richness of both producers and animals remaining approximately the same in all ponds. Instead, regional γ richness increased by 2–3 times with high productivity. Only a subset of the total number of species were able to survive in low-productivity pools.

This is an interesting phenomenon, but at present there are too few studies to determine whether it is a general one across natural systems. In addition there is no accepted mechanism. Further studies, both experimental and theoretical, will be required.

15.5 Latitudinal gradients

This leads on to one of the most striking and widespread patterns in biogeography, the latitudinal trend in species richness (often shortened to LDG for latitudinal diversity gradient). Put simply, there are not merely more species in the tropics but vastly more. This was first documented by von Humboldt (1808), and in the 200 years since, ever more examples have accumulated (Hillebrand, 2004). It occurs throughout the world, in all regions, for all forms of life, encompassing terrestrial, freshwater and marine systems. It can be detected at scales from local to global and is found irrespective of whether all taxa are counted, solely endemic species or taxonomic levels from species up to families. It is also clearly not the result of sampling bias,

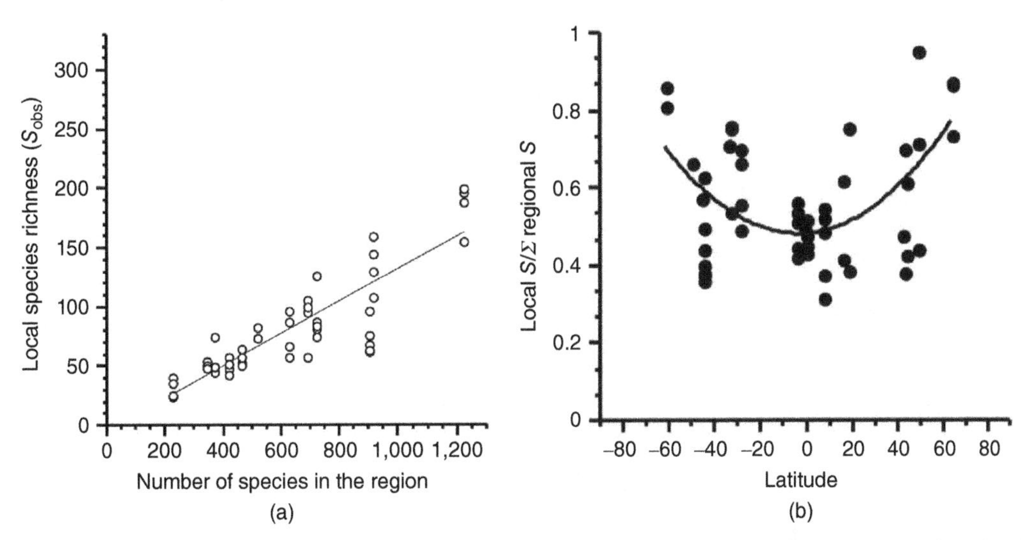

Figure 15.9 (a) Local versus regional species richness of marine epifauna and (b) ratio between local and regional species richness versus latitude. (*Source*: Witman et al. (2004). Copyright (2004) National Academy of Sciences, USA.)

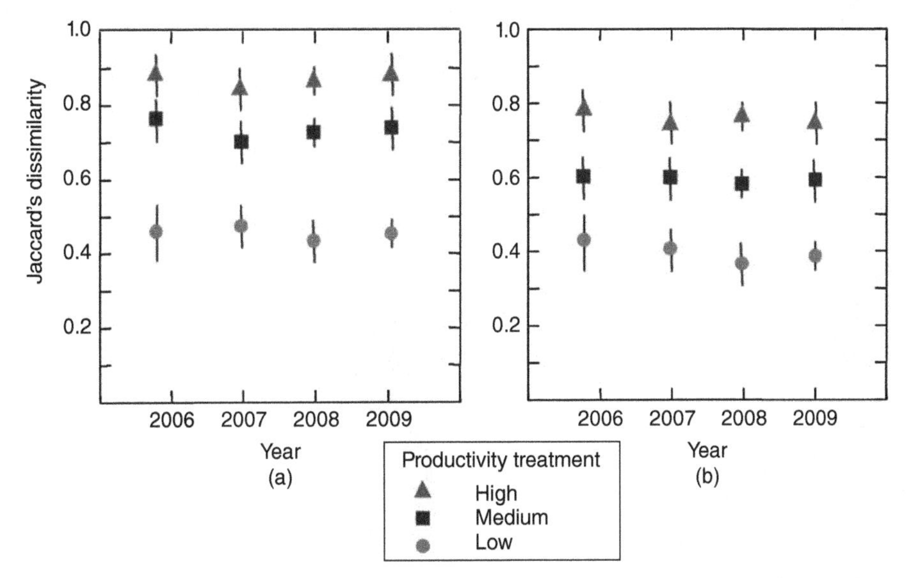

Figure 15.10 Average dissimilarity of (a) producers and (b) animals among experimental ponds within three productivity treatments across four years measured as Jaccard's index, means ± SE. (*Source*: Chase (2010). Reproduced with permission from American Association for the Advancement of Science.)

because the greatest number of species is found in the tropical areas which have been sampled the least, whereas the well-described temperate zone contains markedly fewer.

A deep cause must underlie this pattern since it is so universal, applying to almost every clade that has been studied (with a few telling exceptions). numbers of ethnic groups and languages peaking in the tropics (Gorenflo et al., 2012; Gavin et al., 2013). The gradient is not always perfectly symmetrical (e.g. Dunn et al., 2009), nor does the peak lie exactly on the equator (Powell et al., 2012), but the highest richness normally occurs somewhere within the tropics (Fig. 15.11).

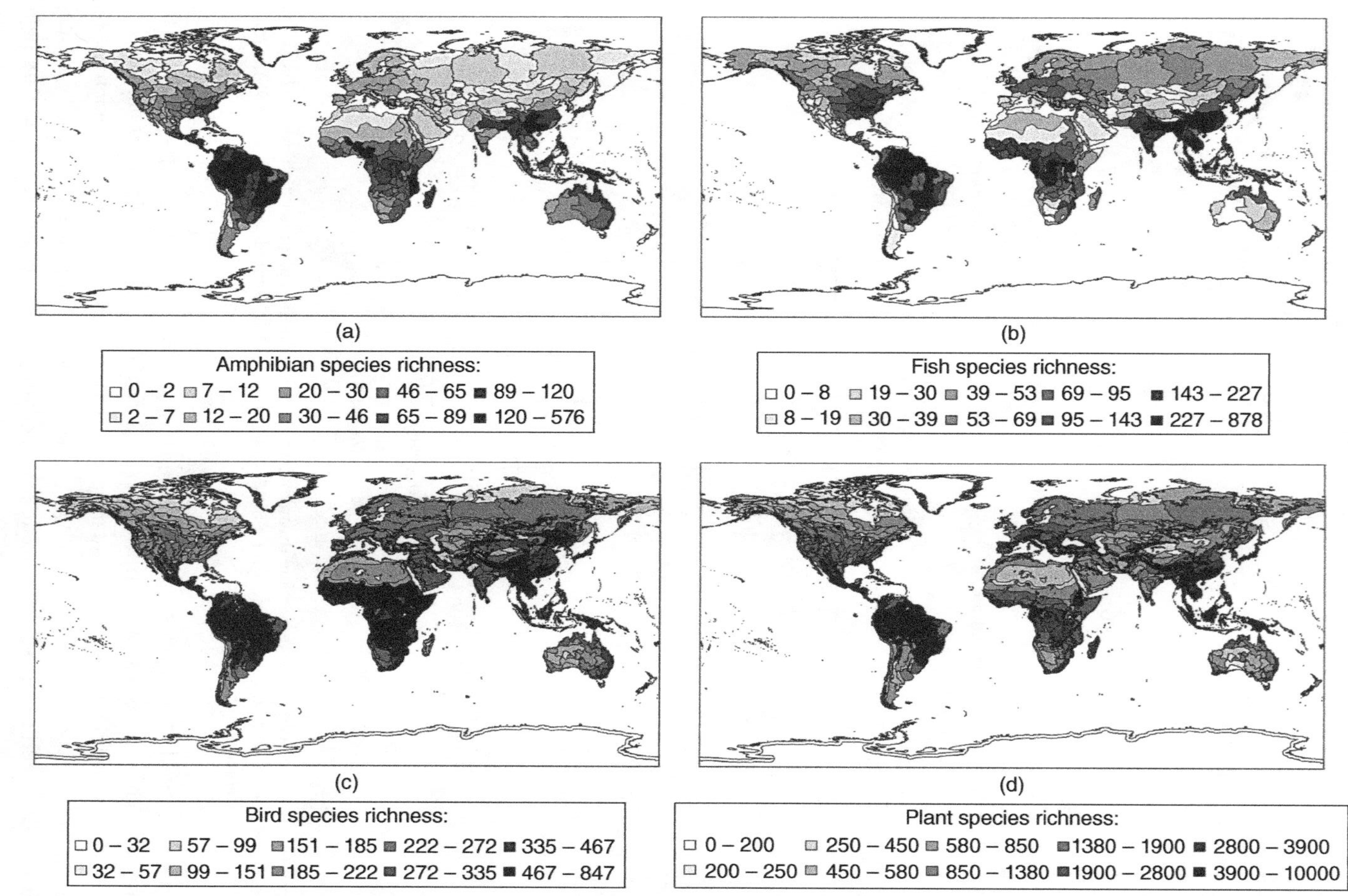

Figure 15.11 Global patterns in species richness of (a) amphibians and (b) fish in freshwater ecoregions; (c) birds and (d) plants in terrestrial ecoregions. Ecoregions are ranked by species richness and split into equal-sized categories. (*Source*: Data provided by The Nature Conservancy; figure prepared with help from Tim Newbold.)

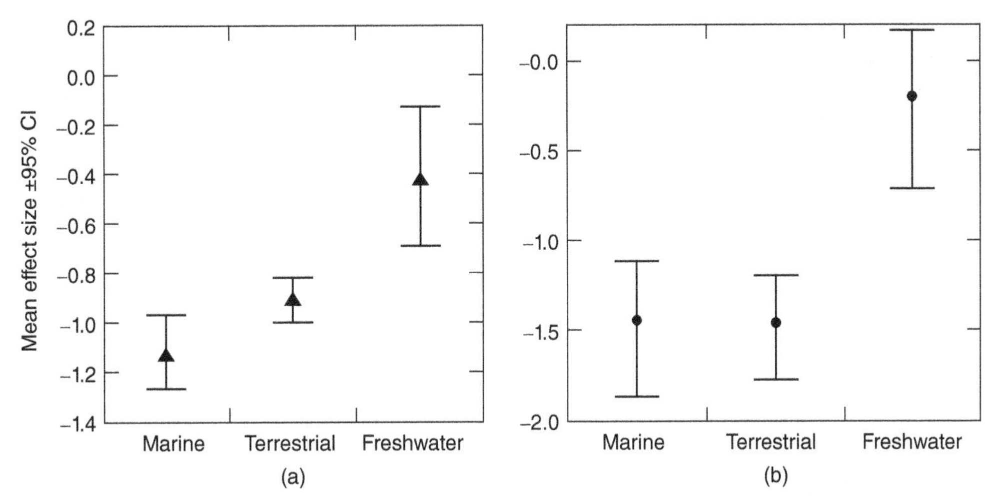

Figure 15.12 (a) Strength and (b) slope of c. 600 latitudinal gradients in species richness. (*Source*: Hillebrand (2004). Reproduced with permission from University of Chicago Press.)

A vast collection of gradients was assembled by Hillebrand (2004), who compared the LDGs exhibited by different environments. He found that the strength of the relationship, in other words its explanatory power, was greatest in marine and terrestrial systems and weakest in freshwater (Fig. 15.12). The slope of the line was also steepest in marine and terrestrial environments and, while negative for freshwater, was not significantly different from zero. Once again freshwater systems are peculiar (see Section 14.7), but the overall pattern is robust elsewhere.

In general Hillebrand (2004) found that LDGs measured at regional scales were stronger than those at local scales. The pattern was usually symmetrical on both sides of the equator. Its strength increased with trophic level and body mass, and larger species also tended to show steeper slopes.

Examples of the LDG even occur in the fossil record (Mittelbach et al., 2007; Alroy et al., 2008), though these mostly arose or at least strengthened within the last 30 My (Mannion et al., 2014). There are also counterexamples. As will be discussed in the next chapter, dinosaur richness peaked outside the tropics (Mannion et al., 2012), and it may be that equatorial peaks only occur during relatively cold periods of earth history, that is, when the poles are covered in ice (Mannion et al., 2014).

The two powerful trends seen so far, the link between climate and species richness, and the LDG, must be connected somehow, because equatorial regions are warmer and have higher NPP.

15.6 Centres of origin

No trend is without its exceptions, and these can often be revealing when attempting to unveil the underlying causes. In the oceans another pattern interferes with the LDG. There are three areas, referred to as centres of origin, which act as sources of the species found elsewhere (Briggs, 2003). These occur in the East Indies Triangle, the cool North Pacific and the cold Antarctic. All three possess higher than expected species richness and buck the trend of the LDG.

Their influence is pervasive; assessing reef fish in 70 locations throughout the Indian and Pacific Oceans, spanning two thirds of the circumference of the world, Mora et al. (2003) found that 86% of the species in all sites were also found in the East Indies. Species richness fell with distance from the East Indies (Fig. 15.13). This implies that the majority of species are found there and have spread out to as far as the east coast of Africa and the western shores of America. Any explanation for global patterns in richness must be able to build in these trends as well.

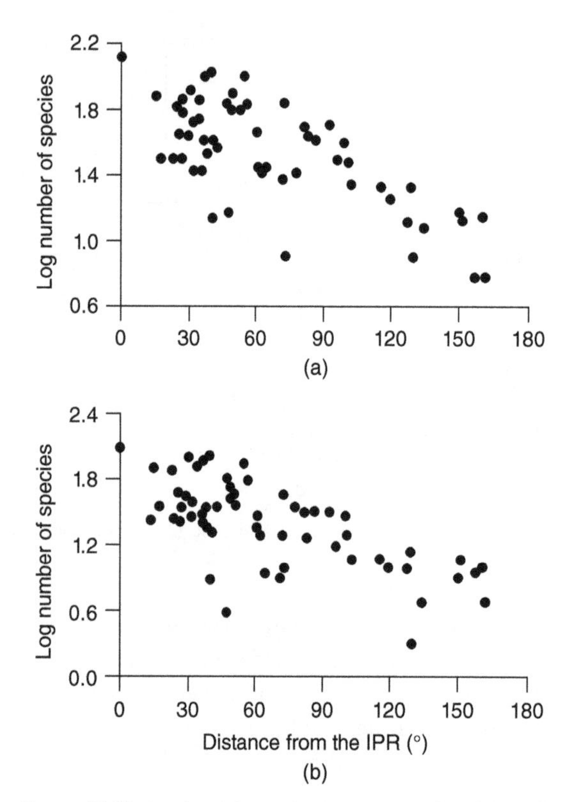

Figure 15.13 Species richness for two groups of reef fish, (a) Labridae and (b) Pomacentridae, with increasing distance from the Indonesian and Philippine region (IPR) in the East Indies. (*Source*: Mora et al. (2003). Reproduced with permission from Nature Publishing Group.)

The idea that centres of origin covering many different taxa also exist on land has fallen out of favour in recent years, although there is a general trend for any particular group to be most diverse in the area where it first arose. It may be that a consistent set of processes account for marine patterns as well, but at present the centres of origin remain puzzling anomalies.

15.7 Regional species–area relationships

Back when considering patterns of local species richness, a universal law was recognised, known as the Species–Area Relationship (SAR; Fig. 7.1). This gives a straight line when plotted on logarithmic axes. At larger scales, however, something strange happens, and the line begins to curve upwards (Fig. 15.14). This holds true across five continents for amphibians, birds and mammals. Note that this method incorporates all species found within the area shown on the x-axis, regardless of whether they are found elsewhere. If the number of endemics is taken instead, that is, the species only found within a given area, the straight line is recovered (Storch et al., 2012). This latter relationship is therefore known as the endemics–area relationship (EAR).

On one level this is an encouraging finding, as it suggests that a general law applies across many taxa in all parts of the world, with great potential for making predictions for unstudied taxa or areas. As for why the SAR and EAR differ in shape, the reasons are complex and remain a subject for further investigation.

15.8 Confounding effects

Never forget that documented patterns of diversity are always best guesses. No single location can ever be completely sampled, let alone a large array of sites. Sometimes apparent trends are caused by a confounding effect. For example, more than 50% of variation in the global diversity of monogonont rotifers (tiny aquatic animals) can be explained by the distribution of specialist taxonomists able to identify them, completely obscuring other potential drivers (Fontaneto et al., 2012). These problems tend only to occur when examining restricted taxonomic groups and should not distract from the consistent overall pattern.

Another issue worth bearing in mind is that all trends discussed so far have referred solely to species richness, not any of the other elements of diversity outlined in Section 5.7. To some extent this is due to the available data; lists of species for regions are easy to obtain, whereas data on relative abundances, functional traits and phylogenies are seldom collected on global scales. Stuart-Smith et al. (2013) examined patterns of reef fish diversity, discovering that although species richness peaks in the

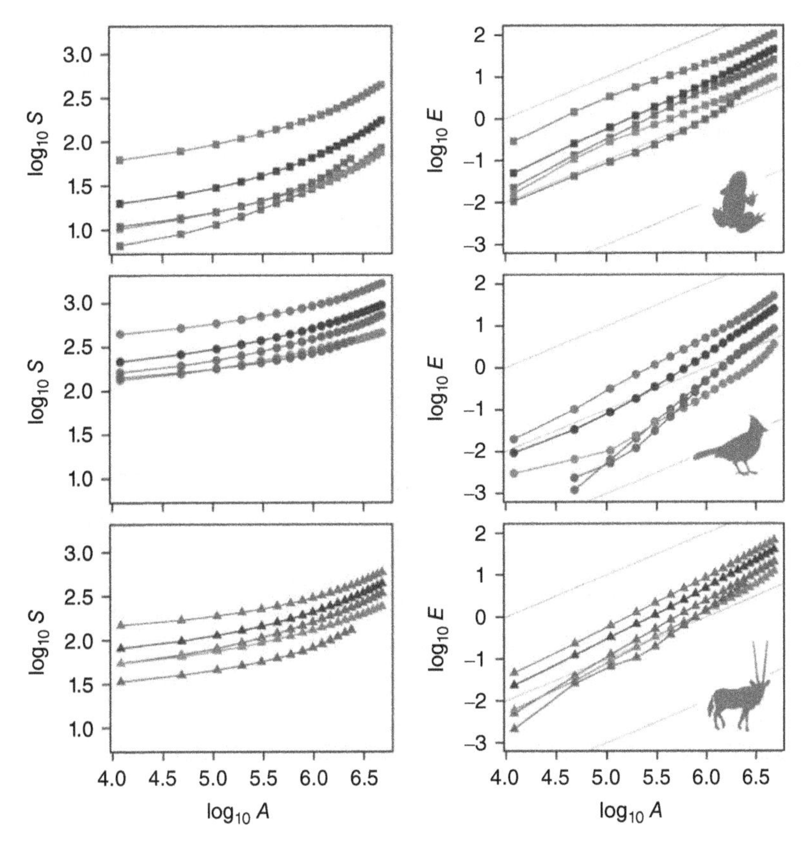

Figure 15.14 Species- and Endemics-Area Relationships (SARs and EARs) for amphibians (squares), birds (circles) and mammals (triangles). Data from five continents (blue: Africa, red: Eurasia, yellow: North America, purple: South America, green: Australia). Grey lines correspond to a power law with a slope of 1; that is, proportionality between area and the number of species. S is the mean number of species, E is the mean number of endemics, A is the area in km^2. (*Source*: Storch et al. (2012). Reproduced with permission from Nature Publishing Group.) *(See colour plate section for the colour representation of this figure.)*

tropics, species evenness is in fact greater at higher latitudes, while hotspots for functional diversity are spread irregularly throughout the oceans. How other diversity metrics vary globally, and the implications for ecosystem processes, remains uncertain.

15.9 Conclusions

Having found in Section 7.3 that regional species richness is the main determinant of local patterns, this merely moved the puzzle up to a higher level. There is pronounced variation in the number of species found in different regions of the world. The main correlate of this is temperature, which together with water availability (and their interaction) accounts for the majority of inter-regional variation. These environmental factors also drive NPP, giving rise to a strong positive correlation between regional species richness and productivity which was not found at local scales. Other trends are relatively minor in explanatory power.

Species richness peaks in the tropics, and the decline seen with distance from the equator is known as the Latitudinal Diversity Gradient. It occurs on both land and in the sea, though is weaker for freshwater systems. There can also be modulating factors, and in the oceans the centres of origin are outliers to the general trend. There must be some

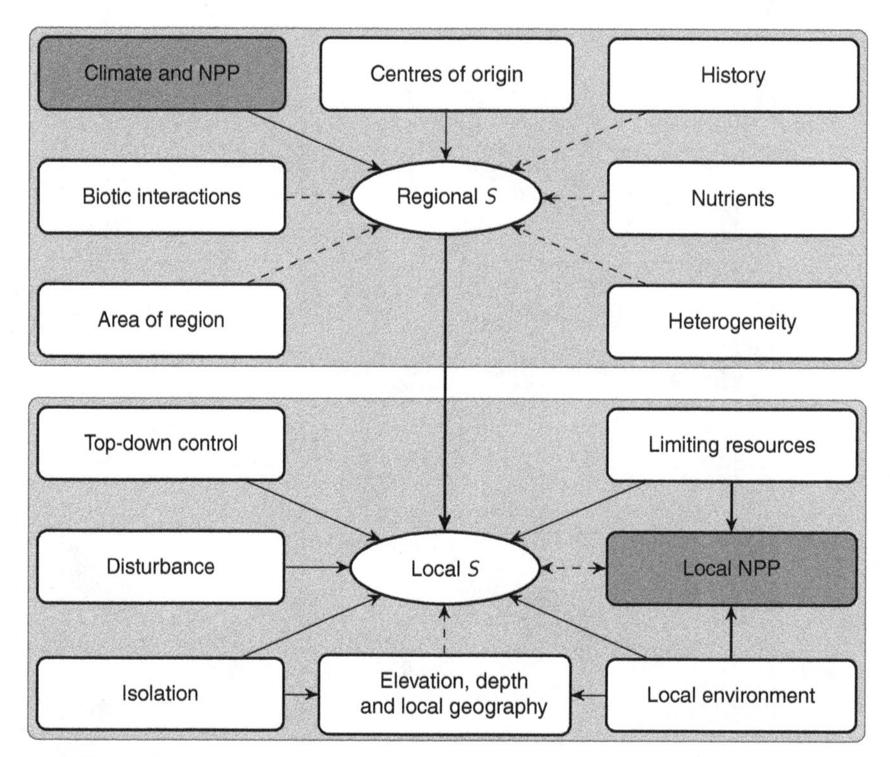

Figure 15.15 Determinants of regional and local species richness S for a given area. Direct links in bold, indirect with dashed lines. Dark grey is used to indicate the distinction between productivity at regional and local scales.

connection among climate, latitude and species richness, though the processes have yet to be accounted for in our scheme (Fig. 15.15) and will be the focus of the next chapter.

15.9.1 Recommended reading

Gaston, K. J., 2000. Global patterns in biodiversity. *Nature* 405:220–227.

Hillebrand, H., 2004. On the generality of the latitudinal diversity gradient. *American Naturalist* 163:192–211.

McGill, B. J., 2010. Matters of scale. *Science* 328:575–576.

15.9.2 Questions for the future

- Why are some ecological processes influential at local scales, while others at large scales, and how do we set the boundaries?
- What mechanism(s) underlie the strong positive correlation between available energy and rainfall and species richness?
- Are apparent centres of origin genuinely sources of species or do they reflect overlap between regions?

References

Adams, J. M. and F. I. Woodward, 1989. Patterns in tree species richness as a test of the glacial extinction hypothesis. *Nature* 339:699–701.

Adler, P. B., E. W. Seabloom, E. T. Borer, H. Hillebrand, Y. Hautier, A. Hector, W. S. Harpole, L. R. O'Halloran, J. B. Grace, T. M. Anderson, J. D. Bakker, L. A. Biederman, C. S. Brown, Y. M. Buckley, L. B. Calabrese, C.-J. Chu, E. E. Cleland, S. L. Collins, K. L. Cottingham, M. J. Crawley, E. I. Damschen, K. F. Davies, N. M. DeCrappeo, P. A. Fay, J. Firn, P. Frater, E. I. Gasarch, D. S. Gruner, N. Hagenah, J. H. R. Lambers, H. Humphries, V. L. Jin, A. D. Kay, K. P. Kirkman, J. A. Klein, J. M. H. Knops, K. J. L. Pierre,

J. G. Lambrinos, W. Li, A. S. MacDougall, R. L. McCulley, B. A. Melbourne, C. E. Mitchell, J. L. Moore, J. W. Morgan, B. Mortensen, J. L. Orrock, S. M. Prober, D. A. Pyke, A. C. Risch, M. Schuetz, M. D. Smith, C. J. Stevens, L. L. Sullivan, G. Wang, P. D. Wragg, J. P. Wright, and L. H. Yang, 2011. Productivity is a poor predictor of plant species richness. *Science* 333:1750–1753.

Alroy, J., M. Aberhan, D. J. Bottjer, M. Foote, F. T. Fürsich, P. J. Harries, A. J. W. Hendy, S. M. Holland, L. C. Ivany, W. Kiessling, M. A. Kosnik, C. R. Marshall, A. J. McGowan, A. I. Miller, T. D. Olszewski, M. E. Patzkowsky, S. E. Peters, L. Villier, P. J. Wagner, N. Bonuso, P. S. Borkow, B. Brenneis, M. E. Clapham, L. M. Fall, C. A. Ferguson, V. L. Hanson, A. Z. Krug, K. M. Layou, E. H. Leckey, S. Nürnberg, C. M. Powers, J. A. Sessa, C. Simpson, A. Tomašových, and C. C. Visaggi, 2008. Phanerozoic trends in the global diversity of marine invertebrates. *Science* 321:97–100.

Boucher-Lalonde, V., J. T. Kerr, and D. J. Currie, 2014. Does climate limit species richness by limiting individual species' ranges? *Proceedings of the Royal Society Series B* 281:20132695.

Briggs, J. C., 2003. Marine centres of origin as evolutionary engines. *Journal of Biogeography* 30:1–18.

Buckley, L. B., A. H. Hurlbert, and W. Jetz, 2012. Broad-scale ecological implications of ectothermy and endothermy in changing environments. *Global Ecology and Biogeography* 21:873–885.

Chase, J. M., 2010. Stochastic community assembly causes higher biodiversity in more productive environments. *Science* 328:1388–1391.

Chase, J. M. and M. A. Leibold, 2002. Spatial scale dictates the productivity-biodiversity relationship. *Nature* 416:427–430.

Davies, R. G., C. D. L. Orme, D. Storch, V. A. Olson, G. H. Thomas, S. G. Ross, T.-S. Ding, P. C. Rasmussen, P. M. Bennett, I. P. F. Owens, T. M. Blackburn, and K. J. Gaston, 2007. Topography, energy and the global distribution of bird species richness. *Proceedings of the Royal Society Series B* 274:1189–1197.

Dias, M. S., T. Oberdorff, B. Hugueny, F. Leprieur, C. Jézéquel, J.-F. Cornu, S. Brosse, G. Grenouillet, and P. A. Tedesco, 2014. Global imprint of historical connectivity on freshwater fish biodiversity. *Ecology Letters* 17:1130–1140.

Dunn, R. R., D. Agosti, A. N. Andersen, X. Arnan, C. A. Bruhl, X. Cerdá, A. M. Ellison, B. L. Fisher, M. C. Fitzpatrick, H. Gibb, N. J. Gotelli, A. D. Gove, B. Guenard, M. Janda, M. Kaspari, E. J. Laurent, J.-P. Lessard, J. T. Longino, J. D. Majer, S. B. Menke, T. P. McGlynn, C. L. Parr, S. M. Philpott, M. Pfeiffer, J. Retana, A. V. Suarez, H. L. Vasconcelos, M. D. Weiser, and N. J. Sanders, 2009. Climatic drivers of hemispheric asymmetry in global patterns of ant species richness. *Ecology Letters* 12:324–333.

Evans, K. L., P. H. Warren, and K. J. Gaston, 2005. Species-energy relationships at the macroecological scale: a review of the mechanisms. *Biological Reviews* 80:1–25.

Field, R., B. A. Hawkins, H. V. Cornell, D. J. Currie, J. A. F. Diniz-Filho, J.-F. Guégan, D. M. Kaufman, J. T. Kerr, G. G. Mittelbach, T. Oberdorff, E. M. O'Brien, and J. R. G. Turner, 2009. Spatial species-richness gradients across scales: a meta-analysis. *Journal of Biogeography* 36:132–147.

Fontaneto, D., A. M. Barbosa, H. Segers, and M. Pautasso, 2012. The 'rotiferologist' effect and other global correlates of species richness in monogonont rotifers. *Ecography* 35:174–182.

Francis, A. P. and D. J. Currie, 2003. A globally-consistent richness-climate relationship for angiosperms. *American Naturalist* 161:523–536.

Gaston, K. J., 2000. Global patterns in biodiversity. *Nature* 405:220–227.

Gavin, M. C., C. A. Botero, C. Bowern, R. K. Colwell, M. Dunn, R. R. Dunn, R. D. Gray, K. R. Kirby, J. McCarter, A. Powell, T. F. Rangel, J. R. Stepp, M. Trautwein, J. L. Verdolin, and G. Yanega, 2013. Toward a mechanistic understanding of linguistic diversity. *Bioscience* 63: 524–535.

Gorenflo, L. J., S. Romaine, R. A. Mittermeier, and K. Walker-Painemilla, 2012. Co-occurrence of linguistic and biological diversity in biodiversity hotspots and high biodiversity wilderness areas. *Proceedings of the National Academy of Sciences of the United States of America* 109:8032–8037.

Grime, J. P., 1973. Control of species density on herbaceous vegetation. *Journal of Environmental Management* 1:151–167.

Hawkins, B. A., R. Field, H. V. Cornell, D. J. Currie, J.-F. Guègan, D. M. Kaufman, J. T. Kerr, G. G. Mittelbach, T. Oberdorff, E. M. O'Brien, E. E. Porter, and J. R. G. Turner, 2003. Energy, water, and broad-scale geographic patterns of species richness. *Ecology* 84:3105–3117.

Hillebrand, H., 2004. On the generality of the latitudinal diversity gradient. *American Naturalist* 163:192–211.

von Humboldt, A., 1808. *Ansichten de Natur mit Wissenschaflichen Erlauterungen*. J. G. Cotta.

Jenkins, D. G. and R. E. Ricklefs, 2011. Biogeography and ecology: two views of one world. *Philosophical Transactions of the Royal Society* 366:2331–2335.

Jetz, W., H. Kreft, G. Ceballos, and J. Mutke, 2009. Global associations between terrestrial producer and vertebrate consumer diversity. *Proceedings of the Royal Society Series B* 276:269–278.

Kreft, H. and W. Jetz, 2008. Global patterns and determinants of vascular plant diversity. *Proceedings of the National Academy of Sciences of the United States of America* 104:5925–5930.

Mannion, P. D., R. B. J. Benson, P. Upchurch, R. J. Butler, M. T. Carrano, and P. M. Barrett, 2012. A temperate palaeodiversity peak in Mesozoic dinosaurs and evidence for Late Cretaceous geographical partitioning. *Global Ecology and Biogeography* 21:898–908.

Mannion, P. D., P. Upchurch, R. B. J. Benson, and A. Goswani, 2014. The latitudinal biodiversity gradient through deep time. *Trends in Ecology & Evolution* 29:42–50.

McGill, B. J., 2010. Matters of scale. *Science* 328:575–576.

Medellín, R. A. and J. Soberón, 1999. Predictions of mammal diversity on four land masses. *Conservation Biology* 13:143–149.

Mittelbach, G. G., D. W. Schemske, H. V. Cornell, A. P. Allen, J. M. Brown, M. B. Bush, S. P. Harrison, A. H. Hurlbert, N. Knowlton, H. A. Lessios, C. M. McCain, A. R. McCune, L. A. McDade, M. A. McPeek, T. J. Near, T. D. Price, R. E. Ricklefs, K. Roy, D. F. Sax, D. Schluter, J. M. Sobel, and M. Turelli, 2007. Evolution and the latitudinal diversity gradient: speciation, extinction and biogeography. *Ecology Letters* 10:315–331.

Mora, C., P. M. Chittaro, P. F. Sale, J. P. Kritzer, and S. A. Ludsin, 2003. Patterns and processes in reef fish diversity. *Nature* 421:933–936.

Pärtel, M., 2014. Community ecology of absent species: hidden and dark diversity. *Journal of Vegetation Science* 25:1154–1159.

Pianka, E. R., 1966. Latitudinal gradients in species diversity: a review of concepts. *American Naturalist* 100:33–46.

Powell, M. G., V. P. Beresford, and B. A. Colaianne, 2012. The latitudinal position of peak marine diversity in living and fossil biotas. *Journal of Biogeography* 39:1687–1694.

Qian, H. and R. E. Ricklefs, 2008. Global concordance in diversity patterns of vascular plants and terrestrial vertebrates. *Ecology Letters* 11:547–553.

Stein, M., K. Gerstner, and H. Kreft, 2014. Environmental heterogeneity as a universal driver of species richness across taxa, biomes and spatial scales. *Ecology Letters* 17:866–880.

Storch, D., R. G. Davies, S. Zajíček, C. D. L. Orme, V. Olson, G. H. Thomas, T.-S. Ding, P. C. Rasmussen, R. S. Ridgely, P. M. Bennett, T. M. Blackburn, I. P. F. Owens, and K. J. Gaston, 2006. Energy, range dynamics and global species richness patterns: reconciling mid-domain effects and environmental determinants of avian diversity. *Ecology Letters* 9:1308–1320.

Storch, D., P. Keil, and W. Jetz, 2012. Universal species-area and endemics-area relationships at continental scales. *Nature* 488:78–81.

Stuart-Smith, R. D., A. E. Bates, J. S. Lefcheck, J. E. Duffy, S. C. Baker, R. J. Thomson, J. F. Stuart-Smith, N. A. Hill, S. J. Kininmonth, L. Airoldi, M. A. Becerro, S. J. Campbell, T. P. Dawson, S. A. Navarette, G. A. Soler, E. M. A. Strain, T. J. Willis, and G. J. Edgar, 2013. Integrating abundance and functional traits reveals new global hotspots of fish diversity. *Nature* 501:539–542.

Tittensor, D. P., C. Mora, W. Jetz, H. K. Lotze, D. Ricard, E. V. Berghe, and B. Worm, 2010. Global patterns and predictors of marine biodiversity across taxa. *Nature* 466:1098–1101.

Vázquez-Rivera, H. and D. J. Currie, 2014. Contemporaneous climate directly controls broad-scale patterns of woody plant diversity: a test by a natural experiment over 14,000 years. *Global Ecology and Biogeography* 24:97–106.

Waide, R. B., M. R. Willig, C. F. Steiner, G. Mittelbach, L. Gough, S. I. Dodson, G. P. Juday, and R. Parmenter, 1999. The relationship between productivity and species richness. *Annual Review of Ecology, Evolution, and Systematics* 30:257–300.

Wang, Z. H., J. H. Brown, Z. Y. Tang, and J. Y. Fang, 2009. Temperature dependence, spatial scale, and tree species diversity in eastern Asia and North America. *Proceedings of the National Academy of Sciences of the United States of America* 106:13388–13392.

Whittaker, R. J., K. J. Willis, and R. Field, 2001. Scale and species richness: towards a general, hierarchical theory of species diversity. *Journal of Biogeography* 28: 453–470.

Witman, J. D., R. J. Etter, and F. Smith, 2004. The relationship between regional and local species diversity in marine benthic communities: a global perspective. *Proceedings of the National Academy of Sciences of the United States of America* 101:15664–15669.

CHAPTER 16

Latitudinal gradients

16.1 The big question

So far we have found that local species richness is strongly linked to patterns at the regional scale, then discovered that regional richness correlates with climate and productivity. This leads to the latitudinal diversity gradient (LDG), which peaks in the tropics. Thus far in our account, however, no satisfactory mechanism accounting for these trends in diversity has been identified. Latitude is merely a measure of position on the Earth and cannot in itself be a cause of anything. The positive relationship between climate (particularly measures of energy and water availability) and species richness across many forms of life is strong, but correlative approaches are unsatisfactory if we are to fully understand natural systems. Ultimately we need a mechanism.

Numerous theories have been advanced, with varying degrees of supporting evidence. The fundamental issues are where species come from, and where they go; in other words what drives the processes of speciation, extinction and dispersal. How are these influenced by climate and earth history? The first step towards understanding global patterns of diversity is to recognise that it is the temperate zones, rather than the tropics, that are unusual. The question we should be asking is not why there are so many species in the tropics, but rather why there are fewer species elsewhere. What limits species richness?

16.2 Hypotheses

A large variety of hypotheses have been advanced to explain the LDG. The top six, at least in terms of plausibility and generality, and in no particular order, are as follows:

- Area effects
- Climatic stability
- Productivity
- Niche sizes
- Evolutionary speed
- Out of the tropics

Each of these will be evaluated in turn. There are others, and several comprehensive reviews have considered them in more detail than is possible here (e.g. Willig et al., 2003; Mittelbach et al., 2007). Some might apply to particular groups and not others. Remember that the crucial question is how each proposed mechanism influences speciation, extinction and dispersal. Correlative explanations for the LDG are not satisfactory—we cannot pass the buck any further having reached the scale of the entire planet.

16.3 Geographic area

The tropics are defined as falling between approximately 23.4°N and 23.4°S of the equator. The logic of the geographic area hypothesis for greater tropical species richness is straightforward: the

Natural Systems: The organisation of life, First Edition. Markus P. Eichhorn.

tropics are larger. A greater number of species might occur in this climatic zone purely as a result of the species–area relationship. If you're unconvinced that the tropics are indeed much larger than other climate zones, note that conventional rectangular projections of the globe stretch out the poles and compress tropical regions. On some maps Greenland appears almost as large as the entire continent of Africa; in fact it is around one fifteenth the size (2 million vs 30 million km^2). Moreover, the tropics form a contiguous single belt, whereas the other climatic zones are split in two with a region on each hemisphere. It is not worth getting mired in the dispute over what is an appropriate map projection; suffice to say that the standard Mercator is misleading, and the tropics include around 40% of the surface area of the world (though due to the distribution of the continents, there is actually more land in the northern hemisphere—see Fig. 16.4).

The tropics have also historically been larger, with tropical-type climates prevailing over much of the planet for most of the history of multicellular life. In the Palæogene there was a warm period from 59 to 50 Mya known as the Eocene Thermal Maximum. At this point the entire world was substantially warmer, such that tropical biomes were present throughout almost the entire globe, with rain forests present as far north as present-day London and the American Midwest. In the oceans warm surface waters occurred even in the Arctic, which was entirely ice-free. During this period the latitudinal gradient in temperature was much weaker, and the climate was overall less seasonal. The gradual cooling trend which brought us to the present day began around 45 Mya (Zachos et al., 2001). Due to the vagaries of preservation, it's difficult to reconstruct ancient diversity patterns with any great resolution, but evidence from dinosaurs suggests that their richness peaked at midlatitudes, corresponding to the regions with the largest land area (Mannion et al., 2012).

More recent changes in the distribution of biomes have occurred due to the repeated glaciations of the past 2.6 My (see Section 17.6). We are currently in a remarkably warm phase for the Quaternary, even without the influence of humans on global climate. Perhaps the species richness of biomes reflects their previous extent? This was considered by Fine and Ree (2006). They chose three forested biomes (tropical, temperate and boreal) then examined whether there was a correlation between their area and plant species richness. For the present day there was none—the boreal zone is vast yet contains hardly any tree species. Next they turned to the last glacial maximum (LGM), 20,000 years ago, and looked at the extent of biomes at that time. The analysis assumed niche conservatism (i.e. species keep the same niche through time and do not change their environmental tolerances) and no restrictions on species' ability to disperse with a changing climate. The clarity of the result was surprising: tree species richness per biome correlates better with the size of that biome at the LGM (Fig. 16.1). This should be tempered slightly with a realisation that the biomes at the LGM were very different in both composition and abundance of species from those we would recognise in the modern world, which means that the analysis is not necessarily comparing like-with-like (Williams et al., 2004).

Area can be shown to account for many patterns in species richness, but is it responsible for the LDG? It is certainly part of the story, but also easy to find counterexamples. In the New World the peak of mammal species richness is ironically found at the very thinnest point, around the isthmus of Panama, declining further north into the broader expanses of the United States and Canada. Likewise there is no clear trend with area moving southwards.

What biological mechanism would lead to a correlation between regional area and its species richness? Here the reasoning becomes harder to sustain than at local scales. While larger areas might contain more species, it is hard to see how this leads to greater species density, that is, α richness. Perhaps larger areas allow for larger populations, increasing the likelihood of speciation and reducing the chance of extinction? This should increase species richness, yet would automatically split populations and reduce

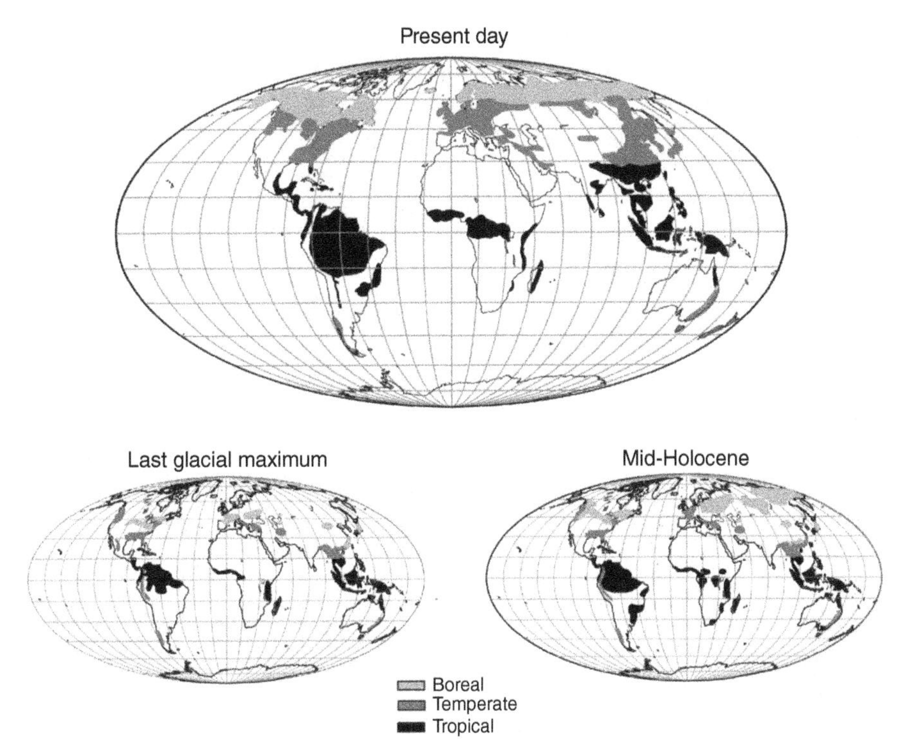

Figure 16.1 Extent of three forested biomes in the present day (upper) and at the last glacial maximum (20,000 ya) and mid-Holocene (6,000 ya). (*Source*: Fine and Ree (2006). Reproduced with permission from the University of Chicago Press.) *(See colour plate section for the colour representation of this figure.)*

species' ranges, making them more vulnerable to extinction. This logic cannot account for greater α diversity in the tropics.

The greatest limitation of the geographic area hypothesis is that it remains untestable at the necessary scale. It cannot be assumed that local-scale richness trends will apply in the same way once evolutionary dynamics are involved. Area alone is unlikely to be the root cause of the LDG.

16.4 Climatic stability

One of the patterns evident from Fig. 16.1 is that, while the boreal and temperate forests have shifted in latitude through time, the tropical forests have stayed in the same place, only expanding and contracting as climate allows. This means that, in effect, tropical habitats are older, having remained stable and in the same place for much longer. It's also likely that the ice ages had less impact on equatorial systems. They will therefore have had more time for speciation to take place and richness to accumulate. The idea dates back to Wallace (1878) and has been a common explanation for the LDG ever since. On top of their historical stability, there is also a more consistent climate in the tropics over shorter timescales, with reduced seasonality. Many regions stay warm and wet throughout the year rather than cycling through summers and winters. Even at the timescale of a single day there are fewer temperature extremes, with generally consistent warmth and no frost except at high altitudes.

The problem is that the facts don't quite add up. The fossil record shows that the LDG predates the Quaternary period, and therefore it cannot be the result of ice ages (Mannion et al., 2014). Nor did the tropics escape glaciations unscathed; the whole climate of the globe changed, meaning that most humid tropical forests contracted and became drier, even at the equator. As ice retreated, those areas behind the ice sheets accumulated plant species rapidly, showing no signs of any lag as the climate warmed (Vázquez-Rivera and Currie, 2014).

It's also quite wrong to assume that life in the tropics is somehow easier and that the more congenial climate leads to fewer extinctions. There are plenty of species that would find tropical conditions intolerable (e.g. penguins), and many species have been able to adapt to harsher, more seasonal climates. Moreover, the intense competition for resources in tropical environments (e.g. for light and nutrients in tropical rain forests) makes them exceedingly difficult places to live. Tropical climates often experience strong seasonality, especially in rainfall, and there is no evidence that populations remain more stable in the tropics (Vázquez and Stevens, 2004).

Most of all, the climatic stability hypothesis fails to provide a satisfactory explanation for where all the tropical species came from. The lack of a convincing account of how speciation and extinction fit into the overall picture means this narrative, while appealing, lacks a convincing biological mechanism at its core.

16.5 Productivity

The last chapter demonstrated that species richness at regional scales is frequently correlated with climatic variables (especially temperature and rainfall) and thereby also productivity. The tropics are warmer because the intensity of solar radiation increases towards the equator. This leads to a much higher NPP; lianas in tropical rain forests can grow at rates up to 1 m per day in full sunlight.

A positive and linear relationship between species richness and energy at regional scales exists, but teasing out what the mechanism behind this might be is difficult. It is certainly not a simple case of more energy in total being available and therefore divided among a greater number of species. One of the best-studied natural systems in the world is the Hubbard Brook Ecosystem.[1] Energy flows between trophic levels have been carefully determined. As energy moves through the food web, only 0.17% of NPP ends up in birds (Gaston and Blackburn, 2000), yet there is a positive correlation between productivity and bird species richness. Such a tiny proportion of NPP cannot be the driving force. Indeed, on global scales, prey richness is a more important driver of predator richness than productivity (Sandom et al., 2013).

Perhaps greater productivity increases the density of individual organisms, permitting larger populations and allowing for greater rates of speciation to occur (as for area)? This is also unsustainable as an argument, since, for example, the density of trees is approximately the same within temperate and tropical forests and only falls near the poles (Enquist and Niklas, 2001). Yet even with the same density of individuals, tropical forests contain vastly greater species richness. For birds in North America, there is an increase in abundance with productivity, but this alone cannot explain variation in species richness (Hurlbert, 2004).

Another possible mechanism, as seen in Chapter 15.4, might be increased stochasticity in community assembly, with higher productivity leading to a greater number of alternative stable states and higher β diversity (Chase, 2010). Some have argued however that β diversity will automatically rise in larger species pools, with no need to invoke community assembly (Kraft et al., 2011). This still leaves us trying to explain how greater regional species richness comes about in the first place.

Perhaps one answer is to step back from assessing productivity alone and examine once again the factors determining productivity at regional scales, which are energy and water. For plants at least, a greater amount of resource space might open up more opportunities for specialisation on different parts of a gradient of availability (Lavers and Field, 2006). Habitat heterogeneity would also be increased, along with the diversity of plant resources available for consumers. It is therefore possible to argue that productivity is a side issue, and the strong correlations between productivity and richness actually reflect a more fundamental link to the

[1] http://www.hubbardbrook.org/.

partitioning of resources. Once again, however, at present we lack the means with which to test this hypothesis rigorously.

16.6 Niche size

Another traditional explanation for the LDG dates back to Dobzhansky (1950). He proposed that tropical species might have narrower niches, allowing more species to coexist in the same place. The logic for this is eminently reasonable; temperate zones have greater climatic variability, with seasonality and wider extremes, meaning that species have to maintain broad tolerances (Janzen, 1967). Hence temperate species should have large ranges, a prediction known as Rapoport's rule. (A related hypothesis suggests that temperate species will have greater dispersal abilities to track changing environmental conditions and therefore larger ranges (Jocque et al., 2010).) Contrast this with tropical species, which, thanks to greater climatic stability, combined with stronger resource competition, would be forced to specialise on a smaller niche. There should therefore be a greater number of endemics with restricted ranges in the tropics.

There is some support for the contention that tropical niches are smaller. Dyer et al. (2007) sampled communities of Lepidoptera (butterflies and moths) on plants in eight locations from Canada to Ecuador. They found evidence that β diversity between host plants was greater at low latitudes (Fig. 16.2). This suggests a greater turnover of caterpillar species between plant species in the tropics, that is, that insects are more specialised. Furthermore, when examining individual species of caterpillar, diet breadth measured as the number of potential host plant genera declined with latitude (Fig. 16.3).

Before taking this evidence at face value, it should be noted that others have found no trend when examining exactly the same interaction. Novotny et al. (2006) compared insect herbivore host specificity between a forest in central Europe and one in Papua New Guinea. In contrast to Dyer et al.

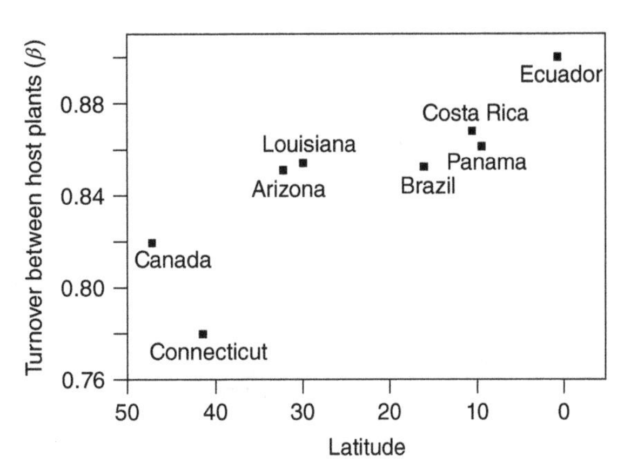

Figure 16.2 Modified Whittaker's β diversity for turnover of lepidopteran species across host plants. Higher levels imply a greater degree of herbivore specialisation. (*Source*: Dyer et al. (2007). Reproduced with permission from the Nature Publishing Group.)

(2007), trends in plant phylogenetic diversity were controlled for. Temperate and tropical tree species supported similar numbers of insect folivore species (29.0 ± 2.2 and 23.5 ± 1.8 per 100 m^2 foliage), and the diet breadths of these herbivore faunas did not differ. Another common interaction is between plants and soil fungi, yet in this case the ratio of plant to fungal species in communities increases towards the equator, implying that these associations become less specialised (Tedersoo et al., 2014).

Overall, the evidence for greater niche breadth is mixed, and it may apply only in limited circumstances (Vázquez and Stevens, 2004). Likewise, there is no evidence that the structure of interaction networks varies consistently with latitude, suggesting that their assembly follows common rules throughout the world (Morris et al., 2014). There is no sign that food webs contain a greater proportion of specialist links in the tropics.

Rapoport's rule has also been subjected to close scrutiny. Orme et al. (2006) took the breeding ranges of all bird species and checked whether these increased at higher latitudes (Fig. 16.4). The pattern occurs in the nothern hemisphere, especially in the Old World, where ranges decline towards the equator. The problem is that they keep declining further south. Bird range size correlates more closely with

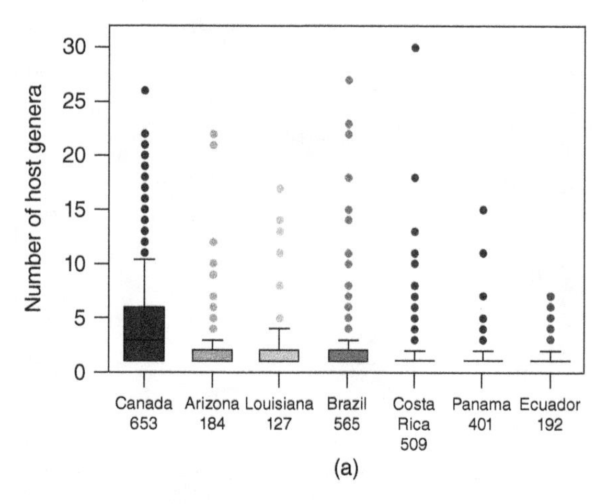

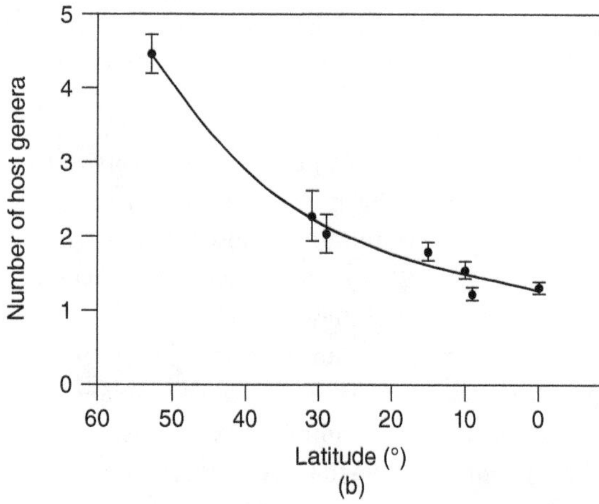

Figure 16.3 (a) Number of host plant genera per caterpillar species along a latitudinal gradient, with number of species listed below the axis; box plots are bounded by the first quartile, median and third quartile; whiskers are 1.5 times the interquartile range; points outside the whiskers are outliers; (b) regression of diet breadth against latitude; points show mean ± SE. (*Source*: Dyer et al. (2007). Reproduced with permission from the Nature Publishing Group.)

total land area than latitude. A weak tendency for narrower ranges at the equator exists but is certainly not sufficient to drive the LDG. The evidence for narrower niches in the tropics is therefore, at best, mixed.

Many authors have posited other biotic interactions than niche division that might lead to higher species richness at low latitudes. These include rates of predation, parasitism, the variety and impact of diseases, Janzen–Connell processes and mutualisms, though this is by no means an exhaustive list. While many of these have merit for particular groups, and all can enhance the trend, none has sufficient generality to fully account for the LDG. All depend on the assertion that 'diversity begets diversity'; this is circular and fails to explain why the cycle of diversification began in the first place. It is likely that any trends in biotic interactions are a symptom rather than the cause of the LDG.

16.7 Evolutionary speed

As yet no theory has given a convincing account of why so many species evolved in the tropics. Perhaps evolution happens faster there? This is not as crazy as it might sound at first. The idea was first put forward by Fischer (1960) and reinvigorated by Rohde (1992). The logic is simple:

1. More energy is available in the tropics.
2. This increases growth rates and reproductive output of organisms.
3. Generation times are thus reduced.
4. There could possibly be a greater rate of mutation due to the effect of temperature on chemical reactions.
5. Selection pressures will be stronger as populations build up more rapidly.

This is a very appealing argument as it not only explains the origins of species but is able to link together the effects of energy, time and area. There are intriguing hints that something is behind this, and the source of the best evidence to date is the evolution of plankton. These are often well preserved in the marine fossil record, with their distinctive hard shells allowing the emergence and spread of particular taxa to be tracked through time. This approach was taken by Allen and Gillooly (2006) for three groups of plankton (Fig. 16.5). After controlling for area and sampling effort, they found that for calcareous plankton (Foraminifera and Nannoplankton) there were more first origins of species in the tropics, matching the LDG.

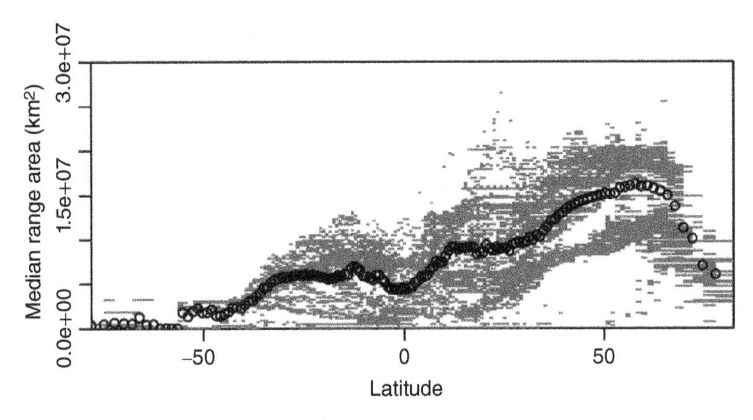

Figure 16.4 Median geographic range area (grey dots, km^2) and land area (black circles) with latitude for all bird species. (*Source*: From Orme et al. (2006). Used under CC-BY-SA 2.5 http://creativecommons.org/licenses/by-sa/2.5/.)

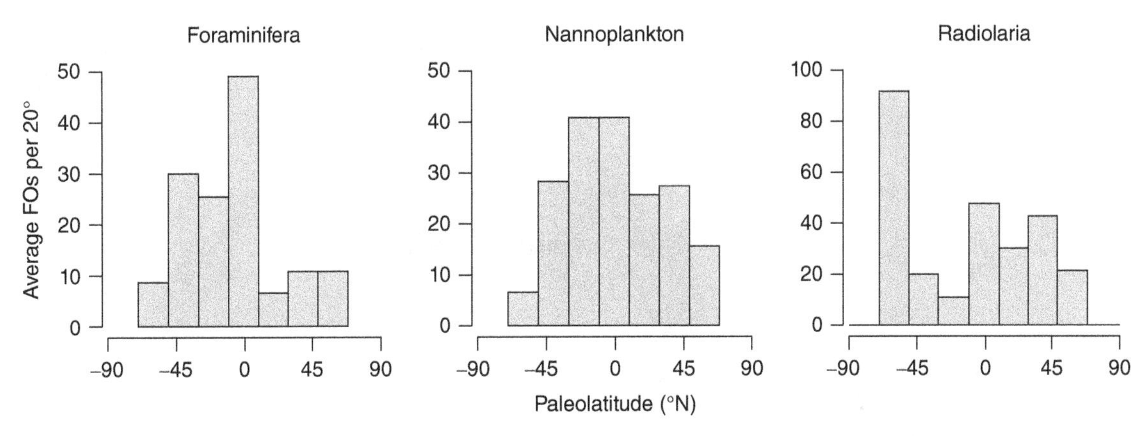

Figure 16.5 First originations (FO) in the fossil record per 20° latitudinal band for species in three groups of plankton. (*Source*: Allen and Gillooly (2006). Reproduced with permission from John Wiley & Sons Ltd.)

One group breaks the trend. For Radiolaria plankton, whose shells are silicaceous rather than calcareous, there was a higher rate of Antarctic origins. Once again, however, this matches the global pattern in their diversity, because Radiolaria are more species-rich in the Antarctic. Here the exception proves the rule, and diversity gradients match speciation rates. There is even evidence that current evolutionary rates of plankton are higher in the tropics. Allen et al. (2006) examined the rate of nucleotide base substitutions (i.e. genetic changes) in plankton, corresponding to neutral evolution, and found that it correlated with temperature (Fig. 16.6).

Another example is the swallowtail butterflies, which also exhibit clear latitudinal gradients in species richness in all parts of the world apart from a predictable dip in the Sahara (Fig. 16.7). Condamine et al. (2012) examined their phylogeny and reconstructed the history of the group. Perhaps surprisingly they evolved in the northern hemisphere, though at a point in the Eocene when tropical conditions extended to high latitudes. They have since spread southwards, in the process splitting into two main subfamilies. The Papilioninae (480 species) are almost exclusively tropical, while the Parnassiinae (70 species) are extratropical. These clades both arose at about the same time, yet the

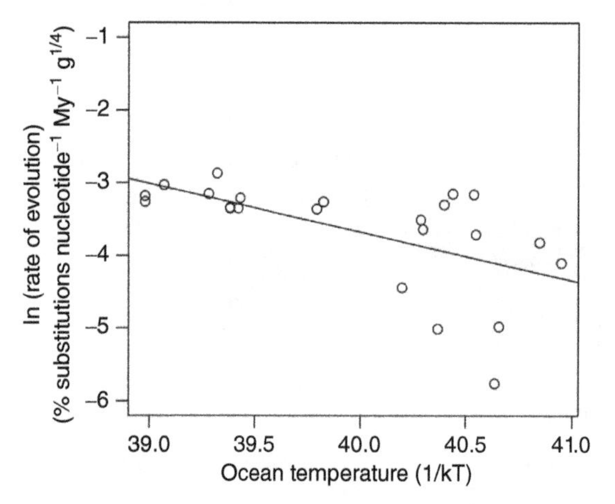

Figure 16.6 Rate of neutral molecular evolution (nucleotide substitutions) in planktonic foraminifera with ocean temperature. Note inverse temperature scale—lower values indicate higher temperatures. (*Source*: Allen et al. (2006). Copyright (2006) National Academy of Sciences, USA.)

phylogeny shows that the tropical group speciated faster. The story is a little less straightforward than this implies because some of the additional tropical speciation was due to shifts between host plants followed by radiations. Evolutionary speed was not the sole cause of patterns, but it was one of the major underlying factors.

These examples from invertebrates are backed by signs that the rate of plant evolution is twice as fast in the tropics (Wright et al., 2006). Elsewhere evidence is more conflicted, with the fossil record of marine invertebrates suggesting not only higher diversification rates in warm periods of Earth history but also higher simultaneous extinction rates, though the overall outcome is for global species richness to increase with temperature (Mayhew et al., 2012). Some studies have found no evidence of latitudinal

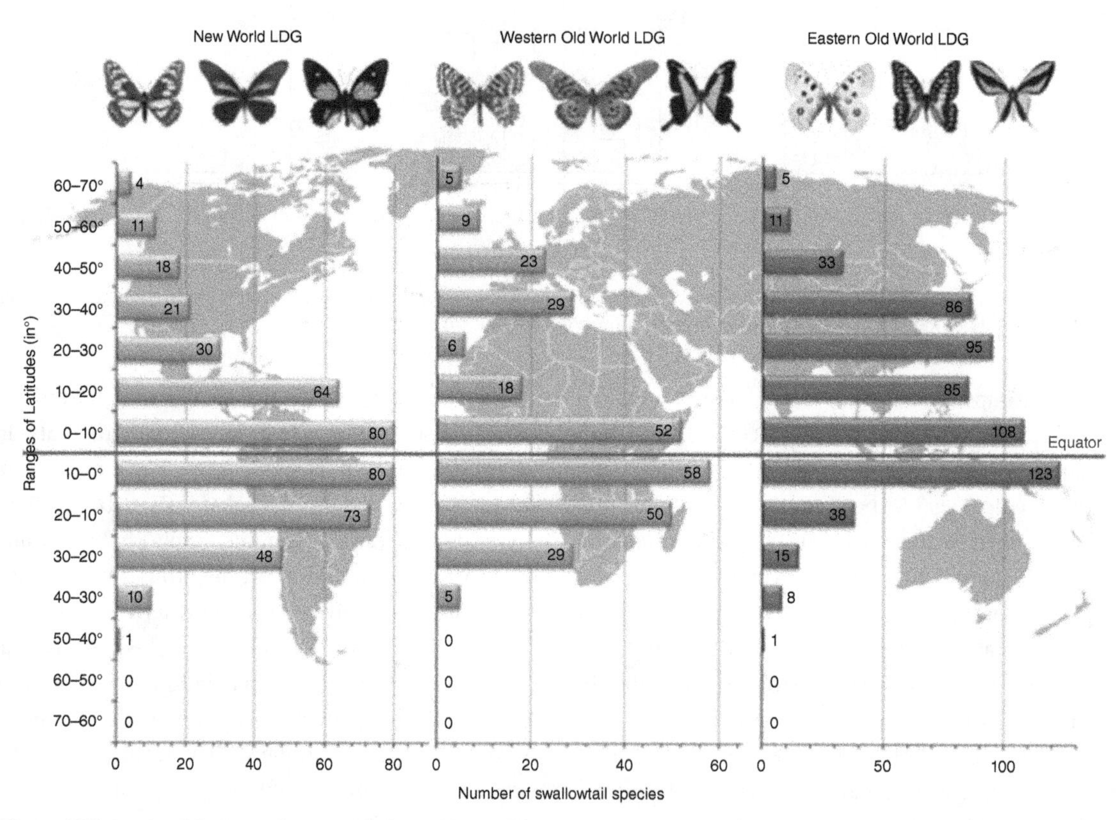

Figure 16.7 Species richness gradients in swallowtail butterflies. (*Source*: Condamine et al. (2012). Reproduced with permission from John Wiley and Sons Ltd.)

variation in speciation rates for either mammals or birds (Jetz and Fine, 2012; Soria-Carrasco and Castresana, 2012), though it may not be coincidental that these are warm-blooded and therefore perhaps less susceptible to effects of external temperature on biochemical processes.

The primary weakness of the evolutionary speed theory is that it is not possible to do any experiments to test it. It ought to apply more to ectotherms than endotherms, yet large animals show stronger relationships between richness and latitude (Hillebrand, 2004). Perhaps the theory works for ectotherms, which in turn drive patterns in endotherms, though evidence for this remains to be found. There is no reason to believe that evolutionary speed will be the sole, unique cause of latitudinal gradients, and it may interact with other processes.

16.8 Out of the tropics

This brings us to a related attempt to explain the LDG, known as the 'out of the tropics' (OTT) hypothesis, which proposes that most groups originate in the tropics then spread out into the rest of the world. Assuming that evolving the ability to survive in cooler or more seasonal climates is difficult, this means that an increasingly small minority of species will make it to higher latitudes.

A number of related processes can generate the observed pattern (Fig. 16.8). It might be that speciation rates are higher in the tropics than outside, making the tropics a cradle of diversity. Alternatively, global speciation rates could be exactly the same, but if extinction rates are higher outside the tropics (perhaps due to the ice ages), then the tropics would instead be a museum of diversity. The OTT model combines both these effects and predicts not only greater speciation rates in the tropics but high extinction rates outside, with the possibility of species sometimes moving between zones, usually from the tropics to elsewhere. Jablonski et al. (2006) found this pattern of origination and extinction to be well supported by the fossil record for marine bivalve molluscs (Fig. 16.9). Most species first occur in the tropics, where the origination rate of taxa is approximately double that at other latitudes. Genera are then often able to spread outside the

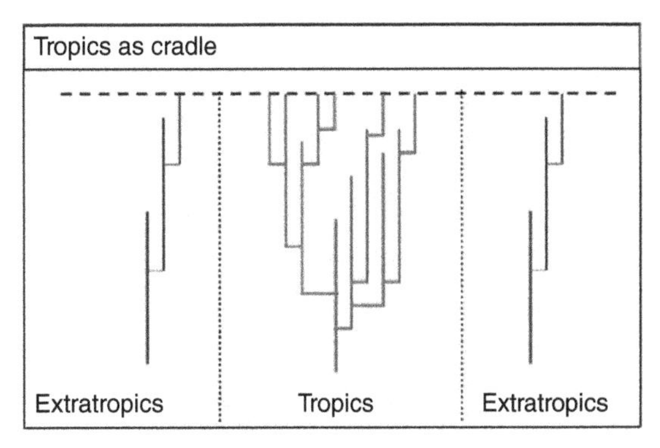

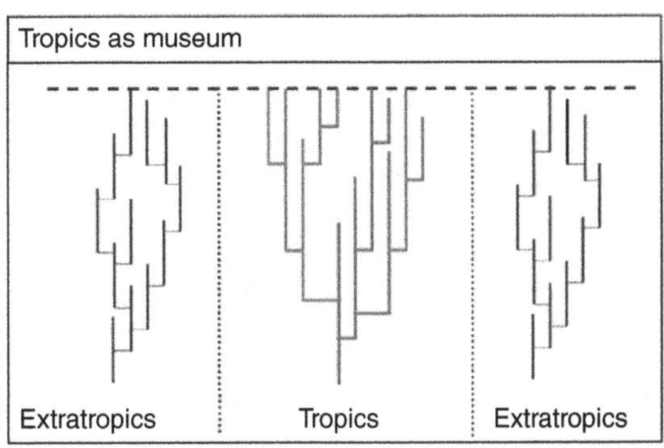

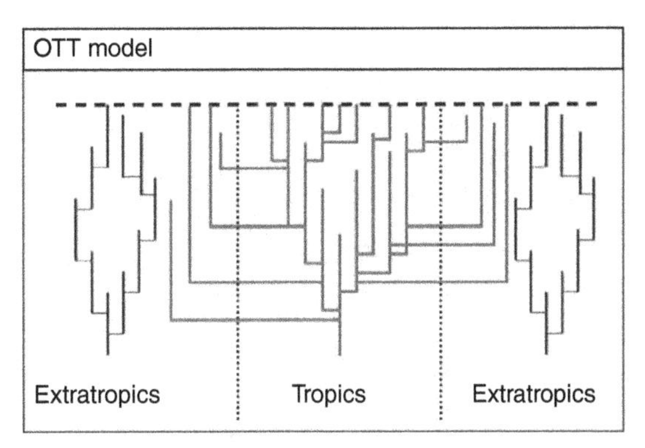

Figure 16.8 Scenarios illustrating the cradle, museum and out of the tropics (OTT) models. Light grey lines are lineages originating in the tropics; dark grey are those originating outside the tropics. Horizontal lines connecting lineages represent expansion of tropical lineages into extratropical regions; dashed horizontal lines indicate present-day richness patterns. (*Source*: Jablonski et al. (2006). Reproduced with permission from the American Association for the Advancement of Science.)

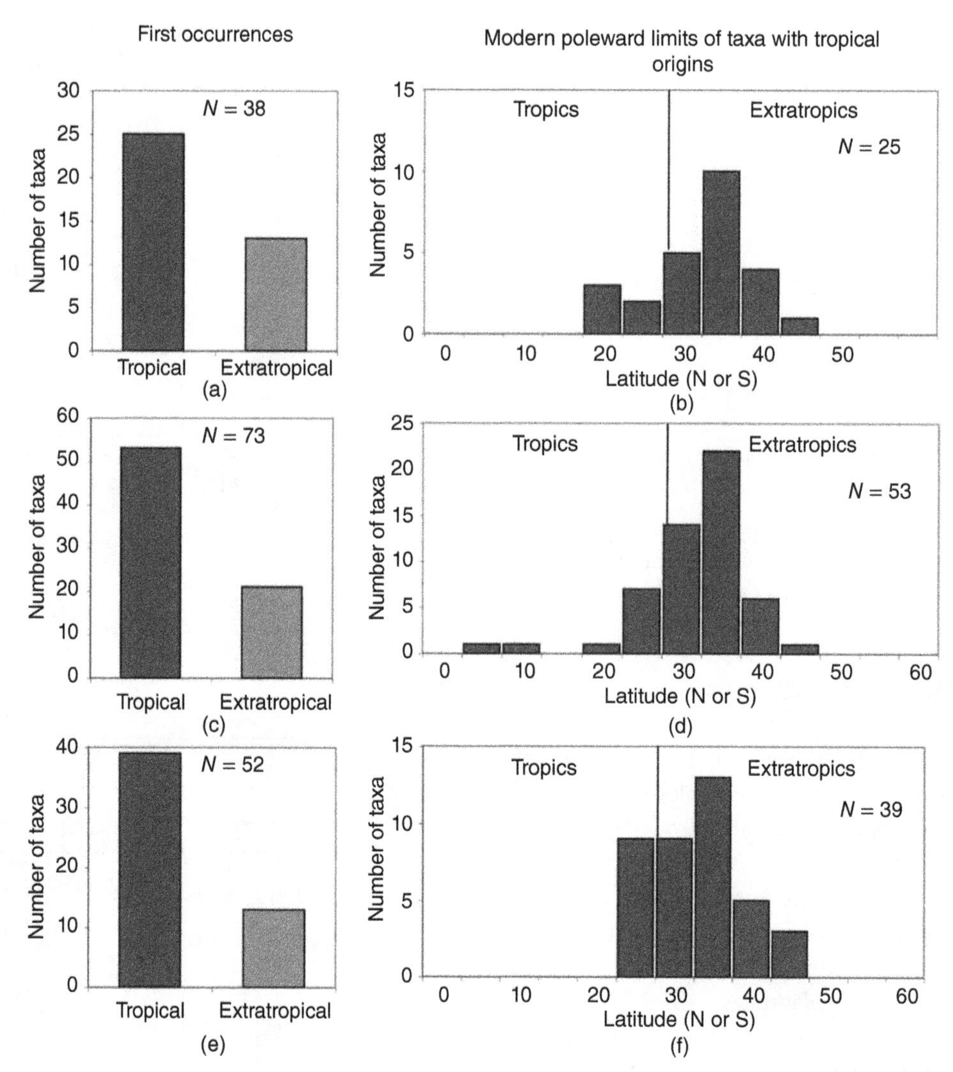

Figure 16.9 Frequency of originations inside and outside the tropics (left) and present-day range limits of marine bivalve genera first occurring in the tropics (right). Genera first appearing in (a and b) the Pleistocene, (c and d) the Pliocene and (e and f) the late Miocene; these represent three approximately even time periods of increasing age over the last 11 My. (*Source*: Jablonski et al. (2006). Reproduced with permission from the American Association for the Advancement of Science.)

tropics to new areas. This is still rare, happening only once every 5 My or so within genera (Jablonski et al., 2013). The same pattern is seen across a wide range of marine invertebrate groups. No latitudinal variation in extinction rates is required to generate it (Martin et al., 2007).

The OTT predictions have also received strong support in mammals, whose speciation rates are higher in the tropics and extinction rates lower and with more evidence of range expansion from than into the tropics, entirely consistent with the theory (Fig. 16.10). In amphibians, high tropical diversity has arisen due to a combination of faster speciation and remarkably low extinction rates in the tropics compared to elsewhere, and few cases of lineages dispersing to higher latitudes. In addition, there is a trend for speciation rates to slow down faster in temperate zones, implying that they fill

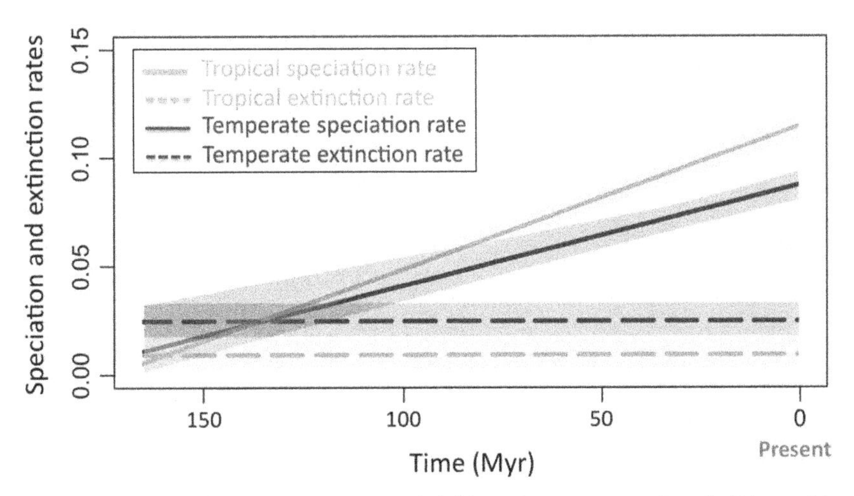

Figure 16.10 Mammalian speciation and extinction rates across clade history in temperate and tropical biomes. Lines represent mean estimates and shaded areas 95% credibility intervals (Rolland et al., 2014).

up with species sooner (Pyron and Wiens, 2013). This assumes that there is a limit to species richness which is set by the climate (Brown, 2014).

One prediction of the OTT model is that the average age of lineages will decrease with latitude, and this appears to generally be the case. The tropics contain not only many more young taxa, which have recently evolved, but also a greater variance in taxonomic ages because they also preserve older groups which have not gone extinct. This applies at the level of both genera and families and has been found for organisms including corals, molluscs, bats and birds (Mittelbach et al., 2007). Among marine bivalve molluscs, younger families have steeper latitudinal gradients, which supports the idea that they have only recently evolved and begun to spread outside the tropics. It also appears that their extinction rates are greater at the poles (Jablonski et al., 2006). The pattern is not universal though (e.g. Pereira and Palmeirim, 2013; Pyron and Wiens, 2013).

Many taxa occurring in temperate regions are actually nested within tropical clades (Mittelbach et al., 2007). Among flowering plants, 50% of families only occur in the tropics, while those occurring in temperate regions are usually subsets of families which are much larger in the tropics. Typically those that extend beyond the tropics only evolved in the last 34 My since global cooling began at the end of the Eocene (Kerkhoff et al., 2014). There are some groups of organisms where the peak of species richness occurs outside the tropics; these include seabirds, hoverflies, aquatic macrophytes and polypore fungi. Yet even these exceptions prove the rule because they tend to be lower taxonomic levels that have recently evolved. Perhaps, like the Papilionidae, their long-term future will see greater diversification in the tropics.

The time that a group has been in a region can sometimes be the most important correlate of species richness patterns. This has the potential to confuse our interpretation because it can be hard to separate from the OTT hypothesis. Hylid frogs in the New World are known to have initially evolved in South America, but some clades spread north, though only those which evolved to tolerate cooler conditions. Having arrived in a new region, the length of time they have been there dictates their species richness (Fig. 16.11).

The ability of clades to escape the tropics might then act as a constraint on diversity at higher latitudes (Wiens et al., 2006). This is the basis of another related hypothesis for the LDG, the tropical niche conservatism (TNC) hypothesis, which proposes that the former extent of the tropics and lengthy time for evolution, combined with the difficulty of

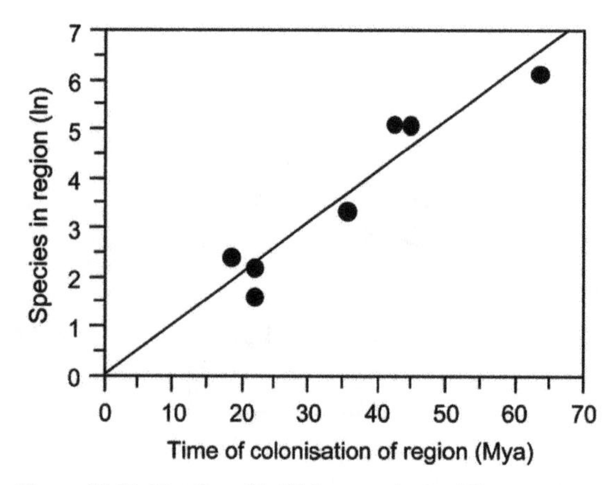

Figure 16.11 Number of hylid frog species in different regions and the time when each region was colonised. (*Source*: Wiens et al. (2006). Reproduced with permission from the University of Chicago Press.)

species spreading into cold regions, has led to present gradients (Wiens and Donoghue, 2004).

Such analyses are contentious as if there are limits on diversity (i.e. the number of species does not increase linearly with time), then simply dividing the age of a clade by the number of species will give a misleading estimate of the speciation rate (Rabosky, 2009). A large number of clades show no correlation between their age and diversity. The OTT and TNC models are also challenged by anecdotal evidence from the Americas, where the two continents met around 3 Mya, with fauna and flora spreading across the new land bridge formed by the Isthmus of Panama in what is referred to as the Great American Biotic Interchange. This was not bidirectional however; groups of animals from the more temperate north were successful in spreading into South America, whereas very few southern lineages from the tropics penetrated northwards (Smith et al., 2012). Even within hylid frogs, while there is evidence that only some groups managed to escape the tropics, their species richness is more closely correlated with precipitation than latitude (Algar et al., 2009). Species composition and richness are not necessarily determined by the same factors and niche conservatism does not always lead to a latitudinal gradient.

It is easy to get confused by various phylogenetic models for the origin and maintenance of the LDG because they overlap in many of their core predictions. It is therefore worth examining them together and asking which predictions are met more or less often (Table 16.1). The hypotheses are not strict alternatives and are better thought of as a set of nested ideas with further assumptions being required for the more complex models.

In a review of over 100 phylogenetic studies, Jansson et al. (2013) found that the majority supported the Out-of-The-Tropics hypothesis. Overall most clades originated in the tropics, contained the most species close to where they arose and moved most commonly from the tropics into the temperate zone. There was no evidence that, on average, speciation rates were higher in the tropics. It's important in this context to realise that it's not necessarily the rate of speciation that matters but the net diversification rate. In some cases the rate of speciation can be

Table 16.1 Expected patterns in phylogenetic trees assuming four different mechanisms for generating and maintaining the latitudinal diversity gradient. ES, evolutionary speed; OTT, out of the tropics; ET, evolutionary time; TNC, tropical niche conservatism.

Hypothesis	Geographic origins of clades	Dispersal between latitudinal zones	Diversification rates
ES	Anywhere	No prediction	Higher in tropics
OTT	Mostly tropical	Tropical to temperate	No latitudinal trend
ET	Mostly tropical	No prediction	No latitudinal trend
TNC	Mostly tropical	All rare	No latitudinal trend

Source: Jansson et al. (2013). Reproduced with permission from the Ecological Society of America.

greater outside the tropics, but when combined with greater extinction rates (e.g. in squamate reptiles; Pyron, 2014), net diversification remains low and turnover high.

The four hypotheses are not mutually exclusive; each might be more appropriate for some clades than others or even in different times and places. Birds and mammals have near-identical patterns of species richness across the Earth, but the phylogenetic origins of their diversity are very different (Hawkins et al., 2012). The general principle remains the same—it is the rates of speciation, extinction and dispersal that drive patterns, and fundamentally it is only through understanding these that we can explain the distribution of species richness across the world.

16.9 Conclusions

The tropics are remarkably more species rich than the rest of the planet, and the Latitudinal Diversity Gradient is one of the most pervasive and universal patterns in biology. There are many interacting processes at work, though underlying all of them are the rates of speciation, extinction and dispersal. Understanding how these rates vary in different parts of the world is the key to solving the riddle of the LDG. Tantalising evidence suggests that most clades originate in the tropics and spread from there, with extinction rates increasing towards the poles. Some puzzles remain, however, and it is likely that the story for each taxon will be slightly different. Remember that the temperate zone is unusual and

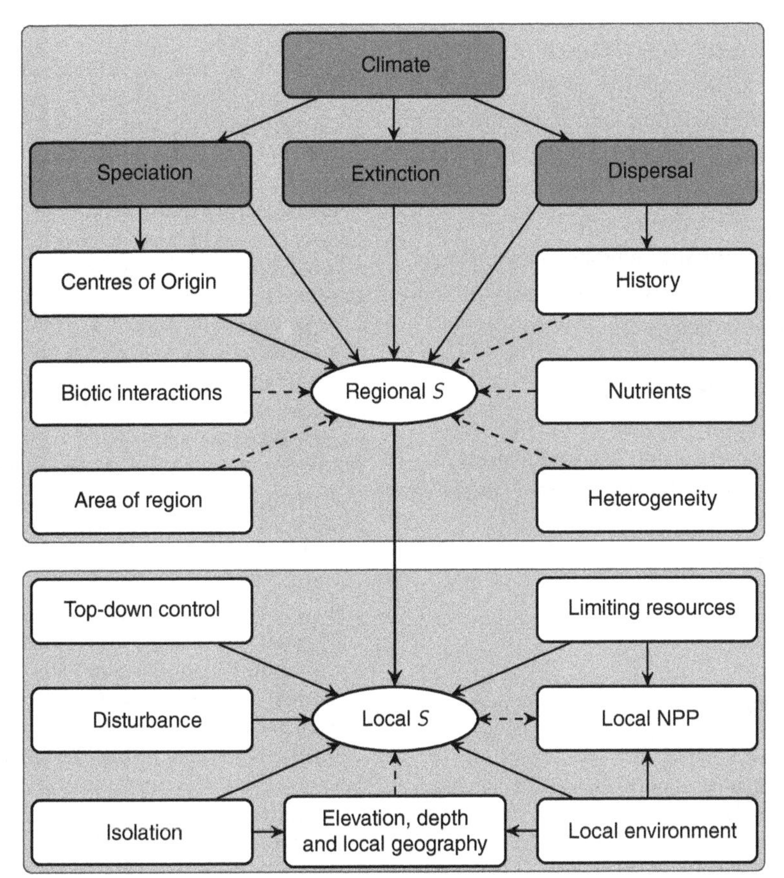

Figure 16.12 Determinants of regional and local species richness S for a given area. Direct links in bold, indirect with dashed lines. Dark grey is used to indicate the main biogeographical drivers of regional species richness.

aberrant, not the tropics. The taxa that reach higher latitudes are a select subset of those that evolved in the tropics. Understanding global-scale patterns requires a reversal of the usual temperate world view to see life from a tropical perspective.

At last we are close to completing our schematic plan of the drivers of local and regional species richness (Fig. 16.12). There remain two branches to be discussed though, which are the roles of history and dispersal at biogeographical scales.

16.9.1 Recommended reading

Brown, J. H., 2014. Why are there so many species in the tropics? *Journal of Biogeography* 41:8–22.

Mittelbach, G. G., D. W. Schemske, H. V. Cornell, A. P. Allen, J. M. Brown, M. B. Bush, S. P. Harrison, A. H. Hurlbert, N. Knowlton, H. A. Lessios, C. M. McCain, A. R. McCune, L. A. McDade, M. A. McPeek, T. J. Near, T. D. Price, R. E. Ricklefs, K. Roy, D. F. Sax, D. Schluter, J. M. Sobel, and M. Turelli, 2007. Evolution and the latitudinal diversity gradient: speciation, extinction and biogeography. *Ecology Letters* 10:315–331.

Willig, M. R., D. M. Kaufman, and R. D. Stevens, 2003. Latitudinal gradients of biodiversity: pattern, process, scale and synthesis. *Annual Review of Ecology, Evolution, and Systematics* 34:273–309.

16.9.2 Questions for the future

- How do speciation rates change with temperature, and does this depend on the taxa under study?
- Are patterns in endotherms driven by those in ectotherms?
- Do biotic interactions cause or strengthen the LDG?

References

Algar, A. C., J. T. Kerr, and D. J. Currie, 2009. Evolutionary constraints on regional faunas: whom, but not how many. *Ecology Letters* 12:57–65.

Allen, A. P. and J. F. Gillooly, 2006. Assessing latitudinal gradients in speciation rates and biodiversity at the global scale. *Ecology Letters* 9:947–954.

Allen, A. P., J. F. Gillooly, V. M. Savage, and J. H. Brown, 2006. Kinetic effects of temperature on rates of genetic divergence and speciation. *Proceedings of the National Academy of Sciences of the United States of America* 103:9130–9135.

Brown, J. H., 2014. Why are there so many species in the tropics? *Journal of Biogeography* 41:8–22.

Chase, J. M., 2010. Stochastic community assembly causes higher biodiversity in more productive environments. *Science* 328:1388–1391.

Condamine, F. L., F. A. H. Sperling, N. Wahlberg, J.-V. Rasplus, and G. J. Kergoat, 2012. What causes latitudinal gradients in species diversity? Evolutionary processes and ecological constraints on swallowtail biodiversity. *Ecology Letters* 15:267–277.

Dobzhansky, T., 1950. Evolution in the tropics. *American Scientist* 38:209–221.

Dyer, L. A., M. S. Singer, J. T. Lill, J. O. Stireman, G. L. Gentry, R. J. Marquis, R. E. Ricklefs, H. F. Greeney, D. L. Wagner, H. C. Morais, I. R. Diniz, T. A. Kursar, and P. D. Coley, 2007. Host specificity of Lepidoptera in tropical and temperate forests. *Nature* 448:696–699.

Enquist, B. J. and K. J. Niklas, 2001. Invariant scaling relationships across tree-dominated communities. *Nature* 410:655–660.

Fine, P. V. A. and P. H. Ree, 2006. Evidence for a time-integrated species-area effect on the latitudinal gradient in tree diversity. *American Naturalist* 168:796–804.

Fischer, A. G., 1960. Latitudinal variation in organic diversity. *Evolution* 14:64–81.

Gaston, K. J. and T. M. Blackburn, 2000. *Pattern and Process in Macroecology*. Wiley-Blackwell.

Hawkins, B. A., C. M. McCain, T. J. Davies, L. B. Buckley, B. L. Anacker, H. V. Cornell, E. I. Damschen, J. A. Grytnes, S. Harrison, R. D. Holt, N. J. B. Kraft, and P. R. Stephens, 2012. Different evolutionary histories underlie congruent species richness gradients of birds and mammals. *Journal of Biogeography* 39:825–841.

Hillebrand, H., 2004. On the generality of the latitudinal diversity gradient. *American Naturalist* 163:192–211.

Hurlbert, A. H., 2004. Species-energy relationships and habitat complexity in bird communities. *Ecology Letters* 7:714–720.

Jablonski, D., C. L. Belanger, S. K. Berke, S. Huang, A. Z. Krug, K. Roy, A. Tomasovych, and J. W. Valentine, 2013. Out of the tropics, but how? Fossils, bridge species, and thermal ranges in the dynamics of the marine latitudinal diversity gradient. *Proceedings of the National Academy of Sciences of the United States of America* 110:10487–10494.

Jablonski, D., K. Roy, and J. W. Valentine, 2006. Out of the tropics: evolutionary dynamics of the latitudinal diversity gradient. *Science* 314:102–106.

Jansson, R., G. Rodríguez-Castañeda, and L. E. Harding, 2013. What can multiple phylogenies say about the latitudinal diversity gradient? A new look at the tropical conservatism, out of the tropics, and diversification rate hypotheses. *Evolution* 67:1741–1755.

Janzen, D. H., 1967. Why mountain passes are higher in the tropics. *American Naturalist* 101:233–249.

Jetz, W. and P. V. A. Fine, 2012. Global gradients in vertebrate diversity predicted by historical area-productivity dynamics and contemporary environment. *PLoS Biology* 10:e1001292.

Jocque, M., R. Field, L. Brendonck, and L. De Meester, 2010. Climatic control of dispersal-ecological specialization trade-offs: a metacommunity process at the heart of the latitudinal diversity gradient? *Global Ecology and Biogeography* 19:244–252.

Kerkhoff, A. J., P. E. Moriarty, and M. D. Weiser, 2014. The latitudinal species richness gradient in New World woody angiosperms is consistent with the tropical conservatism hypothesis. *Proceedings of the National Academy of Sciences of the United States of America* 111:8125–8130.

Kraft, N. J. B., L. S. Comita, J. M. Chase, N. J. Sanders, N. G. Swenson, T. O. Crist, J. C. Stegen, M. Vellend, B. Boyle, M. J. Anderson, H. V. Cornell, K. F. Davies, A. L. Freestone, B. D. Inouye, S. P. Harrison, and J. A. Myers, 2011. Disentangling the drivers of β diversity along latitudinal and elevational gradients. *Science* 333:1755–1758.

Lavers, C. and R. Field, 2006. A resource-based conceptual model of plant diversity that reassesses causality in the productivity-diversity relationship. *Global Ecology and Biogeography* 15:213–224.

Mannion, P. D., R. B. J. Benson, P. Upchurch, R. J. Butler, M. T. Carrano, and P. M. Barrett, 2012. A temperate palaeodiversity peak in Mesozoic dinosaurs and evidence for Late Cretaceous geographical partitioning. *Global Ecology and Biogeography* 21:898–908.

Mannion, P. D., P. Upchurch, R. B. J. Benson, and A. Goswani, 2014. The latitudinal biodiversity gradient through deep time. *Trends in Ecology & Evolution* 29:42–50.

Martin, P. R., F. Bonier, and J. J. Tewksbury, 2007. Revisiting Jablonski (1993): cladogenesis and range expansion explain latitudinal variation in taxonomic richness. *Journal of Evolutionary Biology* 20:930–936.

Mayhew, P. J., M. A. Bell, T. G. Benton, and A. J. McGowan, 2012. Biodiversity tracks temperature over time. *Proceedings of the National Academy of Sciences of the United States of America* 109:15141–15145.

Mittelbach, G. G., D. W. Schemske, H. V. Cornell, A. P. Allen, J. M. Brown, M. B. Bush, S. P. Harrison, A. H. Hurlbert, N. Knowlton, H. A. Lessios, C. M. McCain, A. R. McCune, L. A. McDade, M. A. McPeek, T. J. Near, T. D. Price, R. E. Ricklefs, K. Roy, D. F. Sax, D. Schluter, J. M. Sobel, and M. Turelli, 2007. Evolution and the latitudinal diversity gradient: speciation, extinction and biogeography. *Ecology Letters* 10:315–331.

Morris, R. J., S. Gripenberg, O. T. Lewis, and T. Roslin, 2014. Antagonistic interaction networks are structured independently of latitude and host guild. *Ecology Letters* 17:340–349.

Novotny, V., P. Drozd, S. E. Miller, M. Kulfan, M. Janda, Y. Basset, and G. D. Weiblen, 2006. Why are there so many species of herbivorous insects in tropical rainforests? *Science* 313:1115–1118.

Orme, C. D. L., R. G. Davies, V. A. Olson, G. H. Thomas, T.-S. Ding, P. C. Rasmussen, R. S. Ridgely, A. J. Stattersfield, P. M. Bennett, I. P. F. Owens, T. M. Blackburn, and K. J. Gaston, 2006. Global patterns of geographic range size in birds. *PLoS Biology* 4:e208.

Pereira, M. J. R. and J. M. Palmeirim, 2013. Latitudinal gradients in New World bats: are they a consequence of niche conservatism? *PLoS ONE* 8:e69245.

Pyron, R. A., 2014. Temperate extinction in squamate reptiles and the roots of latitudinal diversity gradients. *Global Ecology and Biogeography* 23:1126–1134.

Pyron, R. A. and J. J. Wiens, 2013. Large-scale phylogenetic analyses reveal the causes of high tropical amphibian diversity. *Proceedings of the Royal Society Series B* 280:20131622.

Rabosky, D. L., 2009. Ecological limits and diversification rate: alternative paradigms to explain the variation in species richness among clades and regions. *Ecology Letters* 12:735–743.

Rohde, K., 1992. Latitudinal gradients in species diversity: the search for the primary cause. *Oikos* 65:514–527.

Rolland, J., F. L. Condamine, F. Jiguet, and H. Morlon, 2014. Faster speciation and reduced extinction in the tropics contribute to the mammalian latitudinal diversity gradient. *PLoS Biology* 12:e1001775.

Sandom, C., L. Dalby, C. Fløgaard, W. D. Kissling, J. Lenoir, B. Sandel, K. Trøjelsgaard, R. Erjnæs, and J.-C. Svenning, 2013. Mammal predator and prey species richness are strongly linked at macroscales. *Ecology* 94:1112–1122.

Smith, B. T., R. W. Bryson Jr., D. D. Houston, and J. Klicka, 2012. An asymmetry in niche conservatism contributes to the latitudinal species diversity gradient in New World vertebrates. *Ecology Letters* 15:1318–1325.

Soria-Carrasco, V. and J. Castresana, 2012. Diversification rates and the latitudinal gradient of diversity in mammals. *Proceedings of the Royal Society Series B* 279:4148–4155.

Tedersoo, L., M. Bahram, S. Põlme, U. Kõljalg, N. S. Yorou, R. Wijesundera, L. V. Ruiz, A. M. Vasco-Palacios, P. Q. Thu, A. Suija, M. E. Smith, C. Sharp, E. Saluveer, A. Saitta, M. Rosas, T. Riit, D. Ratkowsky, K. Pritsch, K. Põldmaa, M. Piepenbring, C. Phosri, M. Peterson, K. Parts, K. Pärtel, E. Otsing, E. Nouhra, A. L. Njouonkou, R. H. Nilsson, L. N. Morgado, J. Mayor, T. W. May, L. Majuakim, D. J. Lodge, S. S. Lee, K.-H. Larsson, P. Kohout, K. Hosaka, I. Hiiesalu, T. W. Henkel, H. Harend, L.-D. Guo, A. Greslebin, G. Grelet, J. Geml, G. Gates,

W. Dunstan, C. Dunk, R. Drenkhan, J. Dearnaley, A. De Kesel, T. Dang, X. Chen, F. Buegger, F. Q. Brearley, G. Bonito, S. Anslan, S. Abell, and K. Abarenkov, 2014. Global diversity and geography of soil fungi. *Science* 346:1256688.

Vázquez, D. and R. D. Stevens, 2004. The latitudinal gradient in niche breadth: concepts and evidence. *American Naturalist* 164:E1–E19.

Vázquez-Rivera, H. and D. J. Currie, 2014. Contemporaneous climate directly controls broad-scale patterns of woody plant diversity: a test by a natural experiment over 14,000 years. *Global Ecology and Biogeography* 24:97–106.

Wallace, A. R., 1878. *Tropical Nature and Other Essays.* Macmillan.

Wiens, J. J. and M. J. Donoghue, 2004. Historical biogeography, ecology and species richness. *Trends in Ecology & Evolution* 19:639–644.

Wiens, J. J., C. H. Graham, D. S. Moen, S. A. Smith, and T. W. Reeder, 2006. Evolutionary and ecological causes of the latitudinal diversity gradient in hylid frogs: treefrog trees unearth the roots of high tropical diversity. *American Naturalist* 168:579–596.

Williams, J. W., B. N. Shuman, T. J. Webb, P. J. Bartlein, and P. L. Leduc, 2004. Late-Quaternary vegetation dynamics in North America: scaling from taxa to biomes. *Ecology* 74:309–334.

Willig, M. R., D. M. Kaufman, and R. D. Stevens, 2003. Latitudinal gradients of biodiversity: pattern, process, scale and synthesis. *Annual Review of Ecology, Evolution, and Systematics* 34:273–309.

Wright, S., J. Keeling, and L. Gilman, 2006. The road from Santa Rosalia: a faster tempo of evolution in tropical climates. *Proceedings of the National Academy of Sciences of the United States of America* 103:7718–7722.

Zachos, J., M. Pagani, L. Sloan, E. Thomas, and K. Billups, 2001. Trends, rhythms, and aberrations in global climate 65 Ma to present. *Science* 292:686–693.

CHAPTER 17
Earth history

17.1 The big question

Earth has changed substantially during the time within which it has been occupied by life. A realisation of the dynamic history of the planet caused a revolution in our understanding of the natural world. Evidence has accumulated over the past century of massive movements of the continents, occurring over hundreds of millions of years. More recently, the last few million years have witnessed dramatic and cyclical shifts in the global climate. Finally, the emergence of humans and their subsequent colonisation of all major land masses (except Antarctica) brought the world into what is becoming known as the Anthropocene—a period when our species has begun to alter the global environment in terms of climate change, nutrient dynamics, water flows and even the structure of whole landscapes.

Life has persisted during all these changes, but biogeographical patterns have been altered in every corner of the planet. Can we detect the signals of these events in the distribution of life? Understanding how natural systems have moved, changed and adapted in the past is the key to predicting how they might respond in the future.

17.2 Geological history

Modern surveys only record a brief snapshot taken from billions of years of life (so far), evolving in response to changes in the environment and layout of the Earth on timescales that can be hard to comprehend. Understanding where things are now depends upon learning about unfamiliar subjects. Ecologists sometimes presume that they ought to concentrate on living organisms and can safely ignore the alien fields of geology and stratigraphy, but this is naïve. Knowing about the deep history of life is essential in order to fully appreciate the present.

Geologists divide the history of Earth into a number of phases. Classically there are four eons, beginning with the Hadean (4.6–4.0 Bya), named for the hellish volcanic conditions at the time. Simple single-celled forms of life appeared in the subsequent Arcæan, the name coming from the Greek for 'beginning'. The atmosphere was almost devoid of oxygen. From 2.5 Bya the Proterozoic eon (meaning 'earlier life') saw a gradual accumulation of atmospheric oxygen, with the first definitive multicellular life forms appearing right at its close during the Ediacaran period (635–541 Mya). From our perspective the most interesting things began to happen in the current Phanærozoic eon (meaning 'visible life') around 541 Mya, when multicellular life forms proliferated and initiated the Cambrian explosion of diversity (see Section 3.3).

The next split within the Phanærozoic eon is into three eras: the Palæozoic, Mesozoic and current Cenozoic (Table 17.1). The divisions reflect two of the most dramatic mass extinction events: the

Natural Systems: The organisation of life, First Edition. Markus P. Eichhorn.

Table 17.1 Eras and periods within the Phanærozoic eon.

Era	Period	Start (Mya)
Cenozoic	Quaternary	2.6
	Neogene	23
	Palæogene	66
Mesozoic	Cretaceous	145
	Jurassic	201
	Triassic	252
Palæozoic	Permian	299
	Pennsylvanian	323
	Mississippian	359
	Devonian	419
	Silurian	443
	Ordovician	485
	Cambrian	541

Modified from www.stratigraphy.org.

Table 17.2 Cenozoic periods and epochs. Starting times given in millions of years ago (Mya).

Period	Epoch	Start (Mya)
Quaternary	Holocene	0.01
	Pleistocene	2.6
Neogene	Pliocene	5.3
	Miocene	23
Palæogene	Oligocene	34
	Eocene	56
	Palæocene	66

Modified from www.stratigraphy.org.

Permian/Triassic boundary, when life on Earth was all but wiped out, and the Cretaceous/Palæogene transition (the K–T boundary), when the dinosaurs went extinct.

In describing the history of life, reference is most often made to the periods, which further subdivide the eras according to notable transitions in the fossil record. These are then split into epochs (Table 17.2). In recent years there have been some revisions to the Cenozoic periods and epochs,[1] which means that older sources might refer to a Tertiary period beginning 65 Mya (covering both the Palæogene and Neogene) and the Quaternary commencing 1.8 Mya. The present day is the Holocene epoch, which began 11,000 years ago when the last glacial cycle ended and a rapid phase of global warming occurred. There has been a recent vigorous debate over whether we should rename the current epoch the Anthropocene to register the substantial changes made to the biosphere by mankind (Corlett, 2015). This is largely motivated by political impact rather than geological evidence, and it is uncertain when an appropriate start time would be.

17.3 Continental drift

Over geological timescales the major land masses have moved around the earth continuously. For an idea which is now considered so central to an understanding of Earth history and modern geology, it is remarkable to think that it was only in the latter half of the last century that continental drift theory gained widespread acceptance. It was proposed by Wegener (1915) who, while not being the first to notice that the continents tessellated and could be fit together into a single land mass, was the first to assemble a serious body of evidence for this proposition. His argument was based on the direction of scars left by glaciers and the concordance of fossil deposits on different continents. It faced severe criticism on first publication, the backlash arising partly from resistance from the geological establishment to being subverted by an outsider (Wegener was originally a climatologist). It should also be acknowledged however that the early theory was flawed and the first edition of Wegener's book contained many errors, some of which were later corrected. The main problem was the lack of any known mechanism, and that no single line of evidence exclusively supported it. Wegener tragically died on an expedition to Greenland in 1930, through which he hoped to collect more supporting data, and never received the acclaim he deserved.

Much more is now known than Wegener was able to discern. At the most basic level, there are two classes of rock which compose the Earth's crust.

[1] See http://www.stratigraphy.org for the very latest consensus view.

Continental rocks are predominantly composed of silica and rich in aluminium and are hence known by the acronym sial. The rocks of the ocean floor, by contrast, contain more manganese and are known as sima. Both float on a layer of molten rock called the mantle. Continental rock remains on the top, while oceanic rock is constantly being created at spreading plate boundaries and drawn back into the mantle at subduction zones.

The crust is split into 16 major tectonic plates (Fig. 17.1) which move around the Earth propelled by internal forces. Rates of movement vary both between plates and through time but in the present day average around 5–10 cm per year. The outlines of the continents have remained largely unchanged through time, though orogenesis (mountain formation) has modified some parts.

During the Mesozoic all the major continental blocks were joined in a single mega-continent known as **Pangæa** (Fig. 17.2). This has since split into many parts as the result of gradual forces caused by radioactive heating rather than catastrophic cleavage. Pangæa was not the first time that the continents united, nor will it be the last; as the Atlantic widens, the Pacific shrinks, and eventually the two sides will meet. It is therefore central to our understanding of the patterns of life to acknowledge that the world looked very different when particular groups of organisms evolved and began to spread.

Pangæa formed at the beginning of the Permian, around 300 Mya, but was only a temporary phase. By 200 Mya it had begun to split into two halves, the northern part known as Laurasia, the southern as Gondwana, with the Tethys Sea separating them. By 120 Mya a complete break had formed.

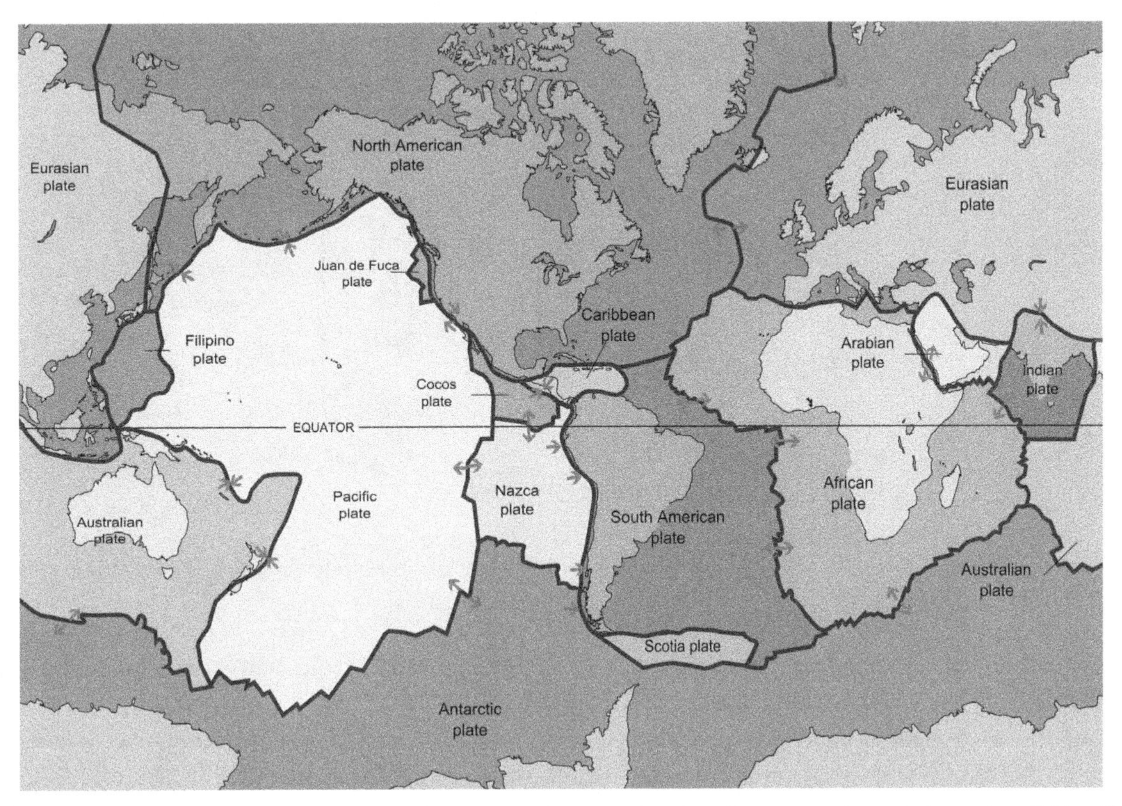

Figure 17.1 Location of continental plate boundaries and predominant direction of movement (arrows). Image from Wikimedia commons based on an original at http://pubs.usgs.gov/gip/dynamic/slabs.html.

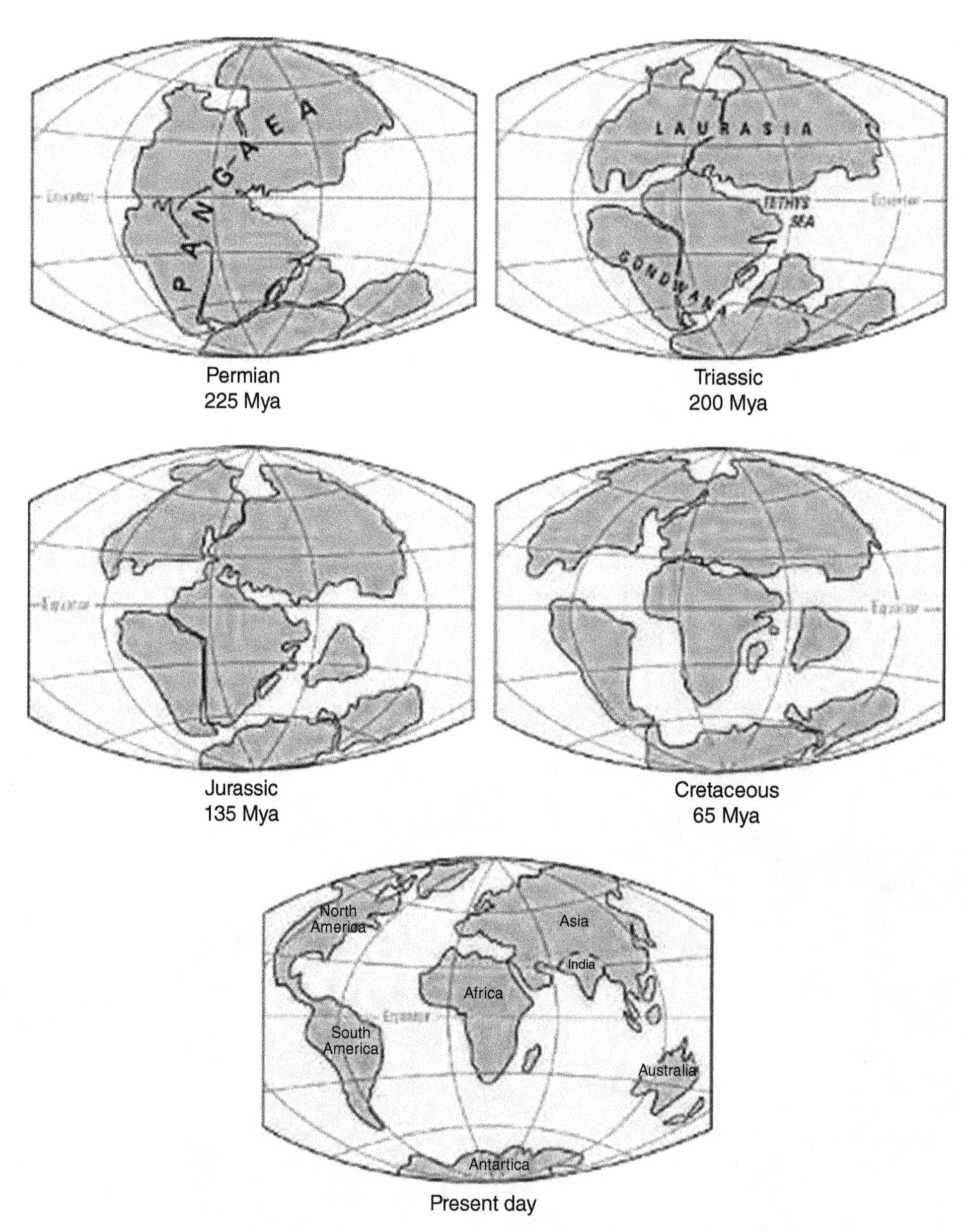

Figure 17.2 Break-up of Pangæa from 225 Mya. From http://pubs.usgs.gov/gip/dynamic/historical.html. *(See colour plate section for the colour representation of this figure.)*

The Tethyan Seaway created a continuous strip of ocean around the equator. This was blocked again around 60 Mya by Africa moving northwards.

Note that formation of a supercontinent meant that there was also only one global ocean in the Permian, known as Panthalassa. The remains of this now make up the shrinking Pacific, while the Atlantic has grown wider.

The theory is crucial for biogeography because the pattern of tectonic plates in Fig. 17.1 bears a striking similarity to the boundaries between the major biogeographical regions (Fig. 14.4). Wegener himself recognised the power of his idea to account for the distribution of organisms. *Lystrosaurus* was a mammal-like reptile that lived around 250 Mya. Its fossils can be found in Antarctica, the Panchet deposits of South India and in the Karoo region of South Africa. The simplest way to explain such a disjunct distribution is that these areas were connected at the time when *Lystrosaurus* existed. Likewise the *Glossopteris* flora is a distinctive band of fossil deposits left by arborescent gymnosperms (pine-like trees) with characteristics suggesting that they grew in temperate climates: they were deciduous, and the wood contains growth rings. Their fossils are found on many continents (including Antarctica), at latitudes which are no longer temperate, but can be matched together as a contiguous band on a single supercontinent.

Be wary of reading too much into distribution patterns though. As will be shown in the next chapter, the ability of organisms to move across the globe can be surprising. In the past over-exuberant biogeographers have tried to use biotic distributions to contest the narrative provided by geologists, proposing land bridges for which no other evidence exists. This should be avoided as it places us on very unsafe (often fictitious) ground; instead biogeographers should accept what geology tells them and use it to infer what happened to natural systems.

17.4 Echoes of Pangæa

The repercussions of the splitting of the supercontinent can be witnessed in the modern world, with the major plates coinciding neatly with the major zoological kingdoms identified by Wallace on the basis of mammalian distributions, as well as the earlier observation by Buffon that each continent possessed different large animals. It accounts for the absence of placental mammals in Australia, which separated from Gondwana prior to their emergence. Australia instead became dominated by marsupials, making it distinct from every other continent on earth (Fig. 17.3). Madagascar is similarly distinctive and will be considered in detail in the next chapter.

In groups of organisms old enough to date back to Pangæa, such as the cypress family (Cupressaceae), which arose in the Triassic, relationships among taxa are consistent with timings of the break-up of the continents (Mao et al., 2012). Likewise among conifers there is a clear difference between northern Laurasian lineages and those from southern Gondwana, dating back to their separation when the tropics became dominated by angiosperms (Procheş, 2013). For bats, which evolved much later, a distinction between the New and Old World faunas illustrates their separation on divided continents (Procheş, 2013).

One particularly ancient group of animals is the velvet worms (Onychophora) which appeared with the first forests of the Devonian around 382 Mya (Murienne et al., 2014). They are the only animal phylum entirely confined to land, particularly to wet forests. They are slow moving and sensitive to desiccation, which means that they avoid sunlight or strong winds. In the modern world most of the nearly 200 species are endemic to very narrow ranges. Fossil data confirms their presence on Pangæa, yet genetic evidence indicates that their diversification occurred long before the supercontinent fragmented. This makes a very important point—Pangæa was not a homogeneous land mass. It comprised a variety of climate zones and was intersected by mountain ranges, and much speciation took place within it.

A focus on taxonomic differences among the continents should not, however, be allowed to obscure the evidence for consistent ecological forces. Tropical rain forests in Southeast Asia, Africa and South America have been separated for many millions of years. There is no such thing as 'The Tropical Rain Forest'—each continent supports forests of entirely different character (Corlett and Primack, 2011). Nevertheless, they share many plant families, a legacy of their former connections. Rather than their communities drifting apart in a similar fashion to their parent land masses, some striking similarities emerge in their composition. For plant taxa shared

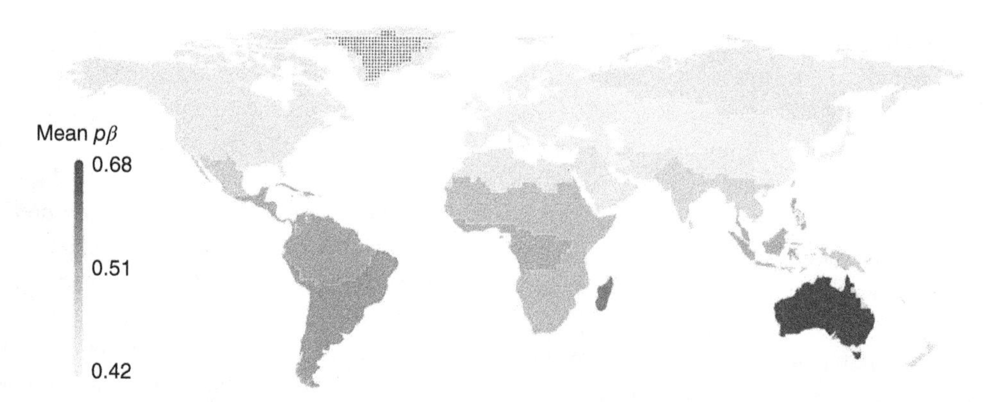

Figure 17.3 Map of evolutionary uniqueness of vertebrate assemblages, calculated as the mean of pairwise phylogenetic β diversity ($p\beta$) comparisons between each focal region and every other; darker shading indicates greater uniqueness. Australia (mean $p\beta$ = 0.68), Madagascar (mean $p\beta$ = 0.63) and South America (mean $p\beta$ = 0.61) have the highest values. (*Source*: Holt et al. (2013). Reproduced with permission from the American Association for the Advancement of Science.)

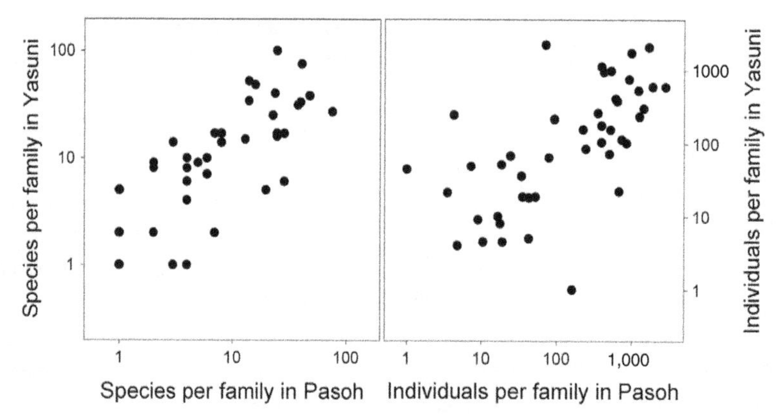

Figure 17.4 Relation between the number of tree species per family and the number of individuals per family in Pasoh, Malaysia, and Yasuni, Ecuador. (*Source*: Ricklefs and Renner (2012). Reproduced with permission from the American Association for the Advancement of Science.)

between rain forest regions, there is a strong correlation in the number of species per family and even number of individuals per family within census plots (Fig. 17.4). This suggests common forces determining the composition of communities despite the geographical distances and environmental differences among them.

Similar rules apply to animals. Even though the dominant mammalian families differ markedly among continents, common syndromes have emerged, implying that there are convergent pathways of evolution leading to the emergence of the same functional types (Fig. 17.5).

What happens when these biotas re-establish contact with one another? One might suspect that, through occupying similar niches, some species would end up being competitively superior to others, leading to rapid displacement. Actually, the opposite is true. Compelling evidence from the fossil record shows that when formerly isolated biotas recombine, their species coexist for over 1 My without any sign of competitive exclusion of one set by another (Tilman, 2011). This has been true for plants as well as animals from molluscs to mammals. For example, around 450 Mya, a wide range of subtidal marine organisms invaded the coast of what would

Figure 17.5 Examples of supposed morphological convergence between unrelated rain forest mammals from (a) tropical America and (b) Africa. The pairs are capybara (*Hydrochaeris hydrochaeris*) and pygmy hippopotamus (*Hexaprotodon liberiensis*); paca (*Agouti paca*) and water chevrotain (*Hyemoschus aquaticus*); agouti (*Dasyprocta* sp.) and royal antelope (*Neotragus pygmaeus*); grey brocket deer (*Mazama gouazoubira*) and yellow-backed duiker (*Cephalophus silvicultor*); giant armadillo (*Priodontes maximus*) and giant pangolin (*Manis gigantea*). (*Source*: Bouliere (1973). Reproduced with permission from Smithsonian Institution.)

later become North America from a western tropical region. This massive influx of species, known as the Richmondian invasion, led to no increase in extinction rates nor the displacement of any native species. Instead the two faunas combined and coexisted for at least 1 My (Patzkowsky, 2007). A similar outcome was seen 3.5 Mya, when the Bering Straits opened and species flooded from the Pacific into the North Atlantic, or at the Great American Biotic Interchange, when the two New World continents were joined around 3 Mya by the Isthmus of Panama. None of these events led to mass extinctions nor the replacement of one set of species by another. In fact, the combination of groups of frogs from North and South America led to a dramatic increase in local species richness (Pinto-Sánchez et al., 2014).

These observations can be generalised into a principle known as the Universal Trade-Off Hypothesis which proposes that all species share a common set of constraints in their ability to evolve new forms (Tilman, 2011). As with niches within communities, there are many possible ways in which species can coexist with one another once they come into contact at the regional scale.

Finally, though the tectonic plates are often thought to only play a role in determining the broad continental patterns of life on earth, the underlying geology of the planet can have more localised effects. In the Indo-Pacific region, species of coral are divided into 11 distinct faunal provinces (Keith et al., 2013). In this complex boundary region, where fragments of plates are moving dynamically, the breaks between marine faunal provinces relate more to the underlying geological features (tectonic plates and tracks of mantle plumes) than to present-day environmental conditions.

17.5 Climatic effects

The movement of continents had profound impacts on the global climate. These occurred due to alterations in oceanic currents, changing the manner in which heat was redistributed. In the early Triassic, equatorial sea surface temperatures rose to around 40 °C, and on land they were much higher. This fierce environment left large gaps in the fossil records: there are no coal deposits in central Pangæa from this time, no fossil tetrapods, almost no fish in tropical seas and no calcareous algae (Sun et al., 2012). Only small invertebrates were able to tolerate such hot conditions.

On the other hand, a concentration of land masses at low latitudes is thought to have triggered glaciations and cold phases in Earth's history (Hoffman and Schrag, 2002). This is due to the increased albedo of land, which reflects more heat than oceans. As ice formed at the poles, it reflected even more solar radiation, leading to a runaway cooling of the planet. Some models have even suggested that all (or most) of the planet was at some points covered in ice, a state known as 'snowball Earth', though this occurred in the Precambrian and therefore would not have troubled multicellular life forms. As ice has formed or melted, it has radically altered the sea level, which has frequently been substantially higher than the present day.

These climatic changes were not without repercussions for the diversity, composition and distribution of life. The species richness of Neotropical plants has varied considerably throughout the Cenozoic (Fig. 17.6a). Diversity increased steadily through the early Eocene stabilized and then fell, with a particular decline at the onset of the Oligocene. This broadly corresponds to the patterns of global temperature, suggesting (as in Section 15.2) that there might be a link between energy and species richness, though it is hard to separate trends in temperature from those in CO_2, which could also account for the patterns (Royer and Chernoff, 2013). The later sharp drop in diversity coincides with the onset of glaciation following formation of Antarctic ice sheets.

Another pattern shown in Fig. 17.6c is turnover of the main plant species, with two substantial shifts occurring. One group of species dominated in the early Palæocene, which were then displaced by another set that persisted until the end of the Eocene, before themselves being replaced. As well as gradual changes within communities, there appear to have also been dramatic transitions that affected entire regions.

An underappreciated aspect of global climatic change is that the biomes outlined in Chapter 14 have not always existed. The earliest tropical rain forests were post-Cretaceous, following the extinction of the dinosaurs (Morley, 2000). Tropical savannahs and grasslands are even younger, with C4 grasses appearing 30 Mya, and only spreading over large areas in the last 8–9 My (Bond and Parr, 2010). While long predating humans, this is relatively recent in the history of life.

When trying to explain the relative numbers of species in each biome, it is therefore important to not only understand their present size and climate but also how they have changed through time. Jetz and Fine (2012) showed that around 80% of

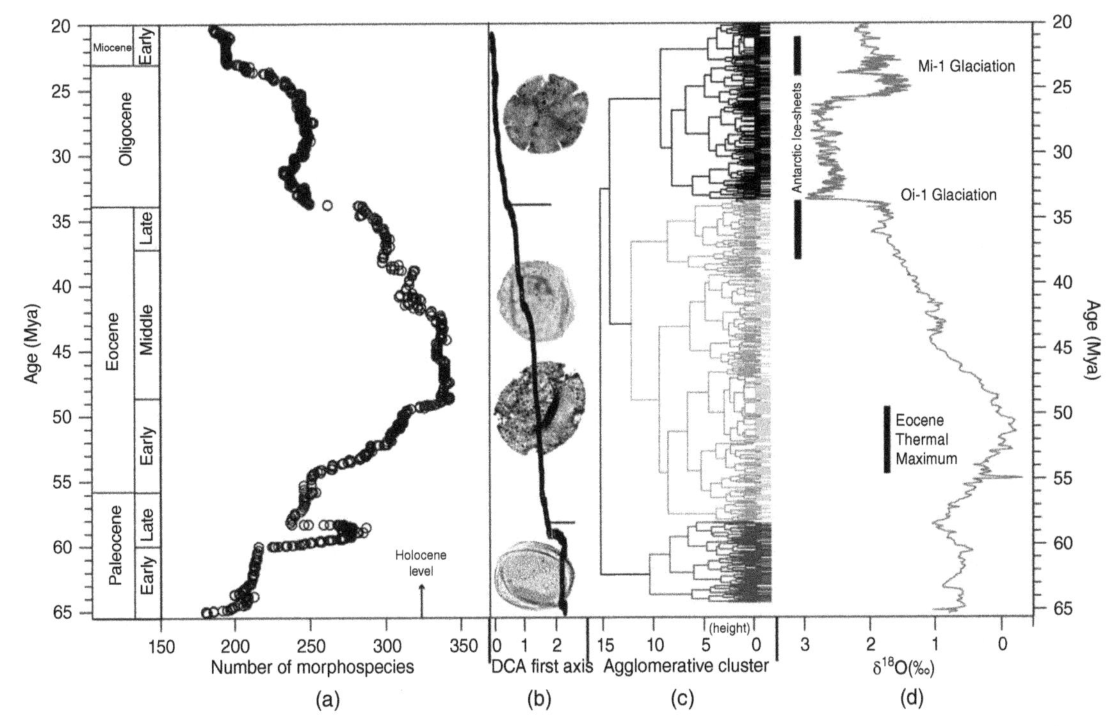

Figure 17.6 Floral diversity and composition in the Neotropics during the early to mid-Cenozoic. (a) Pollen and spore diversity; a Holocene (present-day) standard is marked. (b) First axis of a detrended correspondence analysis that explains 45.9% of the total variance in species composition along the entire profile. Palaeocene palynofloras are clearly different from Eocene to Miocene palynofloras. Characteristic pollen types of each flora are shown. (c) Cluster analysis, showing three distinct palynofloras: Palæocene, Eocene and Oligocene to early Miocene. (d) Global oxygen isotope curve acting as a proxy for average global temperature. (*Source*: Jaramillo et al. (2006). Reproduced with permission from the American Association for the Advancement of Science.)

the vertebrate species richness of biomes could be accounted for by a single model integrating biome area and productivity over time, plus mean temperature (Fig. 17.7). This is best explained by example. The Mediterranean bioregion in Southern Africa has an area of 4.8×10^5 km^2, which is multiplied by its age (55 My) and total productivity over that time period (3×10^{17} kg carbon) to predict its species richness. In large, warm, productive biomes with long histories, such as tropical rain forests, incredible species richness has developed (e.g. the Neotropics, with 7×10^{14} km^2 years and 6.63×10^{20} kg C). This links several of the hypotheses from Chapter 16 for the latitudinal diversity gradient and suggests a role for time, productivity, area and temperature in the accumulation of species.

A common assumption in studies of species interactions with climate is **phylogenetic niche conservatism**, which can be described as the observation that it is easier for species to move than to evolve to cope with new environments (Donoghue, 2008). This was seen in the last chapter, particularly the study by Fine and Ree (2006) in Section 16.3. In a study of over 11,000 plant speciation events, Crisp et al. (2009) found that those within biomes outnumbered those between biomes by a ratio of 25:1, indicating how hard it is to evolve a new set of tolerances.

Sometimes the difficulty of adapting to environmental extremes can lead to sharp biogeographical divisions. Along the border between Malaysia and Thailand, a clear boundary in the floras is known as the Kangar–Pattani Line. This marks the transition

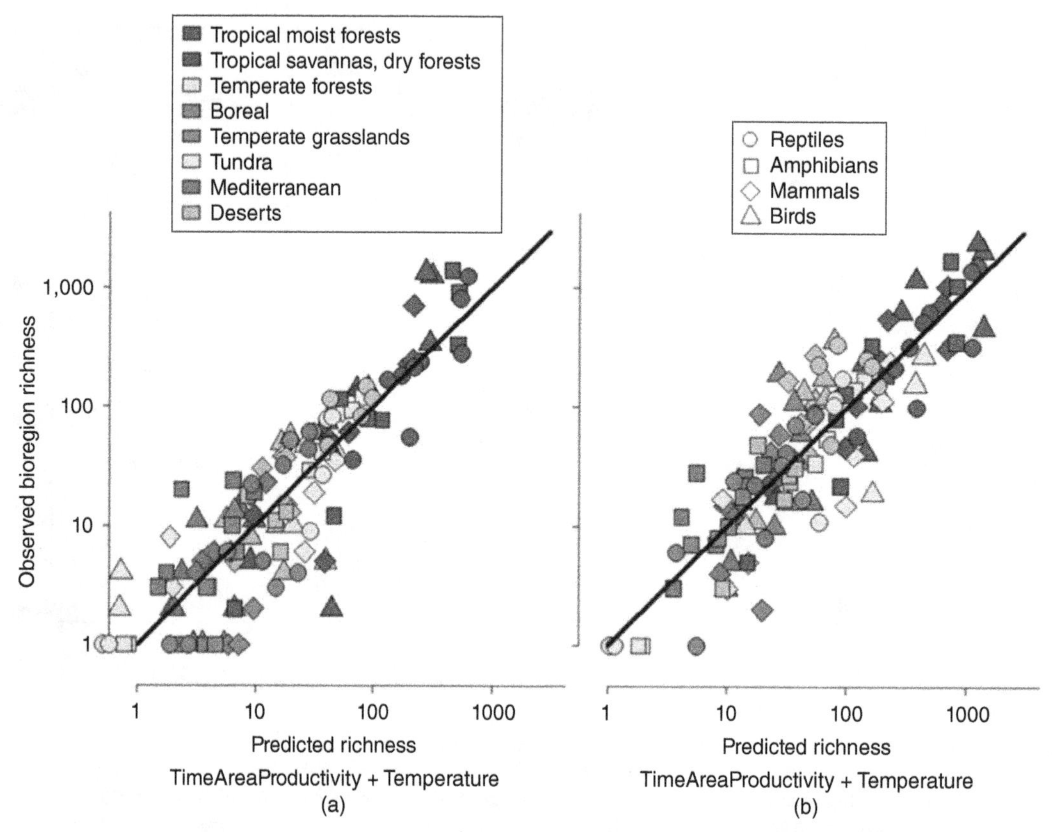

Figure 17.7 Observed versus predicted species richness of terrestrial vertebrates in 32 bioregions and four taxonomic groups (a, endemic species; b, resident species). The model includes two parameters: TimeAreaProductivity and Temperature; lines indicate least squares fit of regressions (r^2 [endemic] = 0.78, r^2 [resident] = 0.78, N = 128). See text for further details. (*Source*: From Jetz and Fine (2012). Used under CC-BY-SA 2.5. http://creativecommons.org/licenses/by-sa/2.5/.) *(See colour plate section for the colour representation of this figure.)*

from evergreen forests to seasonal and sets the northern limit of the distributions of over 500 plant genera (Whitmore, 1984). Once they emerged, each biome has since developed largely independently, with relatively few movements among them.

17.6 Ice ages

Since the beginning of the Pleistocene 2.6 Mya, there have been at least 20 cycles of glacial formation and retreat. These are largely driven by periodicity in Earth's orbit, known as Milanković cycles, caused by slight wobbles in the planet's axis as it spins, as well as minor alterations to its elliptical orbit around the sun. The cycles generate long-term climatic trends on timescales from 20,000 to 100,000 years. At the height of glaciation events, ice covered a third of all land, at depths of up to 2–3 km. The effects were global. Regions now thought of as arid were much wetter, while the equatorial tropics were drier, owing to alterations in global circulation patterns. There was also a greatly reduced seasonality, both in the temperate zone and the tropics. Monsoons were probably weaker than the present day.

In the present interglacial period, it is easy to imagine the ice ages as somehow unusual. In fact the opposite is true; over the Pleistocene as a whole, interglacials make up less than 10% of the epoch, while current conditions are abnormally warm, even

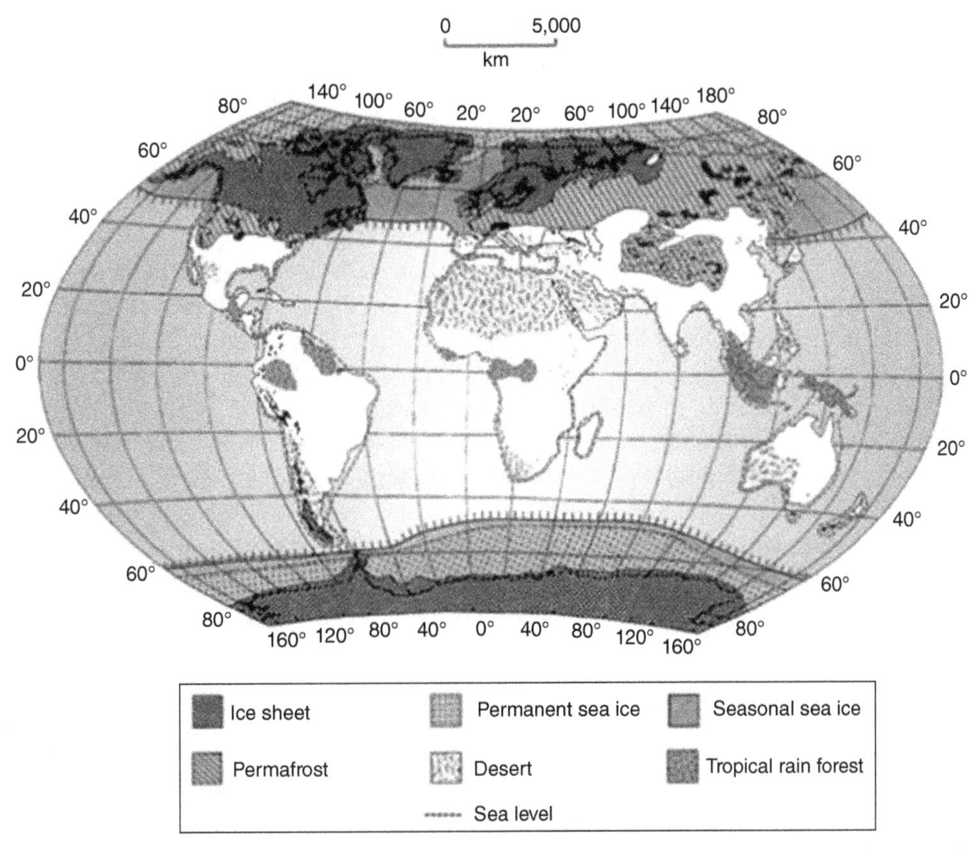

Figure 17.8 Maximum extent of ice and permafrost at the last glacial maximum c. 20,000 years ago. Note changes in the shape of continents owing to lowered sea levels. Reproduced from Hewitt (2000) with permission from the Nature Publishing Group. (*Source*: Modified from Williams et al. (1998) and reproduced with permission from Taylor and Francis in Hewitt (2000).) *(See colour plate section for the colour representation of this figure.)*

without taking into account human-induced climatic change. Some debate continues over whether the Earth should now be warming or cooling; what is certain is that human-induced climate change is occurring at a much faster rate than is possible through orbital effects.

As a result of spreading ice sheets, whole biomes were displaced towards lower latitudes (Fig. 17.8). The consequences for regional species richness differed depending on the layout of land masses. In Europe, severe glaciation covered a large proportion of the continent in ice, with the Alps both preventing southwards movement of taxa and providing another source of glaciers. Many species went extinct. This was less the case in North America, and in East Asia there was little impediment to movement because an unbroken line of vegetation stretched all the way south to the tropical forests allowing scope for survival of species.

Whole biomes did not move *en masse* as the glaciers retreated 11,000 years ago; instead species moved idiosyncratically, not always at the same pace or even in the same direction (Williams et al., 2004). Entire communities have been reshuffled, and in many cases the combinations of species that occurred at the last glacial maximum (LGM) no longer exist, their members now distributed independently. LGM communities share less than a third of species with their modern counterparts, upsetting any belief in communities as obligate associations of species. They are referred to as **no-analogue communities**, communities which do not resemble anything found on Earth in the present day and which occupied climatic conditions that no longer exist (Williams and Jackson, 2007). We are rooted in the present, so this is hard for us to imagine, but a modern ecologist would fail to recognise many of the systems that were widespread after the LGM, even if the species were largely the same.

Moreover, as species continue to move, there is no sign that they have yet reached equilibrium with postglacial changes, never mind anthropogenic climate change. It is owing to this evidence that the superorganism concept of communities (Clements, 1916) was superceded by the impression that species behave largely independently and often simply happen to be found in the same place together (Gleason, 1926). This confirms the judgement that was made in Section 10.2

Communities frequently broke up because the migration rates of species vary. At the LGM, boreal and temperate trees were confined to refugia where glaciers failed to reach. In Europe these are supposed to have been in the Iberian peninsula, Italy and the Balkans (Hewitt, 2000), although little direct fossil evidence has been found. There may have been some smaller refugia in central Europe around the Carpathians where local climates allowed some species to persist (Provan and Bennett, 2008), while elsewhere in the world geothermal activity might have maintained warm spots (e.g. Fraser et al., 2014). The modern distribution of forests therefore reflects the ability of species to spread from these retreats, and their diversity to the ability of species to survive glaciations. Some species, such as the oak *Quercus robur*, have been able to expand rapidly and occupy all the areas in which they are capable of growing, otherwise known as their **climate envelope** (Fig. 17.9). Beech (*Fagus sylvatica*) has yet to reach the limits of its potential range, whereas the Turkey oak *Quercus cerris* remains restricted, as its name suggests, to south-eastern Europe. This is despite their ability to grow in many areas further north, which has been well demonstrated through deliberate planting by humans in gardens and parks throughout the continent. Remember however that interglacials are the exception in the Quaternary; whether tree species survived in Europe may depend as much on their survival through warm periods as on their persistence during glacial cycles (Bennett et al., 1991).

This variable limitation in postglacial dispersal gives rise to patterns of species richness which can be described by a simple model. Imagine a line just south of the Alps below which trees were able to survive at the LGM (Fig. 17.10). Svenning and Skov (2007) found that species richness in European trees bore no resemblance to modern climate, but instead that distance from putative refugia explained 78% of the pattern. European forests remain a long way from reaching equilibrium with present conditions. The same can be said of the European fauna; wide-ranging reptile and amphibian species have reached the current 0°C isotherm (where annual mean temperatures are at freezing point), which acts as a northern limit to their ranges, whereas those with narrow ranges remain limited to where the isotherm lay at the LGM (Araújo et al., 2008). Among European beetles, the gradient of species richness is steeper for poorly dispersed (e.g. wingless) groups (Baselga et al., 2012). Many species are still catching up with the glaciers.

Dispersal of species from refugia following glacial retreat was not solely a terrestrial phenomenon. The red seaweed *Palmaria palmata* occurs along shores on both sides of the North Atlantic, though it probably spent the last glacial period largely confined to a shallow depression on the floor of what is now the English Channel (Provan et al., 2005). Sea level changes also profoundly altered the physical geography of tropical oceans, shifting the distribution of coral reefs. Coral reef fish diversity is closely associated with proximity to reefs that remained stable throughout the Quaternary rather than present-day environmental variables, providing a possible explanation for the existence of marine centres of origin (see Section

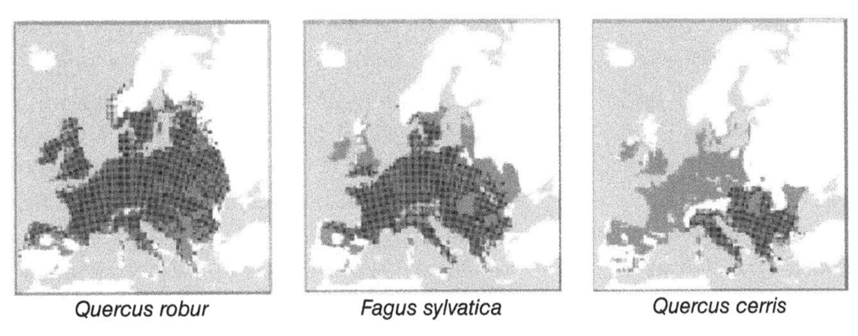

Figure 17.9 Suitable environmental space (grey shading) and observed distribution (black dots) for three European tree species. (*Source*: Svenning and Skov (2007). Reproduced with permission from John Wiley & Sons Ltd.)

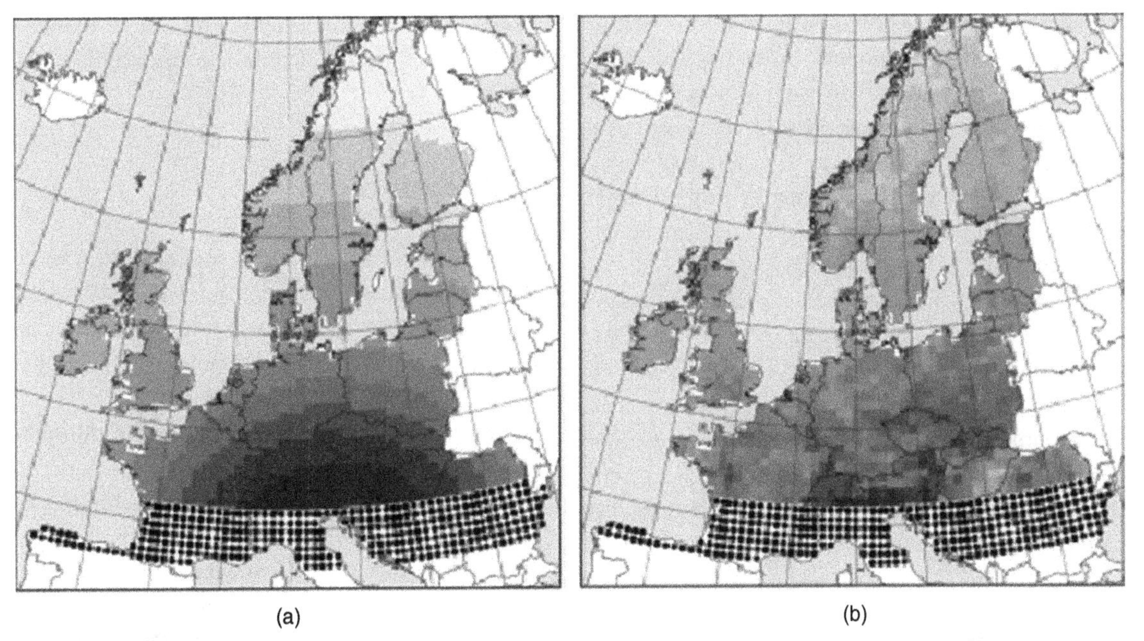

Figure 17.10 (a) Accessibility of areas across Europe predicted solely as distance from refuges in Southern Europe (black dots 43-46°N, and (b) currently observed species richness for 60 species in total, with 2–31 species per cell. Shadings indicate 10 categories from low (light) to high (dark) accessibility and species richness, respectively. (*Source*: Svenning and Skov (2007). Reproduced with permission from John Wiley & Sons Ltd.)

15.6; Pellissier et al., 2014). Some oceanic regions were cut off by land barriers that emerged as sea levels fell and may have lost species as a result.

At the same time as species spread out of refugia, cold-adapted species such as the arctic fox (*Alopex lagopus*) found their ranges contracting towards the poles. Instead of migrating polewards, their ranges contracted through extinction of the southernmost populations (Dálen et al., 2007). The species which now inhabit the far north of the planet were much more widespread for most of the Quaternary.

Increased seasonality of the climate during the Quaternary made it particularly hard for organisms in higher latitudes to tolerate cold winters. This is likely to be the origin of seasonal migrations in many bird species (Winger et al., 2014). Rather than being tropical species which learnt to exploit summer conditions in the temperate zone, as has often been assumed, in the Americas the evidence suggests that long-distance migrants arose in higher latitudes prior to the ice ages and stretched their winter ranges southwards while keeping their breeding ranges in the same place.

There is often an assumption that glaciations only affected those polar areas that were directly covered in ice, but this is an oversimplification. The palms (Arecaceae) are an ancient plant lineage, dating back to the Late Cretaceous (94–89 Mya), long before the tropical rain forests first formed (Couvreur and Baker, 2013). Their distribution reflects their origins on Pangæa. They are present in all tropical zones of the world but limited to such climates due to their soft water-rich tissue, lack of dormancy and inability to tolerate frost. In the modern world their species richness is reduced in tropical areas where temperatures at the LGM were markedly lower than the present day (Kissling et al., 2012). Within Madagascar, palm species richness peaks in areas which had higher precipitation at the LGM, suggesting that patterns have yet to respond to the rapid changes that took place in the Holocene (Rakotoarinivo et al., 2013). This demonstrates that both ancient geological processes and more recent climatic shifts are crucial in understanding global patterns of diversity.

17.7 Sea level

As ice formed and melted, it had global effects on sea levels, which led to land masses emerging or later becoming submerged. Falls in sea level during the ice ages caused the formation of land bridges. The name suggests narrow connecting strips, but the reality is that many islands which are now isolated were subsumed into the continents. The British Channel simply did not exist more than 7,000 years ago—Great Britain was part of the European mainland.

In Wallacea the previous distribution of land is even more telling (Fig. 17.11). The disjunction observed in Section 14.2 between the Indo-Chinese and Australian faunas has a geological basis. At the LGM lower sea levels exposed the Sunda shelf, which grew to twice its present land area. It was possible for taxa to spread southwards on land from mainland Asia, but only as far as Bali. Similarly, the islands off Australia and New Guinea were either connected or the distance between them reduced, allowing easier spread of species as far as Lombok. The narrow gap between Bali and Lombok contains a deep water trench at the boundary between two tectonic plates, meaning they have never been connected. Likewise the neighbouring large islands of Borneo and Sulawesi maintain distinct compositions of species.

17.8 Extinctions

Since the LGM a wave of extinctions has travelled across the world. The megafauna of North America 13,000 years ago was diverse and spectacular, much like modern Africa. There were abundant large herbivores: mastodons, mammoths, camels, llamas, horses, tapirs, ground sloths and numerous ungulates. Large predators included hyenas, dire wolves and other canids, lions and sabre-toothed tigers. There were also giant birds, including *Aiolornis incredibilis*, which with a 5 m wingspan was probably the largest bird ever to take flight and was only one of a number of lost raptor species. Other species of scavenger have vanished, as well as whole groups such as the cave bears. Their fossils can be seen in natural history collections across the continent.

The timing of their extinctions coincides with the spread of people through each region. This is known as the overkill hypothesis (Martin, 1967). Extinctions were seldom triggered by the first humans to arrive. Early human populations were often at low density and confined to coastal regions. Instead the advent of new technologies for hunting allowed a new wave of human migration which drove the majority of other large species (more than 10 kg in size) to extinction (Fig. 17.12). Fossil evidence suggests that

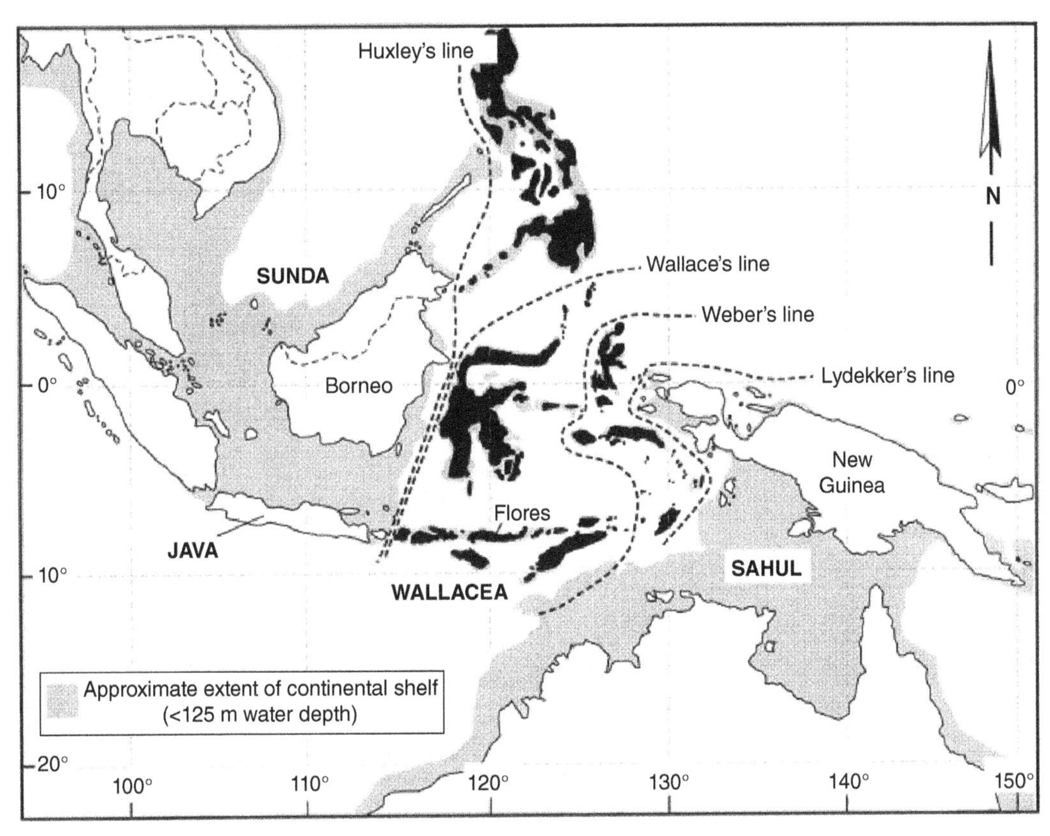

Figure 17.11 Southeast Asia with the division between zoogeographic regions as defined by Huxley, Wallace, Weber and Lydekker, each based on different groups of organisms. (*Source*: Storm (2001). Reproduced with permission from Elsevier.)

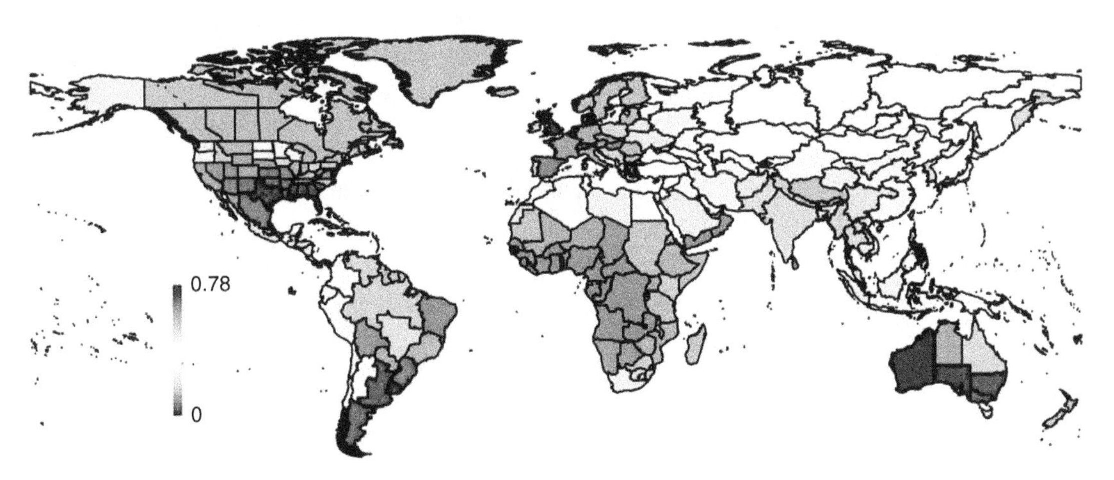

Figure 17.12 Severity of Quaternary large mammal extinctions measured as the proportion of large mammals (≥10 kg) per country which went extinct between 132,000 and 1,000 years ago. Regions in dark grey were excluded from this study. (*Source*: From https://creativecommons.org/licenses/by/4.0/.) *(See colour plate section for the colour representation of this figure.)*

these species only briefly coexisted with humans. The only continent where fewer extinctions occured was Africa, perhaps as a result of those species having evolved alongside humans and being more resistant to hunting. There remains some debate over whether humans were entirely responsible, as perhaps concurrent rapid climate change played a role (Prescott et al., 2012). While both are likely to have been contributors, humans were the dominant cause (Sandom et al., 2014). Nevertheless, most of these extinctions took place long before historical records began. The only place where we can track the loss of large animals through time is in Ancient Egypt, where their disappearance is marked by the gradual loss of representations in archaeological artefacts (Yeakel et al., 2014).

A similar wave of extinctions passed through the Pacific islands as Polynesian peoples settled them around 3,500 ya. Between their first occupation and the arrival of Europeans, around 1,300 species of birds went extinct, accounting for around 10% of global avian species richness, along with perhaps two thirds of all island bird populations (Duncan et al., 2013).

Figure 17.13 The giant kangaroo *Sthenurus*, which went extinct along with 54 other mammal species > 10 kg in size shortly after the arrival of humans in Australia. (*Source*: McGlone (2012). Reproduced with permission from the American Association for the Advancement of Science.)

Stripping the world of most of its megafauna is bound to have had cascading effects on species lower in the food web and even on the structure and composition of plant communities or ecosystem processes. Evidence from Australia suggests that the arrival of humans led indirectly to the replacement of rain forest by drier, sclerophyllous vegetation (Rule et al., 2012). The sequence of events can be traced through fossil deposits, particularly of spores from a fungus that depends on ingestion by herbivores to reproduce because it sporulates in their dung. More spores in deposits imply a greater abundance of large herbivores such as the 150 kg giant kangaroo *Sthenurus* (Fig. 17.13). Around 41,000 years ago, these spores vanish—coincident with the arrival of humans. At the same time, charcoal deposits appear, showing an increase in fire. The two events are related as lower browsing pressure allows fuel loads to build up and increases fire intensity, a pattern evident across the world. Pollen records document a transition from rain forest conifers to grasses and *Eucalyptus*, the landscape most commonly associated with modern-day Australia. The wild bush is, at least in part, an artefact of human impacts.

The effects of humans did not end there; the extinction of the Australian megafauna left behind only the Tasmanian devil and thylacine cat as the dominant carnivores. These disappeared from the mainland by 3,000 years ago, with some debate surrounding whether this was caused by the introduction of dingos (which are actually descendents of domesticated Asian wild dogs) or increasing human populations (for evidence implicating the latter, see Prowse et al., 2014). Regardless of the root cause, it triggered a second regime shift. Where dingos are controlled there are cascading changes in the populations of smaller animals and vegetation structure (Colman et al., 2014).

Figure 17.14 Pre-Columbian raised fields in Savane Corossony, showing the small, round mounds (1–1.5 m diameter). (*Source*: McKey et al. (2010).)

These transformations were followed by dramatic alterations to landscapes as agriculture spread across the world. This is a candidate event with which to mark the beginning of the Anthropocene, the proposed epoch in which the predominant influence on natural systems is no longer climate or geology but mankind. It is tempting to believe that this is only a recent phenomenon, but in fact there are signs that ancient civilisations had similar impacts. In Amazonia, long before Europeans arrived in 1492, farmers on the Guinas coast built systems of mounds on which to grow their crops (Fig. 17.14). These were then colonised by ecosystem engineers—ants, termites and earthworms—that have maintained them as a self-perpetuating stable system ever since (McKey et al., 2010). In cases like this the boundaries between human and natural systems become blurred.

Recent changes have been even more dramatic. Ellis et al. (2010) reconstructed the distribution of life on Earth over the last few centuries, and the pace of change has been remarkable. First they mapped the distribution of anthromes, defined as anthropogenic biomes, that is, habitats that would not exist without their creation by humans. Over the last 300 years these have expanded dramatically, such that in most habitable regions of the world there has been at least some modification. In 1700, 50% of the terrestrial biosphere remained 'wild', or largely unaltered by man, at least in terms of the dominant forms of vegetation (Fig. 17.15). Of the remainder, 45% was still semi-natural, while only 5% had been completely transformed, and only in some centres of civilisation (e.g. Europe, northeast China, northern India). By 2000 only 23% of land could be considered 'wild', mostly restricted to uninhabitable regions, while 20% was semi-natural. More than half the land surface area exists in a state that has been created by humans, with the majority of the change occurring since 1900. Over the entire history of life, the pace of global change has seldom been so rapid.

The rate of species loss has been unremitting in the last few centuries and has lately accelerated dramatically. It is accompanied by major shifts in the abundances and distributions of species, a process masked by the raw numbers of extinctions. While 322 terrestrial vertebrate species have gone extinct in the last five centuries, those that remain have seem marked reductions in their numbers and ranges, which is likely to have had equally profound impacts on natural systems (Dirzo et al., 2014). The candidates for future extinctions are also decreasing in body size. The overkill hypothesis accounted for the prehistorical loss of larger species. In the modern world, ever smaller species are becoming threatened with extinction.

17.9 Conclusions

Over the long history of life on Earth, there have been dramatic and ongoing changes to the layout of land and the climate. Continental drift has caused vicariance, the splitting apart of land masses and their biota, which appears to account for many of the broadscale disjunctions between biogeographical kingdoms and realms, as well as phenomena such as Buffon's Law. Continental movements, along with alterations in Earth's orbit, have led to climate

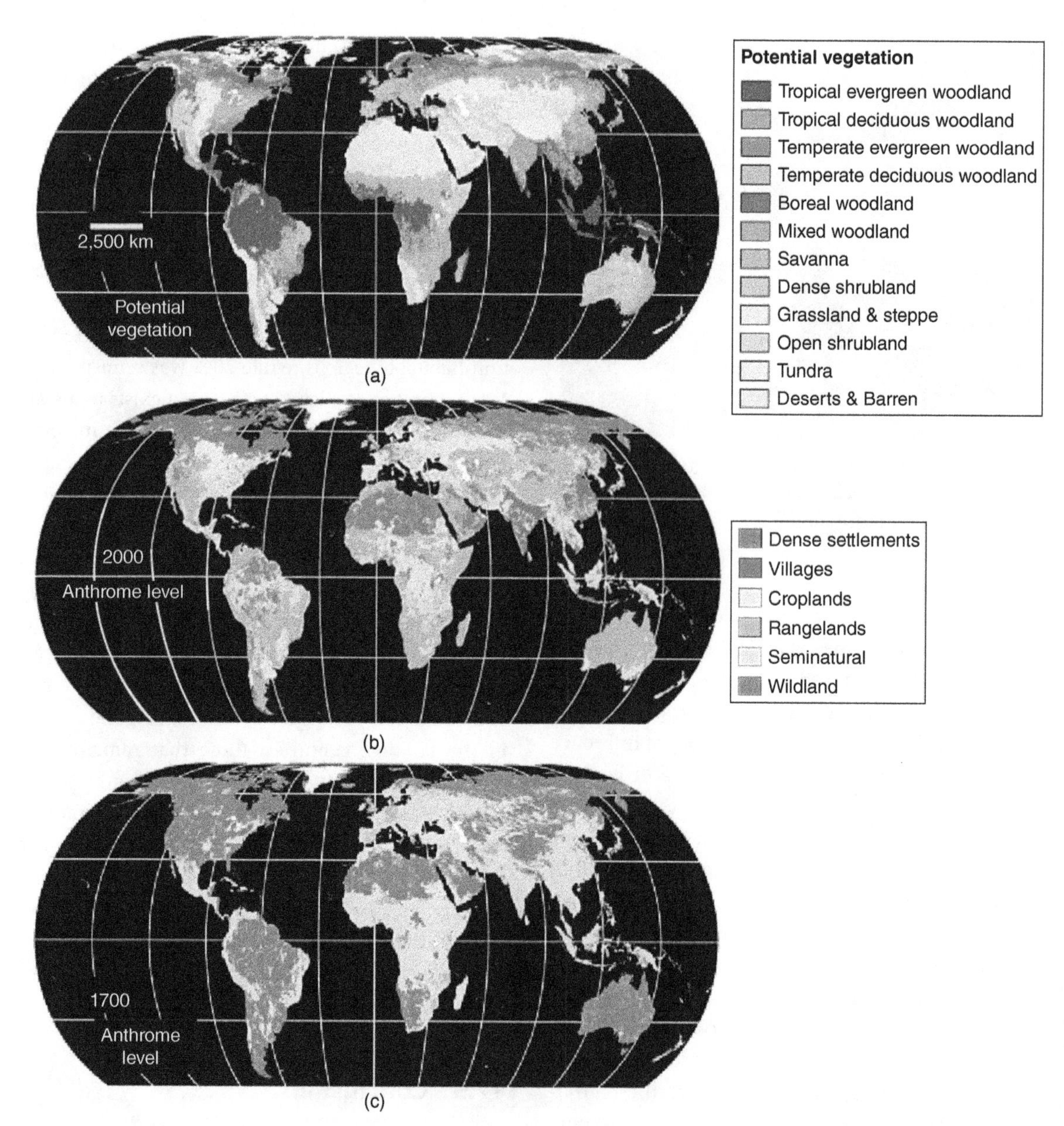

Figure 17.15 (a) Potential natural vegetative biomes with their state in (b) 2000 and (c) 1700. (*Source*: Ellis et al. (2010). Reproduced with permission from John Wiley and Sons Ltd.) *(See colour plate section for the colour representation of this figure.)*

change over geological timescales. More recently, in the Pleistocene, waves of glaciation have forced the movement of species and entire biomes. Communities have reassembled and modern species combinations may have had no analogues at the Last Glacial Maximum; likewise many communities which existed then have since disintegrated. Ice age conditions were not peculiar; in fact the present interglacial is unusually warm and dry for this epoch, and species distributions are still catching up with the last 11,000 years of change. There is therefore no such thing as a 'normal' climate on which to base any judgements about where particular species or biomes ought to be. Finally, the impacts of humans have been profound and ancient, occurring before recorded history, and accelerating in the present day, causing widespread extinctions and habitat change. With this perspective on the changes in life on Earth through deep time, you should now be asking: what do we mean when we call something a natural system?

17.9.1 Recommended reading

- Ellis, E. C., K. K. Goldewijk, S. Siebert, D. Lightman, and N. Ramankutty, 2010. Anthropogenic transformation of the biomes, 1700 to 2000. *Global Ecology and Biogeography* 19:589–606.
- Lohman, D. J., M. de Bruyn, T. Page, K. von Rintelen, R. Hall, P. K. L. Ng, H.-T. Shih, G. R. Carvalho, and T. von Rintelen, 2011. Biogeography of the Indo-Australian Archipelago. *Annual Review of Ecology, Evolution, and Systematics* 42:205–226.
- Tilman, D., 2011. Diversification, biotic exchange and the universal trade-off hypothesis. *American Naturalist* 178:355–371.

17.9.2 Questions for the future

- Can species disperse rapidly enough to keep pace with human-induced climate change, and what does the past tell us about the outlook for a world that could be 4 °C warmer by 2100?
- When did humans become the dominant force determining patterns of life?
- What will happen when Australia finally collides with Asia and the continents come back together?

References

Araújo, M. B., D. Nogués-Bravo, J. A. F. Diniz-Filho, A. M. Haywood, P. J. Valdes, and C. Rahbek, 2008. Quaternary climate changes explain diversity among reptiles and amphibians. *Ecography* 31:8–15.

Baselga, A., J. M. Lobo, J.-C. Svenning, P. Aragón, and M. B. Araújo, 2012. Dispersal ability modulates the strength of the latitudinal richness gradient in European beetles. *Global Ecology and Biogeography* 21:1106–1113.

Bennett, K. D., P. C. Tzedakis, and K. J. Willis, 1991. Quaternary refugia of North European trees. *Journal of Biogeography* 18:103–115.

Bond, W. J. and C. L. Parr, 2010. Beyond the forest edge: ecology, diversity and conservation of the grassy biomes. *Biological Conservation* 143:2395–2404.

Bouliere, F., 1973. The comparative ecology of rainforest mammals in Africa and tropical America; some introductory remarks. In B. J. Meggers, E. S. Ayensu, and W. D. Duckworth, editors, *Tropical Forest Ecosystems in Africa and South America*, pages 279–292. Smithsonian Press.

Clements, F. E., 1916. *Plant Succession: Analysis of the Development of Vegetation*. Carnegie Institute of Washington.

Colman, N. J., C. E. Gordon, M. S. Crowther, and M. Letnic, 2014. Lethal control of an apex predator has unintended cascading effects on forest mammal assemblages. *Proceedings of the Royal Society Series B* 281:20133094.

Corlett, R. T., 2015. The Anthropocene concept in ecology and conservation. *Trends in Ecology & Evolution* 30: 36–41.

Corlett, R. T. and R. B. Primack, 2011. *Tropical Rain Forests: An Ecological and Biogeographical Comparison*. Wiley-Blackwell, second edition.

Couvreur, T. L. P. and W. J. Baker, 2013. Tropical rain forest evolution: palms as a model group. *BMC Biology* 11:48.

Crisp, M. D., M. T. K. Arroyo, L. G. Cook, M. A. Gandolfo, G. J. Jordan, M. S. McGlone, P. H. Weston, M. Westoby, P. Wilf, and H. P. Linder, 2009. Phylogenetic biome conservatism on a global scale. *Nature* 458:754–756.

Dálen, L., M. Nyström, C. Valdiosera, M. Germonpré, M. Sablin, E. Turner, A. Angerbjörn, J. L. Arsuaga, and A. Götherström, 2007. Ancient DNA reveals lack of postglacial habitat tracking in the arctic fox. *Proceedings of the National Academy of Sciences of the United States of America* 104:6726–6729.

Dirzo, R., H. S. Young, M. Galetti, G. Ceballos, N. J. B. Isaac, and B. Collen, 2014. Defaunation in the Anthropocene. *Science* 345:401–406.

Donoghue, M. J., 2008. A phylogenetic perspective on the distribution of plant diversity. *Proceedings of the National Academy of Sciences of the United States of America* 105:11549–11555.

Duncan, R. P., A. G. Boyer, and T. M. Blackburn, 2013. Magnitude and variation of prehistoric bird extinctions in the Pacific. *Proceedings of the National Academy of Sciences of the United States of America* 110:6436–6441.

Ellis, E. C., K. K. Goldewijk, S. Siebert, D. Lightman, and N. Ramankutty, 2010. Anthropogenic transformation of the biomes, 1700 to 2000. *Global Ecology and Biogeography* 19:589–606.

Fine, P. V. A. and P. H. Ree, 2006. Evidence for a time-integrated species-area effect on the latitudinal gradient in tree diversity. *American Naturalist* 168:796–804.

Fraser, C. I., A. Terauds, J. Smellie, P. Convey, and S. L. Chown, 2014. Geothermal activity helps life survive glacial cycles. *Proceedings of the National Academy of Sciences of the United States of America* 111:5634–5639.

Gleason, H. A., 1926. The individualistic concept of the plant association. *Torrey Botanical Club Bulletin* 53:7–26.

Hewitt, G., 2000. The genetic legacy of the Quaternary ice ages. *Nature* 405:907–913.

Hoffman, P. G. and D. P. Schrag, 2002. The snowball Earth hypothesis: testing the limits of global change. *Terra Nova* 14:129–155.

Holt, B. G., J.-P. Lessard, M. K. Borregaard, S. A. Fritz, M. B. Araújo, D. Dimitrov, P.-H. Fabre, C. H. Graham, G. R. Graves, K. A. Jønsson, D. Nogués-Bravo, Z. Wang, R. J. Whittaker, J. Fjeldså, and C. Rahbek, 2013. An update of Wallace's zoogeographic regions of the world. *Science* 339:74–78.

Jaramillo, C., M. J. Rueda, and G. Mora, 2006. Cenozoic plant diversity in the Neotropics. *Science* 311:1893–1896.

Jetz, W. and P. V. A. Fine, 2012. Global gradients in vertebrate diversity predicted by historical area-productivity dynamics and contemporary environment. *PLoS Biology* 10:e1001292.

Keith, S. A., A. H. Baird, T. P. Hughes, J. S. Madin, and S. R. Connolly, 2013. Faunal breaks and species composition of Indo-Pacific corals: the role of plate tectonics, environment and habitat distribution. *Proceedings of the Royal Society Series B* 280:20130818.

Kissling, W. D., W. J. Baker, H. Balslev, A. S. Barfod, F. Borchsenius, J. Dransfield, R. Govaerts, and J.-C. Svenning, 2012. Quaternary and pre-Quaternary historical legacies in the global distribution of a major tropical plant lineage. *Global Ecology and Biogeography* 21:909–921.

Mao, K., R. I. Milne, L. Zhang, Y. Pang, J. Liu, P. Thomas, R. R. Mill, and S. S. Renner, 2012. Distribution of living Cupressaceae reflects the breakup of Pangea. *Proceedings of the National Academy of Sciences of the United States of America* 109:7793–7798.

Martin, P. S., 1967. Prehistoric overkill. In P. S. Martin and H. E. Wright Jr., editors, *Pleistocene Extinctions: The Search for a Cause*, pages 75–120. Yale University Press.

McGlone, M., 2012. The hunters did it. *Science* 335: 1452–1453.

McKey, D., S. Rostain, J. Iriarte, B. Glaser, J. J. Birk, I. Holst, and D. Renard, 2010. Pre-Columbian agricultural landscapes, ecosystem engineers, and self-organized patchiness in Amazonia. *Proceedings of the National Academy of Sciences of the United States of America* 107:7823–7828.

Morley, R. J., 2000. *Origin and Evolution of Tropical Rain Forests*. John Wiley & Sons, Ltd.

Murienne, J., S. R. Daniels, T. R. Buckley, G. Mayer, and G. Giribet, 2014. A living fossil tale of Pangaean biogeography. *Proceedings of the Royal Society Series B* 281:20132648.

Patzkowsky, M. E., 2007. Diversity partitioning of a late Ordovician marine biotic invasion: controls on diversity in regional ecosystems. *Paleobiology* 33:295–309.

Pellissier, L., F. Leprieur, V. Parravicini, P. F. Cowman, M. Kulbicki, G. Litsios, S. M. Olsen, M. S. Wisz, D. R. Bellwood, and D. Mouillot, 2014. Quaternary coral reef refugia preserved fish diversity. *Science* 344:1016–1019.

Pinto-Sánchez, N. R., A. J. Crawford, and J. J. Wiens, 2014. Using historical biogeography to test for community saturation. *Ecology Letters* 17:1077–1085.

Prescott, G. W., D. R. Williams, A. Balmford, R. E. Green, and A. Manica, 2012. Quantitative global analysis of the role of climate and people in explaining late Quaternary megafaunal extinctions. *Proceedings of the National Academy of Sciences of the United States of America* 109:4527–4531.

Procheş, S., 2013. Latitudinal and longitudinal barriers in global biogeography. *Biology Letters* 2:69–72.

Provan, J. and K. D. Bennett, 2008. Phylogeographic insights into cryptic glacial refugia. *Trends in Ecology & Evolution* 23:564–571.

Provan, J., R. A. Wattier, and C. A. Maggs, 2005. Phylogeographic analysis of the red seaweed *Palmaria palmata* reveals a Pleistocene marine glacial refugium in the English Channel. *Molecular Ecology* 14:793–803.

Prowse, T. A. A., C. N. Johnson, C. J. A. Bradshaw, and B. W. Brook, 2014. An ecological regime shift resulting from disrupted predator-prey interactions in Holocene Australia. *Ecology* 93:693–702.

Rakotoarinivo, M., A. Blach-Overgaard, W. J. Baker, J. Dransfield, J. Moat, and J.-C. Svenning, 2013. Palaeo-precipitation is a major determinant of palm species richness across Madagascar: a tropical biodiversity hotspot. *Proceedings of the Royal Society Series B* 280:20123048.

Ricklefs, R. E. and S. S. Renner, 2012. Global correlations in tropical tree species richness and abundance reject neutrality. *Science* 335:464–467.

Royer, D. L. and B. Chernoff, 2013. Diversity in neotropical wet forests during the Cenozoic is linked more to atmospheric CO_2 than temperature. *Proceedings of the Royal Society Series B* 280:20131024.

Rule, S., B. W. Brook, S. G. Haberle, C. S. M. Turney, A. P. Kershaw, and C. N. Johnson, 2012. The aftermath of megafaunal extinction: ecosystem transformation in Pleistocene Australia. *Nature* 335:1483–1485.

Sandom, C., S. Faurby, B. Sandel, and J.-C. Svenning, 2014. Global late Quaternary megafauna extinctions linked to humans, not climate change. *Proceedings of the Royal Society Series B* 281:20133254.

Storm, P., 2001. The evolution of humans in Australasia from an environmental perspective. *Palaeogeography, Palaeoclimatology, Palaeoecology* 171:363–383.

Sun, Y., M. M. Joachimski, P. B. Wignall, C. Yan, Y. Chen, H. Jiang, L. Wang, and X. Lai, 2012. Lethally hot temperatures during the Early Triassic greenhouse. *Nature* 338:366–370.

Svenning, J.-C. and F. Skov, 2007. Could the tree diversity pattern in Europe be generated by postglacial dispersal limitation? *Ecology Letters* 10:453–460.

Tilman, D., 2011. Diversification, biotic interchange, and the Universal Trade-Off Hypothesis. *American Naturalist* 178:355–371.

Wegener, A., 1915. *Die Entstehung der Kontinente und Ozeane*. Vieweg.

Whitmore, T. C., 1984. *Tropical Rainforests of the Far East*. Oxford University Press, second edition.

Williams, D., D. Dunkerley, P. DeDeckker, P. Kershaw, and M. Chappell, 1998. *Quaternary Environments*. Arnold.

Williams, J. W. and S. T. Jackson, 2007. Novel climates, no-analog communities, and ecological surprises. *Frontiers in Ecology and the Environment* 5:475–482.

Williams, J. W., B. N. Shuman, T. J. Webb, P. J. Bartlein, and P. L. Leduc, 2004. Late-Quaternary vegetation dynamics in North America: scaling from taxa to biomes. *Ecology* 74:309–334.

Winger, B. M., F. K. Barker, and R. H. Ree, 2014. Temperate origins of long-distance seasonal migration in New World songbirds. *Proceedings of the National Academy of Sciences of the United States of America* 111:12115–12120.

Yeakel, J. D., M. M. Pires, L. Rudolf, N. J. Dominy, P. L. Koch, P. R. Guimarães Jr., and T. Gross, 2014. Collapse of an ecological network in Ancient Egypt. *Proceedings of the National Academy of Sciences of the United States of America* 111:14472–14477.

CHAPTER 18
Dispersal

18.1 The big question

Widespread acceptance of continental drift theory may have initially been slow, but it swiftly became the default explanation for all biogeographical patterns, often to the exclusion of other processes. It was believed to be more plausible that natural divisions would arise through the erection of physical barriers or **vicariance** than through alternative mechanisms. Yet it remained possible that species could also disperse among regions, either along corridors, across land bridges that had since vanished or by exploiting rare opportunities. Until recently no hypothesis was irrefutable because processes that occurred in the past cannot be directly witnessed or reconstructed.

The challenge posed to biogeographers was to explain **disjunctions**, where a species or group is found in two disconnected regions. Harvestman spiders (superfamily Zalmoxoidea) are found on both sides of the Pacific, even though the basal species is thought to have arisen in Amazonia or Central America (Fig. 18.1). Molecular dating places the split between the two groups at 92 Mya—around 80 My too late for them to share a common ancestor on Gondwana. It is, however, consistent with the direction of oceanic currents at the time. Could they have crossed the Pacific? How many patterns in the natural world are due to dispersal rather than geological mechanisms?

In the past, arguments over which explanations are more plausible often became fierce and intractable. Now, however, with the arrival of molecular genetic techniques, it is possible to piece together the evolutionary history of species and settle some disputes.

18.2 Range expansion

All species disperse. Whether it involves a plant releasing seeds or a chick leaving the nest to find a new territory, dispersal is a feature of all organisms. Most movements within the lifetime of an organism are restricted though, and long-distance dispersal is rare at the individual level. At the population level, however, the combined movements of many individuals can lead to rapid spread. When viewed on a biogeographical scale, there are three processes that give rise to movement across entire regions. These are known as jump dispersal, diffusion and secular migration, and occur at different rates.

The first method of long-distance movement, *jump dispersal*, is characterised by single, dramatic extensions of a species' range. Though any single event sounds improbable, there can be no doubt that it occurs. If you dig a new pond in your garden, it will be colonised within the year by a range of species that are restricted to aquatic habitats, despite having no connection to other waterbodies.

Other evidence emerges at the most dramatic of scales. When the Krakatau volcano erupted in 1883,

Natural Systems: The organisation of life, First Edition. Markus P. Eichhorn.

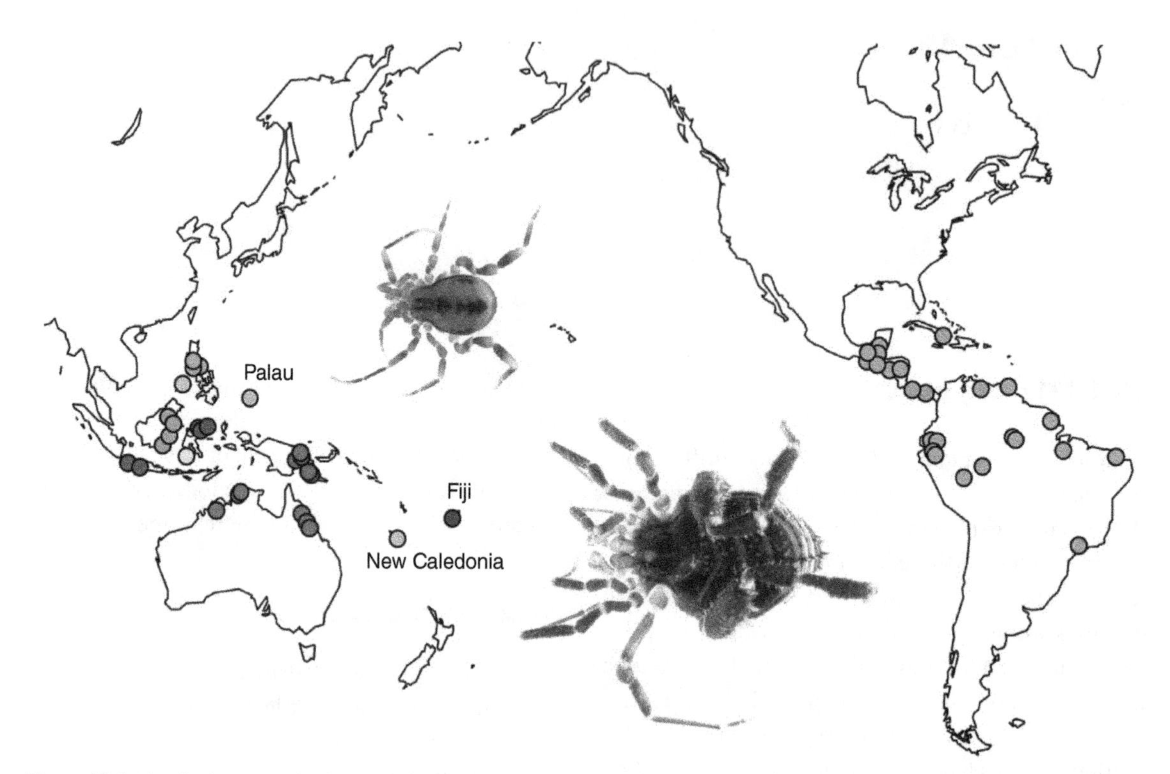

Figure 18.1 Distribution map of Zalmoxoidea samples with two illustrative species. Shading corresponds to recognised biogeographical clusters. (*Source*: Sharma and Giribet (2012). Reproduced with permission from Royal Society.)

the remnant islands were devoid of any life; yet by 1886 there were already 24 plant species, rising to around 300 in 1932 (Whittaker et al., 1989). These communities have been tracked by subsequent trips, and diversity continues to increase through time (Thornton, 1997). Clearly the distance to the nearest mainland, around 40 km away on Java or Sumatra, has posed few problems for the dispersal of these species (even the nearest other small island is 12 km away). Other volcanic islands such as the Galápagos and Hawaii are paradigms of how even distant patches of land can acquire impressive numbers of species through dispersal, despite starting empty. Islands are special cases and will be considered in more detail in the following chapters.

Range expansion can also happen incrementally on a generation-by-generation basis through the process of *diffusion*. In this case a species expands its range to fill the available space more gradually. Invasive **alien species** provide helpful and well-documented case studies. The European starling was introduced to the Americas in 1896 and spread to cover most of the United States, though slowed as it reached the limits of its environmental tolerance. The red fox similarly expanded through Australia following its introduction in 1860. In both cases jump dispersal was necessary to get them onto each continent in the first place, enabled by deliberate human introductions.

New arrivals seldom exhibit immediate population explosions. Typically establishment is slow, requiring a large number of colonists and sufficient time for them to adapt to local conditions. Only once they have done so is the rapid spread characteristic of invasive species seen. Often, if a large area is habitable, the increase in range size is exponential and symmetrical on all its edges, though eventually the spread is arrested once environmental limits or natural barriers are reached. Classically it was believed that climatic tolerances limit species' ranges at their colder edges, while biotic factors are more important at the warm edge (Dobzhansky, 1950; MacArthur, 1972), but more recent evidence suggests that it is usually the environment alone that places limits on the final extent of a species' range (Cahill et al., 2014).

As species reach the limits of their tolerance, diffusion slows and can cease completely, at least until the species is able to evolve new capabilities to survive beyond its present niche. This can lead to a process of *secular migration* whereby a lineage diverges and adapts to each new environment as it spreads. The rate of this is slow but over long timescales allows groups of organisms to colonise large areas. Camels, for example, entered North America via the Bering Straits and then spread southwards across the land bridge between North and South America. There they gave rise to a set of species including the modern llama, guanaco and vicuña, while the linking species in North America subsequently went extinct.

18.3 Mechanisms of dispersal

Often dispersal of organisms is active and deliberate. It is important to distinguish dispersal, referring to individuals spreading apart in space, from **migration**, in which mass directional movement of entire populations takes place. These include the seasonal migrations of many volant species, such as the golden plover, a small bird of only 150 g in size which nonetheless travels each year from a breeding range in Eurasia and North America down to overwintering territories in Southeast Asia and South America. Migrants occur in different ranges in each season rather than expanding into new areas. The same applies even to species that undertake multigenerational migrations. The annual painted lady butterfly migration between Africa and Europe involves a total of six generations, moving across up to 60° latitude and thousands of kilometres, but finishes in roughly the same areas each year (Stefanescu et al., 2013).

Nevertheless, migrations can lead to organisms ending up in unexpected places; this is known as **vagrancy**. Migrants occasionally get lost, perhaps due to strong winds or other freak events, and are unable to return. Hoary bats, *Lasiurus cinereus*, migrate every year from the northern United States down to the Neotropics. There is also a sister species that occurs in Hawaii—perhaps this arose when a population became misdirected? There is evidence from birds that migratory species diversify more often than sedentary ones, and do so by creating new sedentary sister species (Rolland et al., 2014). Migratory behaviour has evolved in birds many times, perhaps in response to climatic changes (Winger et al., 2014). It also appears to be associated with lower extinction rates, which may reflect the greater flexibility of migratory species in actively choosing where to reside (Rolland et al., 2014).

Even large animals undertake dramatic trips for mysterious reasons. There have been numerous accounts of elephants found swimming out at sea (Johnson, 1980). In one case a mother and calf were spotted 50 km off the coast of Sri Lanka. They were followed as they made a return trip to a neighbouring island for unknown purposes. Perhaps this is how elephants first colonised Sri Lanka, swimming out from India? The elephants of Borneo are a more controversial case. The story most frequently told is that they were introduced by the Sultan of Sulu from animals presented to him as a gift in 1750. Genetic evidence, however, suggests that they reached the island during the Pleistocene, when low sea levels exposed the Sunda shelf (Fernando et al., 2003). This still leaves them 1300 km beyond the known historical core range of the species. Could it also be possible that they swam or were swept there by ocean currents?

Other organisms have mechanisms for passive dispersal, whether by wind (anemochory), water (hydrochory), ingestion by animals (endozoochory) or carried by animals externally (epizoochory) (Carlquist, 1967). Still others employ a deliberate hitch-hiking strategy known as phoresy, of which parasites are the prime example.

Seeds often possess adaptations which promote dispersal by wind, sticking to animals or being consumed in fruit. Tempting an animal to eat fruit in order that they deposit seeds elsewhere prompted the evolution of angiosperms, 'angio' coming from the Greek for vessel and 'sperm' meaning seed, referring to plants that enclose their seeds within something. Employing an animal as a dispersal agent enables long-distance movement. There are at least 30 species of vascular plants with bipolar distributions, occurring in both extreme high and low latitudes. In the case of species in the genus *Empetrum*, genetic evidence suggests the southern species split only 1 Mya, making long-distance transport by migratory birds the most likely cause (Popp et al., 2011).

Plants are not the only organisms to utilise passive dispersal. Brine shrimps (*Artemisia* spp.) live in saline pools that frequently dry up. As a result they have evolved the ability to produce encysted eggs which are protected from desiccation and can survive for a considerable time, blown around in the wind. Once they find themselves in water they hatch and are able to colonise the new habitat. Some species do this quite deliberately; money spiders, on windy days, climb to high points of vegetation and spin a silk parachute that allows them to be whisked to new locations (Goodacre et al., 2009).

In the oceans many benthic organisms have a pelagic phase, allowing them to be moved around by currents, even if they are subsequently attached to a rock for the remainder of their lives. One might expect planktonic organisms to be so well mixed as to make the oceans relatively uniform, but as demonstrated in Section 14.6, oceanic algae can be separated into biogeographical regions (Longhurst, 1998). This appears to be because even though currents rapidly move organisms around the world, high rates of evolution outstrip even this constant dispersal, meaning that the oceans are not as well mixed as one might assume (Hellweger et al., 2014).

Darwin was baffled by the presence of species with poor dispersal capacities on isolated oceanic islands. How could there be land snails on Hawaii? His speculations, including that they had been carried on the feet of birds, were subjected to mockery. Any explanation for long-distance dispersal needs to breach this credibility gap. In his search for evidence, Darwin washed the feet of birds and recovered all manner of plant seeds and small invertebrates. One possibility that he perhaps did not consider was that some snails survive being eaten by birds, which has since been demonstrated experimentally (Wada et al., 2012). There is evidence that tropical marine snails crossed the isthmus of Central America between the Pacific and Atlantic over land at least twice within the last million years, almost certainly carried by migratory shorebirds (Miura et al., 2012). One of the most embarrassing examples of epizoochory was the demonstration that invertebrates were transported between continents by biogeographers who had attended a conference field trip, still attached to their muddy boots (Valls et al., 2014).

No one mechanism of passive dispersal is necessarily superior to others. Among the plants of the Galápagos archipelago, around a third possess adaptations for dispersal by wind or water and another third by animals, while the remaining third have no obvious traits for enabling long-distance movement. This highlights that gaps remain in our understanding of how organisms disperse long distances (Vargas et al., 2012).

18.4 Barriers

Obstacles are often placed in the way of dispersing individuals. Such barriers can make dispersal so improbable as to be effectively impossible (though a small chance is still a chance!). The hurdle presented by a barrier differs among species. Janzen (1967) suggested that mountains were more difficult to cross for tropical organisms than temperate; in effect passes were higher in the tropics. His logic was that seasonal climates in the temperate zone necessitated that species possess a wider range of tolerances. Because the climatic range at the top of a temperate

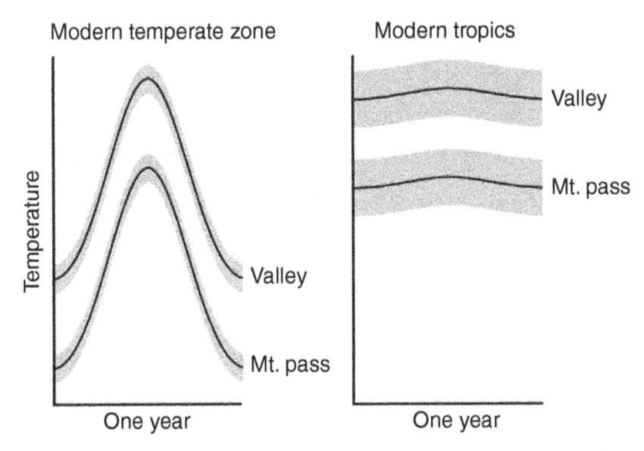

Figure 18.2 Janzen's (1967) hypothesis regarding mean (black line) and daily fluctuations (grey envelopes) of temperature throughout the year in valleys and mountain passes. (*Source*: Archibald et al. (2013). Reproduced with permission from Elsevier.)

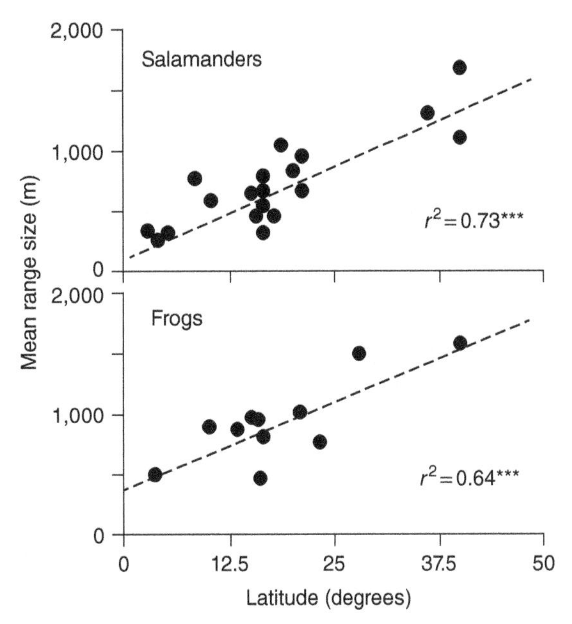

Figure 18.3 Relationships between elevational range size and latitude for salamanders and frogs. (*Source*: McCain (2009). Reproduced with permission from John Wiley & Sons Ltd.)

mountain overlaps that at the base, species should be equipped with the capability to cross them (Fig. 18.2). In the tropics this would not necessarily be the case; a more stable, aseasonal climate would make mountains impassable because lowland species will never encounter similar conditions. This would cause tropical species to have more restricted ranges and potentially make vicariance a more powerful force in tropical biogeography.

It took many years for Janzen's hypothesis to receive detailed attention, but the evidence is supportive. McCain (2009) assessed 16,500 vertebrate species across the world and found that their elevational ranges indeed increased with latitude, as would be expected if temperate species had broader tolerances (Fig. 18.3). Ectotherms showed a stronger response than endotherms, exactly as predicted due to their reliance on external heat sources.

What makes a barrier for one species can turn out to pose no dispersal constraints to another. Most freshwater species are unable to cope with the high salt content of seawater. Some freshwater fauna however are amphidromous—the adults live in freshwater habitats, but their larvae require saline water to develop. These include many species of fish, molluscs and crustaceans. When they possess an oceanic juvenile phase, it is not difficult for species to find their way across large marine barriers. Across the Caribbean, from Puerto Rico to Panama to Trinidad, Page et al. (2013) found that such species showed no spatial genetic structure, even when populations were separated by over 2,000 km of open sea. Similarly mosses, which are dispersed by wind-blown spores, are likely to have little difficulty reaching isolated oceanic islands, while larger-seeded plants are unable to do the same (e.g. Hutsemékers et al., 2011).

Barriers also exist at the global scale between entire climatic regions. Beyond the tropics there are analogous climate zones in both hemispheres, yet they share very few species; hence the failure of polar bears to eat penguins except in unfortunate zoo accidents. The limitation even applies to apparently good dispersers. In the south, penguins and petrels occupy effectively the same roles as auks and skuas in the north. These are clearly convergent species with parallel adaptations, all strong fliers (or swimmers in the case of penguins); yet the tropics pose an insurmountable barrier.

Movement or biotic exchange between areas can occur along corridors which permit dispersal of many taxa. Usually a form of diffusion takes place, whereby a net movement occurs from an area of high species richness to one with fewer. Corridors can also act as dispersal filters, selecting which species make it through. Across Polynesia, the species present on each island are a subset of those from mainland Asia and Australia, increasingly filtered further eastwards (e.g. plant-feeding insects; Hembry et al., 2013).

18.5 Case studies

Having discussed the most important processes in biogeography—speciation, extinction and dispersal—it is time to look at two case studies to examine the role that each has played in determining the composition of regional species pools. One example is taken from the temperate zone, New Zealand, and another from the tropics, Madagascar. Each has been subject to intensive study by biogeographers in recent years. Both were long believed to be classic examples of patterns caused by vicariance, but it is necessary to critically assess the evidence for this.

18.5.1 New Zealand

The islands of New Zealand are of extremely ancient origin, having split from the edge of the southern mega-continent Gondwana some 80 Mya. They have since drifted away from land masses to which they were formerly connected, ending up some 2,000 km from Australia (Fig. 18.4). They were also the last major land mass to be settled by humans, which makes the history of habitation relatively short. Māori peoples reached the islands less than 800 years ago, while Europeans have only known of them for some 200 years. As a result of this, combined with a rugged and impenetrable topography, extinction rates are below those of other Pacific islands (Duncan et al., 2013). This has led to their possessing a unique and distinctive biota; Diamond (1990) declared that New Zealand was 'as close as you will get to... life on another planet'.

Along with geological vicariance, New Zealand has witnessed dramatic environmental changes as a result of glaciations. At the LGM the islands were joined together due to lower sea levels. Despite the increased land area, forests would nonetheless have been confined to the very northern tip, replaced elsewhere by glacial ice or grassy tundra habitats. Finally, in the last few centuries, the arrival of Polynesian Māori peoples led to the loss of 40% of the native forests through wildfires, and then European colonists converted much of the remaining forest and shrubland into pasture (McWethy et al., 2010). This was accompanied by the introduction of numerous alien species.

All these events must have had lasting impacts on the biota of the islands. There has been a prolonged isolation from other land masses, and though it is possible that a link was retained to New Caledonia in the north via a chain of islands which have now sunk beneath the surface of the ocean (the Norfolk ridge), New Caledonia itself was probably submerged for 20 My, only emerging in the Oligocene (Grancolas et al., 2008).

The result of such isolation is that there are many missing groups which never reached the islands. New Zealand lacks native terrestrial mammals and snakes, while monotremes and crocodilians went extinct. In their place arose numerous endemic species found nowhere else on Earth. Examples include the tuatara, a reptile in its own order, only distantly related to lizards elsewhere. The velvet worm *Peripatus* resembles a furry, legged caterpillar and is often misleadingly referred to as a 'living fossil' (the term is unhelpful as it implies a lack of further evolution, which is never true). The giant moas, relatives of the ostrich, went extinct through human hunting but were a distinctive element of the fauna. Among plants, the southern beech *Nothofagus solandri* is a dominant and defining feature of local forests.

It was formerly assumed that these unusual fauna and flora were relicts from Gondwana which had

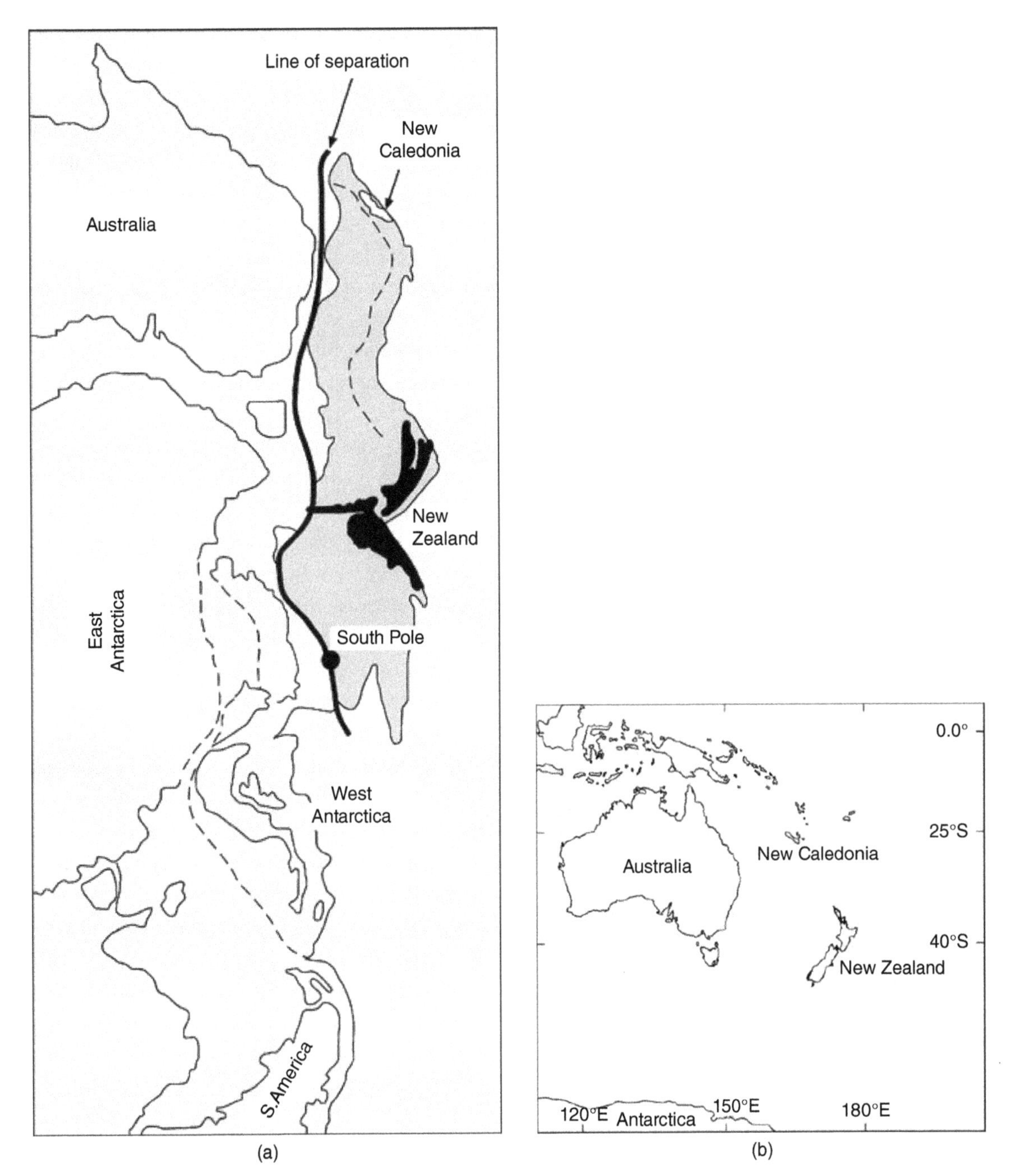

Figure 18.4 (a) Cretaceous position of New Zealand plateau (c. 80 Mya) and (b) modern location in the South Pacific. (*Source*: McDowall (2008). Reproduced with permission from John Wiley & Sons Ltd.)

been preserved in New Zealand while their relatives elsewhere went extinct. This was due to the initial bias towards explanations involving vicariance. But is it really plausible? Further evidence from the history of the islands questions this. During the Oligocene, sea level rises would have placed at least 80% of the islands underwater, with complete immersion remaining a distinct possibility (the highest mountains of the Southern Alps are geologically more recent in origin). Then, during the severe glaciations of the Pleistocene, all warm-temperate elements of the flora went extinct, making it likely that the same was true of their associated animals. Given this sequence of events, the survivors on the islands have to be considered exceptional, and it is likely that most of the biota arrived more recently.

Phylogenetic evidence has accumulated in support of the latter narrative (McDowall, 2008). Charismatic animals thought to be endemics of ancient origin, such as the flightless North Island brown kiwi (a small ratite bird) or the kōkako (wattle bird), which struggles to get airborne, turn out to have ancestral links that considerably post-date the separation of New Zealand. The same is true for endemic freshwater fish like the southern flathead galaxias, some of whose relatives are marine tolerant, and even the tree *Nothofagus fusca*. Such findings were resisted by those who found it galling that genetic evidence implied that New Zealand's species were less special than had been believed (or—even worse—Australian).

There are still some species for which a Gondwanan origin remains credible. The koura genus (*Paranephrops*), which includes two species of freshwater crayfish endemic to New Zealand, split from its closest relatives a considerable time ago. Similarly kauri (conifers in the genus *Agathis*) were widespread during the Jurassic but now remain only in Southeast Asia, Australasia and the South Pacific, reflecting their ancestry. The velvet worms of New Zealand diverged from their nearest mainland ancestor over 22 Mya, before the posited drowning of the islands (Murienne et al., 2014). Yet even in these cases, although the evidence is consistent with vicariance, it is not conclusive. Perhaps speciation took place within Gondwana itself, rather than as a direct result of the separation? An alternative is that there were formerly many more relatives involved in a process of secular migration which have since gone extinct.

Meanwhile, for the majority of species, their closest extant relative is on the Australian mainland. Very few have relatives in South America to which New Zealand was also connected (Fig. 18.4). This is therefore inconsistent with an early split 80 Mya, and it is likely that most species of both flora and fauna reached the islands through dispersal and evolved into the endemics seen today through a process of invasion and subsequent speciation. Even Australia, the most common source of colonists, has a flora which combines both Gondwanan relics and species which arrived later through oceanic dispersal (Crisp and Cook, 2013).

A final dramatic set of changes took place once humans colonised the islands. The arrival of Māori c. 750 years ago led to rapid changes in the vegetation—nearly half of all forests were converted to open vegetation (McWethy et al., 2010). This occurred rapidly as humans brought fires with them to which the native flora was not well adapted, causing a shift of the vegetation into an alternative stable state in which burning was more frequent (McWethy et al., 2014). Only a small number of colonists with no access to modern technology were capable of altering the whole landscape of the islands.

Multiple processes need to be invoked in order to fully understand the biota of New Zealand. It is much more than a Gondwanan ark. Vicariance, climate change, dispersal, speciation and extinction are all part of the story. Next we turn to an even more complicated case.

18.5.2 Madagascar

Madagascar is often described as a large island, though in fact its size, geological origins and variety of habitats mean it is better thought of as a mini-continent. Its 750,000 km^2 make up 0.4% of the total land surface area. It is also isolated, being a little over 400 km from Africa, but much further from India (4,000 km), Antarctica (5,000 km) and Australia (6,000 km), with all of which it was formerly contiguous in the middle of Gondwana (Fig. 18.5). Despite being relatively close to Africa, the land masses are separated by the

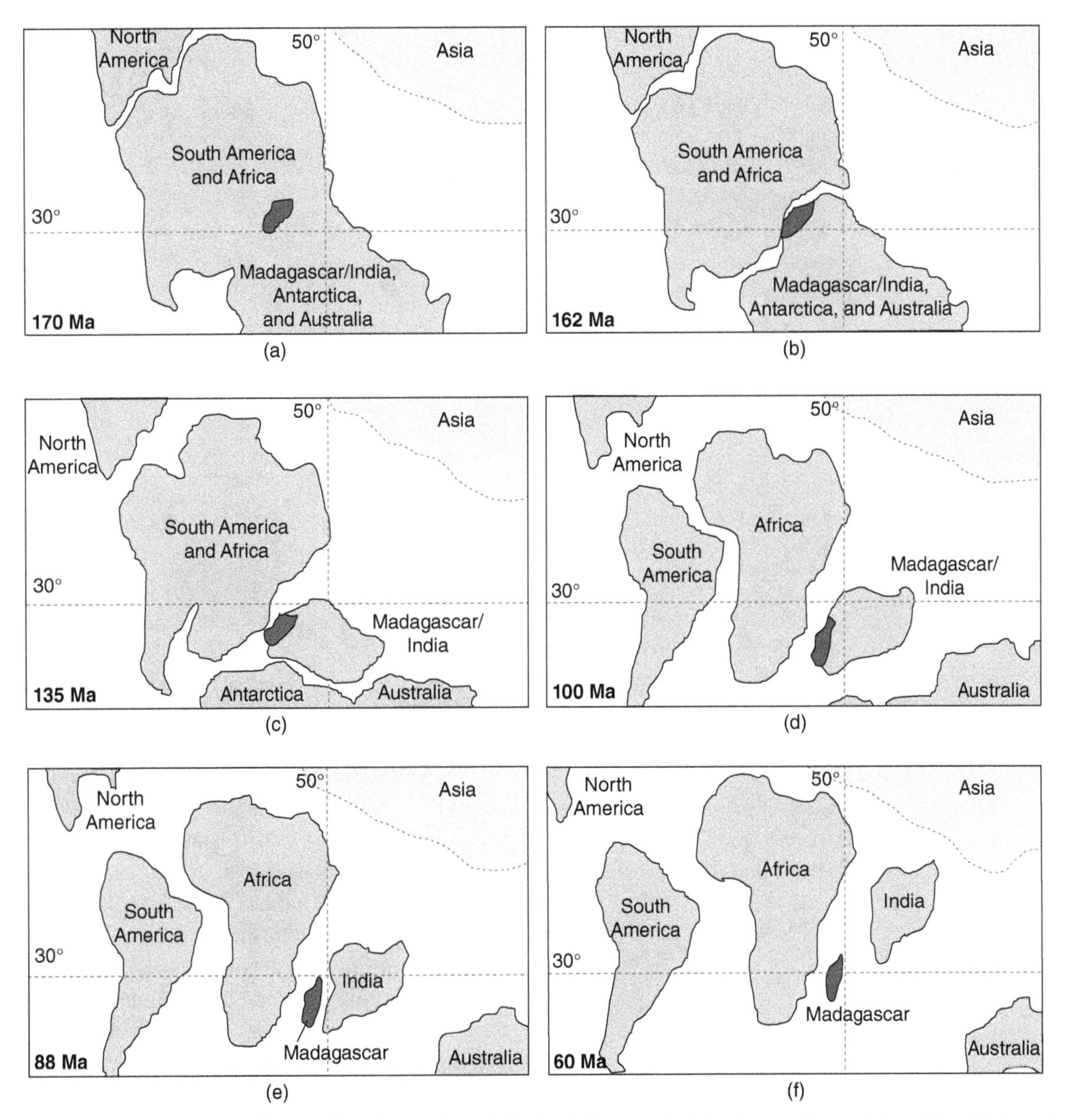

Figure 18.5 Separation of Madagascar from the continental blocks which comprised Gondwana. (*Source*: Adapted from Yoder and Nowak (2006). Chris Scotese, Director, PALEOMAP Project.)

Mozambique channel, which at a depth of 3,000 m has ensured that sea level change was never sufficient to reconnect Madagascar to the African mainland.

The initial split from Africa took place from 162 Mya (Fig. 18.5). At this point Madagascar was still connected to modern India, which then split from Antarctica and Australia. While continents realigned around it, Madagascar remained rooted in approximately the same spot. By 88 Mya it had separated from India, though this episode of cleavage may have taken 30 My to complete. India then drifted northwards, colliding with Laurasia and in the process creating the Himalayas, while Madagascar remained almost exactly in its current position.

The predominant wind direction is from the east; as a result, there is a pronounced rainfall gradient across Madagascar (Fig. 18.6). The environmental heterogeneity created by the eastern mountain range gives rise to a wide spectrum of biomes. The eastern coast is dominated by verdant evergreen rain forest, though this is currently subject to rapid degradation. The highlands have their own distinctive forests, with a further ericoid scrub community on the highest mountains. The west is covered by dry deciduous forest. The greatest contrast is between the northern point, where the Sambirano rain forest receives a relatively high annual rainfall (albeit with a strongly seasonal distribution), and the southern point, where an arid spiny bushland community dominates (Grubb, 2003).

The range of habitats is one reason behind Madagascar's incredible species richness. The 18th century French naturalist Philibert Commerson wrote:

> May I announce to you that Madagascar is the naturalist's promised land? Nature seems to have retreated there into a private sanctuary, where she could work on different models from any she has used elsewhere. There you meet bizarre and marvellous forms at every step...

He was hardly exaggerating—the level of endemism is spectacular (for full details see Yoder and Nowak, 2006). Among the fauna, 95% of

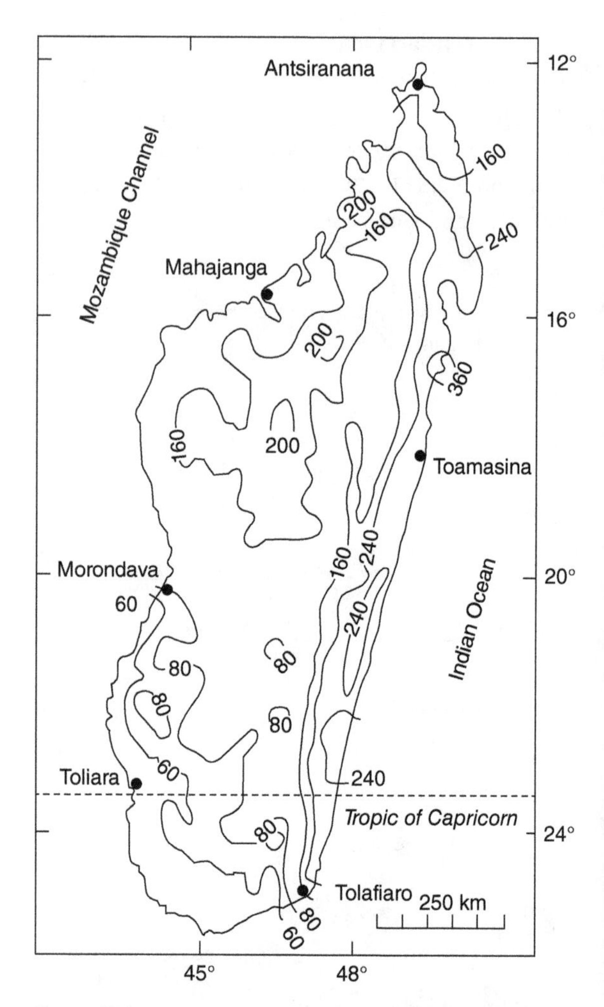

Figure 18.6 Distribution of annual rainfall (cm) across Madagascar, from Koechlin et al. (1974).

reptiles, 99% of amphibians and all native mammals (except for bats) are endemic to Madagascar. It is the global hotspot for chameleons, with new species continuing to be discovered. It contains 15% of all extant primate species, particularly lemurs, an entirely endemic group. In addition, over 80% of plant species are found nowhere else. Explaining how all these species came into existence is more difficult and depends on a complex combination of diversification processes (Brown et al., 2014).

Almost as striking as the bewildering array of unique species is that many other groups are absent. There are no salamanders, vipers, varanid lizards or caecilians, which have an otherwise pantropical distribution. Among frogs, there are 28 families worldwide, of which only three can be found on Madagascar. Even so, these three families have given rise to over 300 endemic species, making up 4% of all amphibians globally.

The fossil record reveals yet more peculiarities. The elephant bird *Aepyornis* is thought to have been the largest bird ever to exist, reaching 2–3 m in height and up to 275 kg in weight. As a ratite, a relative of the ostrich, it was also flightless. It appears to have gone extinct relatively recently, probably shortly after being discovered by humans. The origin of the ratites was on Gondwana in the late Cretaceous. Ancient DNA from elephant birds indicates that its closest living relative is in fact the similarly flightless kiwi of New Zealand, providing a rather unexpected connection between these two regions, and one which cannot be attributed to a common Gondwanan origin because the two land masses were never directly connected (Mitchell et al., 2014).

How have all these endemics arisen? The obvious answer at first seemed to be through vicariance, with Madagascar maintaining the descendents of Gondwanan taxa. If this were so, the age of most clades should be around 88 My or more, with links to other elements of the great land mass. The advent of phylogenetics allows this to be tested and surprisingly reveals that most clades are actually of recent origin. Although there are a few groups for which vicariance remains a plausible mechanism for their presence, the majority arrived within the last 20 My. This is particularly true of plants, none of which appear to have been present for the full 88 My and are generally more recent arrivals.

Bees in the genus *Braunsapis* provide a typical example (Fuller et al., 2005). All the species in Madagascar can be traced back to two common ancestors, 13 and 3 Mya. The likelihood is therefore that there was a minimum of two dispersal events which successfully brought colonists over from Africa. (This genus is also found in South Asia and tropical Australia, indicating that it has dispersed even further.) Among the mammals there are four endemic lineages: tenrecs, lemurs, nesomyine rodents and carnivorans. Each of these is monophyletic; in other words it arose from a single common ancestor, thence radiating in the Cenozoic. A single colonisation event for each is the implication.

In the majority of cases that have been examined, dispersal from Africa is the most plausible source of colonists to Madagascar. This is true for a range of species including frogs, most lizards and reptiles, geckos, tortoises, some freshwater fish and birds (even including the notable endemic Vangidae family). A number of these arrived less than 4 Mya, and in all cases no more than a few colonisation events were required. The overwhelming majority of vertebrates arrived after the final separation from Africa 130 Mya (Crottini et al., 2012).

That's not to say that vicariance can be entirely excluded as a hypothesis; after all, a full biota must have been present at separation, and some of their descendents would be expected to remain. These may include particular families of freshwater fish (rainbowfish, killifishes, cichlids), some lizards and reptiles, boine snakes, podocnemid turtles and iguanid lizards. It remains difficult however to categorically distinguish vicariance from ancient dispersal.

How did so many species manage to cross 400 km of notoriously treacherous seas to reach Madagascar? The most plausible explanation is rafting, and here the cartoon film Madagascar (2005) shows a surprising proximity to the likely scenario (apart from the penguins). Following storms in tropical forests, a large amount of vegetation is swept into streams and rivers, coalescing into substantial aggregations. Records from ancient mariners document these being found far out at sea, with some even claiming to have seen standing trees among them, along with mammalian passengers (Van Duzer, 2004).

This is also the probable route of dispersal for primates around the world, especially because they evolved in Africa after the break-up of Gondwana had already taken place. Rafting took them not only to Madagascar but also to South America, where they again radiated into a wealth of species. A trace of the process of colonisation can be found in the observation that most Neotropical primates are red–green colour-blind, all possessing a non-functional copy of the red light receptor protein, which may be the result of inbreeding among a small founder population.

Other groups followed the same route. Rowe et al. (2010) present convincing evidence that hystricognath rodents crossed the Atlantic 50 Mya, far too late for vicariance to be responsible. These ancestors gave rise to the South American guinea pigs, capybaras, chinchillas, coypu and agoutis, among many other important groups. Even species strongly associated with humans have followed this route. Bottle gourds are widely found in archaeological sites in South America and are still used to make containers and musical instruments, yet no wild populations are known. One might assume that they were brought over by colonists from Asia; in fact, genetic evidence points towards an ancient African origin and another trans-Atlantic dispersal (Kistler et al., 2014). The process involved here is *sweepstake dispersal,* a particularly extreme form of jump dispersal. These events seem staggeringly unlikely. Yet if the odds of dispersal happening are once in a million years—or even 10 million—then on geological timescales it becomes almost inevitable. There is no need to invoke stepping stones or lost continents now submerged, because multiple oceanic crossings are even less credible.

One outstanding puzzle is that the four Madagascan mammal groups all arrived prior to 20 Mya (Fig. 18.7), and since then there has been a decline

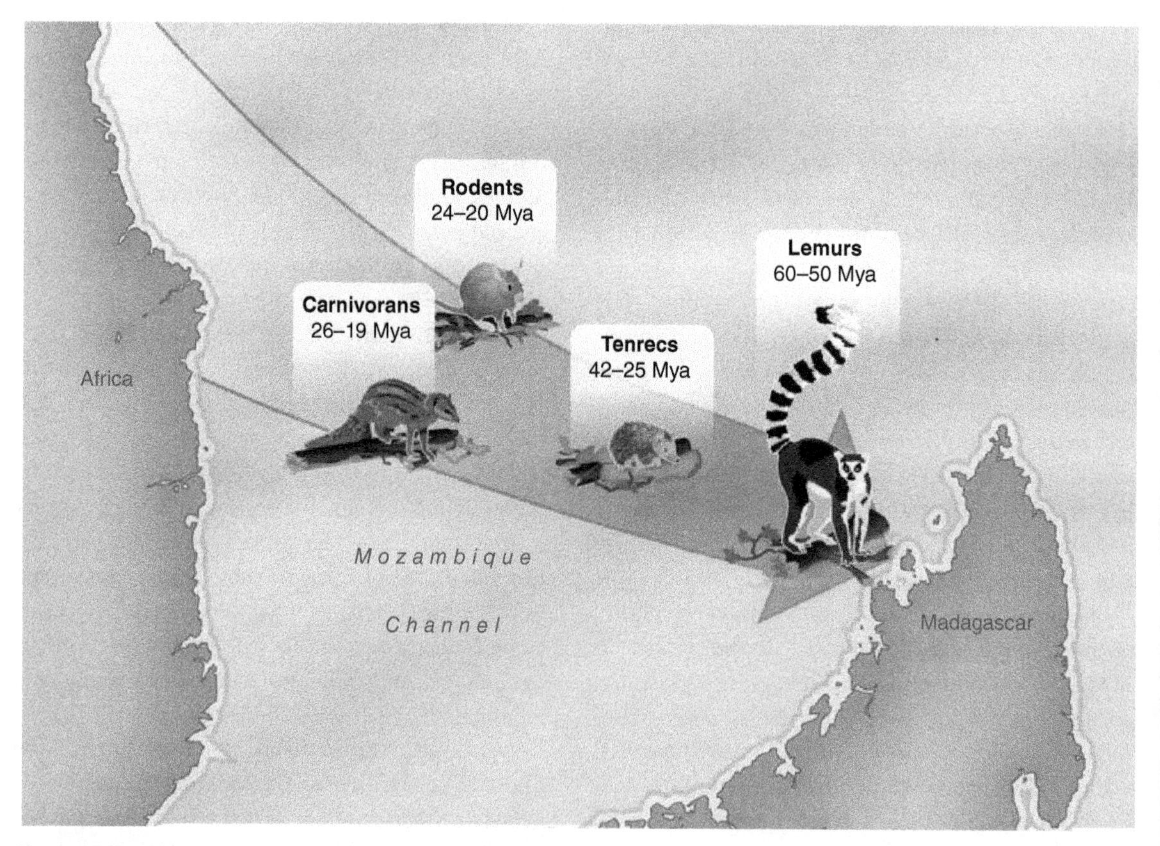

Figure 18.7 Likely arrival times of the ancestors of mammal groups on Madagascar by transoceanic dispersal from the African mainland. (*Source*: Krause (2010). Reproduced with permission from Nature Publishing Group.)

in the number of new groups for which transoceanic dispersal is the most likely origin. Evidence for why was discovered by Ali and Huber (2010), who found that in the Palæogene, strong surface currents would have travelled from present-day Mozambique and Tanzania towards Madagascar, providing a direct route for colonists. Since then, Madagascar has shifted slightly northwards, resulting in a change in ocean currents that now travel predominantly westwards. It is now much more likely that species will move from Madagascar to Africa.

This is an interesting case of how a region can be both a target and a source of colonists, depending on the predominant pathways for dispersal. Chameleons probably first evolved in Africa, with two separate colonisations of Madagascar (Tolley et al., 2013). The Comoros Islands are volcanic in origin and 300 km away from Madagascar, yet contain two sister species of the Malagasy chameleons, both of which are endemic to particular islands. *Furcifer cephalolepis* is found on Moheli, which is only 130,000 years old, whereas *F. polleni* occurs on Grande Comore, much older at 5.4 My. Neither island was ever connected to the mainland, so oceanic dispersal from Madagascar is the only route by which they can have arrived (Raxworthy et al., 2002).

It's not only vegetation that forms rafts. Volcanic eruptions emit vast amounts of floating pumice; in the case of the 2006 eruption of the Home Reef volcano in Tonga, an estimated 2,500,000,000,000 clasts were emitted, each a potential opportunity for long-distance dispersal (Fig. 18.8). It wasn't even a particularly large eruption. Over 80 species were discovered on floating clasts, travelling over 5,000 km in 7–8 months.

New genetic techniques are providing fresh insights into the frequency of transoceanic dispersal events. In the first ever study of its kind, Fraser et al. (2011) showed that kelp wrack deposited on a beach in New Zealand had travelled there from distant islands, in one case spending at least two months at sea, covering around 500 km. Much more is likely to be learnt about the frequency and distribution

Figure 18.8 Epibiont colonisation on Home Reef pumice with coin (2 cm diameter) for scale. (*Source*: From Bryan et al. (2012). Used under CC-BY-SA 2.5 http://creativecommons.org/licenses/by-sa/2.5/.)

of these dispersal events over the coming years. Before completely forgetting about tectonic plates and ancient vicariance however, a cautionary note should be sounded regarding the reliability of genetic dates (Wilf and Escapa, 2015). In some cases taxa which are clearly identified in Gondwanan fossils have been given molecular dates which make them appear more recent. This strikes back against those who favour explanations based only on dispersal and suggests that the debate is far from over.

18.6 Conclusions

All species disperse. The presence of particular species or taxa in different parts of the world is the result of this occurring on geological timescales. Dispersal can be active or passive and deliberate or accidental and may involve jumps, diffusion or secular migrations. Most importantly, on the timescales in which biogeographical patterns are assembled, rare events are bound to happen eventually, and no hypothesis to explain disjunctions can be discounted merely because the occurrence of any single dispersal event is unlikely. Sweepstake dispersal gives every species the chance, albeit small, to travel anywhere.

The debate between proponents of vicariance or dispersal as the main determinant of biogeographical patterns has begun to rest on the side of the latter. This is certainly true in two classic examples first championed by vicariance biogeographers—New Zealand and Madagascar. That said, the two processes are intimately connected; after all, vicariance is dispersal limitation. Even when continents split, there is a prolonged period during which they are close together, increasing the likelihood of dispersal. Taxa vary in their response to barriers, meaning that vicariance for one group can have no impact on others that maintain connectivity. The distinction between the two hypotheses is therefore not as great as is often made out.

18.6.1 Recommended reading

- Gillespie, R. G., B. G. Baldwin, J. M. Waters, S. I. Fraser, R. Nikula, and G. K. Roderick, 2011. Long-distance dispersal: a framework for hypothesis testing. *Trends in Ecology & Evolution* 27:47–56.
- McDowall, J. M., 2008. Process and pattern in the biogeography of New Zealand—a global microcosm. *Journal of Biogeography* 35:197–212.
- Yoder, A. D. and Nowak, M. D., 2006. Has vicariance or dispersal been the predominant biogeographical force in Madagascar? Only time will tell. *Annual Review of Ecology, Evolution, and Systematics* 37:405–431.

18.6.2 Questions for the future

- Some authors (e.g. Donoghue, 2011) think the pendulum of current opinion in biogeography has swung too far in favour of exceptional dispersal events. Perhaps vicariance is harder to detect than dispersal due to confounding effects (see Upchurch, 2008). Is this true?
- Is it easier for species to move to new locations than to evolve to new conditions?
- What determines whether dispersing organisms are able to establish when they arrive in a new location?

References

Ali, J. R. and M. Huber, 2010. Mammalian biodiversity on Madagascar controlled by ocean currents. *Nature* 463:653–656.

Archibald, S. B., D. R. Greenwood, and R. W. Mathewes, 2013. Seasonality, montane beta diversity, and Eocene insects: testing Janzen's dispersal hypothesis in an equable world. *Palaeogeography, Palaeoclimatology, Palaeoecology* 371:1–8.

Brown, J. L., A. Cameron, A. D. Yoder, and M. Vences, 2014. A necessarily complex model to explain the biogeography of the amphibians and reptiles of Madagascar. *Nature Communications* 5:5046.

Bryan, S. E., A. G. Cook, J. P. Evans, K. Hebden, L. Hurrey, P. Colls, J. S. Jell, D. Weatherley, and J. Firn, 2012. Rapid, long-distance dispersal by pumice rafting. *PLoS ONE* 7:e40583.

Cahill, A. E., M. E. Aiello-Lammens, M. C. Fisher-Reid, X. Hua, C. J. Karanewsky, H. Y. Ryu, G. C. Sbeglia, F. Spagnolo, J. B. Waldron, and J. J. Wiens, 2014. Causes of warm-edge range limits: systematic review, proximate factors and implications for climate change. *Journal of Biogeography* 41:429–442.

Carlquist, S., 1967. Biota of long-distance dispersal. V. Plant dispersal to Pacific islands. *Bulletin of the Torrey Botanical Club* 94:129–162.

Crisp, M. D. and L. G. Cook, 2013. How was the Australian flora assembled over the last 65 million years? A molecular phylogenetic perspective. *Annual Review of Ecology, Evolution, and Systematics* 44:303–324.

Crottini, A., O. Madsen, C. Poux, S. Y. Strauß, D. R. Vieites, and M. Vences, 2012. Vertebrate time-tree elucidates the biogeographic pattern of a major biotic change around the KT boundary in Madagascar. *Proceedings of the National Academy of Sciences of the United States of America* 109:5358–5363.

Diamond, J. M., 1990. New Zealand as an archipelago: an international perspective. *Conservation Sciences Publication* 2:3–8.

Dobzhansky, T., 1950. Evolution in the tropics. *American Scientist* 38:209–221.

Donoghue, M. J., 2011. Bipolar biogeography. *Proceedings of the National Academy of Sciences of the United States of America* 108:6341–6342.

Duncan, R. P., A. G. Boyer, and T. M. Blackburn, 2013. Magnitude and variation of prehistoric bird extinctions in the Pacific. *Proceedings of the National Academy of Sciences of the United States of America* 110:6436–6441.

Fernando, P., T. N. C. Vidya, J. Payne, M. Stuewe, G. Davison, R. J. Alfred, P. Andau, E. Bosi, A. Kilbourn, and D. J. Melnick, 2003. DNA analysis indicates that Asian elephants are native to Borneo and are therefore a high priority for conservation. *PLoS Biology* 1:e6.

Fraser, C. I., R. Nikula, and J. M. Waters, 2011. Oceanic rafting by a coastal community. *Proceedings of the Royal Society Series B* 278:649–655.

Fuller, S., M. Schwarz, and S. Tierney, 2005. Phylogenetics of the allopadine bee genus *Braunsapis*: historical biogeography and long-range dispersal over water. *Journal of Biogeography* 32:2135–2144.

Goodacre, S. L., O. Y. Martin, D. Bonte, L. Hutchings, C. Woolley, K. Ibrahim, C. F. G. Thomas, and G. M. Hewitt, 2009. Microbial modification of host long-distance dispersal capacity. *BMC Biology* 7:32.

Grancolas, P., J. Murienne, T. Robillard, L. Desutter-Grandcolas, H. Jourdan, E. Guilbert, and L. Deharveng, 2008. New Caledonia: a very old Darwinian island? *Philosophical Transactions of the Royal Society Series B* 363:3309–3317.

Grubb, P. J., 2003. Interpreting some outstanding features of the flora and vegetation of Madagascar. *Perspectives in Plant Ecology, Evolution, and Systematics* 6:125–146.

Hellweger, F. L., E. van Sebille, and N. D. Fredrick, 2014. Biogeographic patterns in ocean microbes emerge in a neutral agent-based model. *Science* 345:1346–1349.

Hembry, D. H., T. Okamoto, G. McCormack, and R. G. Gillespie, 2013. Phytophagous insect community assembly through niche conservatism on oceanic islands. *Journal of Biogeography* 40:225–235.

Hutsemékers, V., P. Szövényi, A. J. Shaw, J.-M. González-Mancebo, J. Muñoz, and A. Vanderpoorten, 2011. Oceanic islands are not sinks of biodiversity in spore-producing plants. *Proceedings of the National Academy of Sciences of the United States of America* 108:18989–18994.

Janzen, D. H., 1967. Why mountain passes are higher in the tropics. *American Naturalist* 101:233–249.

Johnson, D. L., 1980. Problems in the land vertebrate zoogeography of certain islands and the swimming powers of elephants. *Journal of Biogeography* 7:383–398.

Kistler, L., A. Montenegro, B. D. Smith, J. A. Gifford, R. E. Green, L. A. Newsom, and B. Shapiro, 2014. Transoceanic drift and the domestication of African bottle gourds in the Americas. *Proceedings of the National Academy of Sciences of the United States of America* 111:2937–2941.

Koechlin, J., J. L. Guillaumet, and P. Morat, 1974. *Flore et Végétation de Madagascar*. Cramer.

Krause, D. W., 2010. Washed up in Madagascar. *Nature* 463:613–614.

Longhurst, A., 1998. *Ecological Geography of the Sea*. Academic Press.

MacArthur, R. H., 1972. *Geographical Ecology: Patterns in the Distribution of Species*. Harper and Row.

McCain, C. M., 2009. Vertebrate range sizes indicate that mountains may be 'higher' in the tropics. *Ecology Letters* 12:550–560.

McDowall, J. M., 2008. Process and pattern in the biogeography of New Zealand—a global microcosm? *Journal of Biogeography* 35:197–212.

McWethy, D. B., C. Whitlock, J. M. Wilmshurst, M. S. McGlone, M. Fromont, X. Li, A. Dieffenbacher-Krall, W. O. Hobbs, S. C. Fritz, and E. R. Cook, 2010. Rapid landscape transformation in South Island, New Zealand, following initial Polynesian settlement. *Proceedings of the National Academy of Sciences of the United States of America* 107:21343–21348.

McWethy, D. B., J. M. Wilmshurst, C. Whitlock, J. R. Wood, and M. S. McGlone, 2014. A high-resolution chronology of rapid forest transitions following Polynesian arrival in New Zealand. *PLoS ONE* 9:e111328.

Mitchell, K. J., B. Llamas, J. Soubrier, N. J. Rawlence, T. H. Worthy, J. Wood, M. S. Y. Lee, and A. Cooper, 2014. Ancient DNA reveals elephant birds and kiwi are sister taxa and clarifies ratite bird evolution. *Science* 344:898–900.

Miura, O., M. E. Torchin, E. Bermingham, D. K. Jacobs, and R. F. Hechinger, 2012. Flying shells: historical dispersal of marine snails across Central America. *Proceedings of the Royal Society Series B* 279:1061–1067.

Murienne, J., S. R. Daniels, T. R. Buckley, G. Mayer, and G. Giribet, 2014. A living fossil tale of Pangaean biogeography. *Proceedings of the Royal Society Series B* 281:20132648.

Page, T. J., L. S. Torati, B. D. Cook, A. Binderup, C. M. Pringle, S. Reuschel, C. D. Schubart, and J. M. Hughes, 2013. Invertébrés Sans Frontières: Large scales of connectivity of selected freshwater species among Caribbean islands. *Biotropica* 45:236–244.

Popp, M., V. Mirré, and C. Brochmann, 2011. A single, Mid-Pleistocene long-distance dispersal by a bird can explain the extreme bipolar disjunction in crowberries *Empetrum*. *Proceedings of the National Academy of Sciences of the United States of America* 108:6520–6525.

Raxworthy, C. J., M. R. J. Forstner, and R. A. Nussbaum, 2002. Chameleon radiation by oceanic dispersal. *Naure* 415:784–787.

Rolland, J., F. Jiguet, K. A. Jønsson, F. L. Condamine, and H. Morlon, 2014. Settling down of seasonal migrants promotes bird diversification. *Proceedings of the Royal Society Series B* 281:20140473.

Rowe, D. L., K. A. Dunn, R. M. Adkins, and R. L. Honeycutt, 2010. Molecular clocks keep dispersal hypotheses afloat: evidence for trans-Atlantic rafting by rodents. *Journal of Biogeography* 37:305–324.

Sharma, P. P. and G. Giribet, 2012. Out of the Neotropics: Late Cretaceous colonization of Australasia by American arthropods. *Proceedings of the Royal Society Series B* 279:3501–3509.

Stefanescu, C., F. Páramo, S. Åkesson, M. Alarćon, A. Ávila, T. Brereton, J. Carnicer, L. F. Cassar, R. Fox, J. Heliölä, J. K. Hill, N. Hirneison, N. Kjellén, E. Kühn, M. Kuussaari,

M. Leskinen, F. Liechti, M. Musche, E. C. Regan, D. R. Reynolds, D. B. Roy, N. Ryrholm, H. Schmaljohann, J. Settele, C. D. Thomas, C. van Swaay, and J. W. Chapman, 2013. Multi-generational long-distance migration of insects: studying the painted lady butterfly in the Western Palaearctic. *Ecography* 36:474–486.

Thornton, I. W. B., 1997. *Krakatau—The Destruction and Reassembly of An Island Ecosystem*. Harvard University Press.

Tolley, K. A., T. M. Townsend, and M. Vences, 2013. Large-scale phylogeny of chameleons suggests African origins and Eocene diversification. *Proceedings of the Royal Society Series B* 280:20130184.

Upchurch, P., 2008. Gondwanan break-up: legacies of a lost world? *Trends in Ecology & Evolution* 23:229–236.

Valls, L., A. Mestre, J. A. Gil-Delgado, and F. Mesquita-Joanes, 2014. The shoemaker's son always goes barefoot: intercontinental dispersal of Ostracoda (Crustacea) by scientists attending an IBS excursion. *Frontiers of Biogeography* 6:89–91.

Van Duzer, C., 2004. *Floating Islands: A Global Bibliography*. Cantor Press.

Vargas, P., R. Heleno, A. Traveset, and M. Nogales, 2012. Colonization of the Galápagos Islands by plants with no specific syndromes for long-distance dispersal: a new perspective. *Ecography* 35:33–43.

Wada, S., K. Kawakami, and S. Chiba, 2012. Snails can survive passage through a bird's digestive system. *Journal of Biogeography* 39:69–73.

Whittaker, R. J., M. B. Bush, and K. Richards, 1989. Plant recolonization and vegetation succession on the Krakatau Islands, Indonesia. *Ecological Monographs* 59:59–123.

Wilf, P. and I. H. Escapa, 2015. Green Web or megabiased clock? Plant fossils from Gondwanan Patagonia speak on evolutionary radiations. *New Phytologist*, 207:283–290.

Winger, B. M., F. K. Barker, and R. H. Ree, 2014. Temperate origins of long-distance seasonal migration in New World songbirds. *Proceedings of the National Academy of Sciences of the United States of America* 111:12115–12120.

Yoder, A. D. and M. D. Nowak, 2006. Has vicariance or dispersal been the predominant biogeographic force in Madagascar? Only time will tell. *Annual Review of Ecology, Evolution, and Systematics* 37:405–431.

CHAPTER 19
Life on islands

19.1 The big question

Put simply, islands are different. Travelling from the mainland to an offshore island often results in a noticeably altered set of species. Why might this be? This peculiar phenomenon accounts for the allure of islands and the central role they have played in the development of both theoretical and experimental ecology. It is no exaggeration to say that the greatest insights in biology have come from the study of islands. From Darwin's visit to the Galápagos that inspired his theory of evolution (Darwin, 1859) to Wallace's work in the Malay archipelago (Wallace, 1860), The Theory of Island Biogeography (MacArthur and Wilson, 1967), observation of the fine-scale operation of evolutionary change (Grant and Grant, 2008) and their current status on the frontline of conservation efforts (Caujapé-Castells et al., 2010), islands have played a core role in the development and testing of many of the main ideas in biology.

Noticing that islands and the species living on them are different from the mainland is only the first step. With such a bewildering degree of variation, it can be difficult to spot common patterns that apply to all islands. Can we invoke common processes to account for the features of their biota?

19.2 Types of island

There are three basic types of islands, in the sense of land masses surrounded by water. These are continental islands, volcanoes and coral atolls. Continental islands are formed of continental rock and have therefore been persistent features of Earth's crust over hundreds of millions of years. They include *fragment* islands where small sections of crust have become separated from the main continental blocks; New Caledonia and New Zealand are good examples. These are typically surrounded by deep water as oceanic rock on the sea floor separates them from neighbouring continents. Fragment islands are distinct from *shelf* islands, which remain attached to continental blocks, but become isolated from them at the surface due to sea level changes. The British Isles and Ireland are such a case. Only shallow seas separate them from Europe, while at the LGM they were fully connected to the European mainland. As sea levels rose following the melting of the glaciers, they were cut off. An antipodean parallel is the island of Tasmania, which is in fact united with the Australian continental block.

From the perspective of biogeographical history, fragment and shelf islands behave very differently.

Natural Systems: The organisation of life, First Edition. Markus P. Eichhorn.

In the former case, vicariance is geological in origin and usually ancient. Shelf islands, in contrast, have been isolated only recently (within the last 20,000 years). Their biota is therefore often influenced by the last land bridge that connected them to the continental mainland. Note that the term land bridge can be misleading if assumed to refer to a narrow connecting strip. The continental islands seen today were fully embedded in larger land masses only recently in geological terms. Remember from Section 17.4 that the present interglacial is unusual for the Quaternary period. What appear to be shelf islands in the present day have actually spent most of the past 2.6 My as hills or mountains on a continuous landmass. Their size, degree of isolation and existence as islands in the first place are determined solely by sea level. About three quarters of present islands were connected to the mainland at the LGM (Weigelt et al., 2013); many other islands have since sunk beneath the sea. In the case of Britain, its biota was mostly present 7,500 years ago, when it wasn't an island in any sense of the word. From the perspective of its ancient inhabitants, Europe was a continuous continent with species shared across it. The 'land bridge' closed when the English Channel formed late in the Pleistocene, with Ireland separating slightly earlier.

Shelf islands are necessarily located close to present continental coastlines. Fragment islands also generally remain close to their parent continents (though there are some exceptions, such as Sulawesi). *Volcanoes*, on the other hand, tend to form remote oceanic islands, their positions being determined by other geological processes. They are also taller on average, with younger volcanoes possessing the classical cone shape, enabling them to form clouds and generate their own rainfall. In doing so they elicit processes that ultimately destroy themselves. Volcanic rock is softer than that of the continents, and therefore these islands gradually erode and subside, sinking into the sea, where they remain thereafter as seamounts. These become important centres of diversity within the ocean (Morato et al., 2010).

The biota of volcanoes is necessarily dominated by dispersal because they can never have had connections to the mainland. The island of Surtsey dramatically emerged from the ocean off Iceland on 14 November 1963. The bare rock was swiftly colonised; within 19 months the common seashore plant *Cakile maritima* grew there. Colonists include around 20% of the total flora and fauna of Iceland (Adsersen, 2013). Perhaps this high rate is unsurprising because Iceland is itself an island, and therefore the species present are likely to already possess strong dispersal potential. Nonetheless the accumulation of species has been rapid.

The final class of islands is *atolls*, created from the calcareous deposits formed by coral reefs. Often they initially appear as fringes around volcanic islands; as the volcano collapses, the corals continue growing, until they are all that remains of the original island. Sea level changes or buckling tectonic plates can then expose them at the surface. This explains the characteristic circular form of atolls. The process by which they formed was speculated upon by Darwin (1842). Remarkably he came up with the above hypothesis in his journal on the voyage of the Beagle before he had even seen an atoll for himself.

Once again their genesis dictates their characteristic features. Atolls typically have a very small land area and are only found in warm regions where growth of reef-forming coral is possible. Larger atolls maintain a lens of freshwater beneath the ground, but many smaller ones cannot and are therefore seldom inhabited. Having little elevation, they are incapable of forming clouds and often have dry climates.

The processes studied on islands have received more general application, such that to an ecologist, an island is any patch of habitat surrounded by a barrier to dispersal, usually a hostile environment. These include forest fragments, mountaintops, lakes (also islands within freshwater bodies), urban parks, coral reefs and seamounts (islands beneath the surface). Even cowpats can be considered an insular habitat, maintaining a community of species dependent upon dung that occurs in discrete patches.

In this context most natural systems are divided into islands, formed by particular communities, which are embedded in a matrix of unsuitable habitats. This was the principle behind the theory of metacommunities (Section 13.3). The degree to which this is true, referred to as **insularity**, has implications for their structure and composition. If there are general rules that apply to oceanic islands, then they should be transferable to islands in general. The remainder of this chapter will however refer to islands in the stricter, more literal sense of land surrounded by water.

19.3 Island biotas

Communities on oceanic islands differ markedly from those on nearby land masses. One of the most consistent features of oceanic islands is that their environment is more polar relative to the mainland; their climates tend to be cooler, wetter and less seasonal (Weigelt et al., 2013). (Desert islands are relatively rare, at least above 1 km^2 in size, which excludes the smallest atolls.) This is at least partly responsible for the striking disparity between island and mainland communities, though not the full story. The turnover of species is greater than would be expected based on either geographical distance or environmental change alone (Stuart et al., 2012). This is what makes islands special to ecologists; something more is going on than patterns within metacommunities.

Islands are typically species poor, though of the species present, a large number can be endemics found nowhere else. As a result, islands contain a disproportionate fraction of global species richness given their area. One interesting repercussion of this is that there are more European species in the overseas territories—the smattering of islands around the world that are the last remnants of empire—than in the whole of continental Europe itself. A large proportion are found in species-rich tropical islands such as New Caledonia and Réunion (both French). The frequency of island endemics is partly due to their unusual environments. Moreover, being surrounded by the ocean buffers island climates, creating a zone of localised environmental stability that allows isolated populations to specialise with little gene flow from the mainland. Isolation also enables species to be preserved on islands even when their habitat has long since disappeared from the continent.

The combination of endemism and small area makes islands a particular concern for conservation efforts. Around 70,000 species of plants are only found on islands, comprising a quarter of global plant species richness, but a third of all threatened species (Kreft et al., 2008). Similar relative proportions are true of other groups of organisms. Island communities are small and therefore intrinsically vulnerable.

19.4 Evolution of endemics

The origins of endemics can be split into two broad categories. One pathway is for an island population to become reproductively isolated, diverging from its ancestors in a process known as *anagenesis*. This is likely to be the main mechanism of speciation on fragment islands, or for species that arrive as colonists on volcanoes. The species richness of the island does not change because the endemics simply replace the original species. A second pathway is *cladogenesis*, whereby two island populations diverge from one another, perhaps in response to vacant niches or the extinction of other species. This increases island species richness through *in situ* speciation. As is so often the case with species delineation, it is worth maintaining some detached scepticism, as there can be a temptation among taxonomists to define island populations as being special. Nevertheless, the phenomenon of extraordinary endemism appears robust, with nearly 10 times as many endemics in island communities as those on the mainland (Kier et al., 2009). An interesting observation is that there tends to be high congruence in endemic richness between taxa on islands, in contrast to patterns on the mainland (see Section 7.5), suggesting that a common factor determines their formation.

The number of endemics that can arise is remarkable. St Helena is a remote volcanic island and unusually ancient with an origin around 14.5 Mya. It is 122 km^2 in size, and being 2,800 km from the nearest mainland made a perfect site for the exile of Napoleon Bonaparte. Its native flora is limited, with only 60 indigenous species, of which 50 are endemics. Sadly typical of islands is that seven of these are already extinct; it highly likely that others were lost before they had been described.

Where archipelagoes provide a series of islands, endemics can arise in great numbers through the process of **adaptive radiation**. Here monophyletic groups, derived from a single common ancestor, rapidly give rise to an array of novel species. Classic case studies have been conducted on Caribbean anoles (Losos, 1996), Hawaiian spiders (Gillespie, 2004) and Galápagos weevils (Sequeira et al., 2000). The process requires a large resource base, split into distinct habitats which are isolated from one another, and little competition from existing species. Often the same ecomorphs evolve on different islands, each taking a similar role but evolving independently. It has been suggested more recently that, at least from the perspective of marine organisms, islands are effectively further apart in the tropics (Brown, 2014), a parallel to Janzen's (1967) hypothesis that mountain passes are higher in the tropics. The logic is that the rate of development of planktonic larvae is much faster in warmer waters, reducing the distance they can travel before they establish as adults. This is a plausible idea but remains to be tested.

One of the most well-studied examples of adaptive radiation is the Hawaiian archipelago. These islands have formed over a mid-oceanic hotspot that has created a line of volcanoes from the youngest (the Big Island, which remains seismically active) down to Ni'ihau, the oldest. Through time there has been a continuous turnover, with each new island emerging then gradually subsiding into the sea; a line of seamounts continues north-west where ancient islands have been submerged (some only recently following postglacial sea level rises). The islands are not only geologically young, up to a maximum of 5.1 My, but remote, with North America the closest continent at 4,000 km distance.

It is therefore without doubt that the biota of Hawaii arose as the result of a series of long-distance dispersal events. Many of these must have been involved, few more than 10 Mya. Any group persisting over such long time periods can only have done so by island-hopping as volcanoes emerged and then sank. As with other isolated regions, many common groups of organisms are missing, replaced by a staggering richness of endemic species. An especially diverse genus is the *Hyposmocoma* moths, of which 400 species occur across the archipelago, inhabiting almost all the main habitats and making up a third of the butterfly and moth species on the islands. Their phylogeny places the age of the group at 15 Mya—far older than the oldest remaining island, but consistent with the age of the remote north-western atolls which are all that remains of former volcanoes (Haines et al., 2014).

The Hawaiian crickets provide another dramatic example of an endemic radiation (Otte, 1994). All the species of cricket can be traced back to four initial colonisation events. The phylogenetic evidence suggests that these founder species were flightless, making transoceanic rafting their most likely origin. They have since radiated to such an extent that there are now more than twice as many cricket species in Hawaii as in the whole of the continental United States. One of the most characteristic groups is the tree crickets (Oecanthinae), derived from an ancestral colonist around 2.5 Mya. Since then they have diversified into 3 genera and 54 species, hopping from island to island as they did so. Other insect groups such as *Drosophila* exhibit similar radiations, along with plants such as the silverswords. These widespread plants are unimpressive weeds elsewhere in the world, but on reaching Hawaii were able to evolve into a range of forms, including dominant forest trees.

Table 19.1 Total species richness *S* in four faunal groups on Pacific Ocean islands, with proportions comprising species shared with the continent or regional and local (island-level) endemics.

Group	*S*	Continental (%)	Regional (%)	Local (%)
Butterflies	285	55	10	35
Skinks	100	21	13	66
Birds	592	25	10	65
Mammals	106	40	6	54

Source: Whittaker and Fernández-Palacios (2007). Reproduced with permission from Oxford University Press.

Island assemblages are therefore made up of a blend of continental species that have been able to colonise and establish, regional species found across a number of islands, and local endemics confined to a single island. The latter class contains the majority of species in some clades (Table 19.1).

19.5 Size changes

A peculiar feature of island endemic animals is that they are often noticeably different in size from their closest mainland relatives. Long after the woolly mammoth had gone extinct across mainland Eurasia, around 9,500 ya, they persisted on Wrangel Island, surviving up to 7,000–4,000 ya (Vartanyan et al., 1993). The mammoths on Wrangel, however, became 30% smaller than their forebears. This pattern is repeated for other species, with the most famous being the hominid known as Flores man, *Homo floresiensis*, which was instantly dubbed by the media as the 'hobbit'. On the island of Flores in Indonesia, this dwarf species of human stood less than 1 m tall (Fig. 19.1). It persisted perhaps until less than 10,000 years ago, and it seems likely that competition from modern man played a part in its eventual demise.

Elsewhere similar patterns emerge. The elks (*Megaloceros* spp.), normally around 1.5 m height in Europe, evolved dwarf forms on Sardinia and Crete. Elephants (*Palaeoloxodon* spp.) which were formerly widespread throughout North Africa and Europe, shrank on the islands of Sicily and Malta, with the smallest species being under 1 m tall. The extinction of these insular endemics surely represents a great loss to the pet trade. Some examples are shown in Fig. 19.2.

Other species evolved in the opposite direction. In the Caribbean, giant frogs appeared, whereas in Minorca (Balearic Islands) a gigantic species of rabbit *Nuralagus rex* weighed 12 kg (Fig. 19.3). Despite its greater size, sense-related areas of its skull were reduced, perhaps indicating a lack of predators.

These trends have been repeated numerous times on islands across the globe. When rates of change are compared to those occurring on the mainland, evolution of body size occurs three times faster in insular habitats (Millien, 2006). Changes appear to be greater on islands which are smaller and more distant from the mainland (Lomolino, 2005), and to increase through time (Lomolino et al., 2013). Similar processes are known to have occurred in habitat islands; in Mauritania, climatic change caused wetlands to contract to only a few small, isolated pockets. These relics are inhabited by a species of miniature crocodile that has evolved from its normal-sized forebears (Brito et al., 2011).

The causes remain somewhat obscure. In the case of *nanism*, when species evolve to become smaller, it has been suggested that release from predation is a factor, because large size is often a defence mechanism in animals. Lack of predators may also be a factor in the well-documented tameness of island animals, which either do not flee from

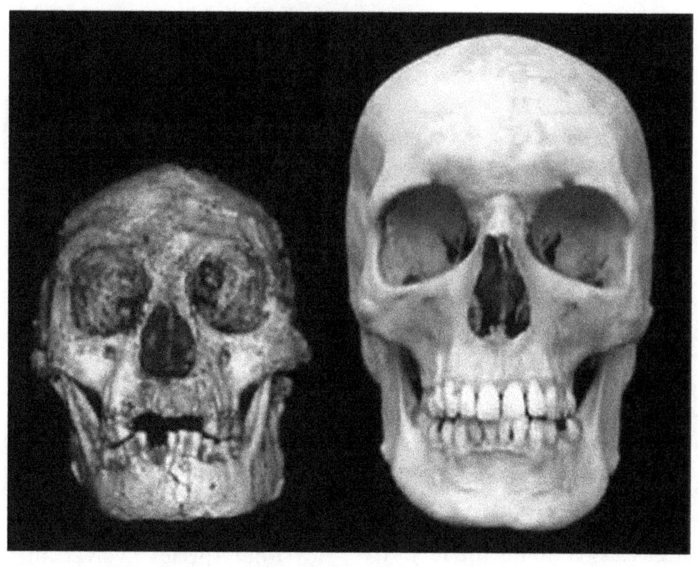

Figure 19.1 Skull of *Homo floresiensis* (left) alongside that of *Homo sapiens*. (*Source*: Brown et al. (2004). Reproduced with permission from Nature Publishing Group.)

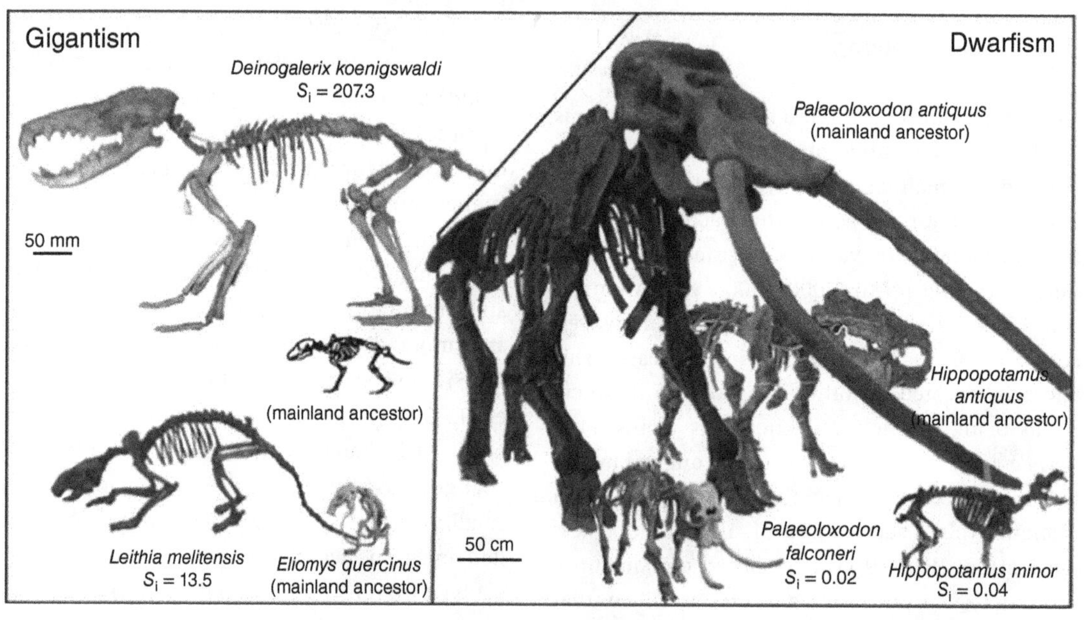

Figure 19.2 Examples of body size evolution in insular mammals, showing the extreme gigantism of erinaceomorph insectivores (e.g. *Deinogalerix koenigswaldi*) and rodents (e.g. *Leithia melitensis*) and the extreme dwarfism of ungulates (e.g. *Hippopotamus minor*) and proboscideans (e.g. *Palaeoloxodon falconeri*). S_i = mass of insular form divided by that of its ancestral or mainland form. (*Source*: Lomolino et al. (2013). Reproduced with permission from John Wiley & Sons Ltd.)

an approaching human or take longer to do so than comparable species on the mainland (Cooper et al., 2014). There may also be an influence of intraspecific competition for limited resources in insular environments. When populations reach their carrying capacity, perhaps those individuals requiring fewer resources to survive to reproduction might be favoured.

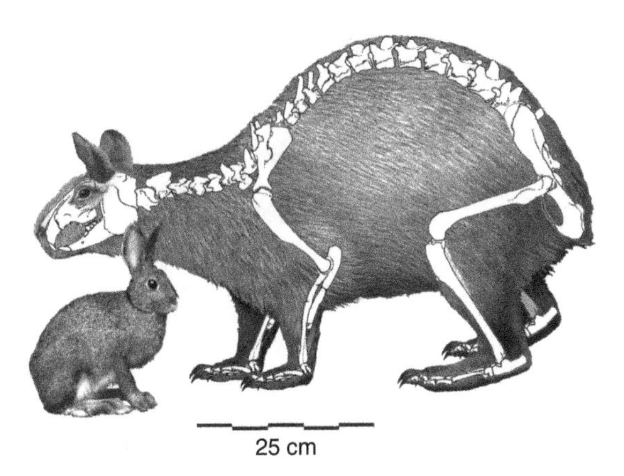

Figure 19.3 *Nuralagus rex* compared to a modern mainland rabbit. (*Source*: Quintana et al. (2011). Reproduced with permission from Society of Vertebrate Paleontology.)

Unfortunately many of the same arguments can be applied in reverse to *gigantism*. It is possible that immigrant selection plays a part, because larger individuals are more likely to possess the resources which will enable them to survive lengthy transoceanic rafting. Once on the island, release from predation and intraspecific competition can again be invoked as explanation, which makes these hypotheses difficult to assess. Of more likely relevance is the lack of larger competitors; in cases such as the Galápagos tortoises (just one of many giant island tortoise species), they have filled the role ordinarily taken by mammalian browsers on the mainland.

It has been suggested that the evolutionary tendency for small species to increase in size, and large to decrease, can be referred to as the island rule (Fig. 19.4; Foster, 1964, Van Valen, 1973). This controversial theory proposes that, at least for mammals, organisms converge on an 'ideal size' once released from selective pressures on the mainland (e.g. predation). The logic is that mammals, and perhaps other groups, will revert to a size that is optimal for energetic or demographic reasons. This is not a universally accepted idea and is likely to prove impossible to test directly. The available evidence suggests that there is some support for changes due to immigrant selection and release from predation (van der Geer et al., 2013), though not for resource limitation due to smaller areas (Lomolino et al., 2012).

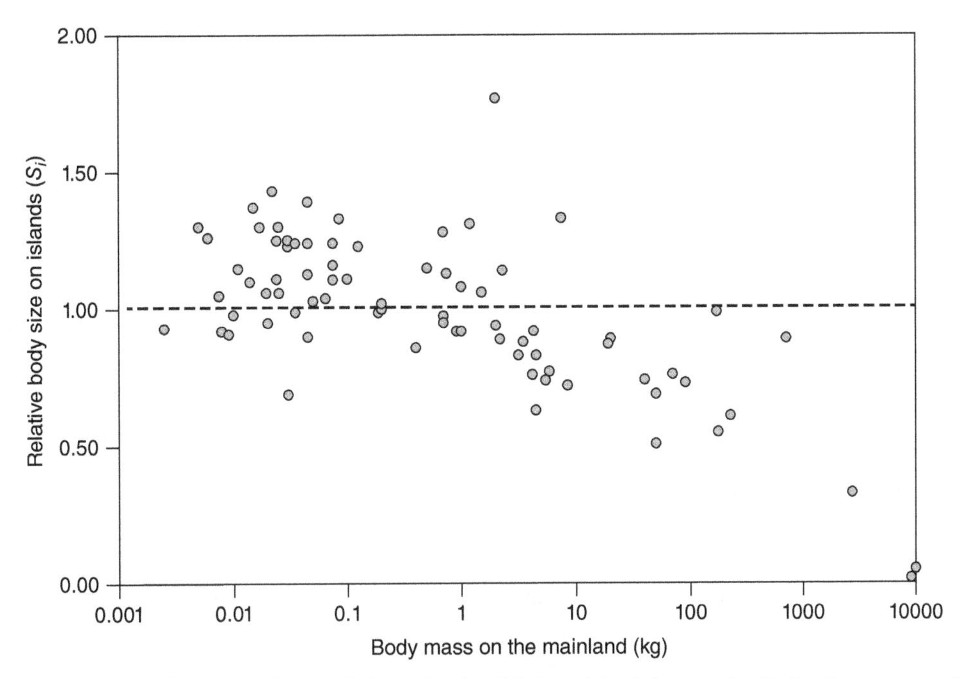

Figure 19.4 Change in size of insular endemics relative to the size of their mainland sister species. Each point represents the average body size for island populations of a particular species, where S_i is the mass in the insular population divided by that in a reference mainland population. (*Source*: Lomolino et al. (2012). Reproduced with permission from John Wiley & Sons Ltd.)

19.6 Reproduction and dispersal

Another trend witnessed among island endemics is that they typically have poor dispersal abilities when compared to their sister species on the mainland. The link between dispersal and speciation is not straightforward. Strong dispersers are more likely to retain connectivity with mainland populations, and even a small amount of gene flow may be sufficient to prevent local speciation occurring. Among plants, there is evidence that seed or diaspore size (a good correlate of dispersal potential) is negatively related to the proportion of island species that are endemics (Adsersen, 1995). Plants with small seeds have given rise to the greatest number of endemics, while those with large seeds seldom reach islands at all. The highest rates of endemism occur within groups possessing intermediate dispersal potential.

Once on islands, many species lose the dispersal mechanisms that allowed them to reach there in the first place. There are a remarkable number of endemic flightless birds on islands, typically also of large body size. Examples include the kakapo in New Zealand, a ground-dwelling descendant of parrots, which are normally strong fliers. The Galápagos cormorant is entirely flightless. Had the dodo retained its ability to fly, it would perhaps not have been so vulnerable to extinction once modern humans reached Mauritius. In such cases a lack of native predators probably removes the necessity of retaining the ability to fly as an escape mechanism. Flight is costly not merely due to the production of wings (which many flightless species nonetheless retain in reduced form) but more as a result of the maintenance of large flight muscles and a constraint on overall body mass relative to wing size.

Many insects illustrate the same syndrome. On the islands of Tristan da Cunha, there are 20 endemic beetle species, of which 18 have reduced wings compared to their ancestors (Williamson, 1981). This pattern is repeated across a wide variety of animal and plant groups (Fig. 19.5). It occurs despite the likelihood that it will be the best dispersers within the source population that first reach the island.

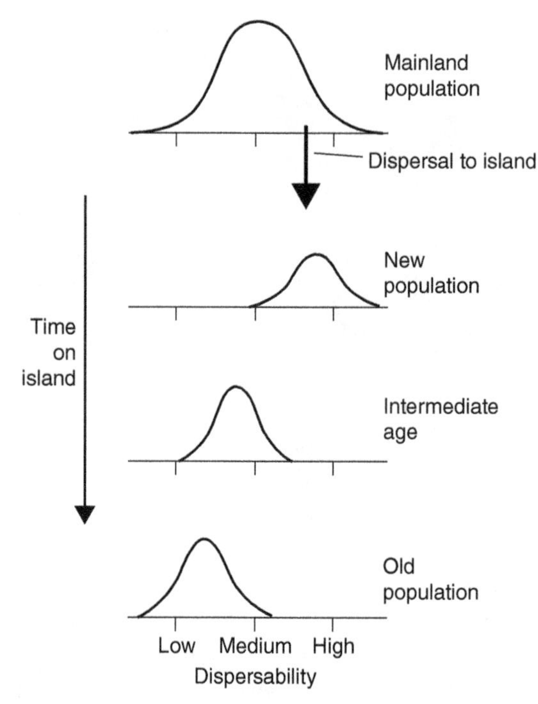

Figure 19.5 Trajectory of dispersal loss during the evolution of insular endemics. (*Source*: Whittaker and Fernández-Palacios (2007). Reproduced with permission from Oxford University Press.)

To understand why this occurs we need to conduct a thought experiment regarding the likely fitness consequences of dispersal (Cody and Overton, 1996). Imagine a species of daisy, which—by some chance—manages to disperse a seed to a previously unoccupied island. Here it grows and reproduces, sending countless seeds into the air. Sadly, however, most of these will land in the sea. Because the probability of landing on the island in the first place was so low, the chance of seeds making it to another patch of land will also be miniscule. Those seeds which are larger or through some mutation lack the pappus that enables them to float in the wind will drop closer to the parent plant and therefore have a higher probability of survival, as they will fall on land. The annual daisy *Lactuca muralis* increased its seed size over the course of only a decade following isolation (Cody and Overton, 1996), and in general island plants have larger seeds than their mainland relatives (Kavanagh and Burns, 2014). As Darwin (1859) put it:

> As with a mariner shipwrecked near a coast, it would have been better for the good swimmers if they had been able to swim further, whereas it would have been better for the bad swimmers if they had not been able to swim at all and had stuck to the wreck.

Over time, plants that evolve to disperse their seeds more locally will have higher fitness. Once the island fills up with daisies, space becomes limited, and there will be further advantages to having large seeds to better enable their offspring to establish in a crowded population. The daisies may also begin to invest more in woody parts than reproduction to ensure that they are able to compete for resources. In due course they will cease to look much like daisies and instead will have become trees.

Across the world, plants normally thought of as weeds or unimpressive herbs have evolved into trees on islands. On the Juan Fernández Islands or in Macaronesia, there are tree lettuces, while tree sunflowers dominate forests in St Helena and the Galápagos. Trees are not a monophyletic group; almost any clade of plants is capable of evolving the tree habit (Petit and Hampe, 2006). Possible reasons why this occurs so often include the competitive advantages offered by greater height and resource storage; a longevity that allows for repeated reproduction events, improving the chances of that population persisting; and the more stable climates of islands that enable ruderal plants normally used to temporary and fluctuating environments to develop specialist adaptations.

Loss of dispersal ability is not inevitable though. Among mosses and ferns, which reproduce via minute spores, very few island endemics have arisen globally (Hutsemékers et al., 2011), though their life histories do adjust to some extent (Patiño et al., 2013). This implies that for those groups to which the sea does not pose a particularly challenging dispersal barrier, they are effectively not on islands.

The numerous flightless endemic birds on islands are examples of the same process. Ironically, while having a high richness across island chains, these species tend to be rather drab and dun-coloured birds with simple song structures. Since there is little need to distinguish themselves from other species, they lose any elaborate morphology or mating behaviours. The Solomon Islands rail is one example of a series of perhaps 800 endemic rails (Galliformes) that were formerly spread across the islands of the Pacific (Trewick, 1997). Being flightless and edible, the overwhelming majority are now extinct, devoured by early Polynesian settlers.

In parallel with the loss of dispersal, species also change their reproductive patterns to adjust to dense populations with limited resources (a process referred to as K-selection). Island birds have lower overall fecundity, laying smaller clutches of eggs, which take longer to develop and require greater investment from the parents (Covas, 2012). This pattern may also be related to the general lack of brood predators on islands, which would normally select against commiting too many resources to any single clutch. Endemic island lizards also lay smaller clutches of larger hatchlings than their nearest similar-sized relatives on the mainland (Novosolov et al., 2013). Islands therefore have predictable impacts on the evolution of life histories.

19.7 Super-generalists

The final common feature of island species is the existence of super-generalists, species that exhibit a broader range of interactions than their mainland relatives. This particularly applies to mutualistic relationships such as pollination and seed dispersal. Although the diversity of interaction networks on islands does not differ from the mainland, they are more asymmetric. In seed dispersal networks a smaller number of frugivores—almost a third of the richness of comparable mainland communities—are responsible for the majority of interactions with plants, whose richness barely differs (Scheuning et al., 2014). This is largely due to a lower number of frugivorous birds, but in some cases may also reflect anthropogenic extinctions.

A number of islands contain species of endemic bees responsible for pollinating a large proportion of the native flora (Olesen et al., 2002). In the high-altitude laurel forests on the Macaronesian island of La Gomera, a single species of bumblebee, *Bombus canariensis*, is the principal pollinator for 48% of the plant species. Even more generalised is the carpenter bee *Xylocopa darwini* which pollinates an astonishing 77% of the Galápagos flora. Plants themselves are often generalised and will accept visits from a range of pollinators. Isolation seems to have made some plants desperate, including a number of bizarre cases where lizards act as the main pollinators (Olesen and Valido, 2003).

19.8 Endemic communities

If species are endemic, then by extension communities can be endemic as well. This is trivially the case for neoendemics, where it is simply impossible for that community to exist elsewhere. Of more interest are palæoendemic communities, where a set of species and their interactions are maintained on islands even after going extinct on the mainland, occasionally recolonising the continent (e.g. Hutsemékers et al., 2011).

In Macaronesia, the volcanic islands off south-western Europe comprising the Azores, Madeira and the Canaries, the relicts can be found of a climatic zone that has since been lost on the mainland. At high altitudes, where the topography encourages the formation of clouds and adequate rainfall, a band of evergreen broadleaf forest dominated by laurels remains. Fossils of similar species are spread throughout Europe, and it was the dominant vegetation of the Mediterranean region around 5–6 Mya. As the climate altered, only the species that managed to reach Macaronesia survived. This is not to say they are the same species today as occurred in the late Miocene; they have no doubt adapted and evolved in their volcanic redoubts, but they are the last remaining lineages of an ancient flora and retain a strong similarity to fossilised species. Islands generally contain a disproportionate global fraction of temperate rain forest, a biome which has become scarce on mainlands worldwide since the LGM (Weigelt et al., 2013).

19.9 Disharmony

Island communities are often described as being disharmonic. This refers to the peculiar composition of species found on islands relative to the mainland species pool. Where species are shared with the mainland, it is often a surprising subset, and their relative abundances can be completely different, with species usually thought of as rare becoming commonplace. Those species present tend to derive from 'weedy' lineages, even if they have subsequently evolved into something quite different. Disharmony is a potentially confusing term, which might be thought to imply that continental biotas are somehow in 'harmony', which is certainly not true.

Unusual combinations of species arise due to the vagaries of which species manage to arrive and establish on a given island. Sometimes this requires careful investigation to discern. For example, Jarak Island is a small forested dot of land, a mere 40 ha in size, resting in the Strait of Malacca over 50 km from mainland Malaysia. A 0.4 ha plot contained 34 tree species: an extraordinary richness compared to a temperate forest, but around a third of the number of species expected in a mainland plot of the same size (Wyatt-Smith, 1953). The most important feature of the forest, however, is that it lacks any trees in the Dipterocarpaceae, the family that dominates rain forests across the entirety of Southeast Asia. Their large seeds are incapable of crossing even narrow oceanic barriers, and their presence throughout the islands of the Sunda Shelf owes more to former land bridges than dispersal. As a result what looks at first sight to be a normal forest turns out to have a very peculiar make-up. Dipterocarps are also notably absent from the Krakatau archipelago, which was discussed in Section 18.2.

Barriers are always relative, permeable to some species while impenetrable to others. This leads to communities being filtered by dispersal ability, which is the main cause of their disharmonic composition. In the South Pacific, most groups of birds originated in Southeast Asia and have since spread across the island chains to varying extents. Corresponding to variation in their dispersal ability, the islands can be divided into a series of nested communities depending on how far each bird group has managed to spread (Fig. 19.6). All species are present in New Guinea, but each subsequent zone is a nested subset of the total species richness. Moving eastwards from New Guinea, an ever smaller proportion of the bird groups have managed to colonise the islands.

Another common cause of disharmony is that small islands are often unable to support higher predators, leading to shorter food chains (Holt, 2010). Higher trophic levels show steeper species–area relationships (SAR), indicating that predators are more sensitive than their prey to changes in available area (Roslin et al., 2014). As a result, the impact of top-down control in communities is diminished on smaller islands, changing their entire structure.

One intriguing result of disharmony is that island communities might be in an unstable state (see Chapter 11), and paradoxically invasions can increase stability through alterations to interaction webs. This was observed for plant–pollinator networks in the Galápagos, where 20% of species were in ivasive aliens but were responsible for 38% of all interactions, increasing overall nestedness

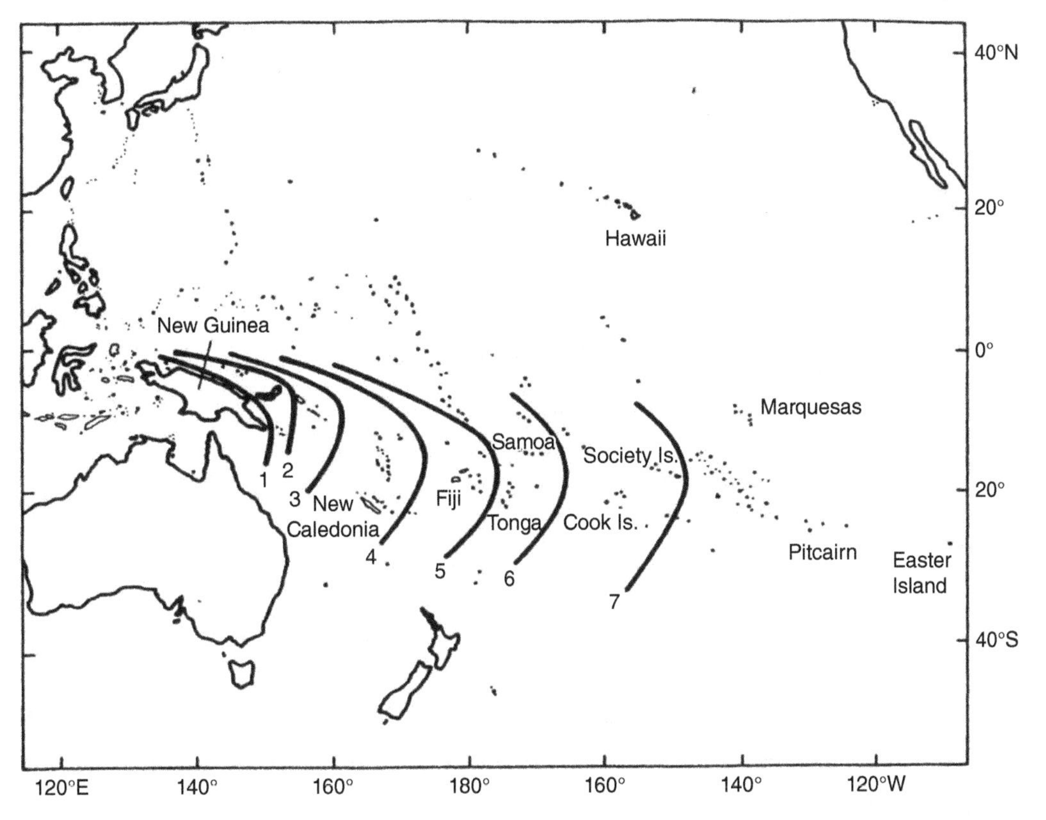

Figure 19.6 Eastern limits in the South Pacific reached by seven families or subfamilies of land and freshwater breeding birds found in New Guinea. Note that there is a concurrent decline in average island size with distance from New Guinea. (*Source*: Williamson (1981). Reproduced with permission from Oxford University Press.)

(Traveset et al., 2013). This could make restoration attempts more difficult once invasive alien species have established themselves.

If a species is fortunate enough to make it to an island where their competitors or predators are absent, they may exhibit another occasional feature of island communities, *ecological release*, which alters the abundance distribution of those species present via density compensation. The same might potentially be true of escape from diseases and parasites if they limit population density on the mainland. On Corsica three native species of tit are present, each of which occurs on both the mainland and the islands. Tit populations on the islands not only occur in a wider range of habitats but also at a greater density (Blondel and Aronson, 1999). Endemic lizards on islands maintain population densities around four times those seen on the mainland (Novosolov et al., 2013). As a note of caution, however, density is notoriously hard to estimate accurately, and it is often difficult to make a fair comparison.

19.10 Assembly rules

Arriving on the shores of an island isn't necessarily the end of the colonisation story. Here it is worth referring back to the metacommunities section of Chapter 13 (a concept that very much originated through studies of islands). Whether a given species manages to establish itself depends not only upon the local environment but also the community that is already present. As Roughgarden (1989) put it:

> A community reflects both its applicant pool and its admissions policies.

Factors that might determine whether or not a particular species establishes include:

- Niche pre-emption. If an incumbent species already occupies a particular role, it may prove impossible to displace.
- Food web levels. Higher trophic levels can only enter when their prey are present.
- Successional stage. Ruderal species require recently disturbed or empty habitats, not those where more competitive species have already arrived or evolved *in situ*. Later successional species may also be unable to colonise at the outset of a primary succession; for example, epiphytes could not colonise Krakatau until closed forest had formed, despite being well dispersed by volant animals.
- Predators. A voracious predator (or potentially a disease or parasite) might prevent vulnerable species from persisting on an island where the opportunities for escape are limited.
- Island size. A species might require a minimum area to allow sufficient territories for a sustainable population, or a maximum area, if islands above a certain size allow in higher trophic levels or dominant competitors.

The implication is that colonisation of islands that are already occupied has as much to do with the resident species as with the chance of arriving and establishing in the first place.

19.11 Island species richness

The final and most intensively studied feature of island communities is their reduced species richness as compared to mainland areas of equal size. The SAR still applies on islands (the island SAR or ISAR; see Section 7.2), but when islands of different size are compared, the steepness of the slope is much greater than is seen in mainland patches. Below a certain size, islands have fewer species than would be expected.

For island plants, the slope of the SAR (the z parameter; Fig. 7.1) is around 0.33, whereas for mainland regions it is only 0.17 (Kreft et al., 2008). This holds even after controlling for distance from the mainland or differences in environment.

As ever greater numbers of datasets have become available, it is now possible to compare island and mainland patches across the entire world. Kreft et al. (2008) assembled species richness estimates

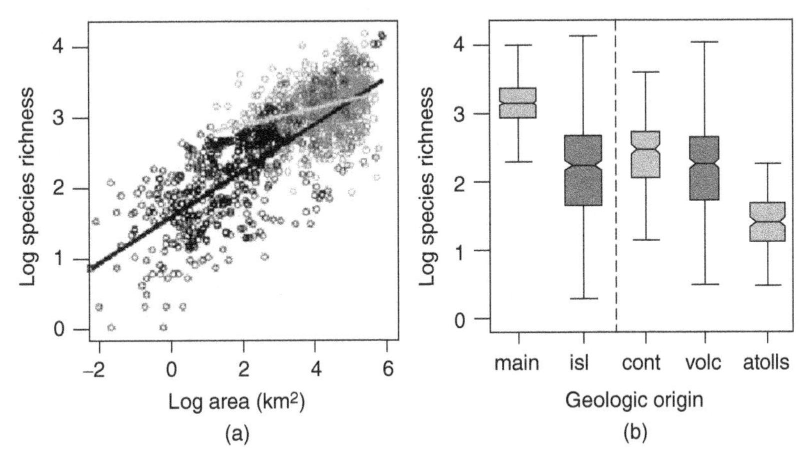

Figure 19.7 (a) Species–area relationship for island (blue) and mainland (brown) plant species richness; (b) variation in plant species richness between mainland areas and islands, dependent on island geological origin (continental, volcanic or atolls). (*Source*: Kreft et al. (2008). Reproduced with permission from John Wiley & Sons Ltd.) (*See colour plate section for the colour representation of this figure.*)

for the native plants of 488 oceanic islands and 970 mainland areas of known size. This demonstrated a number of relationships. Area was the main determinant of island species richness, accounting for 66% of the variation, while for mainland sites it fell to only 25%. The slope of the relationship was also steeper for islands (Fig. 19.7a). Islands consistently contain fewer plant species for a given area, up to about 10,000 km^2 in size. This is important as it sets a limit on when a natural system stops behaving as an island and starts to effectively act as a mainland. Analyses such as this are very sensitive to the choice of islands, and the influence of particular factors is likely to vary between regions, but it provides some rough guidelines.

Once the main effect of island area had been accounted for, the remaining third of the variation in species richness was attributed to a range of more minor influences. These included island isolation, mean annual temperature, precipitation levels and elevation (a good proxy for habitat heterogeneity). That all of these affect species richness should come as no surprise. There was also a difference among islands depending upon their geological histories (Fig. 19.7b). Atolls are particularly depauperate, even given the fact that they are typically smaller. Their flora mostly consists of regionally widespread strand species that will wash up on almost any shore, and very few endemics. A major reason for this is the low altitude of atolls, which means they put up little resistance to disturbances such as hurricanes, which can either strip an atoll of its vegetation or cause it to be submerged beneath even a moderately sized wave. On longer timescales, sea level rises during interglacials are likely to have caused atolls to be serially inundated.

Continental islands have a marginally higher number of species on average than volcanoes. This is again explicable in terms of island history, because all volcanoes need to be colonised, whereas continental islands separate from the mainland with a complete biota already present. The delay imposed by colonisation of remote volcanoes, or evolution of new endemics, accounts for the overall difference.

As usual, the only other group of organisms with global data to match that of plants is birds, and comparable studies have been performed, though not as yet with the same scale and resolution. Kalmar and Currie (2006) collected data from 346 islands around the world. As expected, there was a positive relationship between bird species richness and island area (Fig. 19.8a), along with a negative effect of distance to the nearest continent (Fig. 19.8b). Note that in both cases these are triangular relationships.

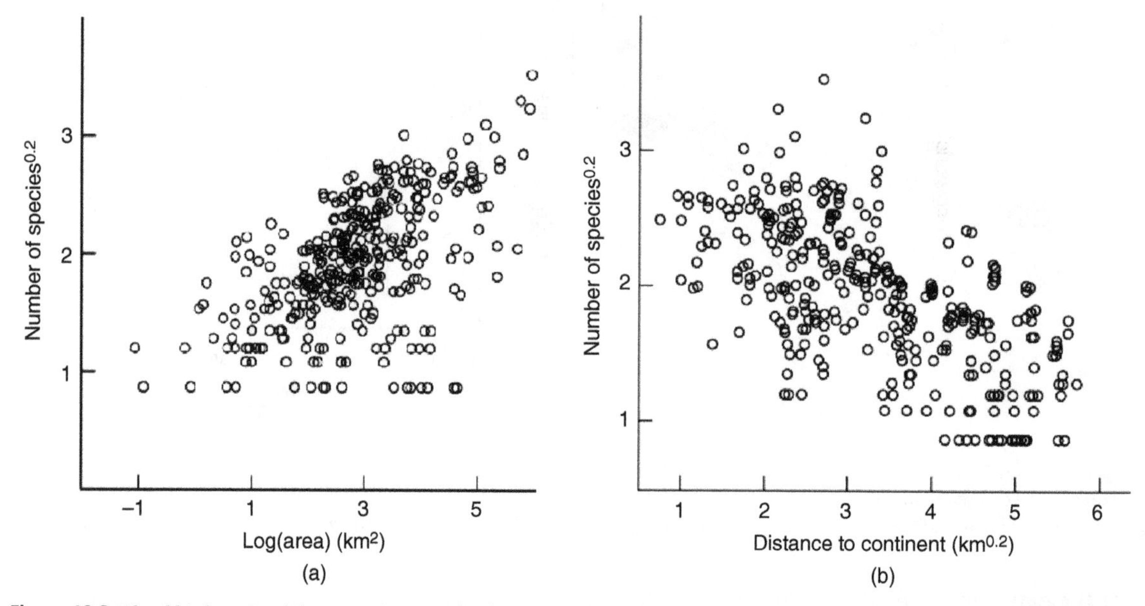

Figure 19.8 Island bird species richness against (a) island area; (b) distance of island from the nearest continental mainland. (*Source*: Kalmar and Currie (2006). Reproduced with permission from John Wiley & Sons Ltd.)

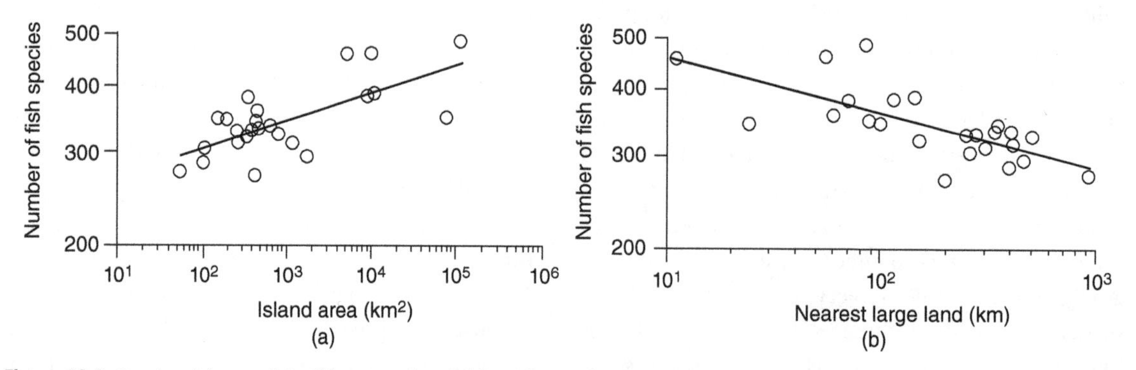

Figure 19.9 Species richness of Caribbean coral reef fish with (a) island area; (b) distance to nearest large island (>5,000 km^2) or mainland. (*Source*: Sandin et al. (2008). Reproduced with permission from John Wiley & Sons Ltd.)

In other words, area and isolation set limits on maximum species richness. It is possible for large or near islands to have relatively few species, but small or distant islands never have many. Similar patterns were seen for coral reef fish in the Caribbean (Fig. 19.9).

Measuring map distance to the nearest major continental land mass may not be the best measure of isolation though; organisms don't tend to reach islands simply 'as the crow flies', even birds that know where they're going. Oceanic currents play a major role in making certain islands more or less accessible to floating organisms, an important consideration that was demonstrated in the context of Madagascar (Ali and Huber, 2010; Section 18.5.2). The source of colonists might depend more on where they are likely to float from than the nearest source pool. With so little information on the frequency and success rate of long-distance dispersal, however, a more realistic measure of island isolation has yet to be formulated, and a combination of parameters might be necessary. For plants, the proportion of land in the area surrounding an island explains a sizeable proportion of variation in species richness, though measures of distance still have some effect (Weigelt and Kreft, 2013). In the modern world, where human-mediated dispersal on ships carries organisms everywhere, the effective isolation of islands is decreasing (e.g. Helmus et al., 2014).

19.12 The equilibrium model of island biogeography

The trends recorded by Kalmar and Currie (2006) and Sandin et al. (2008), showing a decline in species richness with decreasing area and increasing distance, have been known about for almost a century. It was these relationships that inspired MacArthur and Wilson (1967) to write their now classic book, The Theory of Island Biogeography. In order to set the scene, however, a small amount of revision is necessary.

The first reminder is of the Arrhenius SAR (Arrhenius, 1921; Section 7.2), which proposes a straight-line relationship between the area of habitat and the number of species it contains when plotted on logarithmic axes:

$$\log S = \log c + z \log A$$

The second is the observation that species abundance distributions (SADs) from large ecological samples tend to follow a lognormal distribution (Preston, 1948; Section 5.5). Not everyone now agrees on this, and it is seldom adequately tested (what constitutes a large enough sample to be sure?), but it remains a standard assumption. The main implication from an island perspective is that islands are effectively 'samples' of a regional species pool, which means they are expected to exhibit a veil line in their SAD, with the rarer species going extinct.

These were the starting points for the theory set out in a paper by MacArthur and Wilson (1963). They began with the statement that the rate of change of species richness on an island ΔS should be determined by the balance between rates of immigration I, speciation G and extinction E:

$$\Delta S = I + G - E$$

If the logic of these terms seems obvious, this can in large part be credited to the influence of this theory on the whole of ecological thought. To simplify matters, the basic theory ignored speciation G, which was assumed to be rare, and focussed on the assembly of island communities through ecological rather than evolutionary time. The implications of this will be discussed more in the next chapter, but for now we will consider only the fundamental theory.

The easiest way to understand the idea is to imagine a continental shelf island forming as sea levels rise postglaciation. The island begins with a set of species shared with the source pool on the mainland, which has a total species richness P (Fig. 19.10). In other words, the species richness of the island does not start at zero (unlike a volcano), but with some proportion of P.

Since the island has been severed from the mainland, it now has a smaller effective area, which means there are likely to be more species than it can support

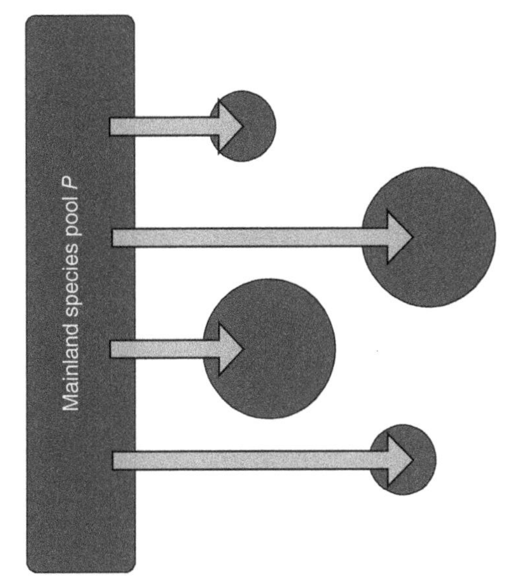

Figure 19.10 Illustration of dispersal from a mainland species pool P onto four islands varying in size and distance.

(as predicted by the SAR). The subsequent loss of species is known as *relaxation*. The probability of extinction decreases with island area because larger islands are predicted to maintain larger and therefore more persistent populations. A greater species richness implies a greater extinction rate (when there are no species, there can be no extinctions). Assuming that the mainland SAD has a lognormal form, the island is effectively a sample of this with a veil line set by its size. Rarer species are likely to go extinct, while common species will maintain stable populations. This gives the extinction rate a curved relationship with species richness (Fig. 19.11).

Species can also colonise the island and increase its species richness S, provided that they are not already present. The probability of a species colonising is dependent on the distance to any given island because more isolated islands are harder to reach. Once all the species P from the mainland are present on an island, the immigration rate will be zero. Assuming once again a lognormal SAD on the mainland, there is a high probability of common species dispersing, whereas rare species are unlikely to make it across. This gives the rate of immigration a curved relationship with island species richness (Fig. 19.11).

The number of species found on a given island at equilibrium can be predicted using the model shown in Fig. 19.11. Note that the x-axis is species richness rather than time; this is a common cause of confusion when interpreting the model. Likewise the y-axis indicates the rates of immigration and extinction per unit time, not the absolute numbers of species. Where the rates of immigration and extinction meet,

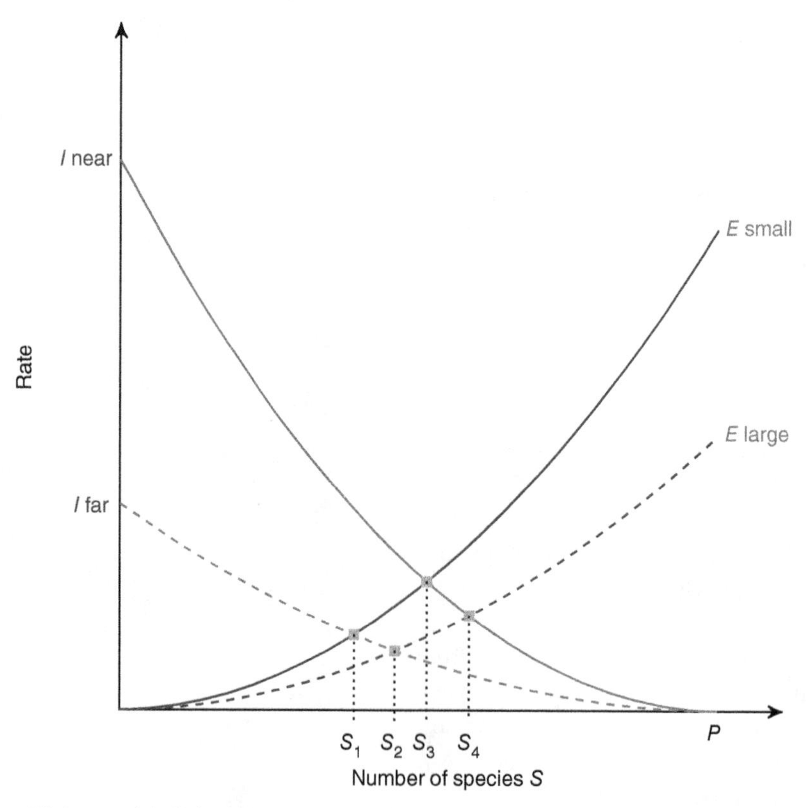

Figure 19.11 The equilibrium model of island biogeography predicts a stable species richness for each island depending on its rates of immigration of species I, modulated by distance from mainland (near vs. far), and rates of extinction E, modulated by island size (large vs. small). The equilibrial turnover rate differs for each island, leading to a particular species richness S_{1-4}, up to the total number of species in the mainland pool P. (*Source*: Adapted from MacArthur and Wilson (1967).)

the species richness will be balanced. This will vary depending on the size and isolation of the island (S_{1-4}) and is read from the bottom of the graph.

The important aspect of this model is that it predicts a stable equilibrium species richness for the island, which is formed by a balance between continuous immigration and extinction. For this reason it is often referred to as the equilibrium model of island biogeography, or EMIB for short. Even when the number of species has reached a stable level, there is still a continuous turnover in the identity of those species.

19.13 Testing the theory

The model is found in every ecology textbook, features on every undergraduate ecology course, and almost any ecologist should be able to sketch out the famous graph from Fig. 19.11. Nevertheless, almost 50 years since its first publication, finding exclusive evidence in support of the theory has proven extremely difficult. Note that finding trends of species richness with island area and isolation as in Kalmar and Currie (2006) does not in itself validate the theory, though this has often been claimed. Recall that MacArthur and Wilson (1963) took these trends as their starting points and proposed the theory to explain them. In order to assess the model's validity, it is necessary to test the specific predictions.

The underlying processes appear obvious—dispersal should decline with distance, and extinction should decrease with area. These are not contentious points. The real difficulty is with the core feature of the theory: that a dynamic equilibrium will form. Evidence for this has proven elusive, which may seem surprising, until you consider what that evidence involves. One would need to not only find all the species present on a particular island but track their colonisation and extinction through time. The trouble is that surveys are always incomplete and biased towards common species. Because the rarest species are those most likely to go extinct (or to have most recently arrived), the records most needed are precisely those that are most difficult to obtain. In the Galápagos, the number of plant species recorded on each island correlates better with the number of visits by botanists, not their area or isolation, a clear indication that sampling effort would have to be considerably increased to detect dynamic turnover (Connor and Simberloff, 1979).

Long-term, comprehensive samples of complete island communities are required. These are almost non-existent—with one happy exception. In the British Isles, a national obsession with birds has led to the offshore islands having rich records of breeding species stretching back many years. Many of these islands have dedicated wardens for this purpose, and some are otherwise uninhabited. With long time series for 13 islands, Manne et al. (1998) evaluated the predictions made by MacArthur and Wilson (1967), in particular the anticipated curvilinear relationships.

They found, as would be expected simply from a sampling effect, that the number of extinctions rose with overall bird species richness. In 9 out of 13 cases these were curved relationships. There was a decline in extinction rate with area, though it was not statistically significant. Immigration rates showed a similar strength of support; 7 of 13 relationships with species richness were curved, and there was the anticipated decline with distance of islands from the mainland. Once again, however, the latter was not statistically significant.

There are therefore two ways to view the paper by Manne et al. (1998), which provides at best equivocal support for the theory. Either one could argue that too little high-quality data has been obtained to reach definitive conclusions, but UK birds at least indicate correspondence with expectations. On the other hand, a more sceptical reader might argue that with the best data available and a set of species that are expected to follow the model predictions closely, in 50 years this is the nearest anyone has come to backing the theory.

What then should be done—keep looking, abandon the theory or try to expand it? This question is the focus of the next chapter. Understanding islands is important because they represent one of the tasks set at the outset of this book: to see what it would take to build a natural system. Islands are themselves microcosms of nature, and finding how they assemble will provide the insights needed to one day perform the task ourselves.

19.14 Conclusions

Island communities are special, or to put it another way, they're odd. Islands differ in terms of their geological origins and history, with particular variation between continental islands, volcanoes and atolls. There is also the broader context of insular habitats occurring wherever some form of dispersal barrier surrounds patches.

The characteristics of island biotas have been a particular source of fascination for ecologists and biogeographers since the emergence of both disciplines. Patterns that continue to intrigue are the presence of numerous endemics, the phenomenon of adaptive radiation and the peculiar features of island taxa when compared to their mainland relatives. Common features of island endemics are size changes, reduced dispersal ability and super-generalism. Endemics are also often from surprising lineages due to the quirks of chance dispersal events.

The presence of endemics implies that the communities containing them are also unique. These exhibit common features, including disharmony and ecological release, but most importantly low species richness relative to the mainland. The latter observation led to The Theory of Island Biogeography, perhaps the most influential model in the history of ecology. Island species richness increases with overall area but declines with distance from the mainland, a puzzle that MacArthur and Wilson (1967) tried to solve by invoking the processes of extinction and dispersal. They predicted an equilibrium species richness to occur when these processes are in balance, maintaining a continual turnover in the species present. This seductively simple theory quickly became one of the cornerstones of modern ecology and lies at the heart of conservation theory, yet after half a century, the existence of a dynamic equilibrium still remains in doubt.

19.14.1 Recommended reading

Kreft, H., W. Jetz, J. Mutke, G. Kier, and W. Barthlott, 2008. Global diversity of island floras from a macroecological perspective. *Ecology Letters* 11:116–127.

Whittaker, R. J. and J. Fernández-Palacios, 2007. *Island Biogeography*. Oxford University Press, second edition.

19.14.2 Questions for the future

- What ecological and environmental forces drive the evolution of the characteristic traits of island endemics?
- Which factors enable (or prevent) adaptive radiations?
- How general is the equilibrium model of island biogeography, and does it deserve its continued prominence?

References

Adsersen, H., 1995. Research on islands: classic, recent and prospective approaches. In P. M. Vitousek, L. L. Loope, and H. Adsersen, editors, *Islands: Biological Diversity and Ecosystem Function*, Ecological Studies, Vol. 115, pages 7–21. Springer-Verlag.

Adsersen, H., 2013. A remarkable anniversary: Surtsey becomes 50 years old. *Frontiers of Biogeography* 5:78.

Ali, J. R. and M. Huber, 2010. Mammalian biodiversity on Madagascar controlled by ocean currents. *Nature* 463:653–656.

Arrhenius, O., 1921. Species and area. *Journal of Ecology* 9:95–99.

Blondel, J. and J. Aronson, 1999. *Biology and Wildlife of the Mediterranean Region*. Oxford University Press.

Brito, J. C., F. Martínez-Freiría, P. Sierra, N. Sillero, and P. Tarroso, 2011. Crocodiles in the Sahara desert: an update of distribution, habitats and population status for conservation planning in Mauritania. *PLoS ONE* 6:e14734.

Brown, J. H., 2014. Why marine islands are farther apart in the tropics. *American Naturalist* 183:842–846.

Jamitko Brown, P., T. Sutikna, M. J. Morwood, R. P. Soejeno, Jatmiko, E. W. Saptomo, and R. A. Due, 2004. A new small-bodied hominin from the Late Pleistocene of Flores, Indonesia. *Nature* 431:1055–1061.

Caujapé-Castells, J., A. Tye, D. J. Crawford, A. Santos-Guerra, A. Sakai, K. Beaver, W. Lobin, F. B. V. Florens, M. Moura, R. Jardim, I. Gómes, and C. Kueffer, 2010. Conservation of oceanic island floras: present and future global challenges. *Perspectives in Plant Ecology, Evolution, and Systematics* 12:107–129.

Cody, M. L. and J. M. Overton, 1996. Short-term evolution of reduced dispersal in island plant populations. *Journal of Ecology* 84:53–61.

Connor, E. F. and D. Simberloff, 1979. The assembly of species communities: chance or competition? *Ecology* 60:1132–1140.

Cooper, W. E. Jr., R. A. Pyron, and T. Garland Jr., 2014. Island tameness: living on islands reduces flight initiation distance. *Proceedings of the Royal Society Series B* 281:20133019.

Covas, R., 2012. Evolution of reproductive life histories in island birds worldwide. *Proceedings of the Royal Society Series B* 279:1531–1537.

Darwin, C., 1842. *The Structure and Distribution of Coral Reefs. Being the First Part of the Geology of the Voyage of the Beagle, under the Command of Capt. Fitzroy, R. N., during the years 1832-36*. Smith, Elder & Company.

Darwin, C., 1859. *The Origin of Species by Means of Natural Selection*. John Murray, first edition.

Foster, J. B., 1964. Evolution of mammals on islands. *Nature* 202:234–235.

van der Geer, A. A., G. A. Lyras, M. V. Lomolino, M. R. Palombo, and D. F. Sax, 2013. Body size evolution of palaeo-insular mammals: temporal variations and interspecific interactions. *Journal of Biogeography* 40:1440–1450.

Gillespie, R., 2004. Community assembly through adaptive radiation in Hawaiian spiders. *Science* 303:356–359.

Grant, P. R. and B. R. Grant, 2008. *How and Why Species Multiply. The Radiation of Darwin's Finches*. Princeton University Press.

Haines, W. P., P. Schmitz, and D. Rubinoff, 2014. Ancient diversification of *Hyposmocoma* moths in Hawaii. *Nature Communications* 5:3502.

Helmus, M. R., D. L. Mahler, and J. B. Losos, 2014. Island biogeography of the Anthropocene. *Nature* 513:543–546.

Holt, R. D., 2010. Toward a trophic island biogeography: reflections on the interface of island biogeography and food web ecology. In J. B. Losos and R. E. Ricklefs, editors, *The Theory of Island Biogeography Revisited*, pages 143–185. Princeton University Press.

Hutsemékers, V., P. Szövényi, A. J. Shaw, J.-M. González-Mancebo, J. Muñoz, and A. Vanderpoorten, 2011. Oceanic islands are not sinks of biodiversity in spore-producing plants. *Proceedings of the National Academy of Sciences of the United States of America* 108:18989–18994.

Janzen, D. H., 1967. Why mountain passes are higher in the tropics. *American Naturalist* 101:233–249.

Kalmar, A. and D. J. Currie, 2006. A global model of island biogeography. *Global Ecology and Biogeography* 15:72–81.

Kavanagh, P. H. and K. C. Burns, 2014. The repeated evolution of large seeds on islands. *Proceedings of the Royal Society Series B* 281:20140675.

Kier, G., H. Kreft, T. M. Lee, W. Jetz, P. L. Ibisch, C. Nowicki, J. Mutke, and W. Barthlott, 2009. A global assessment of endemism and species richness across island and mainland regions. *Proceedings of the National Academy of Sciences of the United States of America* 106:9322–9327.

Kreft, H., W. Jetz, J. Mutke, G. Kier, and W. Barthlott, 2008. Global diversity of island floras from a macroecological perspective. *Ecology Letters* 11:116–127.

Lomolino, M. V., 2005. Body size evolution in insular vertebrates: generality of the island rule. *Journal of Biogeography* 32:1683–1699.

Lomolino, M. V., D. F. Sax, M. R. Palombo, and A. A. van der Geer, 2012. Of mice and mammoths: evaluations of causal explanations for body size evolution in insular mammals. *Journal of Biogeography* 39:842–854.

Lomolino, M. V., A. A. van der Geer, G. A. Lyras, M. R. Palombo, D. F. Sax, and R. Rozzi, 2013. Of mice and mammoths: generality and antiquity of the island rule. *Journal of Biogeography* 40:1427–1439.

Losos, J. B., 1996. Phylogenetic perspectives on community ecology. *Ecology* 77:1344–1354.

MacArthur, R. H. and E. O. Wilson, 1963. An equilibrium theory of insular zoogeography. *Evolution* 17:373–387.

MacArthur, R. H. and E. O. Wilson, 1967. *The Theory of Island Biogeography*. Princeton University Press.

Manne, L. L., S. L. Pimm, J. M. Diamond, and T. M. Reed, 1998. The form of the curves: a direct evaluation of MacArthur & Wilson's classic theory. *Journal of Animal Ecology* 67:784–794.

Millien, V., 2006. Morphological evolution is accelerated among island mammals. *PLoS Biology* 4:1863–1868.

Morato, T., S. D. Hoyle, V. Allain, and S. J. Nicol, 2010. Seamounts are hotspots of pelagic biodiversity in the open ocean. *Proceedings of the National Academy of Sciences of the United States of America* 107:9707–9711.

Novosolov, M., P. Raia, and S. Meiri, 2013. The island syndrome in lizards. *Global Ecology and Biogeography* 22:184–191.

Olesen, J. M., L. I. Eskildsen, and S. Venkatasamy, 2002. Invasion of pollination networks on oceanic islands: importance of invader complexes and endemic super-generalists. *Diversity and Distributions* 8:181–192.

Olesen, J. M. and A. Valido, 2003. Lizards as pollinators and seed dispersers: an island phenomenon. *Trends in Ecology & Evolution* 18:177–181.

Otte, D., 1994. *The Crickets of Hawaii: Origin, Systematics and Evolution*. Academy of Natural Sciences.

Patiño, J., I. Bisang, L. Hedenäs, G. Dirkse, A. H. Bjarnason, C. Ah-Peng, and A. Vanderpoorten, 2013. Baker's law and the island syndromes in bryophytes. *Journal of Ecology* 101:1245–1255.

Petit, R. J. and A. Hampe, 2006. Some evolutionary consequences of being a tree. *Annual Review of Ecology, Evolution, and Systematics* 37:187–214.

Preston, F. W., 1948. The commonness, and rarity, of species. *Ecology* 29:254–283.

Quintana, J., M. Köhler, and S. Moyà-Solà, 2011. *Nuralagus rex*, gen. et sp. nov., an endemic insular giant rabbit from the Neogene of Minorca (Balearic Islands, Spain). *Journal of Vertebrate Paleontology* 31:231–240.

Roslin, T., G. Várkonki, M. Koponen, V. Vikberg, and M. Nieminen, 2014. Species–area relationships across four trophic levels—decreasing island size truncates food chains. *Ecography* 37:443–453.

Roughgarden, J., 1989. The structure and assembly of communities. In J. Roughgarden, R. M. May, and S. A. Levin, editors, *Perspectives in Ecological Theory*, pages 203–226. Princeton University Press.

Sandin, S. A., M. J. A. Vermeij, and A. H. Hurlbert, 2008. Island biogeography of Caribbean coral reef fish. *Global Ecology and Biogeography* 17:770–777.

Scheuning, M., K. Böhning-Gaese, D. M. Dehling, and K. C. Burns, 2014. At a loss for birds: insularity increases asymmetry in seed-dispersal networks. *Global Ecology and Biogeography* 23:385–394.

Sequeira, A. S., A. A. Lanteri, M. A. Scataglini, V. A. Confalonieri, and B. D. Farrell, 2000. Are flightless *Galapaganus* weevils older than the Galápagos Islands they inhabit? *Heredity* 85:20–29.

Stuart, Y. E., J. B. Losos, and A. C. Algar, 2012. The island-mainland species turnover relationship. *Proceedings of the Royal Society Series B* 279:4071–4077.

Traveset, A., R. Heleno, S. Chamorro, P. Vargas, C. K. McMullen, R. Castro-Urgal, M. Nogales, H. W. Herrera, and J. M. Olesen, 2013. Invaders of pollination networks in the Galápagos Islands: emergence of novel communities. *Proceedings of the Royal Society Series B* 280:20123040.

Trewick, S. A., 1997. Flightlessness and phylogeny amongst endemic rails (Aves: Railidae) of the New Zealand region. *Philosophical Transactions of the Royal Society of London, Series B: Biological Sciences* 352:429–456.

Van Valen, L., 1973. A new evolutionary law. *Evolutionary Theory* 1:1–33.

Vartanyan, S. L., V. E. Garutt, and A. V. Sher, 1993. Holocene dwarf mammoths from Wrangel Island in the Siberian Arctic. *Nature* 362:337–340.

Wallace, A. R., 1860. On the zoological geography of the Malay Archipelago. *Journal of the Linnaean Society of London* 4:172–184.

Weigelt, P., W. Jetz, and H. Kreft, 2013. Bioclimatic and physical characterization of the world's islands. *Proceedings of the National Academy of Sciences of the United States of America* 110:15307–15312.

Weigelt, P. and H. Kreft, 2013. Quantifying island isolation—insights from global patterns of insular plant species richness. *Ecography* 36:417–429.

Whittaker, R. J. and J. Fernández-Palacios, 2007. *Island Biogeography*. Oxford University Press, second edition.

Williamson, M. H., 1981. *Island Populations*. Oxford University Press.

Wyatt-Smith, J., 1953. The vegetation of Jarak Island, Straits of Malacca. *Journal of Ecology* 41:207–225.

CHAPTER 20

Reinventing islands

20.1 The big question

At the very outset of the book this challenge was set: if we were to design a functioning natural system, how would we go about it? Islands are compact, self-contained natural systems, and in developing a unified view of how nature works they are the ideal places to start. Understanding how island systems form can perhaps provide lessons about nature as a whole.

As the last chapter showed, however, the processes governing insular species richness and composition are not straightforward. Theories such as the equilibrium model of island biogeography (EMIB) have done much to advance our understanding, but EMIB has proven fiendishly difficult to validate, and additional processes are needed to form a more complete picture. Fortunately these should all be familiar from earlier chapters and include the roles played by niches, species interactions, disturbance, succession and evolution. Can these be combined to generate a unified model of how natural systems are assembled?

20.2 A critique of EMIB

Any model can be criticised for the things it fails to include or account for, but this is lazy science and misses the point of making models in the first place. The purpose of a theoretical model is to reduce systems to the simplest form that can still effectively capture their properties and which makes reliable predictions for how they ought to behave. This explains the power and durability of EMIB. It is a magnificent combination of a few basic processes that describes patterns seen in the real world. In order to reject it, we would need to demonstrate that either:

(a) its assumptions were incorrect, or
(b) it has failed to match the data, or else
(c) its specific predictions were not borne out.

Some of the assumptions can certainly be called into question. The lognormal distribution of abundance is not always present (Dornelas and Connolly, 2008), or as seen in Section 10.10 it can sometimes be more complicated than a simple matter of numbers of individuals (Connolly et al., 2005). Relaxing this assumption will influence the shape of the curves but is by no means fatal for the core theory.

Another questionable assumption is that the effects of area and immigration are independent; in other words, isolation only affects dispersal, and area only affects extinction. This again is unlikely to be true. There could be a target effect of large islands, whereby motile organisms find them more easily or choose them more readily, or else their greater shoreline and surface area might increase the number of species landing on them (Lomolino, 1990). This pattern was seen for birds in the Thousand Island Lake in China, an artificial reservoir created in 1959 for a hydroelectric dam (Si et al., 2014). Larger islands had

Natural Systems: The organisation of life, First Edition. Markus P. Eichhorn.

lower extinction rates, as expected, but also higher colonisation rates, suggesting a target effect; isolation on the other hand affected neither immigration nor extinction. Immigrants may also deliberately leave small islands that do not meet their needs.

Isolation and extinction are likely to be similarly interrelated. Metapopulation theory predicts a rescue effect for near islands, with supplementary immigration maintaining populations of species that would otherwise go extinct in 'sink' habitats (Brown and Kodric-Brown, 1977). On offshore islands of Great Britain and Ireland, extinction rates of bird species actually fall on more isolated islands, leading to patterns of colonisation and extinction which are more consistent with optimal foraging theory than EMIB (Russell et al., 2006). There are also many ways to measure isolation, and distance to the mainland is by no means the best, which may confound analyses (Weigelt and Kreft, 2013).

Evidence for many of these complications was seen in the UK bird data set analysed by Manne et al., (1998). This still implies that tinkering with the model should suffice, rather than total replacement. A more telling omission is the lack of any species traits or interactions. EMIB in its basic form assumes that all species are equal with respect to their ability to survive on islands and that no competition or facilitation takes place (the similarity to the neutral theory of Hubbell (2001) is no coincidence; neutral theory is in fact an extension of EMIB). Yet checkerboard patterns are widespread in metacommunities with a number of potential causes (Section 13.2). Some more recent enhancements have added extra features to the basic model, such as that species can only colonise if their prey are already present, and will go extinct if their food source disappears, which improves the fit to real data (Gravel et al., 2011). Whether adding assembly rules will be sufficient remains an open question.

Another aspect that can be built into the EMIB framework is habitat heterogeneity (Kadmon and Allouche, 2007). The principle is simple; if an island is divided into a greater number of habitats, each will effectively behave as an island within an island. This reduces the area available for species specialised to particular habitats. A humped relationship of species richness with habitat heterogeneity should be the result. The model only works if island species are largely specialists though, and where generalist species dominate, habitat diversity leads to a consistent increase in species richness (Hortal et al., 2009). There is at present no clear evidence for the predictions of this model.

What about EMIB's ability to match real-world systems? Here the difficulty lies in measuring processes such as colonisation and extinction that are inherently difficult to observe. One sign that things might not be working as expected would be if area and isolation failed to account for patterns of island species richness or were superceded by other factors. Power (1972) examined 16 islands in the Bay of California, finding that plant species richness could largely be accounted for by area (58%) and latitude (25%), with only 17% remaining unexplained and apparently unrelated to isolation. Examining bird species richness on the same islands did find an effect of isolation (14%), though the bulk of variation in bird species richness was linked to that of plants (67%). Species richness in plants and birds were determined by different processes and a common model would be inappropriate.

A further complication comes from the observation that very few islands are single, isolated entities. Most form as elements in archipelagos or chains, which share and interchange species amongst themselves. Carstensen et al. (2012) identified four different types of island based on their relative linkage to neighbouring islands (grouped into modules) or the region as a whole (Table 20.1). The classical theory applies best to peripheral islands, but most islands are linked together into networks, sharing species or acting as conduits through which they can disperse across regions. Large, mountainous islands tend to act as hubs and are responsible for the composition of other islands nearby, whereas peripherals behave as sinks, and connectors act as stepping stones.

As an example, Fig. 20.1 divides the islands of the West Indies into module hubs, network hubs,

Table 20.1 Biogeographical roles played by islands within modules and regions.

Module hubs	Network hubs
Many local species Few regional species	Many local species Many regional species
Peripherals Few local species Few regional species	**Connectors** Few local species Many regional species

Source: Carstensen et al. (2012). Reproduced with permission from John Wiley & Sons Ltd.

peripheral and connector islands based on their position within the island network. The map was produced by a statistical method which analyses the role of each island based on the species of birds that it contains. Any given island might play a variety of roles for different taxa, depending on their dispersal abilities and geographical origins.

Finally the EMIB predictions need to be tackled directly, and work here has revolved around the idea of a dynamic equilibrium. In a series of studies which remain some of the most comprehensive experimental tests, Rey (1984, 1985) created artificial islands of the saltmarsh grass *Spartina alterniflora*. These varied in size from 56 to 1023 m^2. At the start of the experiment, islands were fumigated to ensure that they were sterile, after which their arthropod communities were tracked weekly as they recolonised from the mainland pool of species.

What rapidly became clear was that extinction and immigration were not random with respect to species. Turnover was decidedly heterogeneous. A core of species persisted once established, and only

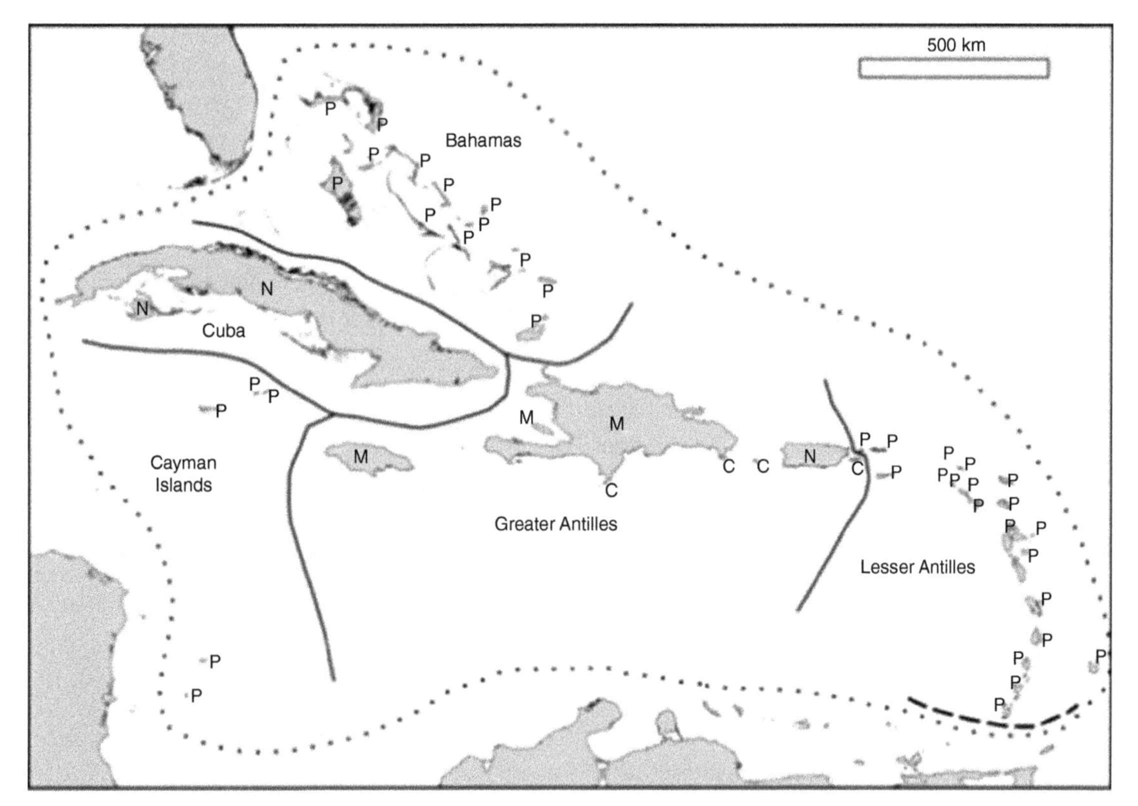

Figure 20.1 Island types within the West Indies archipelago as enclosed by the dotted line (the broken line to the south marks the biogeographical boundary known as Bond's Line). Island roles are P, peripheral; C, connector; M, module hub; N, network hub. (*Source*: Carstensen et al. (2012). Reproduced with permission from John Wiley & Sons Ltd.)

ephemeral species showed any signs of dynamism. Such species were always rare and never remained resident for long. Instead the islands had structured and stable arthropod communities. This should not come as a shock to anyone grounded in basic community ecology, but it causes severe problems for EMIB as it suggests that the bulk of natural systems are at an equilibrium that is anything but dynamic. If the model only applies to minor, transient species, then perhaps it is merely 'true but trivial' (Williamson, 1989).

20.3 Rival hypotheses

It is often perceived by outsiders that science is inherently conservative and unwilling to accept new or contradictory evidence. Nothing could be further from the truth. There is always a ready supply of scientists with their own competing ideas to espouse. There has been no shortage of alternatives to EMIB in the literature, even if few have attracted the same curious devotion.

The simplest explanation for a greater species richness on larger islands is a passive sampling effect, whereby propagules or dispersing animals that spread entirely at random will automatically cause larger areas to receive more colonists (Lomolino, 1990). Any target effect that leads to motile organisms deliberately aiming for larger islands will only serve to enhance this. This does not, however, properly account for the patterns of species richness observed. Larger islands may also enable more species to colonise if their presence requires sufficient space to maintain enough territories for a sustainable population. This would exclude higher trophic levels from many islands. Their entry into the island community might achieve more than simply adding the predators themselves if top-down control were to enable coexistence of multiple prey species (Section 8.6).

Another explanation for species richness variation invokes greater habitat heterogeneity on larger islands, because large islands tend to be taller and therefore provide a greater range of altitudinal and climatic zones for specialisation. Greater numbers of endemics are found at higher altitudes—they are effectively islands within islands, more isolated from areas with similar environments (Steinbauer et al., 2012). When no species adapted to montane environments arrive on an island, they have to evolve *in situ*, as has been the case for Cretan plants (Trigas et al., 2013).

By far the greatest challenge to EMIB thus far came from a paper by Wright (1983). The *species–energy hypothesis* suggests that by multiplying island area by AET (actual evapotranspiration, a measure of available energy), around 70% of variation in plant species richness can be accounted for. Other studies have found its fit to data to be often better than EMIB. Simply fitting a model does not, however, imply that its underlying assumptions or constituent processes are correct. In the case of Wright's model, it rests—like EMIB—on the existence of the lognormal SAD and assumes that all species require roughly the same amount of energy per individual (Fox et al., 2011). The former is doubtful and the latter simply wrong. This does not mean that we should dispense with the idea; a good phenomenological model (one which describes the data well) can still be useful and make accurate predictions, but unlike a mechanistic model (one that explains *why* a pattern occurs), it cannot provide insights into the fundamental workings of natural systems.

Instead of squabbling over which model best fits observed patterns in species richness across islands, we should consider what other processes might be influencing them and how these can be built into our viewpoint.

20.4 Disturbance

One problem faced when attempting to apply or assess the original EMIB is that there is no straightforward way to evaluate how long it will take for species richness to reach an equilibrium. It depends on the rates of immigration and extinction, which are themselves determined by the size of the island, its isolation, the current species richness and the mainland pool from which they are being drawn. Regardless of the number of species on the island when it first forms, it could take an extremely long time for rates to stabilise. It would then only require occasional disturbance for an equilibrium to be deferred almost indefinitely.

In the Caribbean an average of 4.6 hurricanes pass through per year, with any given island usually being struck every few decades (e.g. the return rate in Puerto Rico is c. 21 years). The time between hurricanes is unpredictable. Succession on these islands is perpetually set back by massive disturbance events, which are likely to select which of the mainland species manage to persist, even should they reach the islands.

Other common forms of island disturbance include eruptions of volcanoes, such as the island of Alaid in the Kuril Archipelago, which last erupted in 1996, and is still lacking much in the way of vegetation. Tidal waves strip small islands, leaving a more persistent legacy by infiltrating ground supplies of freshwater. Because atolls have such low elevations, even a strong tropical storm might be enough to flood them entirely from time to time. Circumstantial evidence for the impact of disturbance on species richness was collected by Diamond (1974) who found that islands disturbed by volcanoes or tidal waves in the recent past contained fewer species of bird than expected given their size.

To see what happens when a real island is totally emptied of life and recolonised, there is an ideal and well-studied example: Krakatau. The eruption of this volcano in 1883 was the loudest noise ever recorded, heard on the other side of the world, and tore the original island apart, leaving behind a remnant island (Rakata) and two older neighbouring islands (Panjang and Sertung; Fig. 20.2). The smaller Anak Krakatau emerged as the result of further volcanic activity beginning in the 1920s, an event which also caused large amounts of disturbance on the neighbouring islands.

Little is known about the original vegetation of the islands, but after three months of continuous

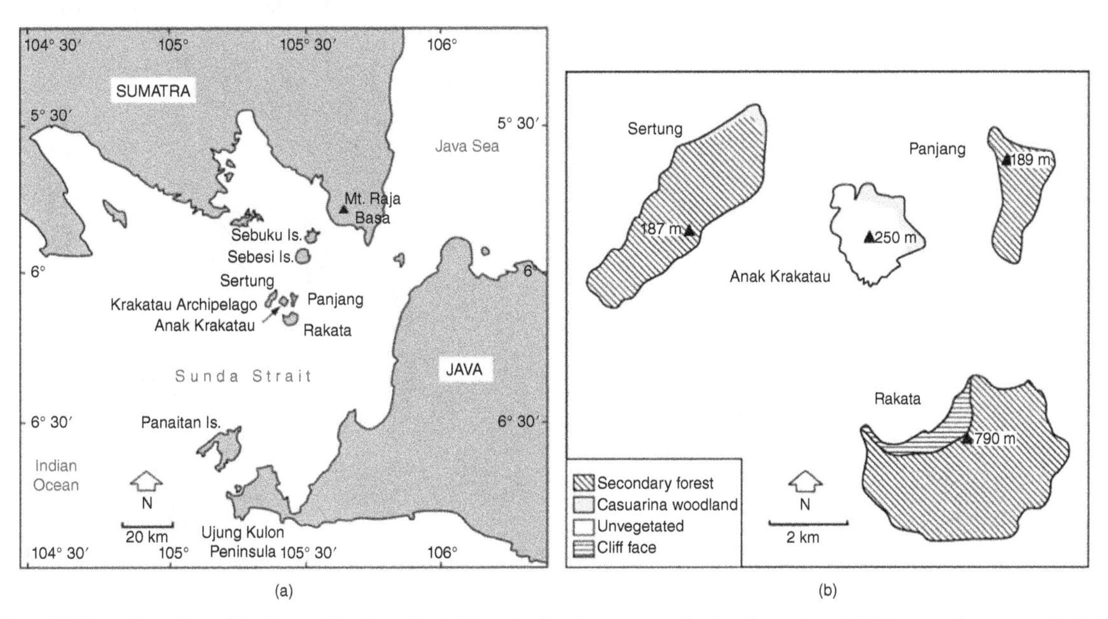

Figure 20.2 (a) Location of Krakatau islands and neighbouring land masses in the Sunda strait and (b) layout of present islands in 1996. (*Source*: Thornton et al. (2002). Reproduced with permission from John Wiley & Sons Ltd.)

eruption, they were utterly devoid of life. An expedition in October 1883 detected no signs of living plants (a detailed account of botanical expeditions to Krakatau is provided by Whittaker et al., 1989). In 1886 an expedition landed on the shores of Krakatau to investigate further. On the coastlines they found the common strand flora composed of species that soon wash up on any beach in the tropics and remain little changed to this day. There were also a scarce few weedy pioneer trees—those with small and well-dispersed seeds carried by wind, water or animals. In the interior were growing 10 species of fern alongside two grass and two daisy species. All of these have tiny, wind-dispersed spores or seeds and therefore might be expected to have arrived quickly, though their colonisation also depended on an ability to grow on unconsolidated ash with no organic matter.

Naturalists continued to visit the islands. In 1897 the interior of Rakata resembled a savannah, with grasses interspersed by small groups of animal-dispersed trees. By the time of the 1906 expedition there were 99 angiosperm species. Between 1919 and 1934 this savannah filled in, creating a mixed woodland which changed in composition through time but remained species poor and dominated by early-successional trees, lacking the tall hard-wooded species such as the Dipterocarpaceae which dominate Southeast Asian forests. The emergence of a new volcanic island in their midst in 1930, Anak Krakatau, not only provided a further opportunity for study but also reset succession on the islands of Sertung and Panjang.

Surveys have continued to the present day and many plant and animal groups have yet to reach the islands. There are still no land mammals (other than rats, which were introduced by humans), though bats have unsurprisingly found a home. Among plants, around 25% of the fern species present on Java have successfully colonised the islands, whereas fewer than 10% of flowering plants have done so (Whittaker et al., 1997). Bird communities have also developed, and though it appeared that their species richness was levelling off in the 1930s, in fact this was due to the replacement of open-habitat specialists by species better adapted to closed forests (Whittaker et al., 1989).

At the same time there have been some apparently natural extinction events, though the majority of these have either been of ephemeral species that were initially rare and never fully established or early-successional species that were later excluded. Almost all of the species were documented within the first 50 years after the eruption. Figure 20.3 compares the early (1883–1934) floras to more recent records (1979–1994) and identifies which species remained extant. Species that were originally common have remained, while those that were rare have more often vanished.

Estimating the rates of colonisation and extinction of groups is necessarily approximate, and tracking them through time is contingent on the availability of survey data, but some informed guesses can be made about the shapes of the relationships for birds, plants and butterflies (Fig. 20.4). It would be difficult to assert that any of these provides unequivocal support for EMIB predictions, although the immigration and extinction rates of plants could be converging. Nevertheless, even this turnover of

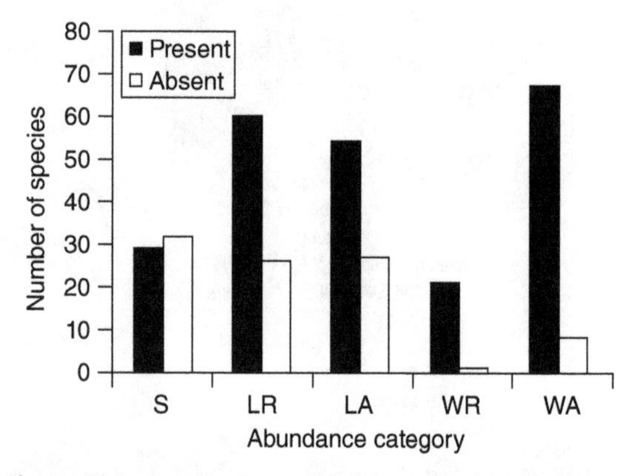

Figure 20.3 Species documented in early plant surveys of Krakatau (1883–1934) and their presence in recent records (1979–1994) against their original abundance (S, singletons; LR, localised and rare; LA, localised but abundant; WR, widespread but rare; WA, widespread and abundant). (*Source*: Whittaker et al. (2000). Reproduced with permission from John Wiley & Sons Ltd.)

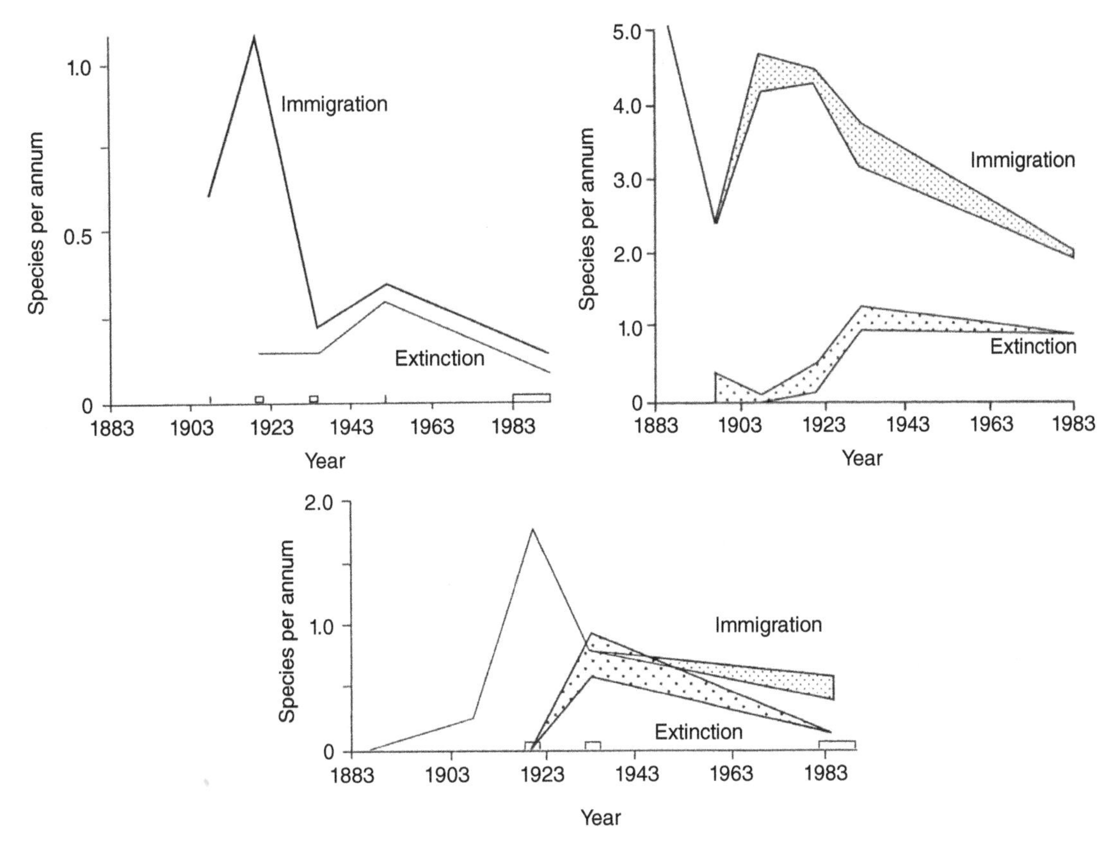

Figure 20.4 Estimated rates of colonisation and extinction rates for birds (a), plants (b) and butterflies (c) on Krakatau. (*Source*: Whittaker and Fernández-Palacios (2007). Reproduced with permission from Oxford University Press.)

species is heterogeneous and dominated by succession processes whereby the forests have been developing an increasingly tall stature. The key to understanding the richness and composition of the island biota is in the history of their disturbances. There is no sign of a dynamic equilibrium, and certainly not one dominated by stochastic colonisation and extinction. Succession on Krakatau looks surprisingly similar to that on the mainland, give or take some disharmonies in species composition, and it is these changes which determine the rates of colonisation and extinction (Whittaker et al., 1989).

Disturbance on islands can also be anthropogenic in origin and leave lasting impacts. On islands in the Stockholm archipelago, off the coast of Sweden, species richness of plants is still elevated on islands which were grazed a century ago, despite this being a relatively mild form of disturbance (Aggemyr and Cousins, 2012). Even if a dynamic equilibrium exists in theory, it will be almost impossible to find in practice if disturbance acts as a more powerful driver of patterns.

20.5 Relaxation

Relaxation on islands may sound like an idyllic topic for scientific enquiry, but in ecological terminology it refers specifically to the changes in species richness that occur when habitat area is reduced (Diamond, 1984). When a continental island is separated from the mainland, it begins with a full complement of species. Rates of extinction due to reduced area are thereafter expected to exceed rates of colonisation, leading to a gradual erosion of overall species richness, until it reaches a new and lower equilibrium.

An impression of what this involves can be gained by looking at the mammals of Tasmania, of which an excellent fossil record exists, allowing their prehistorical species richness to be known with some confidence. As sea levels rose in the present interglacial, Tasmania was cut off from the Australian mainland, and extinctions started to impact upon animal communities.

The State of Tasmania comprises a set of 335 islands, with Tasmania itself the largest. Most mammal species only persisted on the main island, particularly carnivores including the well-known Tasmanian devil and the thylacine (a marsupial carnivore commonly referred to as the Tasmanian tiger despite bearing no relation to true tigers). The same applied to large mammal species, notably humans, and habitat specialists such as the platypus, which requires freshwater. In other words, the animals that went extinct on smaller islands were a non-random selection of the total species pool and shared particular traits that made them more vulnerable.

A similar process can be seen in the Great Basin desert of the United States (Brown, 1971). Covered in forest at the Last Glacial Maximum, this region has undergone a prolonged period of drying, such that tree cover only persists as relicts on isolated mountain tops or the nearby Sierra Nevada plateau. Elsewhere temperature increases have turned the landscape in between into desert. In other words, this is a classic mainland-and-islands situation (Fig. 20.5), though curiously the barrier to dispersal is not water but the lack of it.

The Sierra Nevada plateau retains a full complement of small mammal species in the region, but the mountain-top forests each contain only a proportion of these. The ones that have survived are a predictable subset determined by relative likelihoods of extinction. Small species such as the chipmunk occur on most 'islands', while the largest species

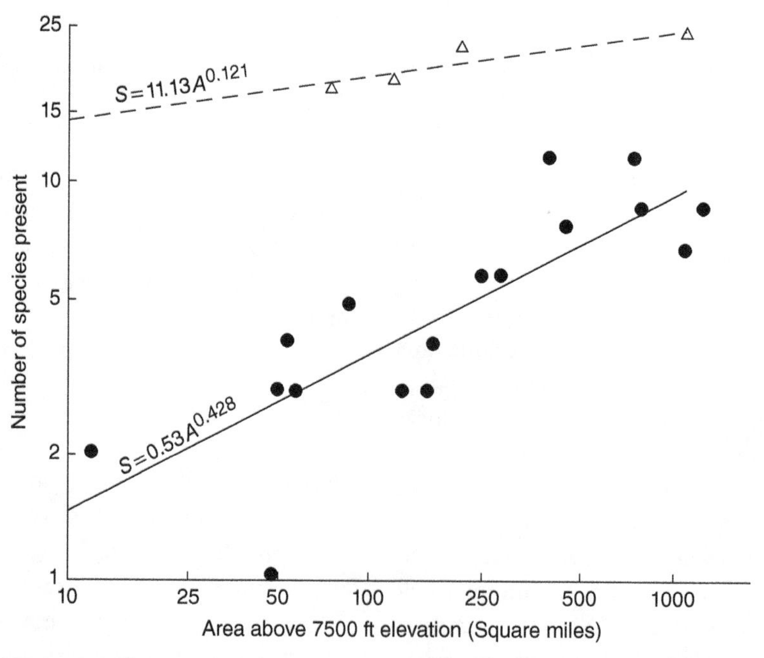

Figure 20.5 Species richness of small mammals with area on the Sierra Nevada plateau (dotted line with triangles) and in remnant montane forests of the Great Basin (solid line with dots). (*Source*: Brown (1971). Reproduced with permission from University of Chicago Press.)

have only remained on mountains with the most forest. Habitat specialists such as the Kaibab squirrel are also confined to the larger blocks. A random turnover of species at equilibrium is not part of the picture.

Similar effects can be seen on shorter timescales, such as the Thousand Island Lake in China described in Section 20.2, where islands in the reservoir date from 1959. Bird communities vary with island size, but are highly nested, such that the species present on small islands are a subset of those on larger ones, which in turn are only a fraction of those from the mainland (Wang et al., 2011). The patterns of species occupancy have been created by selective extinction.

Specialised species on islands are especially vulnerable to extinction, with cascading impacts. On an island in New Zealand, the functional extinction of three species of birds reduced the pollination rates, seed production and density of a shrub species which relied upon them for pollination (Anderson et al., 2011). Given that it is possible to identify individual species as candidates for extinction, in many cases the reorganisation of communities will be predictable as well.

20.6 Extinctions

Exploration of the globe by naturalists led to the discovery of many new species of birds on oceanic islands, and in a number of cases accounts of their extirpation followed soon thereafter. The dodo is the most notorious example, driven to extinction on Mauritius by hunting despite apparently tasting dreadful. Since 1600 there are records of 85 island endemic birds having gone extinct, though doubtless many more occurred prior to this (Duncan et al., 2013; see Section 17.8). These extinctions were also non-random; flightless species were 33 times more likely to go extinct, and large-bodied species were also susceptible. Single island endemics had a 24-fold greater risk of extinction.

What unites all these extinctions is that they were induced by humans, who either introduced new predators (especially rodents, snakes and cats), overhunted the birds or destroyed their habitats. The mere persistence of these species on undisturbed islands for so long prior to occupation gives cause to wonder how applicable the concept of a dynamic equilibrium of species richness can be. These were ancient species that had survived on islands for millions of years in some cases. They were not subject to stochastic turnover or repeated colonisations and extinctions. If dynamic change was really happening, how could these species have arisen?

20.7 Invasions

Another line of evidence to probe for signs of dynamic equilibrium is the effect of invasive alien species on islands. These are colonisation events, albeit human promoted, and can be viewed as experiments, adding new species to insular communities. In an equilibrial community, each naturalised species should lead to the displacement of another. On the other hand, following EMIB logic, a greater immigration rate should increase species richness (it is equivalent to moving the island closer to the mainland), which will in turn increase the extinction rate (Sax et al., 2002).

Insular birds might at first glance appear to have a dynamic equilibrium of species richness, with an apparent 'one-in-one-out' policy operating and numerous extinctions (see Fig. 7.7). Closer inspection upsets this narrative as it turns out that remarkably few extinctions of island vertebrates can be directly attributed to interspecific competition (Sax and Gaines, 2008). Most are instead caused by predation, with intolerant species (such as ground-nesting birds) replaced by those with some means of avoiding introduced carnivores (often rats).

With plants the picture is even more divergent from theoretical expectations. Very few extinctions of island plants have been documented, and in fact their species richness has on average doubled due to naturalised invaders, with no sign of reaching a limit. There is no evidence that replacement of species is

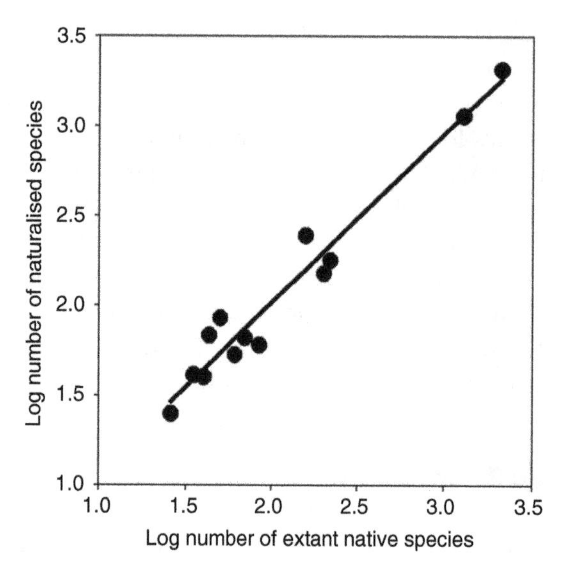

Figure 20.6 Naturalised invasive plant species richness on islands correlated with existing native plant species richness; $r^2 = 0.96$. (*Source*: Sax and Gaines (2008). Copyright (2008) National Academy of Sciences, USA.)

taking place—the number of invasive species that establish is strongly correlated with the number of natives (Fig. 20.6). The more types of plants an island has, the more colonists it is able to accommodate.

What explains the difference? An attempt to reconcile this with equilibrium theory might argue that birds are typically well dispersed, and therefore their rates of immigration are higher, meaning they have reached a stable species richness. In contrast, plants are limited in their dispersal potential, and island plants have further to go before reaching an equilibrium level, though one would be found eventually. This smacks of special pleading and is unsatisfactory in either case.

20.8 A new theory?

Having reviewed a series of case studies, these are beginning to sound less like a collection of exceptions and more like a compelling case against EMIB as a general theory. Can a new theory be assembled that will incorporate the additional processes?

Substantial efforts towards this have been made and point in the direction of a general theory of island biogeography (Whittaker and Fernández-Palacios, 2007). To begin, imagine that islands exist on a continuum between those that are equilibrial, and dominated by the biotic processes of colonisation and extinction, and those that are non-equilibrial, prevented by disturbance from ever reaching a stable state. Another axis of variation runs from islands that are dynamic in species composition, showing continuous turnover, to those that are static and retain the same species indefinitely. The island examples described so far can be mapped out along these axes, and an attempt to synthesise them is made in Fig. 20.7. Each corner can be filled with at least one of the case studies described so far in this chapter.

The equilibrium model envisioned by MacArthur and Wilson (1967) depends upon systems dominated by biotic processes, with relatively little impact of disturbance or variation in the environment among islands. There are likely to be a number of cases where this is true, perhaps only for specific groups of organisms in particular places, but nonetheless it remains a plausible option. The breeding birds on offshore islands studied by Manne et al. (1998) are one such case (see Section 19.13).

On other islands, plenty of turnover takes place, but driven by environmental factors rather than internal biotic processes. Atolls that are frequently denuded will fall into this category, as will islands disturbed by hurricanes, volcanoes or other destructive forces. To understand the species richness and composition of groups of animals and plants in Krakatau, it is more important to consider time since disturbance than the effects of island area and distance from the mainland on the interplay between colonisation and extinction.

Numerous islands are static and unchanging in their composition. Where a set of species have been isolated on a remote island, with little disturbance and plenty of time for adaptation, they are probably resistant to invasion by new colonists and will show limited turnover through time. Remote island birds

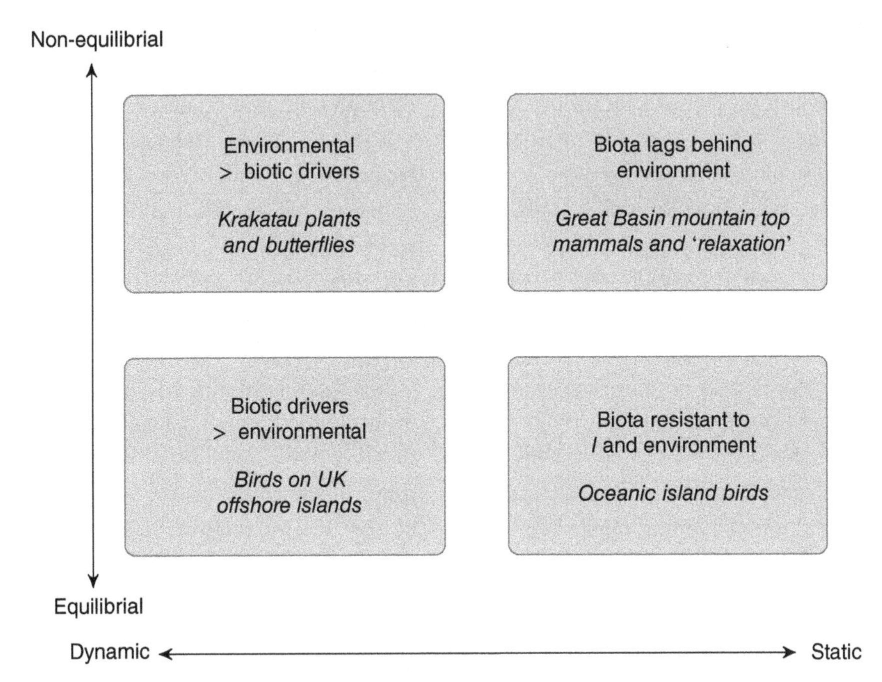

Figure 20.7 Schematic representation of island biogeographical theory along axes from equilibrial to non-equilibrial systems and from dynamic to static. (*Source*: Adapted from Whittaker and Fernández-Palacios (2007).)

are examples of this. Elsewhere, the environment may be changing, but the biota is unable to respond at the same pace. In the Great Basin mountain-top forests, small mammal communities are influenced by the changing climate and shrinking habitat area, a more general process seen in relaxation of the biota on continental islands.

This is beginning to look like a more general theory of island biogeography, capable of encompassing many observations in nature. But there is still something very important missing.

20.9 Evolution

The original book by MacArthur and Wilson (1967) contained plenty of speculation about the potential role that evolution might play in the generation and dynamism of island species, though this extension of the theory received less attention, and it was believed that the rate of speciation would be low relative to the frequency of colonisation and hence less important. This need not be the case, and with new genetic evidence the old assumptions are being overturned (Heaney, 2007). Indeed, among birds, diversification rates on islands far exceed those on the mainland (Jetz et al., 2012).

On remote islands it is likely that local speciation exceeds the rate of immigration of new species. This will be even more true of large islands, particularly those with mountainous topography, because there will be a wide variety of environmental zones generating a range of vacant niches ripe for adaptive radiation to take place. Archipelagos can become factories of diversity, with the subtle differences among islands, combined with just enough dispersal limitation to make inter-island jumps possible but infrequent, allowing all manner of new forms to evolve. The richness of insects on oceanic islands shows little relationship with area but instead correlates positively with island age, for example in the case of Hawaiian crickets (Paulay, 1994). This will not continue indefinitely however; speciation rates of Caribbean lizards decline with time until they reach an equilibrium (Rabosky and Glor, 2010). Presumably at some point the available niche space is filled, whether by colonisation or speciation.

As the potential rate of colonisation falls, the likelihood of neoendemics emerging increases (Fig. 20.8). While islands close to the mainland may have high species richness due to their proximity to a large potential pool, distant islands can compensate for this through evolution of new forms, separated from the gene flow from parent species that would otherwise prevent this. Hence a remote archipelago such as Hawaii can maintain as many species as any offshore continental island. Among plants the effect of isolation on species richness is lower for large islands, implying that speciation compensates for a reduced number of colonists (Weigelt and Kreft, 2013). The same processes apply in lakes as well, which are effectively freshwater islands on land, where speciation of African cichlids is greater in large lakes and increases species richness by an order of magnitude above that which would be present through immigration alone (Wagner et al., 2014).

It would also be a mistake to assume that islands represent the end of the road for colonists. Once a species has reached an island and evolved into something different, there's no reason why it could not then escape and disperse elsewhere, perhaps even invading the mainland (Bellemain and Ricklefs, 2008).

Species can also disperse among islands. In what is probably the longest distance natural dispersal event that we have evidence for, a common Hawaiian tree known as the koa (*Acacia koa*) was shown to be the nearest relative of another island endemic, *Acacia heterophylla*, found on the island of Réunion, 18,000 km away in the Indian Ocean—they are effectively the same species (Le Roux et al., 2014). Botanists had long noted the similarity between the two but assumed either a common origin in Australia or that humans had played some part in

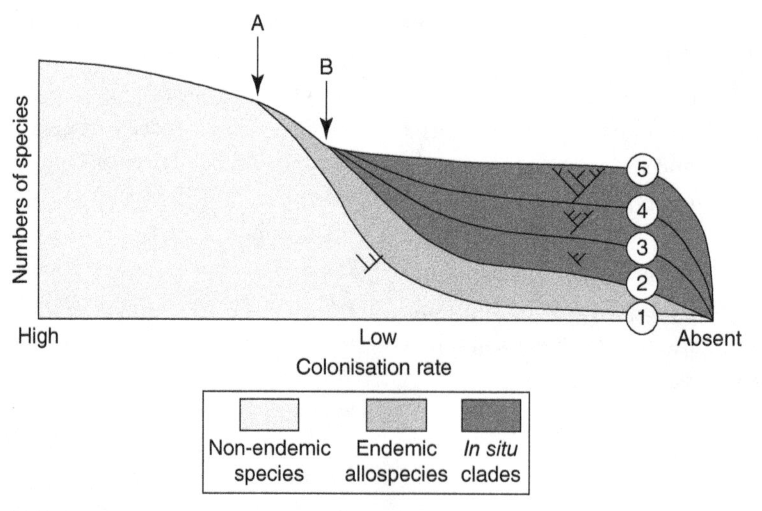

Figure 20.8 Conceptual model to show the relative fraction of species richness on large islands or archipelagos composed of colonists from the mainland versus locally evolved endemics with decreasing rates of colonisation (i.e. increasing isolation). On islands close to a source pool, a high species richness will be present through colonisation, and high rates of gene flow through dispersal of individuals will prevent speciation from occurring. At point A endemic species will begin to develop but remain a minority and of low taxonomic rank (allospecies are species formed through allopatric speciation). From B onwards reduced colonisation generates the potential for endemic clades to grow by sympatric speciation, with their species richness increasing through time (lines 1–5) and partially compensating for decreased number of colonising mainland species. (*Source*: Heaney (2000). Reproduced with permission from John Wiley & Sons Ltd.)

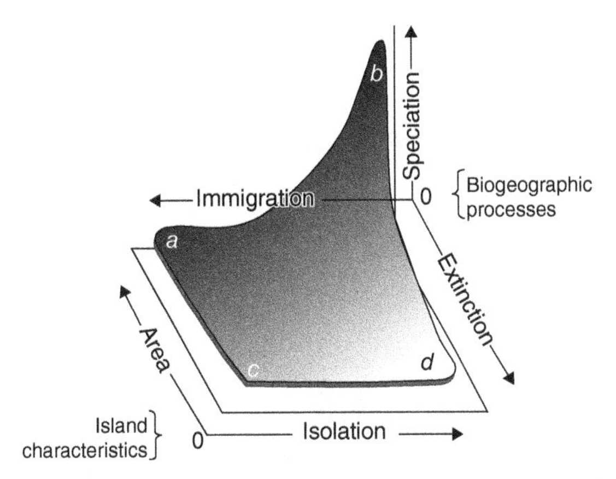

Figure 20.9 Island species richness, represented as intensity of line position shading, as a function of area and isolation, which govern rates of immigration, extinction and speciation Letters match columns in Table 20.2. (*Source*: Lomolino (2000). Reproduced with permission from John Wiley & Sons Ltd.)

moving them. Given that the dispersal event took place 1.4 Mya, humans can have played no part in it. Both trees are restricted to montane habitats, and transportation by birds is a more likely mechanism than floating on ocean currents.

Any attempt to build a new model of island species richness needs to incorporate evolution as a central element, not merely as an afterthought or extension. The likely shape of this was mapped out by Lomolino (2000), represented in Fig. 20.9 as a three-dimensional surface with observed species richness driven by the three processes of immigration, extinction and speciation operating in tandem. While this model remains largely qualitative in form, awaiting the data that might allow the fitting of equations and quantitative predictions, it is a promising line of enquiry that points to the way forward. Some attempts have been made; Rosindell and Phillimore (2011) extended the EMIB by allowing for an increasing rate of speciation on larger and more distant islands. The involvement of the triumvirate of major processes—speciation, extinction and dispersal—is an echo of the determinants of species richness at global level and brings islands back into the greater body of biogeographical theory (Fig. 16.12).

The jigsaw with which this chapter began can now be pieced together. Table 20.2 reviews the variation among islands of different types, allocating them their place in the bigger picture. Compare this means of defining island groups with Table 20.1, which categorised islands based on their connections to other islands within networks, and remember that the terms 'large', 'small', 'near' and 'far' are all relative, both to other islands and to whatever is being considered as the source of colonists.

While it may be overly optimistic to expect that a single mechanistic model will ever be able to cope with the interacting effects of speciation, extinction, dispersal and geological origin, more focussed models might effectively capture the behaviour of particular classes of islands. This was attempted by Whittaker et al. (2008) who took large, remote oceanic islands as their target.

Table 20.2 Features of island biota dependent on their characteristics and origin (adapted from Whittaker and Fernández-Palacios, 2007). Letters in column headings correspond to those in Fig. 20.9.

		Island features			
		Large, near (a)	Large, far (b)	Small, near (c)	Small, far (d)
Process	Immigration	High	Low	High	Low
	Speciation	Low	High	None	Low
	Extinction	Low	Low	High	High
Feature	Species richness	High	High	Moderate	Low
	Turnover	Low	Low	High	Low
Examples	Type	Continental islands	Fragments, volcanoes	Continental islands	Volcanoes and atolls
	Case studies	Tasmania, Britain	New Zealand, Hawaii	Lundy, Skokholm	Easter Island, Ascension

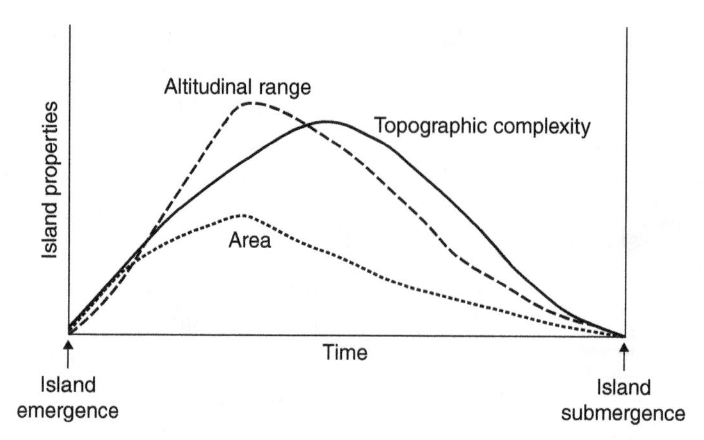

Figure 20.10 Geological trajectory of a remote volcanic island. Initially area and altitude increase while accretion is the dominant process, and then topographic complexity will be generated by erosion, which gradually causes the island to degrade. The time axis of this figure is unlikely to be linear because the early growth phase occurs much more quickly than the decline (Steinbauer et al., 2013). (*Source*: Whittaker et al. (2008). Reproduced with permission from John Wiley & Sons Ltd.)

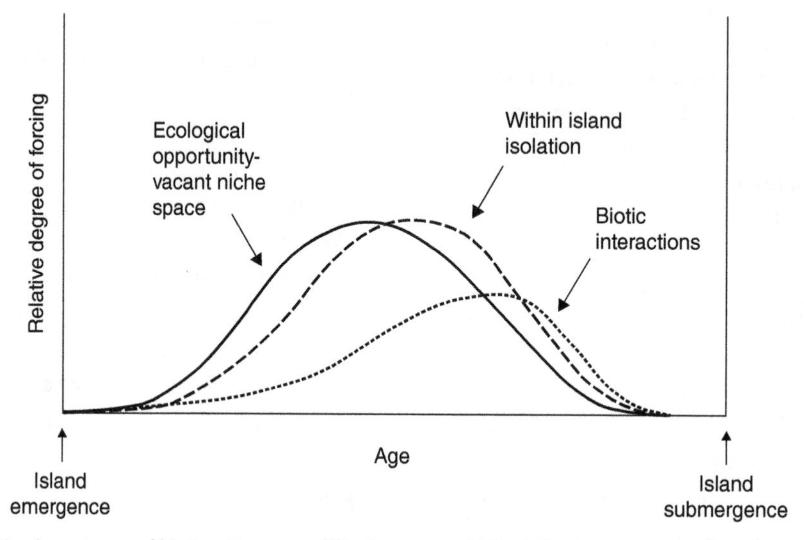

Figure 20.11 Relative importance of biotic processes at different stages of island development. Initially colonisation will dominate (solid line), with later evolution through habitat specialisation (dashed line), while biotic interactions become crucial once established communities form (dotted line). In late stages competitive exclusion and extinction will cause reductions in species richness. Note as for Fig. 20.10 that the age axis is not strictly linear. (*Source*: Whittaker et al. (2008). Reproduced with permission from John Wiley & Sons Ltd.)

Imagine a new volcano emerging from the ocean, perhaps adding to the Hawaiian archipelago (Fig. 20.10). Its geological history will be dominated first by accretion of new material, giving rise to an increase in area and altitude. Next erosion will begin, initially acting to increase topographic complexity, but gradually leading to a long-term decline in area and elevation until at last the island vanishes beneath the waves. Surtsey, for example, has already lost half of its maximum area even in the 50 years since it first emerged (Adsersen, 2013).

These changes in island characteristics will influence the dominant ecological processes during each period (Fig. 20.11). After initially being dominated by colonisation, the biota will increasingly be shaped by speciation over time and as the island gains additional habitat heterogeneity through erosion. Later speciation will be biotic in origin, with interactions such as mutualisms enabling generation of further endemic species. In the final phases of the island's trajectory, certain habitats will be lost (such as higher-elevation zones), competition within declining resource pools will increase, and the overall area will be reduced, leading to extinctions and a reduction in species richness. This narrative account of the dynamics of remote island biotas is known as the general dynamic model (Whittaker et al., 2008).

If we accept the Species–Area Relationship (a reasonable starting point), with endemic richness increasing through evolution over time T, and that there is a humped relationship between species richness S and island age driven by the interaction of geological and biotic processes represented as T^2, then these can be combined into a single model (Fig. 20.12). Simply adding these terms together gives the $A + T + T^2$ model which can account for an impressive proportion of species richness variation among the islands to which it pertains (Whittaker et al., 2008, Steinbauer et al., 2013). The model was consistently supported for patterns of snail species richness across eight archipelagos (Cameron et al., 2013).

It's worth tempering excessive enthusiasm though. While being based on plausible processes, this is still a phenomenological rather than a mechanistic model (unlike e.g. Rosindell and Phillimore, 2011), and the terms within it might need tweaking in the light of appropriate evidence. It also refers to a limited subset of islands, restricting its generality and meaning that it can only be applied and tested within certain prescribed boundaries. This is not intended as a criticism, more as a reminder that the more specific a model, the less transferable it becomes. Perhaps more problematic is that a humped model can fit increases, decreases and peaked curves in species richness, depending on the range of ages of the islands studied. Moreover, any island begins with no species and ends with no species, causing a

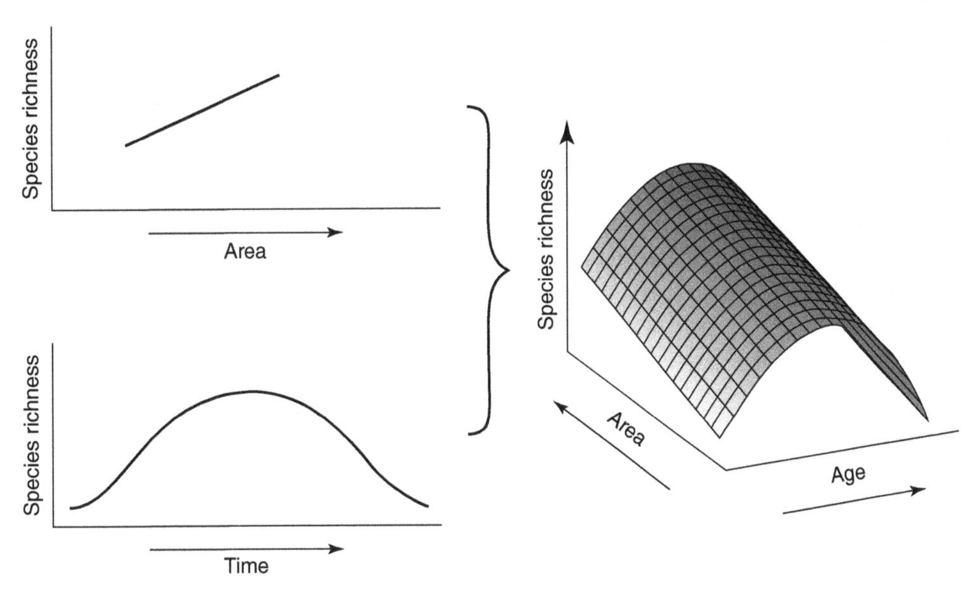

Figure 20.12 Expected linear relationship between species richness and area, and humped with time, combined to give a predictive model incorporating both island size and age. Note as for Fig. 20.10 that the age axis is not strictly linear and species are likely to accrue faster initially than the subsequent rate of loss (Steinbauer et al., 2013). (*Source*: Whittaker et al. (2008). Reproduced with permission from John Wiley & Sons Ltd.)

humped pattern in species richness to be inevitable (Steinbauer et al., 2013). This makes the general dynamic model quite difficult to falsify.

The model might nonetheless point towards a productive way to study islands in future. Armed with a general theory, predictive models might be generated for each of the classes of island in Fig. 20.7, leading potentially to an eventual unification. For example, isolation may have lesser importance for continental islands (Kreft et al., 2008, Weigelt and Kreft 2013), whose dynamics are better explained by relaxation processes related to changes in area. Much research remains to be done, and the necessary data to test new models have yet to be harvested, but islands will continue to play a central role in developing ecological understanding.

20.10 Conclusions

The Equilibrium Model of Island Biogeography has provided useful inspiration to the whole of ecology, but for real-world islands it is not the complete answer. While its processes are reasonable, the main prediction of dynamic turnover has been difficult to support with evidence, and it is likely to apply only to a subset of islands and taxa. Confounding problems obscure its action; these could be artefacts, such as inadequate surveys, but also powerful biological processes including disturbance, succession and assembly rules. Evidence from relaxation (natural extinctions) and invasions (artificial dispersal) also fails to back up EMIB predictions. The basic model omits evolution and thereby fails to account for one of the most striking aspects of island life—the large numbers of endemics. A general framework for island biogeography has emerged in recent years from the familiar themes of speciation, extinction and dispersal that have run throughout this biogeography section. The processes that determine the species richness and composition of island systems are not so different from those elsewhere in the natural world.

As to whether islands provide any lessons for building natural systems, perhaps the most important is that no island is an island. Their communities are not sealed from the rest of the world. They respond dynamically to the arrival of new colonists, through evolution to changing environments, and by extinction of those species that cannot find a place. Even the most remote and isolated of islands do not remain static and unchanging. Whatever the starting point, island communities change and not in entirely predictable directions.

20.10.1 Recommended reading

Heaney, L. R., 2007. Is a new paradigm emerging for oceanic island biogeography? *Journal of Biogeography* 34:753–757.

Lomolino, M. V., 2000. A call for a new paradigm of island biogeography. *Global Ecology and Biogeography* 9:1–6.

Whittaker, R. J., K. A. Triantis, and R. Ladle, 2008. A general dynamic theory of oceanic island biogeography. *Journal of Biogeography* 35:977–994.

20.10.2 Questions for the future

- When can and should the EMIB and its derivatives be used?
- Can multiple models for different types of islands be brought together in a common framework?
- Do systems with lower insularity, such as habitat patches, show similar patterns of variation?

References

Adsersen, H., 2013. A remarkable anniversary: Surtsey becomes 50 years old. *Frontiers of Biogeography* 5:78–.

Aggemyr, E. and S. A. O. Cousins, 2012. Landscape structure and land use history influence changes in island plant composition after 100 years. *Journal of Biogeography* 39:1645–1656.

Anderson, S. H., D. Kelly, J. J. Ladley, S. Molloy, and J. Terry, 2011. Cascading effects of bird functional extinction reduce pollination and plant density. *Science* 331:1068–1071.

Bellemain, E. and R. E. Ricklefs, 2008. Are islands the end of the colonisation road? *Trends in Ecology & Evolution* 23:461–468.

Brown, J. H., 1971. Mammals on mountaintops: nonequilibrium island biogeography. *American Naturalist* 105:467–478.

Brown, J. H. and A. Kodric-Brown, 1977. Turnover rates in insular biogeography—effects on immigration and extinction. *Ecology* 58:445–449.

Cameron, R. A. D., K. A. Triantis, C. E. Parent, F. Guilhaumon, M. R. Alonso, M. Ibáñez, A. M. de Frias Martins, R. J. Ladle, and R. J. Whittaker, 2013. Snails on oceanic islands: testing the general dynamic model of oceanic island biogeography using linear mixed effect models. *Journal of Biogeography* 40:117–130.

Carstensen, D. W., B. Dalsgaard, J.-C. Svenning, C. Rahbek, J. Fjeldså, W. J. Sutherland, and J. M. Olesen, 2012. Biogeographical modules and island roles: a comparison of Wallacea and the West Indies. *Journal of Biogeography* 39:739–749.

Connolly, S. R., T. P. Hughes, D. R. Bellwood, and R. H. Karlson, 2005. Community structure of corals and reef fishes at multiple scales. *Science* 309:1363–1365.

Diamond, J. M., 1974. Colonization of exploded volcanic islands by birds: the supertramp strategy. *Science* 184:803–806.

Diamond, J. M., 1984. 'Normal' extinctions of isolated populations. In M. H. Nitecki, editor, *Extinctions*, pages 191–246. University of Chicago Press.

Dornelas, M. and S. R. Connolly, 2008. Multiple modes in a coral abundance distribution. *Ecology Letters* 11:1008–1016.

Duncan, R. P., A. G. Boyer, and T. M. Blackburn, 2013. Magnitude and variation of prehistoric bird extinctions in the Pacific. *Proceedings of the National Academy of Sciences of the United States of America* 110:6436–6441.

Fox, G. A., S. M. Scheiner, and M. R. Willig, 2011. Ecological gradient theory: a framework for aligning data and models. In S. M. Scheiner and M. R. Willig, editors, *The Theory of Ecology*, pages 283–307. University of Chigago Press.

Gravel, D., F. Massol, E. Canard, D. Mouillot, and N. Moquet, 2011. Trophic theory of island biogeography. *Ecology Letters* 14:1010–1016.

Heaney, L. R., 2000. Dynamic disequilibrium: a long-term, large-scale perspective on the equilibrium model of island biogeography. *Global Ecology and Biogeography* 9:59–74.

Heaney, L. R., 2007. Is a new paradigm emerging for oceanic island biogeography? *Journal of Biogeography* 34:753–757.

Hortal, J., K. A. Triantis, S. Meiri, E. Thébault, and S. Sfenthourakis, 2009. Island species richness increases with habitat diversity. *American Naturalist* 174:E205–E217.

Hubbell, S. P., 2001. *The Unified Neutral Theory of Biodiversity and Biogeography*. Princeton University Press.

Jetz, W., G. H. Thomas, J. B. Joy, K. Hartmann, and A. O. Mooers, 2012. The global diversity of birds in space and time. *Nature* 491:444–448.

Kadmon, R. and O. Allouche, 2007. Integrating the effects of area, isolation, and habitat heterogeneity on species diversity: a unification of island biogeography and niche theory. *American Naturalist* 170:443–454.

Kreft, H., W. Jetz, J. Mutke, G. Kier, and W. Barthlott, 2008. Global diversity of island floras from a macroecological perspective. *Ecology Letters* 11:116–127.

Le Roux, J. J., D. Strasberg, M. Rouger, C. W. Morden, M. Koordom, and D. M. Richardson, 2014. Relatedness defies biogeography: the tale of two island endemics (*Acacia heterophylla* and *A. koa*). *New Phytologist* 204:230–242.

Lomolino, M. V., 1990. The target area hypothesis: the influence of island area on immigration rates of non-volant mammals. *Oikos* 57:297–300.

Lomolino, M. V., 2000. A call for a new paradigm of island biogeography. *Global Ecology and Biogeography* 9:1–6.

MacArthur, R. H. and E. O. Wilson, 1967. *The Theory of Island Biogeography*. Princeton University Press.

Manne, L. L., S. L. Pimm, J. M. Diamond, and T. M. Reed, 1998. The form of the curves: a direct evaluation of MacArthur & Wilson's classic theory. *Journal of Animal Ecology* 67:784–794.

Paulay, G., 1994. Biodiversity on oceanic islands: its origin and extinction. *American Zoologist* 34:134–144.

Power, D. M., 1972. Numbers of bird species on the California islands. *Evolution* 26:451–463.

Rabosky, D. L. and R. E. Glor, 2010. Equilibrium speciation dynamics in a model adaptive radiation of island lizards. *Proceedings of the National Academy of Sciences of the United States of America* 107:22178–22183.

Rey, J. R. J., 1984. Experimental tests of island biogeographic theory. In D. R. Strong, D. Simberloff, L. G. Abele, and A. B. Thistle, editors, *Ecological Communities: Conceptual Issues and the Evidence*, pages 101–112. Princeton University Press.

Rey, J. R. J., 1985. Insular ecology of salt marsh arthropods: species level patterns. *Journal of Biogeography* 12:97–107.

Rosindell, J. and A. J. Phillimore, 2011. A unified model of island biogeography sheds light on the zone of radiation. *Ecology Letters* 14:552–560.

Russell, G. J., J. M. Diamond, T. M. Reed, and S. L. Pimm, 2006. Breeding birds on small islands: island biogeography or optimal foraging? *Journal of Animal Ecology* 75:324–339.

Sax, D. F. and S. D. Gaines, 2008. Species invasions and extinction: The future of native biodiversity on islands. *Proceedings of the National Academy of Sciences of the United States of America* 105:11490–11497.

Sax, D. F., S. D. Gaines, and J. H. Brown, 2002. Species invasions exceed extinctions on islands worldwide: a comparative study of plants and birds. *American Naturalist* 160:766–783.

Si, X., S. L. Pimm, G. J. Russell, and P. Ding, 2014. Turnover of breeding bird communities on islands in an inundated lake. *Journal of Biogeography* 41:2283–2292.

Steinbauer, M. J., K. Dolos, R. Field, B. Reineking, and C. Beierkuhnlein, 2013. Re-evaluating the general dynamic theory of oceanic island biogeography. *Frontiers of Biogeography* 5:185–194.

Steinbauer, M. J., R. Otto, A. Naranjo-Cigala, C. Beierkuhnlein, and J. Fernández-Palacios, 2012. Increase of island endemism with altitude—speciation processes on oceanic islands. *Ecography* 35:23–32.

Thornton, I. W. B., D. Runciman, S. Cook, L. F. Lumsden, T. Partomihardjo, N. K. Schedvid, J. Yukawa, and S. A. Ward, 2002. How important were stepping stones in the colonization of Krakatau? *Biological Journal of the Linnaean Society* 77:275–317.

Trigas, P., M. Panitsa, and S. Tsiftsis, 2013. Elevational gradient of vascular plant species richness and endemism in Crete—the effect of post-isolation uplift on a continental island system. *PLoS ONE* 8:e59425–.

Wagner, C. E., L. J. Harmon, and O. Seehausen, 2014. Cichlid species-area relationships are shaped by adaptive radiations that scale with area. *Ecology Letters* 17:583–592.

Wang, Y., S. Chen, and P. Ding, 2011. Testing multiple assembly rule models in avian communities on islands of an inundated lake, Zhejiang Province, China. *Journal of Biogeography* 38:1330–1344.

Weigelt, P. and H. Kreft, 2013. Quantifying island isolation—insights from global patterns of insular plant species richness. *Ecography* 36:417–429.

Whittaker, R. J., M. B. Bush, and K. Richards, 1989. Plant recolonization and vegetation succession on the Krakatau Islands, Indonesia. *Ecological Monographs* 59:59–123.

Whittaker, R. J. and J. Fernández-Palacios, 2007. *Island Biogeography*. Oxford University Press, second edition.

Whittaker, R. J., R. Field, and T. Partomihardjo, 2000. How to go extinct: lessons from the lost plants of Krakatau. *Journal of Biogeography* 27:1049–1064.

Whittaker, R. J., S. H. Jones, and T. Partomihardjo, 1997. The re-building of an isolated rain forest assemblage: how disharmonic is the flora of Krakatau? *Biodiversity and Conservation* 6:1671–1696.

Whittaker, R. J., K. A. Triantis, and R. J. Ladle, 2008. A general dynamic theory of oceanic island biogeography. *Journal of Biogeography* 35:977–994.

Williamson, M. H., 1989. The MacArthur and Wilson theory today: true but trivial. *Journal of Biogeography* 16:3–4.

Wright, D. H., 1983. Species-energy theory: an extension of species-area theory. *Oikos* 41:496–506.

CHAPTER 21

What is a natural system?

21.1 The big question

This book began with the challenge of trying to work out whether it is possible to create natural systems. What species richness and what composition should they have? Can they operate as sealed, autonomous units, or do they depend on interactions at both local and broader scales? The answers are more complex than either hoped or expected. In truth we have barely scratched the surface. There are endless opportunities for research in this field, and the science of ecology remains in its early phases.

In retrospect, it was naïve of the builders of Biosphere 2 to expect a large glasshouse in the desert to settle into a stable, functioning ecosystem; or rather to expect that it would form a system capable of supporting human life (see Section 1.1). Actually, had it been left to its own devices, it would probably have developed into a stable state, only one some distance from the initial or desired composition of species. As a small, isolated system, it was inevitable that most species would go extinct, beginning with those that were large or specialised in their role. All islands do the same.

If we were to perform the same experiment in space, sending a closed system into the stars and expecting it to maintain itself, the consequences would be equally disastrous for its human dependents. Maintaining ecosystem functions such as photosynthesis and decomposition is not a chemical engineering problem, and simple equations fall apart in the face of the complex dynamics of communities and ecosystems. Natural systems are open and their existence depends upon effects beyond their borders. Perhaps if civilisation were ever to reach a new world and seed it with life, it would be large enough to break free of the scale dependence of small experiments, but there's no guarantee that it would take a direction amenable to ourselves.

By now we are ready to tackle a much more fundamental question, one which has been prompted at several points in the book—what is natural? The systems observed today are the product of millions of years of evolution in a changing world, where the climate and the layout of the continents themselves have interacted with the distribution and composition of life. In more recent years the impacts of humans have been widespread and dramatic; little of the globe can be said to exist in a 'wild' state, and even areas that appear so may only reflect the shifting baselines of our perceptions. As the climate changed since the last glacial maximum and as it continues to do so in the future, communities assemble and disintegrate, and species themselves alter in response through rapid evolution. Novel ecosystems spring up as new combinations of species are thrown together, often made up of alien invasives from all over the world.

In truth, the only thing that can be called natural is constant change. The stability that we might detect in communities is relative, soon perturbed by

Natural Systems: The organisation of life, First Edition. Markus P. Eichhorn.

humans or processes operating on timescales beyond our lifetimes. The balance and continuity of nature is a myth. This prompted the historical ecologist Oliver Rackham to opine that 'natural' is whatever goes on in between human interventions. By definition we can't make something natural nor should we aspire to do so. Whether an island is occupied by native endemics or filled with the alien species introduced by man, it only becomes natural as soon as we pack our bags and set sail.

Is this an argument for non-intervention in nature? By no means. There are many species and habitats that exist only because humans choose to maintain them, often at great expense and effort. We do so because we place an intrinsic value on their continued existence. Usually, however, this involves trying to stem the tide of 'unnatural' processes and their repercussions. Extinction might be a natural process, but the current global crisis is very much anthropogenic in origin. We must learn how to manage natural systems better for the outcomes we desire rather than running away and calling it a conservation strategy.

21.2 Lessons learnt

It is worth stepping back and seeing whether any general rules can be formulated that will apply to all natural systems. There have been a number of attempts to codify a set of laws that would form the foundation of a general theory of ecology; see Scheiner and Willig (2011) or Knapp and D'Avanzo (2010) for some suggestions. The draft laws presented here overlap to some extent with these sources.

21.2.1 Ecological processes are scale dependent

The processes acting at any particular spatial scale in ecology can only be understood and applied *at that scale*. Evidence obtained from plot-level studies cannot be used to infer what might be happening across entire regions nor *vice versa*. For example, competition among species within communities cannot be used to inform us about species distributions across regions (at least not solely) nor do regional-scale evolutionary dynamics matter for local community assembly. This is not to say that they are not all linked together; one aim of this book has been to show the exact opposite. Instead, a cascading hierarchy can be envisaged (as in Fig. 16.12) in which different sets of processes assume primary importance at each scale. Unwise application of ideas at the inappropriate level leads to demonstrably incorrect conclusions such as Gaia theory. The challenge is to determine where the boundaries between levels lie—it is also unlikely that they will be at the same scales for all types and sizes of organism.

21.2.2 All interactions are nested

In physics or chemistry it is easier to separate a system into its component parts and probe how each functions or responds to change. This information can be used to predict the behaviour of composite systems. In ecology this is not merely impossible but also misleading. All ecological interactions and processes are enmeshed in a complex web within which each cannot be understood in isolation. The creation of these complex systems gives rise to emergent properties—unexpected and sometimes unpredictable outcomes that arise from interactions among many components. While this makes it difficult to anticipate the fine-scale behaviour of any single element, the broad patterns remain consistent. Ecologists need to accept what cannot be controlled or predicted and focus on those aspects of nature that can.

21.2.3 There is no such thing as the balance of nature

At the very outset of the book, it was stated that the fabled 'balance of nature' was a myth, and this should by now have been categorically demonstrated. Everything is constantly changing at a range of timescales, from the fluctuations and cycles of individual populations through to the movement and reorganisation of entire biomes as the climate changes and continents rearrange themselves around the globe. This is not to say that there is no such thing as stability, only that it is relative and transient. A lack of change over the course of a human lifetime does not imply perpetual stasis.

21.2.4 Everything is contingent

Adding to the former point is the realisation that the state exhibited by any natural system depends upon its history and starting conditions. Minor changes in the species that happen to arrive and the interactions among them can have lasting consequences for the structure of not only communities but even whole regions of earth. Consider, for example, the consequence of placental mammals having never colonised Australia (until humans transported them). The outcome of natural processes depends on where they begin. This applies as much to the assembly of single communities as to the evolution of entire clades. Rare and chance events can have lasting repercussions.

Given their inherent complexity and contingency, there are limits to our ability to predict the behaviour of ecological systems, even in the medium term (Beckage et al., 2011). Natural systems are intrinsically unpredictable. They may also be computationally irreducible—even if the processes underlying their behaviour were completely understood, one would need to simulate them in their entirety to know the outcome, which means we will only know what will happen when it happens! There is a wry Russian proverb which states that 'the past is unpredictable'—if we had known what was going to happen, we would not have acted in the way we did. Current models struggle to cope with previous events in earth history. How much less can we predict the future?

Some ecologists are nevertheless building models to predict the behaviour of entire ecosystems or even the whole world (Purves et al., 2013, Harfoot et al., 2014). These are necessarily stochastic models, incorporating uncertainty, leading to a range of possible outcomes with varying probability, much like weather forecasting. There are some things that can be predicted with a degree of confidence, such as the shapes of species abundance distributions. The overall trends can be anticipated but not the behaviour of any single component.

21.3 Processes not systems

Ecology has often been criticised for lacking general laws. It is easy to be swept up in the small details, the particularities of individual systems, which can obscure the bigger picture. Some even resist the suggestion that general laws exist or would be helpful if they did. Nonetheless, regardless of the species concerned, a limited number of processes govern how nature works (Vellend, 2010). These can be grouped into just four categories: *selection* (incorporating differences in fitness among individuals due to the environment or biotic interactions), *drift* (random variation), *speciation* and *dispersal*. Between them these control all natural systems. Species enter,

Figure 21.1 Léon Croizat (1894–1982) during the 1950–1951 Franco-Venezuelan Expedition to the sources of the Orinoco River. (*Source*: From http://en.wikipedia.org/wiki/File:CroizatOrinoco.jpghttp://creativecommons.org/licenses/by-sa/3.0/.)

either through dispersal or speciation, and then their abundances vary in response to selection and drift. Every one of the examples detailed in this book can be reduced to one of these processes or several acting in combination.

In closing, I offer the words of Léon Croizat (Fig. 21.1), one of the great names in biogeography, who could fairly be described as both an eccentric and divisive figure. As he put it (Croizat, 1984):

> A process is always more important than any of its by-products.

It is not enough to study natural systems—they are just the outcome, an epiphenomenon. Instead we should be looking for natural processes.

References

Beckage, B., L. J. Gross, and S. Kauffman, 2011. The limits to prediction in ecological systems. *Ecosphere* 2:125.

Croizat, L., 1984. Mayr vs. Croizat: Croizat vs. Mayr—an enquiry. *Tuatara* 27:49–66.

Harfoot, M. B. J., T. Newbold, D. P. Tittensor, S. Emmott, J. Hutton, V. Lyutsarev, M. J. Smith, J. P. W. Scharlemann, and D. W. Purves, 2014. Emergent global patterns of ecosystem structure and function from a mechanistic global ecosystem model. *PLoS Biology* 12:e1001841.

Knapp, A. K. and C. D'Avanzo, 2010. Teaching with principles: toward more effective pedagogy in ecology. *Ecosphere* 1:15.

Purves, D. W., J. P. W. Scharlemann, M. Harfoot, T. Newbold, D. P. Tittensor, J. Hutton, and S. Emmott, 2013. Ecosystems: time to model all life on Earth. *Nature* 493:295–297.

Scheiner, S. M. and M. R. Willig, 2011. *The Theory of Ecology*. The University of Chicago Press.

Vellend, M., 2010. Conceptual synthesis in community ecology. *The Quarterly Review of Biology* 85:183–206.

APPENDIX A

Diversity analysis case study: Butterfly conservation in the Rocky Mountains

To build on the techniques presented in Chapter 5, this appendix is intended to give examples of calculating and interpreting the types of diversity statistics that are commonly used in ecological studies, environmental impact surveys and conservation biology. For a more comprehensive introduction to diversity analysis, see Magurran (2004). A detailed account of the state of the art can be found in Magurran and McGill (2011), though this is more suitable for advanced readers.

The example here involves samples taken from butterfly communities in Colorado, in the Rocky Mountain region of the United States. Butterflies are well-known taxonomically, straightforward to identify and easy to capture in large numbers, making them a widely studied group of organisms. Their dependence on their host plants, and sometimes on interacting mutualists such as ants, makes them useful indicator species of the health of a natural system. At larger scales they are one of the most commonly used groups in studies of biotic responses to global change. If you wish to practise the analyses for yourself, then the full dataset can be downloaded from[1] (Oliver et al., 2006). The data were collected by Sharon Collinge, Jeff Oliver and Katherine Prudic, and for more see Collinge et al. (2003).

In this study surveys were conducted on 66 protected grassland sites. These included remnants of the original tall-grass prairie, lowland hayfields in floodplains and short- and mixed-grass prairies adjacent to woodlands. The hayfields were planted, in contrast to the other three which are native vegetation types. The dataset comprises records of 7,246 individuals from 58 species in 5 families, including the majority of grassland butterfly species present in the region and a number of endangered species. Five surveys were conducted over 2 years. Here the data have been summed within sites to compensate for seasonal and interannual variation in community composition. Sites were sampled on sunny days between 1000 and 1600 using a standardised method. Sampling was uneven between habitats, which creates some problems for comparing their diversity, as will be seen shortly.

The central question is whether these data provide a strong case for conservation of tall-grass prairie, which is a threatened habitat regionally, relative to other forms of grassland management. Which habitat is best for butterflies? In this kind of study, there are no 'right' answers; the task of the ecologist is to consider the statistical evidence and make sure that any answer is backed up by what was actually found, not by prejudices or preconceptions. This is not as easy as it sounds.

A.1 Software resources

A large number of programs are available to calculate diversity metrics, many of which are free. By far

[1] http://www.esapubs.org/archive/ecol/E087/061/

Natural Systems: The organisation of life, First Edition. Markus P. Eichhorn.

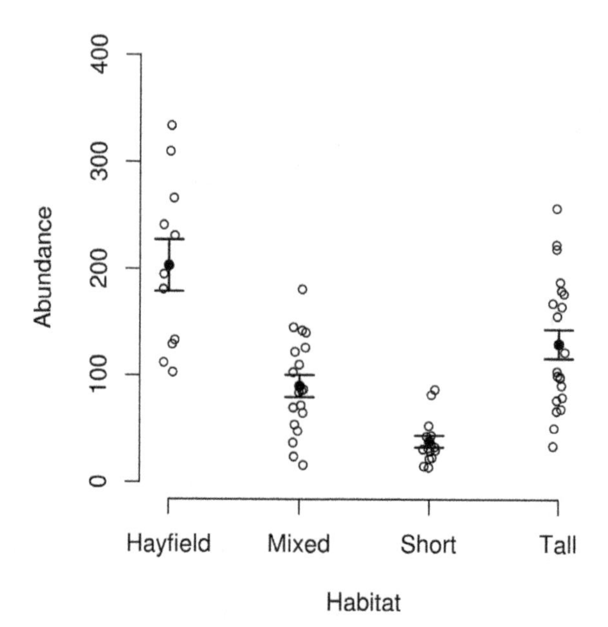

Figure A.1 Abundance of butterflies captured per sample in four grassland habitats in Colorado. Clear points represent individual samples; filled points show means with standard error bars.

the most powerful and versatile approach is to use the statistical computing environment `R` with the `vegan` package[2], which works on all operating systems and has been used in preparing this section. It is strongly recommended to those undertaking academic research and is free, but for straightforward applications there are simpler programs with a less steep learning curve.

All the metrics calculated here can be repeated using the free package SPADE, written by Anne Chao, one of the leaders in developing new techniques for the analysis of diversity[3]. It is up to date with the latest methods. A popular alternative used by many ecologists is EstimateS, which is also free to download[4]. Commercial software performing the same functions is available from Pisces Conservation Ltd.[5], with the SDR program being the most appropriate for these tests, though there are separate programs for community analyses as well. One note of caution is that the precise equations used to estimate species richness or calculate diversity indices may vary between programs, and it is worth checking that each is using the method you expect. Don't be surprised if they provide slightly different answers.

A.2 Calculations

The best place to start is by looking at patterns of abundance. Figure A.1 shows the total number of butterflies captured in each sample. There are significant differences in abundance between habitats (Kruskal–Wallis rank sum test; $\chi^2 = 35.2$, d.f. = 3, $P < 0.001$), with the greatest abundance in hayfields and the lowest in short-grass prairies. Bear in mind, following Chapter 5, that such variation might be an artefact of the sampling if butterflies are harder to catch in some habitats than others and therefore doesn't necessarily reflect genuine ecological differences. It is unlikely to be a problem here though.

What about the number of species? Figure A.2 shows sample-based species accumulation curves for each habitat. These have been created by averaging the value for each number of samples, which

[2] (http://www.r-project.org)
[3] (http://chao.stat.nthu.edu.tw/blog/software-download/spade/)
[4] (http://viceroy.eeb.uconn.edu/estimates)
[5] (http://www.pisces-conservation.com)

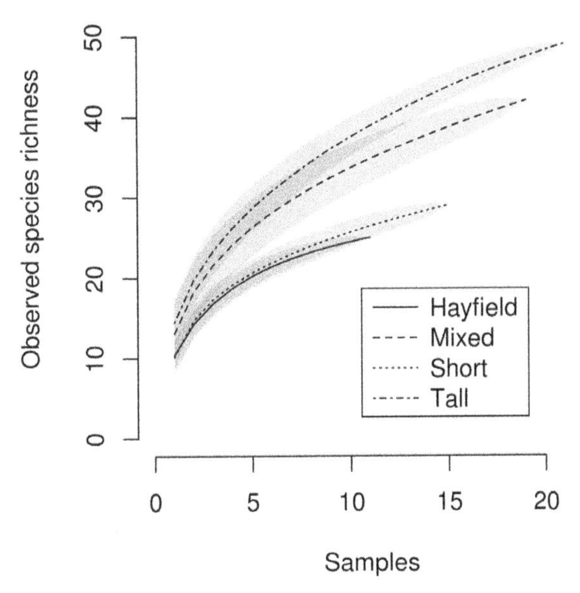

Figure A.2 Accumulation curves for butterfly species richness in three grassland habitats in Colorado; mean ± SE; see text for details.

smoothes the curves and prevents any single sample from having a dramatic effect on the overall trend. They show that the number of species found is continuing to rise with increased sampling, and there is little indication that an asymptote will be reached soon in any habitat. There appears to be some evidence, however, of a lower species richness in short grasslands and hayfields and perhaps more in the tall grasslands; these differences are worth investigating further.

Turning next to the distribution of abundance among the species present, a useful step is to explore the overall pattern using a Whittaker plot (Fig. A.3). Three of the curves are similar in form, but the slope

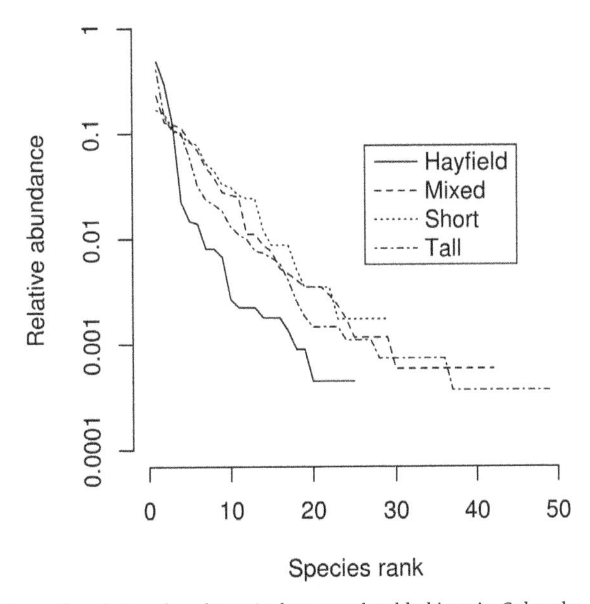

Figure A.3 Whittaker plots for butterfly relative abundance in four grassland habitats in Colorado; see text for details.

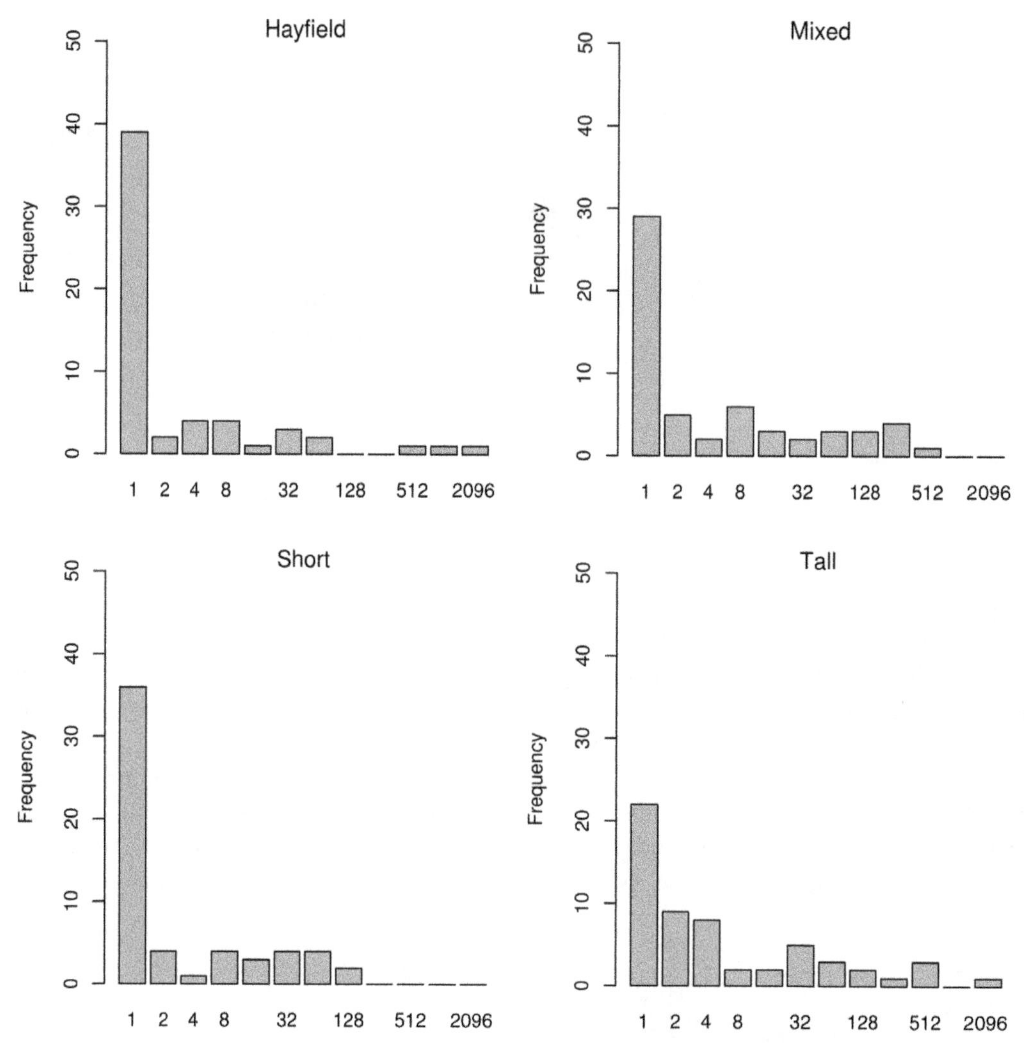

Figure A.4 Preston plots of butterfly relative abundance in four grassland habitats; see text for details.

for the hayfield is markedly steeper, indicating a loss of evenness and therefore overall diversity. A greater total number of species (S_{Obs}) were captured in tall grasslands. Remember that these curves take no account of species identities, only their relative abundance within habitats.

How completely have the butterfly assemblages been sampled? First we inspect the shapes of the Preston plots (Fig. A.4). According to classical theory, if sampling is complete, then these should show a lognormal distribution. In this case they don't, and estimating coverage C is a much more robust criterion for assessing completeness of sampling. For a total sample size N we can use the following estimator which requires you to know the number of species that appear either once (f_1) or twice (f_2) (Chao and Jost, 2012):

$$\hat{C}_N = 1 - \frac{f_1}{N}\left[\frac{(N-1)f_1}{(N-1)f_1 + 2f_2}\right]$$

This gives high values for all the habitats, with a minimum of 0.988 for the short grassland, while the hayfield is at 0.997. These indicate that sampling is not far from being complete, and the majority of

Table A.1 Diversity indices of butterfly samples from four grassland habitats in Colorado; means ± SE.

^qD	Metric	Hayfield	Mixed	Short	Tall
	S_{Obs}	10.4 ± 0.6	13.1 ± 0.9	10.1 ± 0.6	14.4 ± 0.6
^0D	S_{Chao}	13.0 ± 1.3	18.9 ± 2.3	13.4 ± 1.2	19.6 ± 2.1
^1D	$e^{H'}$	3.9 ± 0.3	7.5 ± 0.5	7.0 ± 0.4	6.7 ± 0.5
^2D	$1/D$	2.9 ± 0.2	5.5 ± 0.5	5.5 ± 0.4	4.7 ± 0.4

individuals in the full assemblage (at least 98.8%) are represented by species collected in the samples. Any species still missing are likely to be rare, and we are justified in making estimates of total diversity. More importantly, however, the values are similar, so comparing estimates among habitats will be reasonable.

There is still an excess of singletons (found only once, f_1), which means that there are more rare species still to collect, especially in the hayfields and short grasslands from which fewer samples were taken. All of this evidence implies that we cannot take the observed differences in species richness at face value, and some form of estimation is necessary. Diversity indices should also be calculated to incorporate information on the uneven abundance among species. We will only use the basic formulae here as a demonstration; for more sophisticated and accurate estimation techniques, see Magurran and McGill (2011). The advanced methods would be recommended for a scientific publication.

First we estimate species richness within each habitat using the Chao1 technique, which adds the minimum number of unseen species to the total number observed in the habitat overall (S_{Obs}):

$$S_{Chao} = S_{Obs} + \frac{f_1^2}{2f_2}$$

We can also estimate the exponential of Shannon's H':

$$H' = -\Sigma p_i \ln p_i$$

Finally we can estimate the inverse of Simpson's index D:

$$D = \Sigma p_i^2$$

In Table A.1 the values have been estimated for every sample and then averaged (it is possible to calculate the standard error of an estimate from a single sample, but this requires more complex maths). Once again we can test for differences among them, this time using simple analyses of variance (ANOVAs). There are significant differences in observed species richness ($F_{3,62} = 7.7$, $P < 0.001$), with samples from mixed and tall grasslands containing more species than those from hayfields and short grasslands. The difference remains significant for estimated species richness ($F_{3,62} = 2.8$, $P = 0.047$), but only just, due to the large variance (see the standard errors). This is a consequence of the difficulty of estimating true species richness accurately. From the perspective of the sampling, the difference between S_{Obs} and S_{Chao} indicates there are many species which were not collected, especially as S_{Chao} is thought to give a minimum estimate.

There is significantly lower butterfly diversity in hayfields at both the ^1D levels, which is $e^{H'}$ ($F_{3,62} = 7.9$, $P < 0.001$), and the ^2D level, or $1/D$ ($F_{3,62} = 6.5$, $P < 0.001$). These numbers are often described as 'effective species'. A ^1D score of 3.9 means that these samples have the same $e^{H'}$ as an assemblage with 3.9 equally abundant species. It is sensitive to the numbers of moderately abundant species, while $1/D$ reflects the numbers of highly abundant species. The message is that relative abundance of species is skewed in hayfields, an effect that can be seen most clearly in a plot of the full Hill series, where the line for hayfields falls clearly below the other three (Fig. A.5).

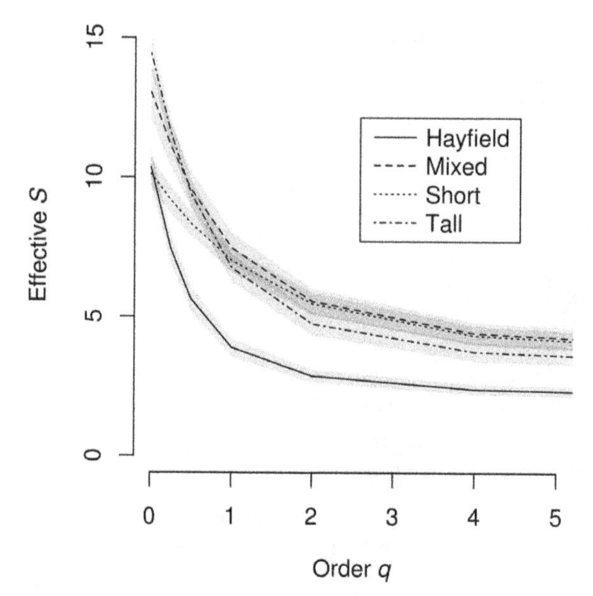

Figure A.5 Hill series plots for butterfly diversity in four grassland habitats in Colorado; means ± SE. Note that here ${}^0D$ shows observed rather than estimated species richness.

Finally we turn to the question of what proportion of species is shared between the three habitat types using Whittaker's index of β-diversity, which is the simplest to calculate and interpret. It assumes that sampling intensity is the same in all habitats, which was not the case here. To account for this, a random selection of 11 samples has been taken from each habitat, given that this was the minimum number (as in hayfields). This means that you might get slightly different answers. There are more sophisticated ways around this problem, which avoid effectively having to throw away some of the data, but for here the basic approach will be sufficient.

The formula for β-diversity is

$$\beta_W = S/\bar{\alpha} - 1$$

For this we need to know the total number of species jointly contained in any two habitats (S) divided by the average number contained in each ($\bar{\alpha}$). Dividing it by ($N - 1$), where N is the number of samples makes it scale from 0 (complete similarity) to 1 (maximum turnover). This gives an answer identical to Sørensen's index of dissimilarity. The results are shown in Table A.2.

Table A.2 Beta diversity (β_W) between butterfly assemblages in four grassland habitats in Colorado.

	Hayfield	Mixed	Short
Mixed	0.29		
Short	0.29	0.24	
Tall	0.35	0.27	0.39

Overall the assemblages share 60–75% of their species, so the β-diversity values are low (closer to 0). There is a suggestion that turnover is greatest between tall and short grasslands, but the differences are marginal.

A.3 Synthesis

All these statistics provide useful information regarding butterfly assemblages in the four habitat types. The first point is that there is a wide variation in butterfly abundance among habitats, with hayfields containing around five times as many individuals as short grassland. If you were to assess purely on the basis of raw numbers, then you might assume that this indicates a healthy butterfly assemblage. The Whittaker plot, however, suggests something different—that the hayfields contain fewer species with greater unevenness in their abundance.

The assemblages have not been completely sampled; this is obvious from inspecting the species accumulation curves, which are still rising, or the large numbers of singletons in the Preston plots, showing that many species were only collected once. After estimating true species richness, differences between the habitats are still not completely clear, but there is good evidence that tall and mixed grasslands contain more species overall. We could use a more sophisticated method than Chao1, such as the abundance coverage estimator (ACE), though the data here suggest that it is unlikely to make a difference. Remember that species richness is one of the hardest properties of a community to estimate accurately, and diversity indices are often more robust when small samples are being used.

A much clearer picture emerges from the diversity indices, with a noticeable fall in $e^{H'}$ and $1/D$ in hayfields. These figures suggest that only half as many species reach high or moderate abundance in these patches, with a few species dominating. The full Hill series plot emphasises that evenness is much lower in the hayfield butterfly assemblages. Turnover among the habitats as measured by β-diversity is not particularly high though; many of the same species are present in all grassland types.

A.4 Conclusions

Do the butterflies make a strong case for preserving the tall grasslands? The answer depends on the measure chosen to represent the relative health of a community. Tall grasslands have moderate abundance of butterflies, and while species richness is the highest of any of the habitats, it is not significantly greater than the mixed grasslands. Tall grasslands are not the most diverse by any other measure nor are its species particularly distinct from other habitats. Neither species richness nor overall abundance are necessarily good indicators, despite being commonly taken as the targets of conservation. Diversity indices provide an alternative set of metrics, though in this case the evidence for singling out tall grasslands for special attention remains equivocal. Mixed grasslands deserve as much consideration on these grounds.

What can be said definitively is that butterfly diversity is lowest in hayfields, which can still give butterflies a role as indicators of ecosystem health. All of these habitats are managed. Skewed butterfly abundance may have implications for ecosystem functioning or indicate similar disruptions in the abundances of food plants or nectar sources.

Nevertheless, problems can arise from assuming that one group of species is an effective proxy for others. The status of butterflies as indicator species is based on the assumption that a diverse butterfly assemblage implies a diverse plant community. This is not necessarily the case, and making predictions about patterns in other groups is inadvisable. Butterflies only give an indirect measure of the other species present. Though the overall congruence of butterfly richness with other taxa is relatively high (around 57% in Gardner et al. (2008)), we cannot pretend that this result could be applied to birds or lizards.

Moreover, even five samples over 2 years have provided only partial descriptions of the butterfly assemblages, with many species still missing. These samples may not be robust enough to draw firm inferences from. Butterflies are highly mobile and assemblages might therefore overlap between locations, or samples could include migratory or transient species. Separating out the resident and specialist species would be a useful next step.

Finally, it might be more worthwhile to focus on individual species of concern rather than the communities as a whole. All the statistics calculated here are *species neutral* in that the identities of the species do not matter. This is seldom true in conservation, and frequently management focusses on particular species of concern. As noted in Collinge et al. (2003), 11 of the 58 butterfly species were only seen in tall grasslands, a fact which does not emerge from any derived index. Most studies require local knowledge and interpretation to put the findings into context.

Sometimes it is hard to know the appropriate level of balance with which to present evidence from diversity statistics. There is a temptation to cherry-pick the values that match your preconceptions (perhaps to promote conservation of tall grasslands) or to focus on the differences rather than the properties that remain unchanged. You should always give the full picture, not merely the results that you might have wanted or expected. It is also wise to remain suspicious when you only see one set of values being presented. After all, had we based our judgements solely on the sheer numbers of butterflies, we might have assumed that hayfields were the superior habitat.

References

Chao, A. and L. Jost, 2012. Coverage-based rarefaction and extrapolation: standardizing samples by completeness rather than size. *Ecology* 93:2533–2547.

Collinge, S. K., K. L. Prudic, and J. C. Oliver, 2003. Effects of local habitat characteristics and landscape context on grassland butterfly diversity. *Conservation Biology* 17:178–187.

Gardner, T. A., J. Barlow, I. S. Araujo, T. C. Avila-Pires, A. B. Bonaldo, J. E. Costa, M. C. Esposito, L. V. Ferreira, J. Hawes, M. I. M. Hernandez, M. S. Hoogmoed, R. N. Leite, N. F. Lo-Man-Hung, J. R. Malcolm, M. B. Martins, L. A. M. Mestre, R. Miranda-Santos, W. L. Overal, L. Parry, S. L. Peters, M. A. Ribeiro-Junior, M. N. F. da Silva, C. da Silva Motta, and C. A. Peres, 2008. The cost-effectiveness of biodiversity surveys in tropical forests. *Ecology Letters* 11:139–150.

Magurran, A. E., 2004. *Measuring Biological Diversity*. Blackwell Publishing.

Magurran, A. E. and B. J. McGill, editors, 2011. *Biological Diversity: Frontiers in Measurement and Assessment*. Oxford University Press.

Oliver, J. C., K. L. Prudic, and S. K. Collinge, 2006. Boulder County Open Space butterfly diversity and abundance. *Ecology* 87:1066.

Glossary

Abundance
The number of individuals of a given taxon. This can be measured in a fixed area to obtain population density.

Adaptive radiation
The process by which a single ancestral species gives rise through speciation to an array of new species in response to a range of environments or resources.

Alien species
Species which are not native to a given locality and owe their presence to introduction by humans.

Allopatric speciation
Speciation that occurs between two populations which are geographically separated. Compare sympatric speciation.

Angiosperm
Vascular flowering plant with seeds enclosed by fruits. Includes most extant plant species.

Anthropocene
A term proposed to describe the most recent phase of geological history in which modern humans have become a dominant global driver of ecosystem and geological processes.

Apparent competition
A negative effect of one species on the population of another, mediated through the action of shared natural enemies rather than direct resource competition.

Assemblage
A collection of species present in a given location, making no assumptions regarding their interactions or interdependence. The term is preferred to **community** when the relationships among species are unknown.

Assembly rules
The set of local processes determining which candidate species are able to enter or persist in a given community.

Asymptote
The maximum value reached by a curved function.

Autotroph
A species capable of producing chemical energy from inorganic materials, e.g. through photosynthesis. Contrast **heterotroph**.

Benthic
Related to the bed of an aquatic habitat (e.g. a sea or lake) or the organisms that occur there.

Biodiversity
According to the Convention on Biological Diversity (1992), the 'diversity wthin species, between species and of ecosystems'. Has no consistent quantitative measure.

Biological control
Reduction of pest populations by actively introducing their natural enemies (e.g. predators, diseases).

Natural Systems: The organisation of life, First Edition. Markus P. Eichhorn.

Biomass
The total mass of live organic material in a system.

Biome
A large geographical region dominated by a particular habitat type, general climate and set of dominant organisms.

Biosphere
The whole earth system.

Bottleneck
A sharp reduction in population size, followed by an increase; this tends to leave a signal of reduced genetic diversity.

Carrying capacity
The maximum size of a population which can be sustained indefinitely at a given supply rate of resources. Often represented in ecological terminology by the parameter K.

Clade
A single branch of a phylogeny, consisting of an ancestral species and all its descendants.

Climate envelope
The range of environmental conditions within which a species is able to persist, projected onto a map to predict its potential distribution.

Coexistence
Co-occurrence of species in space and time despite overlap in resource usage.

Community
A collection of species present in a given location and consistent through time which are linked by interactions (e.g. feeding relationships, mutualisms).

Competition–colonisation trade-off
The common observation that species fall on a continuum between those adapted for high rates of reproduction and dispersal, and those adapted to compete for scarce resources.

Connectance
A statistical measure of networks which describes the ratio between actual and potential links among nodes.

Convergence
The evolution of similar forms or structures among species caused by common environmental drivers rather than through inheritance.

Crown group
Within a phylogeny, all the descendents of the most recent common ancestor of the extant taxa. Contrast **stem group**.

Cryptic species
Species which are morphologically identical (or indistinguishable using normal taxonomic characters) but revealed as distinct through their genetics, behaviour or other characters.

Density dependence
The tendency for many ecological processes (e.g. resource competition) to intensify with increased population density.

Diaspore
A single seed or spore, combined with the associated organs which enable its dispersal (e.g. fruit, wings).

Disharmony
The difference in composition and relative abundance of species on islands (or other isolated areas) relative to the mainland.

Disjunction
The presence of a species in two or more geographical locations separated by uninhabitable regions.

Dispersal
The movement of individual organisms apart in space; distinct from **migration**.

Distance decay
The decrease in similarity of species composition (i.e. higher turnover) between sites with increasing distance.

Distribution
The geographical area within which a taxon is present.

Disturbance
Disruption to a community causing the death of individual organisms.

Diversity index
A composite measure of both the richness and evenness of a sample. See Chapter 5.

Ecomorph
A type of organism with consistent habitat preferences, morphology and behaviour, though constituent species are not necessarily related to one another phylogenetically.

Ecoregion
A geographical region with a broadly consistent set of species, habitats and climatic conditions.

Ecosystem
A linked set of biotic and abiotic components in a given location.

Ecosystem engineer
A species whose activities create or modify the environment they require to survive.

Ecosystem function
A movement or transformation of abiotic material mediated by a biological system.

Ecosystem service
A process generated by an ecosystem which has a measurable role in supporting or sustaining human life and well being.

Ectotherm
An organism which relies on the external environment to regulate its body temperature; cold-blooded.

Edaphic
Related to or caused by soil properties.

Endemic
A species whose entire range falls within a specified area, often a country, ecoregion, island or habitat patch.

Endotherm
An organism which generates metabolic heat in order to maintain a constant body temperature, usually above that of the external environment; warm-blooded.

Eutrophication
Normally applied to aquatic systems, this refers to pollution which takes the form of an excessive supply of nutrients.

Extinction debt
The number of committed extinctions in a patch following a reduction in area and predicted by the species–area relationship. There is usually a time lag between the change in area and the final loss of populations.

Fitness
The contribution of an individual to the next generation relative to the average for the population. In ecological terms it is usually measured by number of offspring which survive to reproductive age, though strictly it refers to a genetic contribution.

Frugivore
A species whose diet primarily consists of fruit.

Functional group
A set of species sharing a similar role within a community or ecosystem, e.g. primary producers, pollinators.

Functional trait
Feature of an organism's phenotype which influences processes at the ecosystem scale.

Gaia
A controversial theory which views the earth as a self-regulating system that acts to maintain conditions conducive to life. See Chapter 11.

Grain size
The minimum unit area in a study of spatial patterns. It can vary in ecological studies from smaller scales relevant to individual interactions (1 m^2 or below) through the organisation of communities (e.g. hectares) up to biogeographical patterns (e.g. 1° grid squares).

Gymnosperm
Vascular plant with seeds not enclosed by fruits. Includes conifers and cycads.

Heterotroph
A species which obtains chemical energy by consuming other organisms or organic material. Includes herbivores, predators, parasites and decomposers. Contrast **autotroph**.

Holocene
The present geological epoch, commencing at the end of the last glaciation.

Individual
A single, reproductively-viable member of a species. Note that this definition is not universally applied; see Chapter 1.

Insularity
A measure of the degree to which a given system resembles an island.

Invasive
A species that has naturalised and spread widely within a region outside its normal range. The term is most often used to refer to species which cause ecological or economic damage.

Janzen-Connell effects
Negative Density- or distance-dependent recruitment in populations due to transmission of natural enemies from adults to offspring.

***K*-Selection**
Directional selection on populations at high density in relatively constant environments and with high levels of resource competition. Contrast **r-selection**.

Keystone species
A species whose influence on a community or ecosystem, measured following removal, exceeds that expected given its abundance or biomass. In practice the definition is hard to apply rigorously.

Lognormal
The shape of a histogram in which the *x* axis is log-transformed and *y* axis values follow a normal distribution.

Megafauna
Relatively large organisms within a given system; a common threshold is > 45 kg in body size.

Metacommunity
A set of local communities spread across a landscape or region and linked by the dispersal of species among them.

Metapopulation
A set of distinct populations spread across a landscape and linked by the dispersal of individuals.

Metazoan
A multicellular life form.

Microcosm
An enclosed, miniaturised system. Often used in experiments as a means of studying small, sealed systems.

Migrant
A species or individual that regularly and deliberately moves from one region to another.

Migration
The mass directional movement of an entire population of organisms; distinct from **dispersal**.

Monophyly
The descent of all individuals from a single common ancestral taxon.

Mutualist
An interacting species with net benefits for both partners in the interaction, e.g. a pollinator.

Native
A species that is naturally capable of occupying a given site within its range.

Naturalised
A non-native species which has established in a site outside its normal range. The use of this term implies that the species has not become invasive, though all invasive species are naturalised.

Necromass
The total mass of dead organic material in a system; contrast **biomass**.

Niche
The joint description of the environmental conditions that allow a local population to persist and the per capita effects on the environment. See Chapter 6.

No-analogue communities
Communities formed of species which are extant today but which in the past occurred in combinations unlike any in the modern world, often occupying regions of climatic space which no longer exist.

Novel ecosystem
An ecosystem which differs in structure and composition from historical analogues as a result of human influence, and which no longer requires human intervention to persist (therefore excluding agricultural or managed landscapes).

Ordination
A statistical procedure in which objects (e.g. ecological samples) are characterised by multiple variables (e.g. abundances of species) and represented figuratively such that the distance between objects in space correlates with their relative similarity. Includes techniques such as PCA and NMDS.

Pangæa
The Mesozoic supercontinent which united all the major continental land masses. It formed around 300 Mya and began to separate from 200 Mya.

Pelagic
Of or related to the open sea.

Photosynthetically Active Radiation (PAR)
Light at wavelengths that plants are able to use in phytosynthesis.

Phylogenetic diversity
A measure of the degree of evolutionary differentiation among a set of species.

Phylogenetic niche conservatism
The assumption that species retain ancestral niche properties through inheritance.

Phylogeny
Evolutionary history of a group of species, usually depicted as a tree.

Pioneer
A species which is the first colonist of empty sites, usually adapted for rapid growth and broad dispersal.

Population
A group of individuals of the same species in a given location which are linked by reproduction (or other interactions).

Precipitation
Falling condensation; includes rain, snow or hail.

Propagule
A life history stage which can give rise to a new organism, typically adapted for dispersal. Includes seeds, spores and the larvae of many benthic marine species.

Pteridophyte
Vascular plant reproducing by means of spores rather than seeds. Includes horsetails, ferns, and several extinct phyla.

Radiation
In evolutionary terms, an increase in the number of species (or types).

***r*-Selection**
Directional selection on populations at low density in unpredictable environments where resources are relatively abundant. Contrast **K-selection**.

R*
The minimum levels of a resource at which a population of a given species is able to maintain itself (zero net growth).

Range
The geographical map space occupied by a species.

Regeneration
The suite of processes which restore the composition of a community to a stable state following disturbance.

Regeneration niche
The environmental conditions required by juveniles of a species to establish; these may differ from those required by adults.

Ruderal
A species adapted to rapid growth and reproduction in transient high-quality habitats or disturbed sites.

Sample
A series of measurements taken from a much larger, unknown set. For example, a small collection of individual organisms taken from a large community.

Singleton
A species (or type) which is only collected once in a sample.

Sink
A site where the growth rate of a population is negative, and therefore the presence of a species is only maintained by continual immigration from elsewhere.

Species
Individual organisms linked by commonality of reproduction, appearance or descent. See Chapter 2.

Species-Abundance Distribution
The relative frequency of abundances across a set of species from an assemblage or community.

Species accumulation curve
A graph which plots the cumulative number of species collected with increasing sample size, frequency, time, effort or numbers of individuals.

Species–Area Relationship
The increase in the number of species found with greater area surveyed. This typically follows a log-linear relationship. See Chapter 7.

Stem group
The inferred lineage leading to a **crown group** from the last common ancestor shared with a sister group.

Stepping stone
A location which acts as a dispersal pathway between two regions divided by a barrier. Most commonly used in the context of an oceanic island between two larger land masses.

Stochastic
Involving or containing one or more random variables drawn from a probability distribution.

Storage effect
Stable coexistence of multiple species through variation in relative fitness through time.

Succession
Change in the composition of a community through time which is nonseasonal, directional and continuous.

Sustainability
In ecological terminology, the ability of a system, process or activity to be maintained in perpetuity given current conditions.

Sympatric speciation
Speciation that occurs between two populations which overlap in their distribution. Contrast **allopatric speciation**.

Synonym
A taxonomic name applied to a species which already has an established name.

Taxon
Plural taxa. A single biological unit within the taxonomic hierarchy (e.g. a family, genus or species).

Trait
A genetically-determined, quantifiable property of an organism. Those that describe the ecological roles of a species are known as **functional traits**.

Transpiration
Evaporation of water from the photosynthetic tissues (usually leaves) of a plant.

Trophic
Related to nutrition, particularly with regard to the consumption of one species by another in a food chain or web.

Trophic cascade
A phenomenon whereby higher trophic levels (e.g. predators) in a food web reduce the abundance or alter the behaviour of their prey, thereby having indirect effects on the next trophic level down.

Turnover
The change in species composition between locations in space or at a single location over time.

Vagrancy
The movement of individuals out of the normal range of a species, defined as the areas normally used for breeding, migration or overwintering.

Veil line
A phenomenon whereby an under-sampled lognormal species abundance distribution appears as though the left-hand side of the curve has been hidden by the *y* axis.

Vicariance
Geographical separation and isolation of subpopulations, resulting in their differentiation into new species or varieties.

Volant
Capable of active flight, e.g. bats and most birds.

Index

Note: Page numbers in *italic* refer to figures and tables
Page numbers in **bold** refer to definitions

Natural Systems: The organisation of life, First Edition. Markus P. Eichhorn.